MOLECULAR BIOLOGY OF HUMAN CANCERS

# Molecular Biology of Human Cancers

An Advanced Student's Textbook

*by*

WOLFGANG ARTHUR SCHULZ

*Department of Urology and Center for Biological and Medical Research,
Heinrich Heine University, Dusseldorf, Germany*

A C.I.P. Catalogue record for this book is available from the Library of Congress.

ISBN 1-4020-3185-8 (HB)
ISBN 1-4020-3186-6 (e-book)

Published by Springer,
P.O. Box 17, 3300 AA Dordrecht, The Netherlands.

Sold and distributed in North, Central and South America
by Springer,
101 Philip Drive, Norwell, MA 02061, U.S.A.

In all other countries, sold and distributed
by Springer,
P.O. Box 322, 3300 AH Dordrecht, The Netherlands.

*Printed on acid-free paper*

All Rights Reserved
© 2005 Springer
No part of this work may be reproduced, stored in a retrieval system, or transmitted
in any form or by any means, electronic, mechanical, photocopying, microfilming, recording
or otherwise, without written permission from the Publisher, with the exception
of any material supplied specifically for the purpose of being entered
and executed on a computer system, for exclusive use by the purchaser of the work.

Printed in the Netherlands

meinen Eltern gewidmet

# TABLE OF CONTENTS

**Preface: How to read this book**.................................................................. xiii
**Acknowledgements** ........................................................................................ xvii

## PART I – MOLECULES, MECHANISMS, AND CELLS

**1 An Introduction to Human Cancers**................................................................ 1
   *1.1 An overview of the cancer problem* ........................................................... 1
   *1.2 Causes of cancer*....................................................................................... 5
   Box 1.1 Reactive oxygen species .................................................................. 10
   *1.3 Characteristic Properties of Cancers and Cancer Cells*............................ 11
   Box 1.2: Hallmarks of Cancer ....................................................................... 18
   *1.4 Characterization and Classification of Cancers in the Clinic* .................. 17
   *1.5 Treatment of Cancer* ............................................................................... 21
   Further reading.............................................................................................. 23

**2 Tumor Genetics** ........................................................................................... 25
   *2.1 Cancer as a genetic disease*..................................................................... 26
   *2.2. Genetic alterations in cancer cells* ......................................................... 27
   *2.3 Inherited predisposition to cancer*........................................................... 37
   *2.4 Cancer genes* .......................................................................................... 42
   *2.5 Accumulation of genetic and epigenetic changes in human cancers* ............ 44
   Further reading.............................................................................................. 45
   Box 2.1 Tumor viruses in human cancers ..................................................... 46

**3 DNA Damage and DNA Repair** ................................................................. 47
   *3.1 DNA damage during replication: base excision and nucleotide excision*
      *repair*..................................................................................................... 48
   *3.2 Nucleotide excision repair and crosslink repair* ..................................... 55
   *3.3 Strand-break repair* ................................................................................ 62
   *3.4 Defects in DNA repair and cancer susceptibility*.................................... 66
   *3.5 Cell protection mechanisms in cancer*..................................................... 68
   Further reading.............................................................................................. 70

**4 Oncogenes**.................................................................................................... 71
   *4.1 Retroviral oncogenes*.............................................................................. 72
   *4.2 Slow-acting transforming retroviruses* ................................................... 75
   *4.3 Approaches to the identification of human oncogenes*............................ 78
   *4.4 Functions of human oncogenes* .............................................................. 84
   Further reading.............................................................................................. 89
   Box 4.1 Carcinogenesis by HTLV-I............................................................... 90

**5 Tumor Suppressor Genes** ........................................................................... 91
   *5.1 Tumor suppressor genes in hereditary cancers*....................................... 92
   *5.2 RB1 and the cell cycle*............................................................................ 97
   *5.3 TP53 as a different kind of tumor suppressor* ...................................... 101

    *5.4 Classification of tumor suppressor genes* ............................................................. *109*
    Further reading ............................................................................................................ 111
    Box 5.1 Human papilloma viruses ............................................................................. 112

## 6 Cancer Pathways ..................................................................................................... 113
    *6.1 Cancer Pathways* ................................................................................................. *114*
    *6.2 MAPK signaling as a cancer pathway* ................................................................ *115*
    *6.3 The PI3K pathway* .............................................................................................. *119*
    *6.4 Regulation of the cell cycle by the MAPK and PI3K pathways* ......................... *123*
    *6.5 Modulators of the MAPK and PI3K pathways* ................................................... *126*
    *6.6 The TP53 network* ............................................................................................... *129*
    *6.7 Signaling by TGFβ factors* ................................................................................. *131*
    *6.8 Signaling through STAT factors* ......................................................................... *132*
    *6.9 The NFκB pathway* .............................................................................................. *135*
    *6.10 Developmental regulatory systems as cancer pathways* ................................... *137*
    Further reading ............................................................................................................ 144

## 7 Apoptosis and Replicative Senescence in Cancer ................................................ 145
    *7.1 Limits to cell proliferation* .................................................................................. *146*
    *7.2 Mechanisms of apoptosis* .................................................................................... *150*
    *7.3 Mechanisms of diminished apoptosis in cancer* .................................................. *156*
    *7.4 Replicative senescence and its disturbances in human cancers* .......................... *159*
    Further reading ............................................................................................................ 164
    Box 7.1: Human aging and cancer .............................................................................. 165

## 8 Cancer Epigenetics ................................................................................................... 167
    *8.1 Mechanisms of epigenetic inheritance* ................................................................ *168*
    *8.2 Imprinting and X-inactivation* ............................................................................ *170*
    *8.3 DNA methylation* ................................................................................................ *174*
    *8.4 Chromatin structure* ........................................................................................... *179*
    *8.5 Epigenetics of cell differentiation* ....................................................................... *182*
    *8.6 Epigenetics of tissue homeostasis* ....................................................................... *185*
    Further reading ............................................................................................................ 191
    Box 8.1 Carcinogenesis by HIV                                        192

## 9 Invasion and metastasis .......................................................................................... 193
    *9.1 Invasion and metastasis as multistep processes* ................................................. *194*
    *9.2 Genes and proteins involved in cell-to-cell and cell-matrix adhesion* ............... *197*
    *9.3 Genes and proteins involved in extracellular matrix remodeling*
        *during tumor invasion* ........................................................................................ *202*
    *9.4 Angiogenesis* ....................................................................................................... *206*
    *9.5 Interactions of invasive tumors with the immune system* .................................... *210*
    *9.6 The importance of tumor-stroma interactions* .................................................... *212*
    Further reading ............................................................................................................ 216
    Box 9.1 Tumor hypoxia and its consequences ........................................................... 217

# PART II - HUMAN CANCERS

**10 Leukemias and Lymphomas** ............................................................ 219
    *10.1 Common properties of hematological cancers* ............................ 221
    *10.2 Genetic aberrations in leukemias and lymphomas* ...................... 223
    *10.3 Molecular biology of Burkitt lymphoma* ..................................... 226
    *10.4 Molecular biology of CML* ......................................................... 232
    *10.5 Molecular biology of PML* ......................................................... 237
    Further reading ....................................................................................... 242

**11 Wilms Tumor (nephroblastoma)** ....................................................... 243
    *11.1 Histology, etiology and clinical behavior of Wilms tumors* ........ 244
    *11.2 Genetics of Wilms tumors and the WT1 gene* ............................. 246
    *11.3 Epigenetics of Wilms tumors and the 'WT2' locus* ..................... 250
    *11.4 Towards an improved classification of Wilms tumors* ............... 252
    Further reading ....................................................................................... 253

**12 Cancers of the skin** ............................................................................ 255
    *12.1 Carcinogenesis in the skin* ........................................................... 256
    *12.2 Squamous cell carcinoma* ........................................................... 260
    *12.3 Basal Cell Carcinoma* .................................................................. 262
    *12.4 Melanoma* .................................................................................... 266
    *12.5 Tumor antigens* ........................................................................... 269
    Further reading ....................................................................................... 270

**13 Colon Cancer** ..................................................................................... 271
    *13.1 Natural history of colorectal cancer* ........................................... 272
    *13.2 Familial Adenomatous Polyposis Coli and the WNT pathway* ... 273
    *13.3 Progression of Colon Cancer and the Multi-Step Model*
        *of Tumorigenesis* ............................................................................. 280
    *13.4 Hereditary nonpolyposis colon carcinoma* .................................. 282
    *13.5 Genomic instability in colon carcinoma* ..................................... 284
    *13.6 Inflammation and colon cancer* ................................................... 285
    Further reading ....................................................................................... 287
    Box 13.1 Positional cloning of tumor suppressor genes in hereditary cancers . 288

**14 Bladder Cancer** .................................................................................. 289
    *14.1 Histology and etiology of bladder cancer* ................................... 290
    *14.2 Molecular alterations in invasive bladder cancers* ..................... 297
    *14.3 Molecular alterations in papillary bladder cancers* .................... 302
    *14.4 A comparison of bladder cancer subtypes* .................................. 304
    Further reading ....................................................................................... 305
    Box 14.1: Tumor suppressor candidates at 9q in bladder cancer ........... 306

**15 Renal Cell Carcinoma** ....................................................................... 307
    *15.1 The diversity of renal cancers* ..................................................... 308
    *15.2 Cytogenetics of renal cell carcinomas* ........................................ 310

*15.3 Molecular biology of inherited kidney cancers* ............ 311
*15.4 Von-Hippel-Lindau syndrome and renal carcinoma* ............ 316
*15.5 Molecular biology of clear cell renal carcinoma* ............ 321
*15.6 Chemotherapy and immunotherapy of renal carcinomas* ............ 324
Further reading ............ 326

## 16 Liver Cancer ............ 327
*16.1 Etiology of liver cancer* ............ 328
*16.2 Genetic changes in hepatocellular carcinoma* ............ 331
*16.3 Viruses in HCC* ............ 336
Further reading ............ 339
Box 16.1 Hepatocellular carcinoma in experimental animals ............ 340

## 17 Stomach Cancer ............ 341
*17.1 Etiology of stomach cancer* ............ 342
*17.2 Molecular mechanisms in gastric cancer* ............ 345
*17.3 Helicobacter pylori and stomach cancer* ............ 348
Further reading ............ 354
Box 17.1: Barrett esophagus and esophageal cancer ............ 355

## 18 Breast Cancer ............ 357
*18.1 Breast biology* ............ 358
*18.2 Etiology of breast cancer* ............ 364
*18.3 Hereditary breast cancer* ............ 365
*18.4 Estrogen receptors and ERBB proteins in breast cancer* ............ 373
*18.5 Classification of breast cancers* ............ 378
Further reading: ............ 382

## 19 Prostate Cancer ............ 383
*19.1 Epidemiology of prostate cancer* ............ 384
*19.2 Androgens in prostate cancer* ............ 389
*19.3 Genetics and epigenetics of prostate cancer* ............ 394
*19.4 Tumor-stroma interactions in prostate cancer* ............ 398
Further reading ............ 402

# PART III - PREVENTION, DIAGNOSIS, AND THERAPY

## 20 Cancer Prevention ............ 403
*20.1 The importance of cancer prevention* ............ 403
*20.2 Primary prevention* ............ 404
*20.3 Cancer prevention and diet* ............ 408
*20.4 Prevention of cancers in groups at high risk* ............ 415
*20.5 Prevention of prostate cancer by screening the aging male population* ............ 420
*20.6 Other types of prevention* ............ 423
Further reading ............ 426

**21 Cancer Diagnosis ........................................................................... 427**
   *21.1 The evolving scope of molecular diagnostics ............................................ 427*
   *21.2 Molecular diagnosis of hematological cancers......................................... 429*
   *21.3 Molecular detection of carcinomas ............................................................ 433*
   *21.4 Molecular classification of carcinomas .................................................... 439*
   *21.5 Prospects of molecular diagnostics in the age of individualized therapy .. 442*
   Further reading.................................................................................................... 447

**22 Cancer Therapy ............................................................................ 449**
   *22.1 Limitations of current cancer therapies ..................................................... 449*
   *22.2 Molecular mechanisms of cancer chemotherapy ....................................... 450*
   *22.3 Principles of targeted drug therapy............................................................ 459*
   *22.4 Examples of new target-directed drug therapies........................................ 464*
   *22.5 New concepts in cancer therapy: Immunotherapy ..................................... 475*
   *22.6 New concepts in cancer therapy: Gene therapy......................................... 479*
   *22.7 The future of cancer therapy ...................................................................... 486*

**KEYWORD INDEX ............................................................................ 489**

# PREFACE: HOW TO READ THIS BOOK

The present book grew out from a lecture course I have taught for more than 5 years, often together with colleagues who covered topics and cancers they are more familiar with than myself. These lectures were mainly attended by biology and medical students well advanced in their curricula, but also by clinical trainees doing cancer research in the lab, and by graduate students and postdocs having entered cancer research from different fields, including chemistry, pharmacology, developmental genetics, physics, and even mathematics. This experience reflects how cancer research, cancer prevention, and even cancer treatment increasingly become interdisciplinary efforts.

While writing the book I had this motley group of people in mind, figuring that they all, with their different backgrounds, could make use of an introduction to the molecular biology of human cancers. Specifically, I felt that a textbook for more advanced students was required to fill a gap between standard textbooks on one hand and specialized reviews or even original research papers on the other hand. Moreover, the textbooks on molecular biology and genetics are usually read by biologists and those on pathology and clinical oncology only by medical students. Accordingly, for biologists and chemists medical terms had to be introduced, and for the readers from the medical profession some general molecular biology had to be explained. So, please do not scoff if some statements in this book are found on the first five pages of the standard textbook in your specialty. However, this is neither a book on biochemical mechanisms nor on clinical oncology. So, if you do not understand medical terms or molecular issues in the book, please borrow a textbook from a student of the other discipline or (better...) ask them to explain.

The book is intended to provide a relatively short overview of important concepts and notions on the molecular biology of human cancers (according to the author's opinion), including many facts essential to find one's way in this field. It is, however, not meant to be comprehensive and probably cannot be, as our knowledge is rapidly growing. A list of other textbooks and handbooks can be found in the 'further reading' section of Chapter 1.

This book is named 'Molecular Biology of Human Cancers' because molecular biology is at its center, although in some places I have attempted to put this specific angle of approach to the cancer problem into a broader, 'real world' perspective, particularly in Part III. 'Human' is supposed to stress that cancers in humans and not in other organisms or in vitro models of cancer are the main issue (most evidently in the central Part II). The plural 'Cancers' is to indicate that the diversity of human cancers is stressed, although common mechanisms are treated in depth.

The subtitle is also programmatic. It reads 'An Advanced Student's Textbook' most of all because the book is supposed to bridge the gap from undergraduate textbooks to the specialized literature. This is only one of several reasons for putting 'Advanced' in the subtitle. Advanced students also understand that knowledge is evolving and must be constantly questioned. So I have not avoided to point out the limits of current knowledge and open questions, as they appear to me. Advanced students have learned that scientific questions are pursued in a real world, which imposes limits on what we understand about human cancers and what we can do about them. I have therefore not avoided to mention limits of that sort either, where appropriate. Advanced students also know how to find literature on a subject, if they want to check statements in the book. For that reason and to limit the size of the volume, I have restricted the references to suggestions for further reading, which follow each chapter. Most references are review articles, and almost all are in English.

I have tried to keep the book to a size that can be read in a couple of weeks (one or two chapters per day) and to write in a style that is not so dry as to make reading a trial in endurance. Hopefully, you will not find it too journalistic. Ideally, you would read the volume from start to finish. If you do so, you will notice that many important facts and ideas appear in several places, often described from different angles. This redundancy is intended to help memorize important issues and link facts and concepts to each other. If you find essential points missing, please follow my request in the last paragraph of this preface.

A second reading strategy is to follow the many cross-references in the book and 'surf' it. This approach is strongly recommended to casual readers and also to those who are already familiar with cancer molecular biology. To help you scan the book and find out what interests you, all chapters (except the first introductory one and the three more essay-style chapters in Part III) contain a short introductory section listing their essential points. Each chapter then proceeds to a more detailed exposition, but also contains additional points that are too specialized (or too speculative) to be of interest to every reader. So, you could browse this book by reading the one or two introductory pages of each chapter and if you think 'I know all this' move on to the next one. Some issues which did not fit well into the general stream of thought are placed in separate boxes. Be aware that these are treated in a particularly cursory fashion.

Part I starts with a brief overview chapter, which serves mainly to define terms and introduce concepts and issues that are treated in more depth and detail in later chapters. The following chapters deal with molecules and mechanisms important in human cancers. Although these chapters contain a lot of information, they present only a fraction of what is known in this regard, since knowledge on individual molecules and their interactions is presently expanding at a particularly rapid pace. For instance, ≈250 genes are now considered oncogenes or tumor suppressor genes in humans, by a strict definition, and at least ten times as many are implicated as important in cancer biology. Some molecules and many mechanisms are common to several human cancers, but the actual mechanisms for each type or even subtype of cancer differ. One could picture this relationship as a large set of oncogenic mechanisms, from which various subsets are relevant in different cancers.

In some cancers, the mechanisms involved in carcinogenesis, cancer progression, or response to therapy are rather well understood, and the role of specific genes and proteins has been well elucidated. In my opinion, the most exciting and promising development in cancer research is an increasing ability to not only identify and list molecular changes, but also to relate them to the characteristic biological properties and clinical behavior of specific human cancers. The cancers discussed in Part II are selected according to this criterion. They are therefore not necessarily the most prevalent or lethal cancers, although many of these are treated. Accordingly, the treatment of each cancer type is not comprehensive either, but focuses on those selected issues and mechanisms that we understand best.

Application of the knowledge presented in Part I and Part II has begun to improve the prevention, diagnosis, and therapy of human cancers. Part III therefore gives a short sketch of the progress, problems, and possible future of the developments in these areas. Clearly, there is a still a long way to go until the insights from molecular biological research are fully translated into a benefit for fellow humans. Hopefully, this part will have to be much longer in future books.

Finally, I have a request to the readers. I have tried, with the help of my colleagues acknowledged in the following acknowledgement section to keep this book as free of mistakes as possible, to deal with the most important issues in cancer molecular biology and to label opinions and speculations as such, but certainly many errors remain. Please, send a short note on any that you find, best by e-mail (wolfgang.schulz@uni-duesseldorf.de), I have asked for an extra-large size mailbox on our university server.

# ACKNOWLEDGEMENTS

Many colleagues from our university have contributed figures and references for the book. They are usually acknowledged in the figure legends. I am particularly grateful to Prof. Claus-Dieter Gerharz for most of the histology figures, to Dr. Georg Kronenwett for expert advice on hematological cancers, and to Dr. Andrea Linnemann-Florl for drawing many figures and improving many of my own. I am further obliged to them and several other colleagues for reading and commenting on individual chapters or large parts of the book, specifically Dr. Aristoteles Anastasiadis, Dr. Sylvia Geisel, Michèle Hoffmann, Prof. Joseph Locker, Prof. Stephan Ludwig, Dr. Dieter Niederacher, Dr. Julia Reifenberger, Dr. Ingo Schmitz, Dr. Valerie Schumacher, and Dr. Hans-Helge Seifert. Ms. Bettina Möller has kindly helped with the formatting and editing of the final version and Ms. Olga Schulz with the index.

I have to apologize to my co-workers in the lab, Andrea Linnemann–Florl, Michèle Hoffmann, Christiane Hader, Andrea Prior, and Marc Cronauer, for not giving them my full attention for more than a year and even more to our lab students and trainees during that period. I am very grateful to all of them for compensating through their own initiative. I am also grateful to our head of department, Prof. Ackermann, for supporting the endeavor.

At Kluwer press (now Springer), Ebru Umar talked me into writing the book, and later Christina dos Santos and Melania Ruiz took care of the project.

If I ever did wonder why book authors feel obliged to thank their families for support, I now know. So, thank you, Geli, Olga, and Edwin!

Düsseldorf, September 2004

WAS

# PART I

# MOLECULES, MECHANISMS, AND CELLS

# CHAPTER 1

# AN INTRODUCTION TO HUMAN CANCERS

## 1.1 AN OVERVIEW OF THE CANCER PROBLEM

What is commonly called 'human cancer' comprises in fact more than 200 different diseases. Together, they account for about one fifth of all deaths in the industrialized countries of the Western World. Likewise, one person out of three will be treated for a severe cancer in their life-time. In a typical Western industrialized country like Germany with its 82 million inhabitants, >400,000 persons are newly diagnosed with cancer each year, and ≈200,000 succumb to the disease. Since the incidence of most cancers increases with age, these figures are going to rise, if life expectancy continues to increase.

If one considers the incidence and mortality by organ site, while ignoring further biological and clinical differences, cancers fall into three large groups (Figure 1.1). Cancers arising from epithelia are called 'carcinomas'. These are the most prevalent cancers overall. Four carcinomas are particular important with regard to incidence as well as mortality. Cancers of the lung and the large intestine (colon and rectum, →13) are the most significant problem in both genders, together with breast cancer (→18) in women and prostate cancer (→19) in men. A second group of cancers are not quite as prevalent as these 'major four' cancers. They comprise carcinomas of the bladder (→14), stomach (→17), liver (→16), kidney (→15), pancreas, esophagus, and of the cervix and ovary in women. Each accounts for a few percent of the total cancer incidence and mortality. Each of them is roughly as frequent as all leukemias or lymphomas (→10) taken together. The most prevalent cancers are those of the skin (→12), not shown in figure 1.1. They are rarely lethal, with the important exception of melanoma. Cancers of soft tissues, brain, testes, bone, and other organs are relatively rare; but can constitute a significant health problem in specific age groups and geographic regions. For instance, testicular cancer is generally the most frequent neoplasia affecting young adult males, with an incidence of >1% in this group in some Scandinavian countries and in Switzerland.

The health situation in less-industrialized countries differs principally from that in the highly industrialized part of the world because of the continuing, recurring or newly emerged threat of infectious diseases, which include malaria, tuberculosis, and AIDS. Nevertheless, cancer is important in these countries as well, with different patterns of incidence and often higher mortalities. Cancers of the stomach (→17), liver (→16), bladder (→14), esophagus, and the cervix are each endemic in certain parts of the world (Figure 1.2). Often, they manifest at younger ages than in industrialized countries. Conversely, of the major four cancers in industrialized countries, only lung cancer has the same impact in developing countries.

# CHAPTER 1

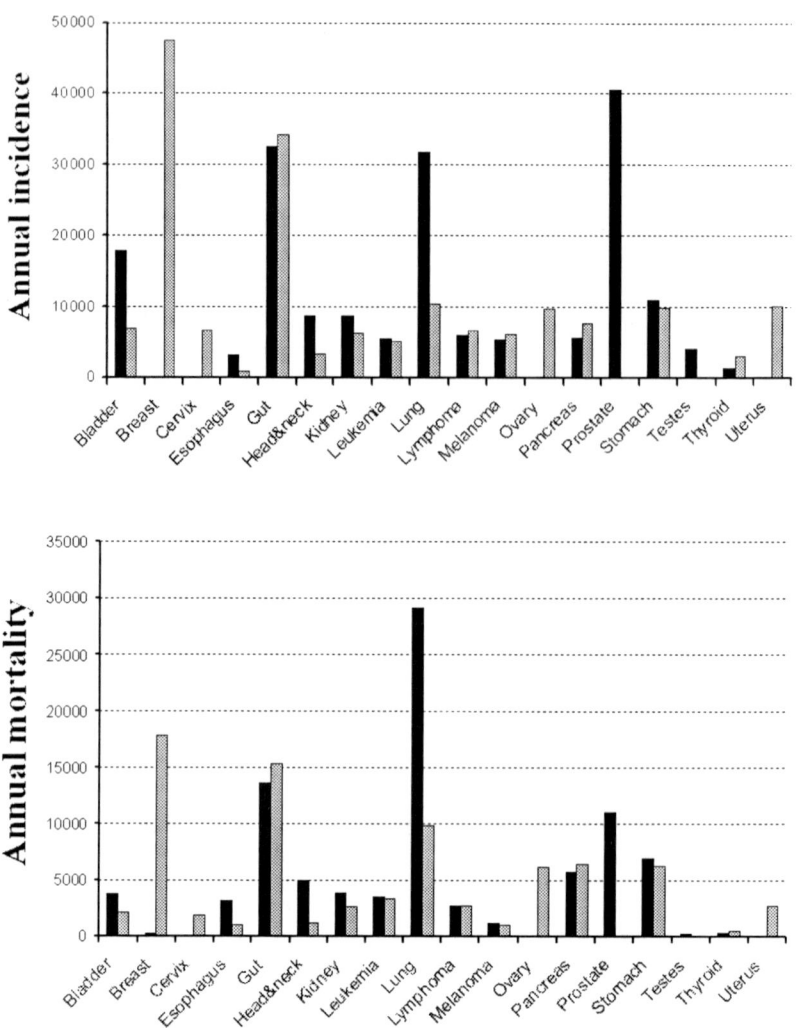

**Figure 1.1** *Incidence (top) and mortality (bottom) of cancers (cases per year) by organ site for females (grey bars) and males (black bars) in Germany in 2000.*
Data are from the Robert Koch Institute (www.rki.de).

This snapshot view of present-day cancer incidence of course conceals changes over time (Figure 1.3). For instance, large-scale industrialization and the spread of cigarette smoking are generally associated with an increased incidence of lung, kidney, and bladder cancer. On the positive side, improvements in general hygiene and food quality may have contributed to the spectacular decrease in stomach cancer incidence that is continuing in industrialized countries (→17.1). On the negative side, prostate and testicular cancer appear to have increased over the last decades. In prostate cancer, a slight increase in the age-adjusted incidence is exacerbated by the overall aging of the population (→19.1).In some regions, the incidence of melanoma has escalated in an alarming fashion. This increase is not related to the aging of the population, but perhaps to life-style factors (→12.1).

One important aim of molecular biology research on human cancers is to understand the causes underlying the geographical and temporal differences in cancer incidence. This understanding is one important prerequisite for cancer prevention (→20). Obviously, the prospects for prevention are brightest for those cancers that exhibit large geographical differences or the great changes over time in their incidences. To give just one example: The incidence of prostate cancer of East Asia residents may be 10-20-fold lower than that of their relatives who grow up in the USA (→19.1). It is easy imagining the potential for prevention, if the causes for this difference were understood.

Unfortunately, overall, neither incidence nor mortality of human cancer have been much diminished by conscious human intervention over the last decades. The mainstay of treatment of the 'big four' cancers and of the carcinomas in the second group outlined above remain surgery, radiotherapy, and chemotherapy, as they were 30 years ago. Surgery and radiotherapy are often successful in organ-confined cases, and chemotherapy is moderately efficacious for some advanced cancers. In general, only modest improvements have been made in cure and survival rates for these. Importantly, the quality of life for the patients is now widely accepted as a criterion

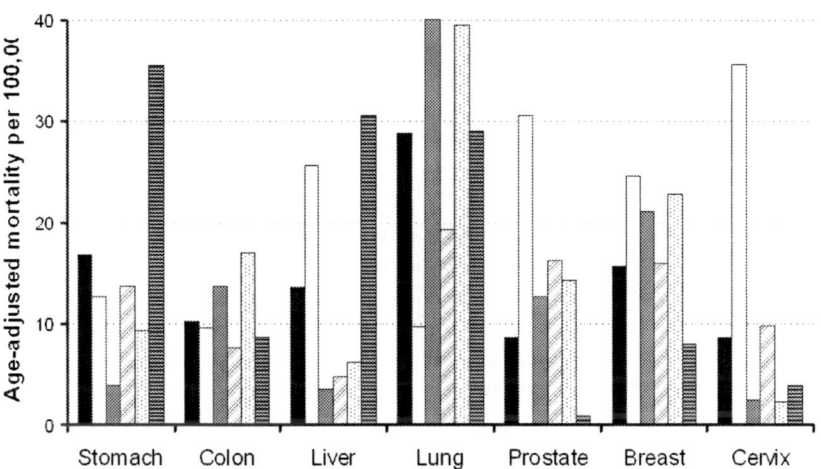

**Figure 1.2** *Mortality of selected cancers by organ site in different regions of the World*
In each group of bars from left to right: World average, Africa, North-America, South-America, North-West Europe, China. Data source: Shibaya et al, BMC Cancer 2, 37ff

for successful therapy. Modern cancer therapy recognizes that not every malignant tumor can be cured by the means presently available. So, treatment needs to be carefully chosen to maximize the chance for a cure while retaining a maximum of life quality. Providing a better basis for this choice will perhaps constitute the most immediate application of new insights on the molecular biology of cancers (→21). In addition, palliative treatments have become more sophisticated and pain medications are less restrictively administered. Nevertheless, the treatment of metastatic carcinomas remains the weakest point of current cancer therapy and a crucial goal of cancer research (→22).

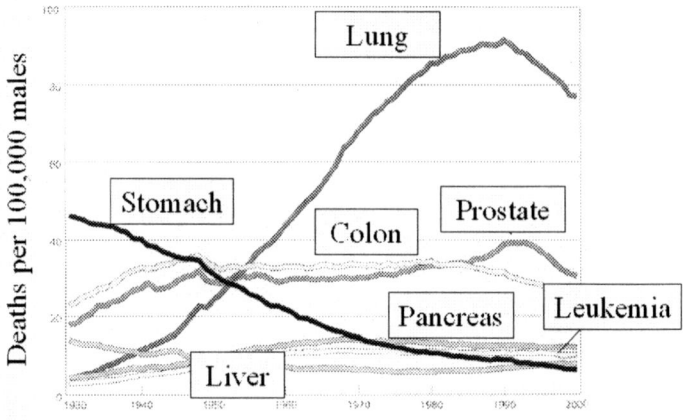

**Figure 1.3** *Trends in the mortality of selected cancers in the USA*
The original data figure is from the American Cancer Society.

Great steps towards successful treatment have been made with specific cancers, unfortunately mostly from the third group above. These improvement have had little effect on the impact of cancer on the overall population, but have helped many individuals, often young people and children. Formerly incurable leukemias and lymphomas can now be successfully treated by chemotherapy and/or stem cell transplantation, particularly in children and young adults. Likewise, the rise in testicular cancer incidence is stemmed by highly efficacious chemo- and radiotherapy, with cure rates exceeding 90%. Obviously, there is a need to understand why these cancers, but not others respond so well to the chemotherapeutic drugs currently available. It is hoped that a better understanding of the molecular and cellular basis underlying this difference will eventually open the door to successful treatment of the major carcinomas, as will the development of novel drugs and novel therapies based on the results of molecular biological cancer research (→22).

## 1.2 CAUSES OF CANCER

Since the genetic constitution of mankind hardly changes within a century and differs only moderately between human populations in different parts of the world, the changes in the incidences of individual cancers over time and their geographical variation to a large extent reflect environmental effects. Cancers are caused by exogenous chemical, physical, or biological carcinogens. They act on humans who, however, vary in their ability to cope with them due to differences in their genetic constitution ($\rightarrow$2.3, $\rightarrow$3.4) and – not to forget - their psychological, social, and economic conditions. Endogenous processes in the human body also contribute to the development of cancer, on their own or by interacting with exogenous agents.

The mechanisms of carcinogenesis in humans are often multifactorial and complex. Different factors may act by different mechanisms and at different stages of tumor development. In experimental animals carcinogens can be applied in a controlled fashion and the individual steps and interactions can therefore be analyzed more precisely. It is, e.g., possible in some cases to distinguish between initiating and promoting agents as well as complete carcinogens, or between carcinogens and co-carcinogens. In these laboratory models, initiating carcinogens are usually mutagens, while promoting agents act by facilitating the expansion of cells with altered DNA.

These distinctions are more difficult to apply in real human cancers. For instance, tobacco smoke is a human carcinogen, without a shade of doubt. In fact, it contains a variety of different carcinogens, some of which may act as initiators and some as promoters, and some as both. Nicotine itself is almost certainly not a direct carcinogen, but a potent alkaloid which acts not only on the central nervous system, but also influences cell signaling and cell interactions in the airways and in the lung. So, it would have to be classified as a co-carcinogen. Similarly complex interactions take place during skin carcinogenesis caused by UV radiation ($\rightarrow$12.1). Moreover, the actions of carcinogens and co-carcinogens are modulated by genetic differences in the human population, which is outbred, unlike many laboratory animals.

As a consequence, it is often difficult in humans to elucidate exactly by which mechanism a potential carcinogen acts, even though it is clearly identified as being associated with a specific cancer by epidemiological data. Attempts at prevention must therefore often be started before the relationship between a carcinogen and cancer development is fully understood ($\rightarrow$20). Nevertheless, precise elucidation of the mechanisms is helpful and insights from molecular biology are beginning to contribute to improved prevention of cancer ($\rightarrow$20).

Many carcinogens have been established as important in human cancer, in one or the other way. Exogenous carcinogens can be classified into chemical, physical, and biological agents. Table 1.1 provides an overview of these classes, with prominent examples for each class. For some carcinogens the evidence is very strong, while for others the notion 'carcinogen' has to be applied in a broader sense. Another type of classification issued by the World Health Organization groups human carcinogens by the level of available evidence. The most important criteria of this classification are listed in Table 1.2.

Chemical carcinogens come from different sources and comprise very different chemicals (Figure 1.4). Inorganic compounds like nickel, cadmium, or arsenic are encountered in the workplace or are present as contaminants in water. Organic compounds acting as carcinogens can be aliphatic, like nitrosamines, which occur in smoked and pickled foods, or trichloro-ethylene, which is used for cleaning. Nitrosamines are thought to contribute to stomach cancer, in particular (→17.1) Aromatic compounds like benzopyrenes and arylamines are generated from natural sources by burning, and are among the many carcinogens in tobacco smoke. They also present a danger in the workplace, e.g. during coal processing and dye production and use, respectively. Arylamines are thought to cause bladder cancer, in particular (→14.1). Polyaromates like benzopyrene are also released into the environment by burning of coal and fuels. Natural compounds produced by plants and molds can be highly carcinogenic. Aflatoxin B1 is implicated as a carcinogen in liver cancer (→16.1) and is the most infamous of many chemically diverse compounds in this group. Medical drugs can be carcinogenic, notably those used in cytostatic tumor therapy like cyclophosphamide, nitrogen mustards, and platinum compounds. Various hormones and hormone-like compounds from natural and pharmaceutic sources also influence the development of cancers in specific tissues, e.g. in the breast (→18.1) and prostate (→19.1). Doubtless, the most abundant exogenous carcinogen is oxygen. The form present in air, dioxygen, is relative inert and, of course, safe when fully reduced towards $H_2O$. However, partially reduced oxygen or dioxygen activated towards its singlet state are highly reactive and can be mutagenic (Box 1.1). Reactive oxygen species are formed at low levels during normal metabolism and are produced at increased rates during certain physiological processes such as immune defense and inflammation. Their concentrations can also be increased during the metabolism of some exogenous compounds, e.g. quinones (→20.2), and by pathophysiological states such as iron overload (→16.1).

**Table 1.1.** *Types and examples of human carcinogens*

| Type of carcinogen | Examples |
|---|---|
| Chemical carcinogens | Nickel, cadmium, arsenic, nitrosamines, trichloro-ethylene, arylamines, benzopyrene, aflatoxins, reactive oxygen species |
| Physical carcinogens | UV irradiation (specifically UVB), ionizing radiation |
| Biological carcinogens | Human papilloma virus (e.g. strain 16), Epstein-Barr-Virus, Hepatitis virus B, Helicobacter pylori, *Schistosoma mansoni* |
| Endogenous processes | DNA replication, metabolic reactions generating reactive oxygen species, chronic inflammation |

**Table 1.2.** *Classification of human carcinogens according to the WHO/IARC*

| Group | Definition |
|---|---|
| Group 1 | The agent is carcinogenic in humans. The exposure circumstance entails exposures that are carcinogenic to humans. |
| Group 2A | The agent is probably carcinogenic to humans. The exposure circumstance entails exposures that are probably carcinogenic to humans. |
| Group 2B | The agent is possibly carcinogenic to humans. The exposure circumstance entails exposures that are possibly carcinogenic to humans. |
| Group 3 | The agent (or exposure circumstance) is not classifiable as to carcinogenicity in humans. |
| Group 4 | The agent (or exposure circumstance) is probably not carcinogenic to humans. |

NB: the group definitions apply to single agents or mixtures.

Physical carcinogens: Any energy-rich radiation can in principle act as a carcinogen, depending on dose and absorption. Visible light is not usually carcinogenic, unless it is absorbed by 'photosensitizing agents' which generate reactive oxygen species. UVB irradiation is an important carcinogen in the skin (→12.1), and its effect is augmented by UVA. In contrast, UVC is strongly absorbed in the non-cellular protective layers of the skin and does not usually act as a carcinogen. $\gamma$-Radiation from natural, industrial, and iatrogenic sources (e.g., used in X-ray diagnostics) can penetrate into and through the body. It is carcinogenic to the extent to which it is absorbed, damaging DNA and cells by direct absorption but also indirectly by generating reactive oxygen species. Damage and carcinogenicity by $\gamma$-radiation therefore depend on the concentration of oxygen and also on the repair capacity (→3.3). Radioactive $\beta$-radiation and specifically $\alpha$-radiation is most dangerous when nuclides are incorporated, e.g. of cesium, uranium, and plutonium. The effect of radioactive isotopes depends also on their distribution in the body. For instance, radioactive iodine is accumulated in the thyroid gland and therefore causes specifically thyroid cancers, whereas radioactive cesium isotopes tend to become enriched in the urinary bladder. The potential carcinogenicity of microwave and radio wavelength electromagnetic radiation are, of course, the subject of public debate.

Biological carcinogens: Certain viruses and bacteria act as biological carcinogens in man. Specific strains of human papilloma viruses (HPV16 and HPV18) are established as causative factors in cervical and other genital cancers (→Box 5.1). They also influence the development of cancers of the skin and of the head and neck, and perhaps others as well. Papovaviruses like SV40 (simian virus 40) cause cancers in animals and partially transform human cells in vitro (→5.3), but whether they actually cause human cancers is a controversial issue. The best evidence exists for mesothelioma, a rare cancer which may be caused by the combined action of SV40 and asbestos. Specific herpes viruses can also act as carcinogens or co-carcinogens, e.g. human herpes virus 8 (HHV8) in Kaposi sarcoma (→ Box 8.1) or Epstein-Barr virus (EBV) in lymphomas (→10.3). The hepatitis B virus (HBV) with its DNA genome is certainly involved in the causation of liver cancer, although in a complex fashion (→16.3), as is the hepatitis C virus (HCV) which has an RNA genome. Human retroviruses such as HIV facilitate the development of cancers mostly by interfering with the immune system, but HTLV1 (human T-cell leukemia virus) causes a rare leukemia by direct growth stimulation of T-cells (→Box 4.1).

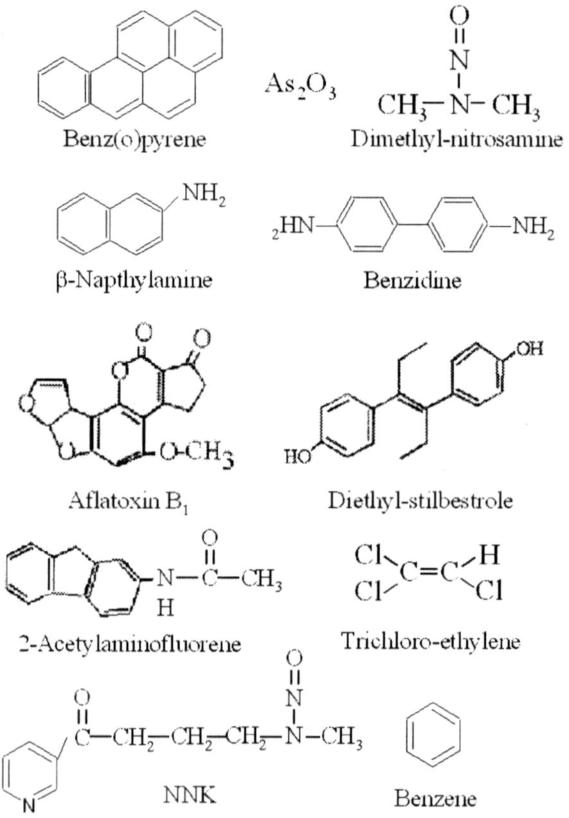

**Figure 1.4** *Some chemical carcinogens*
NNK is 4-(Methyl-nitrosamino)-1-(3-pyridyl)-1-butanone, a nitrosamine like dimethylnitrosamine in the upper right corner.

While there are many speculations on a carcinogenic role of bacteria, definitive evidence exists for a relationship between *Helicobacter pylori* infection and stomach cancer (→17.3). More generally, bacterial infections may contribute to inflammation which can promote cancer development. Schistosoma trematodes also cause cancer in humans, mostly in the urinary bladder (→14.1).

Endogenous carcinogens: How effective exogenous carcinogens elicit cancer in a specific person depends strongly on an individual's exposure, specific responses, and general health. So, endogenous processes are in any case involved in cancer development through modulation of the response to exogenous carcinogens. However, cancers may also be caused by strictly endogenous processes:

➢ Normal metabolism generates carcinogenic compounds such as nitrosamines, aromatic amines, quinones, reactive aldehydes, and – as mentioned above – reactive oxygen species. The concentration of these potential carcinogens may vary depending on factors like diet or physical activity, but a minimum level is associated with any level of metabolic activity and any type of diet. Potent protective and detoxification mechanisms exist for many of these compounds, but they can never be perfect.

➢ In the same vein, damage to cells and specifically DNA occurs at a minimum rate spontaneously and particular during cell proliferation, e.g. by errors in replication or by spontaneous chemical reactions of DNA bases (→3.1). The potentially huge number of such errors is kept at bay by very efficient DNA repair mechanisms specifically directed at typical errors of this kind. In addition, damaged cells are removed by apoptosis and other mechanisms. These protective mechanisms, however, cannot be perfect, either.

➢ There is, consequentially, some evidence that damaged DNA and cells accumulate with age and that protective mechanisms may become less efficient in the course of a human life. These factors may contribute to the increase of cancer incidence with age, although it is not clear, to which extent.

➢ While each of the above processes takes place in any human anytime, the risk of carcinogenesis is certainly higher during specific phases, e.g. when tissues proliferate after incurring damage. A period in human life with particularly high proliferative activity is, of course, fetal development. Genetic and even epigenetic errors occuring during this period may lead to cancer in children (→11), but also favor cancer development much later in life.

➢ Some pathophysiological conditions may increase the risk of cancer development. Chronic inflammation, in particular, is associated with increased cancer risk in many organs, e.g. colon (→13.6) or stomach (→17.1). Several factors are involved, including an increased production of mutagenic reactive oxygen species by inflammatory cells as well as secretion of proteases, cytokines, and growth factors by various cell types in the tissue that favor the growth and spreading of tumors (→9). So, tissue regeneration in general is associated with an increased cancer risk, but particularly, if it involves remodeling as in liver cirrhosis (→16.1) or in cystic kidneys (→15.3). Some carcinogens have, in fact, been proposed to act simply by stimulating tissue growth and rebuilding without being actually mutagenic.

## Box 1.1 Reactive oxygen species

Oxygen is present throughout the human body as the relatively unreactive dioxygen molecule in its triplet state (•O=O•). Singlet oxygen (|O≡O) is much more reactive. It can be generated by energy transfer from photoactivated compounds, e.g. porphyrins, and can be deactivated by transfering its energy to water molecules in a few µs. It also can add to double bonds in biomolecules, e.g. of unsaturated fatty acids, yielding unstable endoperoxidase that initiate chain reaction leading to further toxic and perhaps mutagenic compounds.

By taking up 4 electrons and 4 protons, e.g. in the cytochrome oxidase reaction in the mitochondria, $O_2$ yields $H_2O$. Water is of course innocuous, but all intermediate stages of this reduction are not.

Superoxide ($O_2^{\bullet -}$) is a byproduct of several enzymatic reactions, but in the cell may mostly be formed by leakage of electrons from electron transport chains in the mitochondria and ER. It is highly reactive and initiates chain reactions with coenzymes, nucleotides and thiol groups of enzymes. Like $H_2O_2$, it reacts with metal ions. $H_2O_2$ is a general oxidant, although not quite as reactive as superoxide. However, in Fenton type reactions (→Fig. 16.3), it yields the extremely reactive hydroxyl radical, which has a ns half-life, oxidizing essentially any biomolecule it encounters. It is highly mutagenic.

A certain level of reactive oxygen species and other radical molecules in the cell is normal. In fact, cells use reactive molecules like singlet oxygen and NO for signaling and specialized cells, e.g. macrophages, generate them to kill infectious agents.

Nitrous oxide, NO•, is another radical molecule used for signaling. It has a short half-life, but can also react with multiple biomolecules, in particular with the amino groups of DNA bases and proteins. Specifically, it can combine with superoxide to peroxynitrite ($ONOO^-$). This compound has a longer half-life. So the reaction potentially decreases the toxicity of the parent compounds, but in fact is a mixed blessing, because nitration of some amino groups in the cell and mutagenicity may be enhanced.

The amount of reactive oxygen species in a cell is normally contained by a variety of protective molecules, both low molecular weight and enzymes (→3.5), which remove reactive and prevent radical chain reactions that they initiate. One can regard these as anti-oxidants. If the level of pro-oxidant molecules surpasses that of anti-oxidants, a state of 'oxidative stress' arises. This may lead to cell death by necrosis or apoptosis or, if transient, to damage that can be repaired. In particular, certain DNA repair mechanisms are specifically targeted at the reaction products arising by oxidative stress.

B. Halliwell & J.M.C. Gutteridge (1999) 'Free radicals in biology and medicine', Oxford University Press, 3rd edition.

In summary, both endogenous and exogenous factors can be responsible for human cancers. In many cancers, they interact so intricately that their contributions are difficult to discern. In some cases, the involvement of specific carcinogens can be identified by characteristic mutations (→12.1, →16.1). In other cases, the absence of such 'fingerprints' and the epidemiological evidence point to a predominance of endogenous processes (→19.1). On this background, estimates of 5% of all human cancers being due to occupational carcinogens or 15% being caused by viruses have to be taken with more than one grain of salt. They do provide, however, rough estimates of how much could be achieved by prevention.

## 1.3 CHARACTERISTIC PROPERTIES OF CANCERS AND CANCER CELLS

In spite of their diversity, human cancers share several fundamental properties (Table 1.3). Different cancers display each of these to different extents. Moreover, these properties may be acquired step by step and become evident at various stages during the progression of a cancer. Most of these properties individually are also found in other diseases and some are even exhibited during physiological adaptive responses. However, the combination of uncontrolled cell proliferation, altered differentiation and metabolism, genomic instability, and invasiveness with eventual metastasis is unique to and defines cancer.

<u>Increased and autonomous cell proliferation</u>: The most obvious property of tumors is growth beyond normal measures. In fact, the term 'tumor' when used in a broader sense designates every abnormally large structure in the human body, also including swellings or fluid-filled cysts. More precisely, then, cancers belong to those tumors caused primarily by increased cell proliferation, i.e. a permanent and continuing increase in cell numbers. Increased cell proliferation as such is also observed during tissue regeneration, adaptative tissue growth, and in some non-cancerous diseases. For instance, atherosclerosis can also with some right be regarded as a tumor disease. In general, an increased number of cells in a tissue is designated 'hyperplasia'. Extensive hyperplasia or hyperplasia with additional changes such as altered differentiation ('dysplasia') is considered a 'benign' tumor. Dysplasia often precedes malignant tumors and in such cases is regarded as a 'preneoplastic' change. Substantial alterations in the tissue structure and in particular the presence of tumor cell invasion define a malignant tumor or cancer (Table 1.4). The borderlines between hyperplasia, benign tumors, and cancer are often evident from microscopic or even macroscopic inspection, but in some cases additional criteria or markers have to be employed to make the distinction. Hyperproliferation in cancers is brought about by an altered response to exogenous growth regulatory signals (→6). On one hand, cancers are often hypersensitive to growth-stimulatory signals, and some cancers become largely independent of them. On the other hand, sensitivity to growth-inhibitory signals is usually diminished or abolished. Together, these altered responses result in the growth autonomy that characterizes cancers. Moreover, it typically increases during their progression.

**Table 1.3.** *Characteristic properties of human cancers*

| Property |
| --- |
| Increased cell proliferation (often autonomous) |
| Insufficient apoptosis |
| Altered cell and tissue differentiation |
| Altered metabolism |
| Genomic instability |
| Immortalization (growth beyond replicative senescence) |
| Invasion into different tissue layers and other tissues (with disturbed tissue architecture) |
| Metastasis into local lymph nodes and distant tissues |

Insufficient apoptosis: Cell proliferation in cancers may be caused by a combination of three factors: (1) The rate of cell proliferation is enhanced by an increase in the proportion of cells with an active cell cycle, i.e., a higher 'proliferative fraction', and/or by a more rapid transit through the cell cycle, together resulting in increased DNA synthesis and mitosis. (2) The rate of cell death is often decreased by relatively diminished apoptosis ($\rightarrow$7). In some cancers, this is in fact the driving force for increased proliferation. In others, the rate of apoptosis is enhanced compared to normal tissue; but not sufficiently so as to compensate for the increase in mitotic activity. (3) In normal tissues, successive stages of differentiation are typically associated with progressively decreased proliferative capacity and/or with apoptotic death of the fully differentiated cells ($\rightarrow$7.1). Thus, a block to differentiation is in some cases sufficient to confer an increased proliferation rate.

Altered differentiation: Many cancers consist of cells which resemble precursor cells of their tissue of origin and have not embarked on the normal course of differentiation, whereas others show properties of cells at intermediate stages of differentiation. Some cancers, however, do consist of cells with markers of full differentiation, with the crucial difference that they continue to proliferate. In these cancers, it is not difficult to identify the cell of origin, which is important for diagnosis. Many cancers, however, express markers that do not occur in their tissue of origin ($\rightarrow$12.5). Frequently, cancer cells express proteins which are otherwise only found in fetal cells. Such proteins, e.g. carcinoembryonic antigen in colon carcinoma or alpha-fetoprotein in liver cancer are called 'oncofetal' markers. Other proteins expressed in cancers are never synthesized in the original cell type, e.g. 'cancer testis antigens' in melanoma and various peptide hormones in small cell lung cancer. This phenomenon is called 'ectopic' expression. Some cancers change their phenotype to resemble cells from a different tissue in a process called 'metaplasia'. One might think that this is a clear hallmark of cancer, but metaplasia occurs also in some comparatively innocuous conditions. Metaplasia can, in fact, precede cancer development, e.g. during the development of a specific type of stomach cancer ($\rightarrow$17). In some other carcinomas, metaplasia may be a late event. Other changes of cell differentiation obliterate the original cellular phenotype so strongly that it can be difficult to distinguish from which primary site a metastasis

**Table 1.4.** *Some basic definitions in oncology*

| Designation | Meaning | Remarks |
|---|---|---|
| Tumor | any abnormal increase in the size of a tissue | also used for swellings, unusual for benign hypertrophy or hyperplasia |
| Malignant tumor | a tumor characterized by permanently increased cell proliferation, progressive growth, and invasion or metastasis | corresponding to 'cancer' in everyday language |
| Benign tumor | a tumor lacking growth beyond a circumscribed region within a tissue | |
| Cancer | a malignant tumor | preferentially used for (suspected or verified) systemic disease |
| Neoplasia | a malignant tumor | |
| Leukemia | a malignant tumor formed by cells of the hematopoetic cells and found in the blood | |
| Lymphoma | a malignant tumor formed by cells of the lymphocyte cell lineage | can be restricted to specific lymphoid organs |
| Sarcoma | a solid malignant tumor formed from connective tissue (mesenchymal) cells | |
| Carcinoma | a solid malignant tumor formed from cells of epithelial origin | |
| Adenoma | a benign tumor displaying a glandular structure | often originated from gland tissue |
| Adenocarcinoma | a malignant tumor showing resemblance to glandular structures | often originated from gland tissue |
| Tumor stage | a measure of the physical extension of a (malignant) tumor | different systems are in use, for different (and even the same) cancer types |
| Tumor grade | a measure of the cellular and/or architectural atypia of a tumor | different systems are in use, for different (and even the same) cancer types |

originates. Two such 'generic' cell types are a small epithelial-like cell with a large nucleus to cytoplasm ratio and a spindle-shaped cell resembling a mesenchymal fibroblast. These cell types are end points of cancer progression in some carcinoma cases, typically found in aggressive cases and therefore also in metastases.

So, altered differentiation confers properties to cancer cells that are otherwise found in tissue precursor cells, fetal cells or cells of other tissues. Moreover, altered differentiation is also related to increased proliferation. As pointed out above, the control of proliferation and differentiation are intimately linked in normal tissues. The final stages of differentiation of many normal tissues are associated with an irreversible loss of replicative potential or even with cell death. This process is therefore called 'terminal differentiation' ($\rightarrow$7.1). For instance, differentiated cells in keratinizing epithelia crosslink with each other, dissolve their nuclei, and become filled with structural proteins. This way, a steady state between cell generation and loss is maintained which breaks down, if differentiation fails in a cancer.

Altered metabolism: Cell proliferation, whether normal or abnormal, requires according changes in cell metabolism. Most evidently, DNA synthesis requires deoxynucleotides, so enzymes required for nucleotide biosynthesis, and specifically of deoxynucleotide biosynthesis, are induced and activated in proliferating cells. Further cell components such as membranes and organelles also need to be duplicated. For this reason, lipid biosynthesis is increased in cancer cells, likely because they cannot obtain enough fatty acids, phospholipids and cholesterol from lipoproteins supplied by the gut and liver. As a consequence, expression and activity of key enzymes like fatty acid synthase and hydroxymethylglutaryl-coenzyme A reductase are increased in cancer cells. Porphyrin biosynthesis is also often increased. As Warburg already noted in the 1930's, many tumor cells switch from aerobic to anaerobic glucose metabolism.

A key requirement for cell growth is increased protein synthesis, which is apparent at several levels by enhanced size and number of nucleoli, increased expression of transcription initiation factors and enhanced phosphorylation of ribosomal proteins. A particularly strong boost in protein synthesis may be required during invasion and metastasis ($\rightarrow$9).

Overall, cancer growth poses an enhanced energy demand on the patient, which increases with the tumor load. Moreover, cancers release the waste products of their metabolism, such as lactate, with which the body has to cope. These are, unfortunately, only some of the systemic effects of cancers. Cancers also secrete enzymes and hormones that act on the host, some of which are toxic. In particular, cytokines like tumor necrosis factor $\alpha$ can elicit a general break-down of metabolic function with visible wasting, termed 'cachexia', and suppression of the immune system, thereby facilitating 'opportunistic' infections. Other tumor products, such as FAS ligand ($\rightarrow$7.3), can damage sensitive organs such as the liver, and ectopically produced hormones can interfere with homeostasis. For instance, calcitonin production by small cell lung cancers may cause life-threatening variations in calcium levels. Such indirect disturbances of the body homeostasis by cancers, designated as 'paraneoplastic' symptoms, can be as problematic for the well-being and survival of a patient as the malignant growth per se.

Genomic instability: A clear distinction between cancerous and non-cancerous cell proliferation lies in genomic instability. Cancer cells as a rule contain multiple genetic and epigenetic alterations (→2). Polyploidy, an increase in the number of genomes per cell, can be ascertained by measuring cellular DNA content. Aneuploidy, i.e. a change in the number and structure of individual chromosomes, is revealed by cytogenetic methods. These aberrations are often already revealed upon microscopic observation of tumor tissues by altered size and shape of the nuclei in the cancer cells and aberrant mitotic figures. Other cancers remain diploid or nearly so, but contain point mutations and/or altered DNA methylation patterns.

As cancers progress, the numbers of alterations in their genome tend to increase. Therefore, cancers, even if outwardly homogeneous, usually consist of cell clones that differ at least slightly in their genetic constitution. The variant clones are continuously selected for those proliferating fastest, tolerating adverse conditions best, capable of evading immune responses, etc., with the best-adapted cell clone dominating growth (Figure 1.5). This variation becomes particularly evident during tumor treatment by chemotherapy which exerts a strong selection pressure for those cell clones carrying alterations that allow them to survive and continue to expand in spite of therapy.

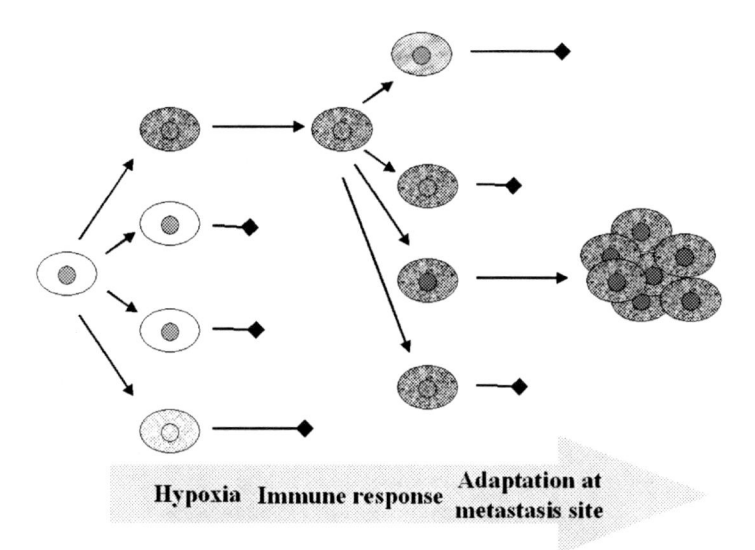

**Figure 1.5** *Clonal selection model of cancer growth.*
Genomic instability in a cancer continuously creates novel clones from the initial tumor (left). These clones are selected according to their ability to proliferate in the face of hypoxia and immune responses, and to adapt at metastatic sites. One (right) or several clones may succeed.

There is some debate, whether some cancers have simply accumulated many mutations during their development or whether all exhibit genomic instability leading to an increased rate of chromosome alterations, point mutations, and/or epigenetic defects (→2.5). This is not an academic question, because cancers with genomic instability will display greater variation and a higher risk of developing resistance. It seems indeed possible that true genomic instability develops in some cancers during progression, e.g. in CML (→10.4). Genomic instability in cancer cells can be derived from several sources, e.g. from defects in DNA repair and in mechanisms checking genomic integrity (→3.4).

Immortalization: Many cancer cells are 'immortalized', which means they are capable of a theoretically infinite number of cell divisions. Most human cells can undergo only a finite number of divisions, likely up to 60-80, before they irreversibly lose their ability to proliferate (→7.4). Obviously, the cells constituting the germ line are exempt from this restriction and so are tissue stem cells. For instance, hematopoetic stem cells in the bone marrow can be successively transplanted across several recipients and still remain capable of reconstituting the entire hematopoetic system, blood and immune cells. Immortality in stem cells is maintained by specific mechanisms such as expression of telomerase (→7.4), which is also found in cancers. Moreover, some human cancer cells can be maintained in tissue culture or as transplants in animals, designated 'xenografts', over many generations, as far as we can tell, infinitely many. On a note of caution, it is not certain that all human cancers are immortalized, since many cannot be grown in tissue culture or as xenografts. Even telomerase expression is not universal. To become life-threatening, however, a cancer does not need to consist of cells with infinite growth potential. Starting out from a single cell, 50 replications would yield up to $2^{49}$ tumor cells, which must be compared to something between $10^{13}$ and $10^{14}$ normal cells in a human. Lethal cancers are much smaller than that.

Invasion and metastasis: A property more directly evident in human cancers is their ability for invasion and metastasis. Invasion and metastasis (→9) are the definitive criteria which distinguish benign from malignant tumors (the expression 'malignant tumor' is synonymous with cancer, Table 1.4). Moreover, invasion and metastasis, with tumor cachexia and immune suppression, account for most of the lethality of human cancers.

During invasion, cancers spread from their site of origin into different layers and parts of the same tissue, eventually growing beyond it and into neighboring structures. Invasion involves multiple steps and often substantial rebuilding of the tissue structure by the tumor cells, by other cells in the tissue responding to signals from the tumor cells, and by immune and inflammatory cells. Typically, in carcinomas, the basement membrane separating epithelium and mesenchyme is destroyed and tumor extensions push through the connective tissue and muscle layers. From some cancers, cells separate and migrate through the neighboring tissues, as single cells, in an Indian file pattern or as small, adherent cell clusters. Invasion is often accompanied by inflammation, so lymphocytes, granulocytes and macrophages are present in the invaded tissue and in the tumor mass.

An important component of malignant growth is neoangiogenesis (→9.4). The nutrient and oxygen supply from preexisting blood vessels is usually not sufficient to support growth of tumors beyond a size of a few mm. Therefore, cancers, but also some benign tumors, induce neoangiogenesis, which comprises the growth of new capillaries, and the rebuilding of existing blood vessels (→9.4). Lymph vessels can also be remodeled or newly formed.

During metastasis, cancer cells separate from the primary tumor and migrate by the blood or lymph to different organs where they form new tumors. Depending on the route, 'hematogenic' metastasis, which usually leads to metastases at distant organ sites, is distinguished from 'lymphogenic' metastasis, which leads initially to the formation of metastases in lymph nodes draining the region from which the cancer emerges. Like invasion, metastasis is really a multistep process. Thus, many more cancer cells enter the blood or lymph than actually form metastases. Important barriers are posed by the necessity to leave the blood stream at capillaries which carcinoma cells cannot pass ('extravasation') and to survive and resume proliferation in the microenvironment of a different tissue. In fact, individual cancer cells or small groups may end up in a different tissue only to survive over long periods without net growth. These 'micrometastases' are not detectable by current imaging techniquew, although they may be biochemically detectable by proteins secreted by the cancer cells. Over time, they may adapt to their new environment and expand to larger metastases that threaten the patient's life. This may occur several years after the primary tumor has been removed. Cancers differ in the extent and the sites to which they metastasize (→9). Generally, preferred organs for metastasis are those with extended microcapillary systems such as liver, lung, and bone.

## 1.4 CHARACTERIZATION AND CLASSIFICATION OF CANCERS IN THE CLINIC

Many properties of cancers described in the previous section are reflected in the terms and methods used in the clinic and in diagnostic histopathology to describe and classify cancers as a prequisite for appropriate treatment and for prognosis. For these purposes, as well as for cancer research using specimens of human cancers, it is mandatory to obtain as exact as possible descriptions of the extension of the tumor, of its degree of malignancy and of its histological subtype.

Staging: The extension of a tumor is described by 'staging'. Prior to surgery or if none is performed, a clinical stage is defined by visual inspection, palpation and various imaging techniques. These techniques use a.o. ultrasound, X-rays, scintigraphy, computer tomography, magnetic resonance, and positron emission tomography. Some imaging procedures detect changes in tissue shape and density, whereas others react to changes in metabolism and blood flow in cancers. If surgery is performed, a more precise delineation of the extension of the tumor can be made by inspection of the tumor site and by histopathological investigation of the specimen. The stage defined in this fashion is called pathological stage. It is denoted by a 'p' prefix to distinguish it from clinical stage, which is denoted by a 'c'.

## Box 1.2: Hallmarks of Cancer

In a millenium issue article which appeared on January 7$^{th}$, 2000 (1), a set of characteristic properties of cancers has been proposed. The traits considered as 'hallmarks of cancer' by Hanahan and Weinberg comprise

- self-sufficiency in growth signals
- insensitivity to anti-growth signals
- evasion of apoptosis
- limitless replicative potential
- sustained angiogenesis
- tissue invasion and metastasis.

These criteria are widely used in experimental cancer research and preclinical development of anti-cancer therapies. A comparison shows that they are not too different from those listed in Table 1.3. They are not identical, though, because the two lists reflect slightly different angles of view. The 'hallmarks of cancer' are based to a greater degree on experience in experimental models and on insights from cell biology research. Each trait can be found and can be analyzed as such in cancer cells and in animal models. In selected models, one or the other of these properties can even be generated or suppressed by genetic manipulation. It is, however, not always possible to ascertain each trait in actual individual human cancers, and, in fact, the individual traits may apply to different human cancers to different degrees.

The properties discussed in section 1.3 are to a greater degree oriented at observations on human cancers. They are consequently more descriptive and their molecular basis is often incompletely understood. Unfortunately, at present, not all human cancers can be studied in adequate models. This is one of the more severe problems in cancer research. The transfer of experimental insights to the clinic, called 'translation', requires good models and an understanding of real human cancers alike (cf. Part III). Indeed, in today's best cancer research, experimental data obtained using cell and animal models are combined with descriptive data on human cancer tissues.

(1) Hanahan D, Weinberg RA (2000) The hallmarks of cancer. Cell 100, 57-70.

Several staging systems are employed. The most widely used and systematic staging system is the TNM classification, while others remain in use for specific cancers. In the TNM system, the extent of the primary tumor is normally described by T1-T4, where increasing numbers describe larger and/or more invasive tumors (Figure 1.6). The system varies for different tumor sites (cf. 21.1). The presence of cancer cells in lymph nodes is denoted by N0, N1, and in some cancers also N2, with N0 meaning none detected. The presence of metastases is indicated by M0 meaning none detected, M1, or in some cancers also M2. After surgery, it is also important to know whether all of the local tumor growth has been removed. This is designated by the R value. R stands for resection margin, so R0 means that the tumor seems to be wholly contained within the removed specimen. In all categories, the affix 'x' is used for 'not determined/unknown'.

Grading: The degree of malignancy of a tumor is further estimated by grading systems. Again, several systems are in use for different tumors and in different countries. To different extents they score the degree of cellular and nuclear atypia and/or the degree of tissue disorganization in tumor sections, biopsies, or even single tumor cells. The most prevalent system is G grading, which usually ranks from G0 to G4 (Figure 1.7). The designation G0 typically denotes normal differentiation and no cellular atypia, as would be found in a benign tumor. At the other end, G4 would be assigned to cancers with a cellular morphology completely different from the normal tissue and pronounced atypia of the cells and nuclei. The grades G1-G3 are called well-differentiated, moderately and poorly differentiated. It is important to be aware that the use of 'differentiation' in this context is different from that in cell or developmental biology. This can be confusing, the more so as in modern pathology grading based on morphology is sometimes supported by staining for specific markers of cell differentiation or of cell proliferation. For instance, immunohistochemical staining against PCNA, a subunit of the DNA replisome mainly expressed in S-phase cells, and Ki67, a protein essentially restricted to actively cycling cells, can be used to estimate the proliferative fraction in a tumor, in addition to estimating the proliferative rate by counting mitotic figures.

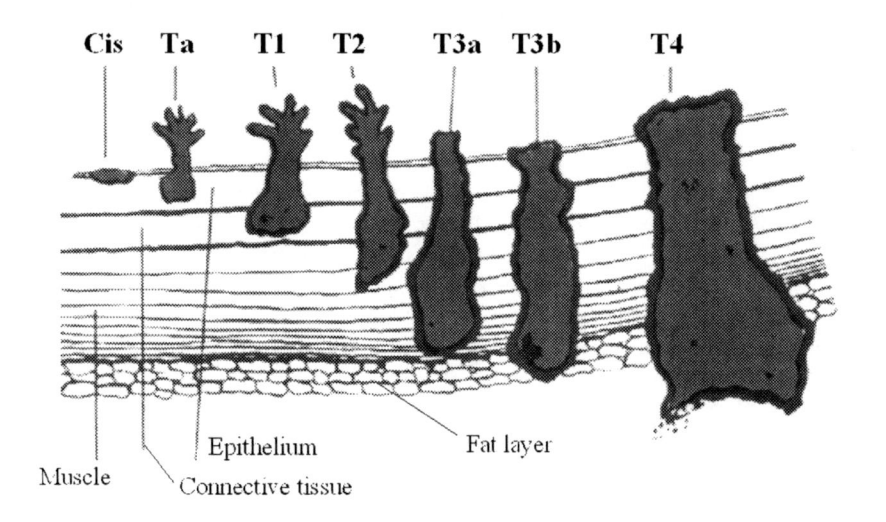

**Figure 1.6** *Tumor staging in bladder cancers*
Carcinoma in situ (CIS) is a severe dysplasia restricted to the epithelial (urothelial) layer. Ta tumors are papillary structures formed by hyperproliferative urothelium growing into the lumen of the urinary tract, but not into the underlying tissue layers. T1 tumors extend into the connective tissue layer, T2 tumors into the muscular layers, and T3 tumors into the fat layer. Suffixes a and b are used to further distinguish the extent of invasion. T4 cancers have grown into neighboring organs, such as the prostate or the uterus.

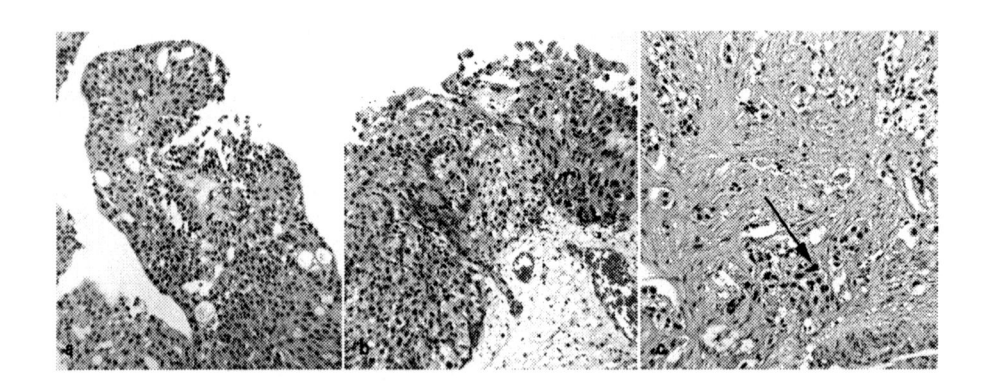

**Figure 1.7** *An example of tumor grading by the G system*
Grading of urothelial cancer: Note increasing nuclear size and pleomorphism from G1 (left) via G2 (center) to G3 (right). The arrow points to examplary markedly enlarged nuclei in the poorly differentiated cancer.

Histological classification: The location of a tumor is the first clue to its classification. It is, of course, not sufficient, since (1) a tumor mass may represent a metastasis or the extension of a cancer originated in a neighboring organ. Moreover, (2) several different kinds of cancer may develop within one tissue, often with very different properties, clinical course and treatment options. Therefore, a tumor must be histologically classified from samples acquired by biopsy or from surgical specimens. Several of the designations used in this context have already been introduced in this chapter and are summarized in Table 1.4. Histological typing of tumors is performed by evaluating their morphology. Routine procedures use a variety of specific stains developed over centuries in anatomy and pathology to highlight particular cell types as well as extracellular structures like basement membranes, fibers or mucous. Increasingly, tumor classification by histopathological investigation is being improved by specific molecular markers. Immunohistochemical staining with antibodies directed against specific antigens of the presumed tissue of origin, e.g. cytokeratins, or tumor-specific antigens, e.g. carcinoembryonic antigen, is often performed. For leukemias, analysis of subtypes can be determined by antibody staining followed by flow cytometry. Analyses at the RNA or DNA level for specific patterns of gene expression or specific genetic alterations are not standard yet, but are employed by specialized institutions (→21). Similarly, cytogenetic techniques are becoming more widely used, particularly for the classification of hematological cancers (→21.2).

Insights into the molecular biology of human cancers have begun to improve staging, grading, and histological classification. These improvements have an immediate impact on cancer therapy, because in most cancers the choice of therapy is contingent on these parameters (→21.5). Even now, a tumor in the kidney, e.g., will be treated quite differently, if it is a malignant renal carcinoma at an early or at an advanced stage, a benign tumor of mesenchymal origin, a melanoma metastasis, or a lymphoma. However, recent and future molecular markers are expected to go far beyond such distinctions. They may reveal differences between cancers that look morphologically the same, but represent different diseases, as do certain leukemias (→21.2). They may reveal previously unrecognized subclasses within one disease, as may be the case with breast cancer (→18.5) and may predict different clinical courses for morphologically similar tumors (→21.4). Moreover, molecular insights should allow selection of certain cancers for treatment with drugs tailored to specific targets and may allow to predict how well a patient tolerates cancer treatment by radiotherapy or cytostatic medication (→21.5).

## 1.5 TREATMENT OF CANCER

In principle, a range of different therapies is available for the treatment of human cancers. Surgery, irradiation or drugs can be employed, or a a combination of these. Which therapy is chosen depends strongly on the classification of the cancer by the criteria described in the previous section.

Surgery or radiation are treatment choices for localized cancers. In contrast, leukemias, lymphomas, and metastatic or locally advanced carcinomas and soft tissue cancers require drug chemotherapy, which is in some cases supplemented by radiotherapy or surgery of primary cancers or metastases. Conversely, surgery can be followed by chemotherapy or irradiation to attack residual local tumor or metastases. This is called 'adjuvant' treatment. Accordingly, chemotherapy applied before surgery to shrink the tumor mass and facilitate its complete resection is often called 'neo-adjuvant' treatment. The standard chemotherapy regime for a cancer is usually designated as 'first-line', if it fails, 'second-line' therapy can be attempted. The efficacies of chemotherapy and radiotherapy are extremely dependent on the tumor type. Some cancers, e.g. some testicular cancers and certain lymphomas, are highly sensitive, whereas others, e.g. renal cell carcinoma (→15.1), appear to be overall less sensitive than many normal tissues.

In cancer chemotherapy a wide range of different drugs are employed. Some drugs are aimed at the cancer itself, but other medications are employed to stabilize specific body functions in the patient or for pain relief. The most important component in the treatment of many cancers is cytotoxic drug therapy, often simply called chemotherapy. In this kind of therapy, chemical compounds are employed that block DNA synthesis, transcription, and/or mitosis in the cancer cells, often driving them into apoptosis (→22.2). A different type of anti-cancer drugs summarized as 'biological agents' bind to receptor molecules in the cancer cells that are not directly involved in DNA replication or mitosis, but regulate them. Examples for this type of drugs are hormones and antihormones used in the treatment of breast cancer (→18.4) and prostate cancer (→19.2) and retinoids used for several cancers, especially a specific acute leukemia, promyelocytic leukemia (→10.5). Since the advent of recombinant DNA biotechnology, cytokines and growth factors can be produced at reasonable cost and sufficient purity to be used in cancer therapy. In some cases, they act directly on the cancer cells, in other cases, they stimulate the immune response against the cancer, and in still another application, they stabilize the hematopoetic system of the patient against the effects of the cancer and the treatment. Interferons and interleukins used in the therapy of leukemias (→10.4) and renal carcinoma (→15.1) are examples for the first two applications. Erythropoetin stimulating erythrocyte production and GM-CSF enhancing hematopoesis more globally are examples for the third type.

Several different types of radiation can be used in cancer radiotherapy and even particles and radioactive isotopes. Most widely used is ionizing radiation in the form of high energy γ-radiation. It damages cancer cells by direct effects on cellular macromolecules or by generating reactive oxygen species, as during carcinogenesis (→1.2). Typically, radicals are induced which initiate chain reactions that lead to DNA double-strand breaks which prevent further cell proliferation or induce apoptosis. The radiation dose that can be applied is limited by its effects on normal tissues. Modern techniques allow an improved focussing of the radiation or use radioactive isotopes implanted into the tumor. However, these improvements are mostly useful for the treatment of localized cancers. In some cases, the tumor cells can be sensitized towards the radiation. For instance, the fact that many tumors have a higher rate of porphyrin biosynthesis than normal tissues is exploited in photodynamic therapy. In this technique, intense light with a wavelength near the absorption maximum of protoporphyrins is applied by a laser beam to generate reactive oxygen. In other cases, the preferential uptake of radioactive isotopes by the cancer can be exploited, e.g. in the therapy of well-differentiated thyroid cancers by radioactive iodine.

Even where the full range of modern cancer therapies is available, many cancers cannot be cured today. In not too few malignant diseases, state of the art therapies are at best palliative, i.e. symptoms caused by the cancer are alleviated, but survival is not or only slightly prolonged. Novel cancer therapies are urgently required. Therefore, cancer therapy is the area, in which expectations are highest for the applications resulting from insights into the molecular biology of cancers. Indeed, novel drugs have been developed based on such insights (→22.4). Moreover, the understanding of established therapies like cytotoxic chemotherapy and radiotherapy has also been deepened (→22.2). These developments are beginning to have a significant impact on cancer therapy. Evidently, it is hoped that completely novel therapies may emerge. Immunotherapy is already being applied in some cancers, albeit with highly variable efficacy. Since the causes for its successes and failures are becoming gradually elucidated, it may be more broadly used in the future (→22.5). Gene therapy of cancer also carries high hopes and several hundred clinical trials have been performed or are underway. At this stage, however, it is a purely experimental therapy (→22.6).

## *Further reading*

Selected textbooks and handbooks on molecular aspects of cancer and related diseases:

Allgayer H, Heiss MM, Schildberg FW (eds.) Molecular Staging of Cancer. Springer, 2003.
Brenner C, Duggan DJ (eds.) Oncogenomics: Molecular Approaches to Cancer. Wiley, 2004
Bronchud MH et al (eds.) Principles of Molecular Oncology. Humana Press, 2003
Casson AG, Ford CHJ (2003) Molecular Biology of Cancer. BIOS Scientific.
Coleman WB, Tsongalis GJ (2001) Molecular Basis of Human Cancer. Humana Press
Cowell JK (2001). Molecular Genetics of Cancer. Academic Press.
Crocker J, Murray P (eds.) Molecular Biology in Cellular Pathology 2$^{nd}$ ed. Wiley, 2003
Epstein RJ (2002) Human Molecular Biology. An Introduction to the Molecular Basis of Health and Disease. Cambridge University Press
Knowles M, Selby P (eds.). Introduction to the Cellular and Molecular Biology of Cancer 4$^{th}$ ed. Oxford University Press, 2004
Lalloo F (2001) Genetics for Oncologists: The Molecular Genetic Basis of Oncological Disorders. REMEDICA Publishing
McKinnell RG, Parchment RE (2005) The Biological Basis of Cancer. Cambridge University Press
Mendelsohn J et al (eds.) Molecular Basis of Cancer 2$^{nd}$ ed. Harcourt Publishers, 2001
Pitot HC (2002) Fundamentals of Oncology 4$^{th}$ ed. Marcel Dekker.
Ross DW (2004) Molecular Biology of Cancer 2$^{nd}$ ed. Springer
Strachan T, Read AP (2003) Medical Molecular Genetics. BIOS Scientific
Strausberg RL (ed.) Cancer Genomics: Disease Markers. IOS Press 2002
Sullivan NF. The Molecular Biology of the Cancer Cell. Cambridge University Press, 2002
Vogelstein & Kinzler (eds.) The genetic basis of human cancer 2$^{nd}$ ed. McGraw-Hill, 2002
Warshawsky D (2002) Molecular Carcinogenesis, CRC Press
www.cancerhandbook.net

# CHAPTER 2

# TUMOR GENETICS

- Cancer cells typically contain multiple alterations in the number and structure of genes and chromosomes.
- The majority of genetic alterations found in cancer cells are acquired by mutations in somatic cells. A few cancers in children and young adults are caused by genetic or epigenetic defects acquired during fetal development. Germ-line mutations underlie familial cancer syndromes. These can be inherited in an recessive or in a dominant fashion.
- The mutations causing inherited cancer syndromes increase the risk of cancers by orders of magnitude. Overall, they are relatively rare. In contrast, less dramatic inherited variations ('polymorphisms') in a large number of genes influence cancer risk only slightly to moderately, but are highly prevalent. Such genes encode, e.g., proteins involved in the metabolism of carcinogens and in protection against cell damage, in the regulation of immunity and inflammation, and in the metabolism of hormones and growth factors.
- Many different genetic alterations are observed in cancer cells. Individual genes display point mutations such as base changes, insertions and deletions, or can be affected by chromosomal translocations or inversions. These changes lead to the expression of altered gene products, to decreased or increased gene expression, or to novel gene products like fusion proteins. Moreover, cancer cells are often aneuploid exhibiting numerical and/or structural alterations of chromosomes. These comprise loss or gain of chromosomes or chromosomal parts as well as rearrangements and recombinations. The consequences for individual genes range from complete loss by homozygous deletion through decreased copy numbers and increased gene dosage to gene amplification. At polymorphic loci, loss of heterozygosity may occur as a consequence of deletions or recombination.
- In certain cancers, infection by DNA viruses and retroviruses alter the composition of the genome, adding new sequences and mutating existing ones as a consequence of insertion or induced deletions and rearrangements.
- These diverse types of genetic alterations occur to different extents in different types of cancers and even in cancers of the same type. So, in some cancers point mutations may prevail, while in others predominantly chromosomal aberrations are found.
- The diverse types of genetic changes result in altered patterns of gene expression and altered gene products in cancer cells. Changes in gene expression are compounded by epigenetic mechanisms. Epigenetics in general designates the stable inheritance of alterations in gene expression without changes in the DNA sequence.

➢ Two particular important classes of genes affected by genetic and epigenetic alterations in cancer cells are oncogenes and tumor suppressor genes. Oncogenes contribute to tumor development by increased or misdirected activity. In the case of tumor suppressors, conversely, insufficient or lost function supports tumor development. Typically, in human cancers activation of oncogenes and inactivation of tumor suppressor genes are both observed.

➢ Multiple genetic as well as epigenetic changes accumulate during the development of malignant tumors, of which many are necessary. To accumulate so many changes, cancer cells may need to acquire a 'mutator phenotype' during tumor initiation or progression. Indeed, cancer cells typically show defects in genome stability which lead to increased rates of fixed point mutations or chromosomal alterations or both.

## 2.1 CANCER AS A GENETIC DISEASE

The characteristic properties of cancer cells (→1.3) are to a large extent the consequences of genetic changes in the tumor cells. Indeed, genomic instability is one of the properties defining cancer and the aberrant structure of the nucleus seen in many cancer cells is an obvious morphological consequence of their altered genome. It is plausible that every cancer cell contains structural or numerical alterations of its genome. The number of alterations is not precisely known, and certainly varies with cancer type and stage of progression. Systematic DNA sequencing has yielded estimates of hundreds to thousands of point mutations in some tumors. Screening by arbitrary PCR has suggested an even higher number of alterations for some cancers. Certainly, 20 or more chromosomal aberrations detectable by cytogenetic techniques are not unusual in an advanced carcinoma.

It is therefore very appropriate to regard cancer as a genetic disease. Still, a few points must be kept in mind:

(1) Only a minority of cancers are caused by mutations inherited in the germ-line. Rather, the vast majority of genetic alterations found in cancers develop during the life of a patient in somatic cells. Thus, cancer is almost always a disease caused by 'somatic mutations'. Even in cancers which are passed on over several generations in a family, the initial inherited mutations are complemented by additional somatic mutations. Likewise, cancers arising in young children or adolescents are often caused by mutations originating de novo in their parents' germ cells or during intra-uterine development. A typical cancer of this kind is Wilms tumor (→11), but similar circumstances apply to testicular cancer and certain childhood leukemias.

(2) The relationship between mutant genotype and disease phenotype is not straightforward in cancer, which per se is not so unusual for genetic diseases. However, in cancer the relationship is extremely complex. Cancer cells as a rule contain many different mutations which each may contribute to various extents to the properties of the tumor.

(3) Not all properties of cancer cells may result from genetic defects. Many stable changes in cancer cells may be set up by regulatory loops without alterations

in the sequence or the amount of DNA. Such changes are designated as 'epigenetic' (→8). They can occur within a cancer cell or concern its interaction with other cell types (→8).

## 2.2. GENETIC ALTERATIONS IN CANCER CELLS

Many different types of genetic alterations can be observed in human tumor cells by molecular or cytogenetic methods (Table 2.1). Some alterations affect single or only a few genes. The DNA sequence of an individual gene can be altered by point mutations, smaller or larger deletions or insertions, or by rearrangements, with a wide array of potential consequences. Larger deletions can affect several genes at once. Rearrangements can lead to the creation of novel genes from others.

Point mutations: Point mutations are due to base exchanges in the DNA. These are categorized as transitions (pyrimidine→ pyrimidine or purine→ purine) or transversions (pyrimidine→purine or purine→pyrimidine). They can have very different effects, even if one considers only those occurring within the coding region of a gene (Figure 2.1). Silent mutations lead to a different codon encoding the same amino acid. They therefore do not change the coding potential of the mRNA, but they may affect its stability. The same effect can be elicited by mutations in the 3'- or 5'-UTR (untranslated region). Missense mutations lead to a change in the amino acid sequence. The resulting altered protein may possess increased, decreased or

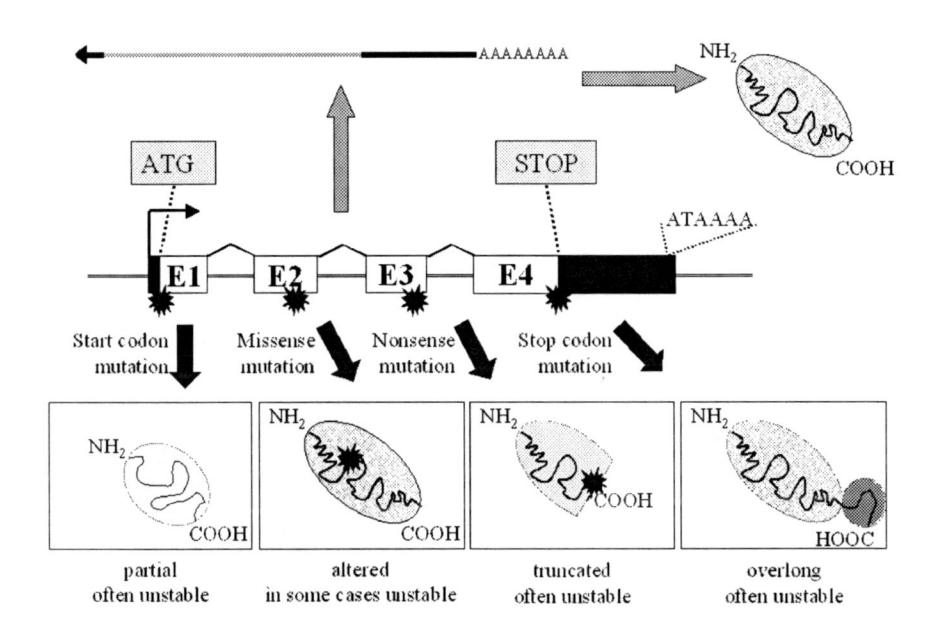

**Figure 2.1** *Effects of point mutations in the coding sequence of a gene*
See text for further explanations.

**Table 2.1.** *Types of genetic alterations in human cancers*

| Type of genetic alteration | | Typical consequences | Examples |
|---|---|---|---|
| Changes in the DNA sequence | base exchange | none, altered coding, altered splicing, protein truncation/read-through, altered mRNA stability, altered regulation | activation of *RAS* proto-oncogenes, inactivation of *TP53* tumor suppressor gene |
| | small insertion or deletion | frameshift mutations with protein truncation, altered splicing, altered gene regulation | expansion/contraction of oligo-A stretch in *TGFBR2* gene |
| | larger deletion | gene loss, protein truncation or shortening by exon loss | deletion of *CDKN2A*, loss of pocket domain in RB1 tumor suppressor gene |
| | larger insertion | gene disruption with loss of protein, altered splicing, altered regulation | inactivation of *APC* tumor suppressor gene by retrotransposon insertion |
| Viral infection | introduction of viral genomes and insertion of viral sequences into the genome | introduction of novel regulatory proteins, altered gene regulation by viral insertion, chromosomal instability at viral integration sites | inactivation of TP53 and RB1 by HPV E6 and E7 proteins, oncogene activation by retroviral LTR |
| Structural chromosomal changes | very large deletions | loss of several genes | 8p deletions in prostate cancer |
| | chromosomal translocation | altered gene regulation, formation of fusion genes encoding fusion proteins | activation of *MYC* oncogene in lymphomas, *BCR-ABL* fusion gene in CML |
| | chromosomal inversion | altered gene regulation, formation of fusion genes | activation of *RET* oncogene in thyroid cancers |
| | gene amplification | increased gene dosage, usually with increased expression | amplification of *MYC* or *ERBB1* oncogenes |
| Numerical chromosomal changes | aneuploidy | altered gene dosage | altered chromosome numbers in many advanced cancers |
| | chromosome gain | altered gene dosage | gain of chromosome 7 in papillary renal cell carcinoma |
| | Chromosome loss | altered gene dosage | loss of chromosome 3p in clear-cell RCC |

even unchanged activity. Therefore, the evaluation of the functional impact of a missense mutation detected in a tumor cell requires a biochemical or cellular assay for the function of the encoded protein. Nonsense mutations are more easy to evaluate since they lead to a truncated, often unstable protein product. Moreover, nonsense mutations occurring between the ATG start codon and the next splice site, i.e. in the first or second exon of a gene, destabilize its mRNA by making it prone to 'nonsense-mediated decay (NMD)', a cellular quality control mechanism. Importantly, point mutations in principle can be activating or inactivating. However, the spectrum of mutations that lead to increased activity of a gene product is normally much narrower than that decreasing its activity. For instance, mutations activating RAS oncogenes are restricted to three codons (→4.3), while mutations inactivating the tumor suppressor gene TP53 are distributed across the gene (→5.4).

Splice mutations: Splicing requires specific consensus sequences at the exon - intron junctions and also in the intron. The efficiency of splicing is also influenced by sequences within the exon. Mutations of sequences required for splicing may alter the protein product or the expression level of a gene more or less subtly (Figure 2.2). Mutations disrupting a 5'-splice site may lead to skipping of an exon, whereas mutations affecting a 3'-splice site may lead to an elongation of an exon until the next recognizable splice site. In both cases, an altered protein may result that lacks

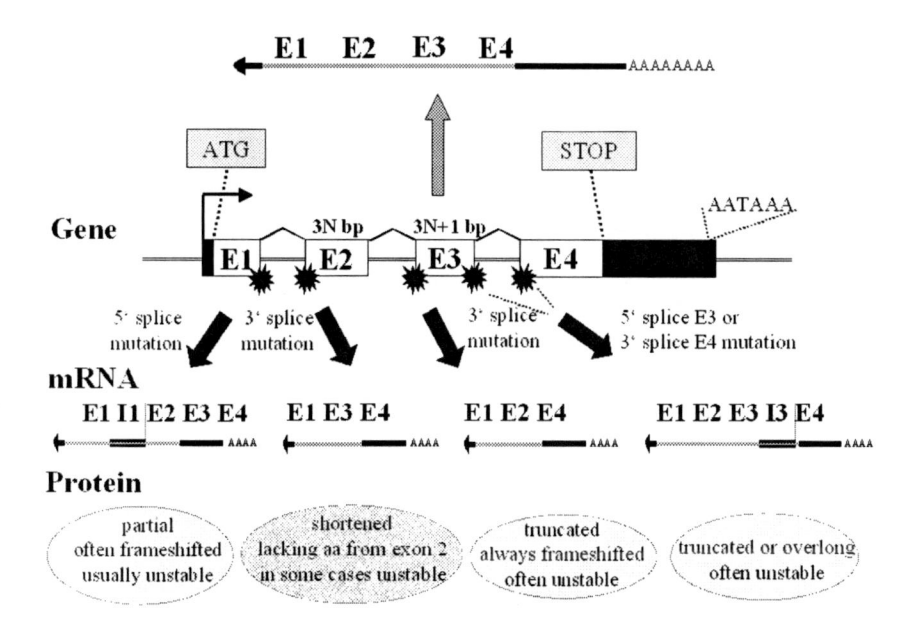

**Figure 2.2** *Potential effects of splice site mutations*
E: exon; I: intron. Note that the effect of a 3' splice site mutation depends on whether the exon has a number of nucleotides divisible by 3. See text for further explanations.

amino acids or contains additional amino acids. If the reading frame in the following exons is changed, the ensuing protein product is often truncated and/or unstable. With similar consequences, point mutations in introns or exons may create novel splice sites from 'cryptic splice sites', i.e. sequences resembling proper splice consensus signals.

Alternative splicing: The issue of altered splicing in human tumors is rendered difficult by the vagaries of splicing in normal mammalian cells. According to current estimates the average human gene generates about five different mRNA variants by differential splicing, alternative promotor usage and use of alternate termination signals. This can make it very difficult to decide, whether altered splicing found in tumors is relevant or not, and to determine whether it is caused by mutations in the affected gene or perhaps by mutations in a splice regulator. Several important genes display variations in splicing in human cancers that are probably functional relevant, but whose causes are unknown. Among them are relatives of the tumor suppressor *TP53* (→5.3), called *TP73L* and *TP73* (Figure 2.3). They encode several different proteins with different functions, the prototypic forms being designated p63α and p73α, respectively.

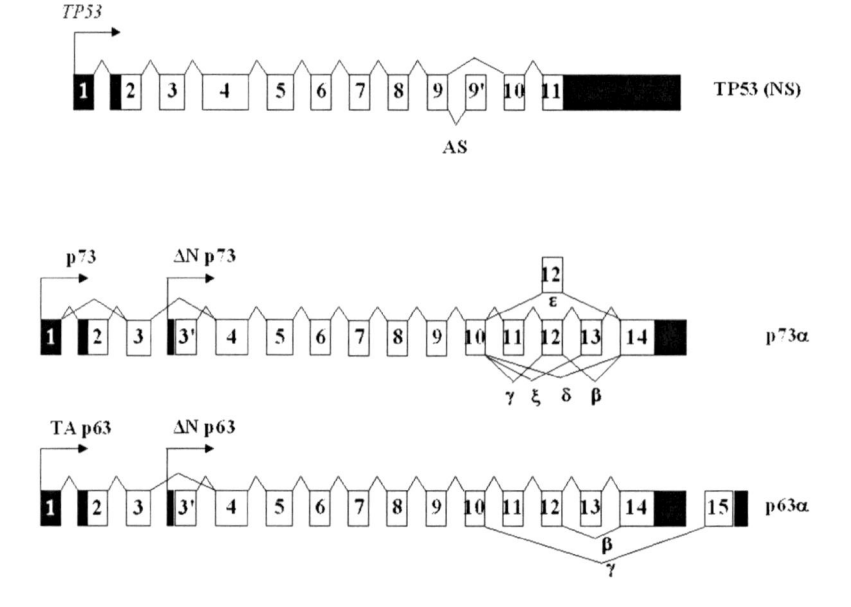

**Figure 2.3** *Splice variants of p63 and p73*
p73 and p63 are homologs of the important tumor suppressor protein TP53, with partially overlapping functions. Multiple protein forms result from alternative splicing (greek letters) and use of different transcriptional start sites, at least two in each gene (flag arrows). Untranslated regions are drawn in black. In cancer cells, the relative frequencies of the different messages is often altered by mechanisms that are not understood. The *TP53* gene likely also expresses additional splice variants whose patterns change in cancers.

Mutations in regulatory sequences: Even more complicated is the issue of mutations in non-coding sequences that regulate transcription. The regulatory sequences for a gene are often spread out over several 10 kb. In addition to the basal promoter, they comprise enhancers and silencers that may be located upstream and downstream of the gene as well as in introns and boundary elements. They are not well characterized for many human genes, and the effect of point mutations or even larger changes can only be judged by tedious experiments. As a consequence, many investigations of human cancers have focussed on mutations in coding sequences only, considering at most exon/intron junctions. It is therefore generally assumed that in many cancers mutations occur outside the coding regions of genes, but their frequency and their functional impact is impossible to assess at present. Moreover, many mutations are documented outside of actual genes, e.g., in microsatellite repeats, whose functional importance is also uncertain.

Deletions: In addition to base changes, smaller or larger deletions or insertions in tumor cells disrupt the coding regions of genes or destroy their regulatory elements. Small deletions become evident in PCR-based analyses, while very large deletions can be detected by cytogenetic methods. Intermediate-sized deletions in the kb range are more easily detected and charted in tumor cell lines than in tumor tissues, because PCR methods amplify residual alleles from tumor cells or non-tumor cells in the tissue. Therefore, it is not quite clear to what extent deletions contribute to loss of gene function in human tumors. In some cases, deletions affect both copies of a gene, e.g. through an internal deletion in one chromosome and loss of the second corresponding chromosome. This case is called 'homozygous deletion'. Charted homozygous deletions in tumor cells sometimes extend across several Mbp.

Insertions: Insertions in the genomes of tumor cells can comprise one to several bps. In the coding sequence of a gene, they typically lead to frame-shift mutations and/or change the stability of its mRNA. Insertions can arise by several mechanisms. Some are in fact duplications, e.g., in short repetitive sequences such as genomic polyA-tracts, caused by unrepaired slipping of DNA polymerases (→3.1). Others may arise during repair of DNA strand breaks (→3.3). Insertions as well as deletions can also be caused by viruses or endogenous retroelements. Somewhat surprisingly, in human cancers, transposition of endogenous retroelements appears to be relatively rare, even though a few active retrotransposons, all from the LINE-1 class, exist in the human genome.

Viral genomes: Viruses add extra genetic material to cells and often change the genome of the host cell. In human cancers, DNA viruses are more prevalent than retroviruses and contribute to the development of several human cancers by expression of specific viral proteins (→5.3). They normally replicate as episomes, but in cancer cells integrates are not unusual. These insertions typically disrupt genes and are often associated with partial losses of the viral genomes and altered expression patterns of the viral genes. Importantly, the integrates are often unstable and therefore induce chromosomal breaks with losses and rearrangements. Retroviral insertions in particular also often affect gene expression near the insertion site substantially, and cause over-expression (→4.2). In human cancer, retroviral insertions seem to be rare, but the DNA virus HBV may sometimes act in a similar manner (→16.3).

Chromosomal translocations: Alterations in the structure or expression of specific genes can also be caused by chromosomal translocations (Figure 2.4), which are most evident in hematological and soft tissue cancers. Obviously, a translocation may simply destroy and inactivate genes at the translocation sites. However, the opposite outcome is not infrequent and is important in human cancers. A translocation may separate inhibitory regulatory elements from the coding region of a gene and/or place it under the influence of activating regulatory elements from

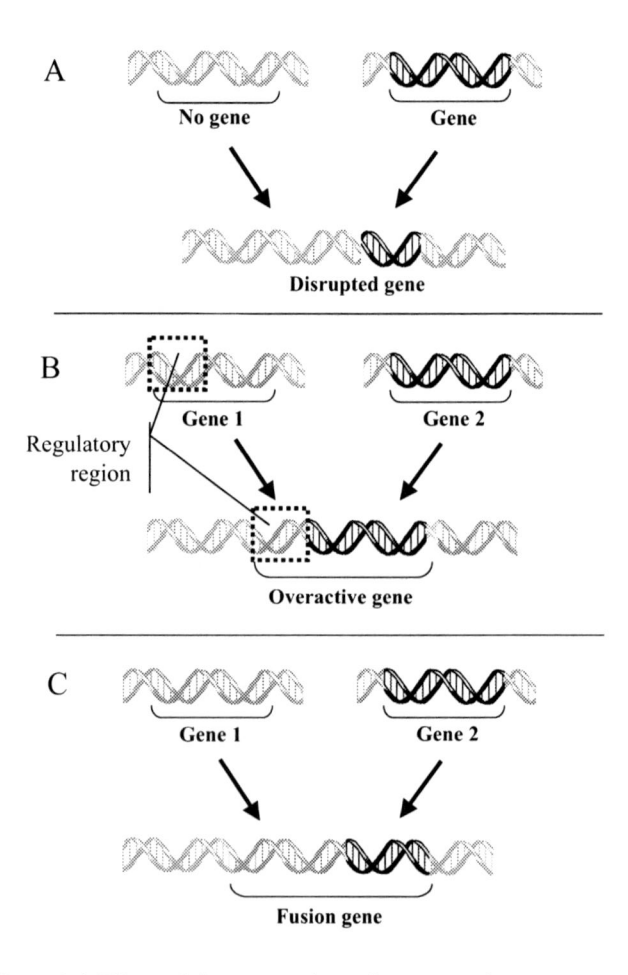

**Figure 2.4** *Effects of chromosomal translocations in human cancers*
In A a gene is inactivated by disruption. In B regulatory regions of gene 1 become appositioned to gene B, deregulating its expression. In C, gene 1 and gene 2 are fused resulting in the expression of a fusion protein controlled by regulatory regions from gene 1.

another gene. In either case, over-expression or deregulation ensue. This mechanism accounts, e.g., for the activation of the MYC gene in Burkitt lymphoma (→10.3). A third possible outcome is the generation of novel genes by translocations appositioning two genes to each other. Typically, the product of the novel gene is a fusion protein which contains N-terminal sequences from one fusion partner and C-terminal sequences from the other. If a balanced translocation occurs, two fusion genes and proteins may be formed. The fusion proteins can possess properties that differ significantly from those of the original proteins. Two infamous proteins of this type are BCR-ABL (→10.4) and PML-RARα (→10.5).

Chromosomal inversions can basically have the same consequences as translocations, since they are essentially translocations within one chromosome. If they are not evident from the karyotype or from the analysis of a particular gene, they are difficult to detect because many molecular or cytogenetic methods in use do not distinguish the orientation of a sequence on a chromosome. Their prevalence is therefore difficult to ascertain. A prominent example concerns the RET gene in thyroid cancers (Figure 2.5). In this case, inversions within the long arm of chromosome 10q lead to activation of RET by fusing it with other genes. These inversions are particularly frequent following exposure to radioactive iodine.

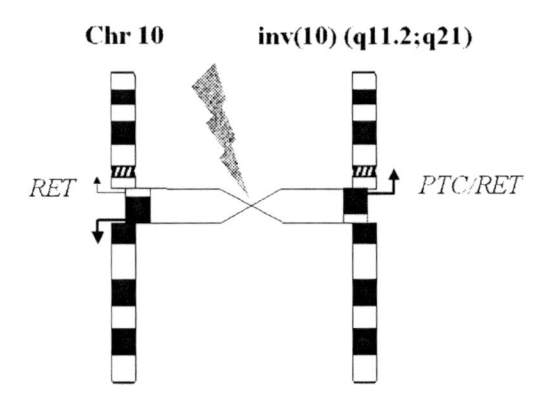

**Figure 2.5** *Activation of the* RET *gene by inversion in thyroid cancers*
Inversion of a segment of chromosome 10 in some cases of thyroid cancers (induced e.g. by ionizing radiation) yields a *PTC/RET* fusion gene with overexpression of the RET receptor tyrosine kinase. In other cases, the RET protein is oncogenically activated by point mutations.

The above alterations typically affect individual genes. However, genetic alterations in most cancers are not restricted to individual genes, but affect the genome at large.

Polyploidy and aneuploidy: While some cancers retain a nearly diploid genome with few alterations in the number and structure of chromosomes, many advanced cancers are aneuploid, exhibiting numerous structural and numerical chromosomal aberrations. According to measurements of DNA content, many carcinomas appear to start out from a near-diploid stage and go through a near-tetraploid stage, from which chromosomes are lost and gained (with losses prevailing) until a final metastable 'pseudo-triploid' state is reached. Indeed, many established carcinoma cell lines show this state with a modal chromosome number ≈70 (Figure 2.6). While this number is close to that of a triploid genome, the 'pseudo' indicates that it often belies a multitude of changes in the number and structure of individual chromosomes. Some leukemias remain near-diploid through most of their course, displaying however distinctive chromosomal changes like specific translocations or losses of particular chromosomal fragments, but then develop multiple and different chromosomal changes in their terminal 'crisis' phase (→10.4). In general, numerical and structural aberrations are distinguished.

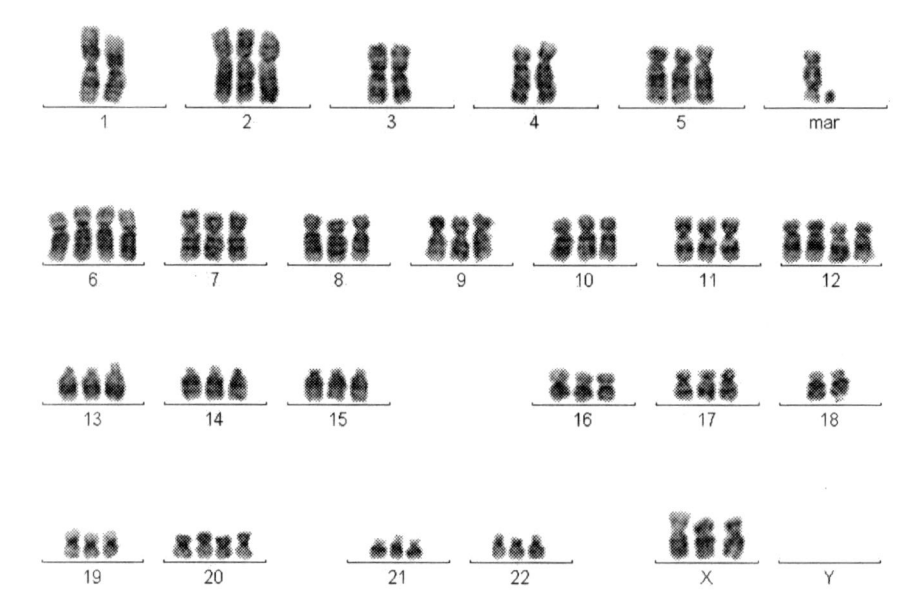

**Figure 2.6** *A pseudo-triploid chromosome set in a human cancer cell line*
Courtesy: Dr. V. Jung and Prof. B. Wullich

Numerical chromosomal aberrations: In aneuploid cells, the numbers of chromosomes or chromosome arms deviate from the total number of complements present, e.g., three chromosomes of one kind may be found in a diploid cell. These numerical aberrrations imply an altered gene dosage for the affected genes. The copy of one gene in a tumor cell therefore may range from zero (i.e. homozygous deletion) to very high. However, even in a tumor cell it is unusual to have more than 5 or 6 chromosomes of one kind.

Gene amplification: Higher copy numbers of genes are reached by amplification of smaller chromosomal regions called amplicons, which may range from several hundred kbs to several Mbp in size. This size implicates that amplifications may contain several genes, of which one or several can be over-expressed and be relevant for the tumor phenotype. If sufficiently large, amplified regions contained in a chromosome can become cytogenetically detectable as homogeneously staining regions (HSR). Amplicons in tumor cells can also be episomal, typically as small DNA circles presenting as small speckles in cytogenetic analyses. These are designated 'double minutes'. Double minutes replicate autonomously, but do not possess kinetochores and are randomly distributed to daughter cells. Accordingly, their numbers vary. Moreover, they may be in a dynamic exchange with corresponding HSR from which they may originate and into which they may re-integrate by recombination. Copy numbers of genes on double minutes can run up to thousands per cell. Gains or losses of whole chromosomes may result from missegregation at mitosis. However, amplifications presuppose structural changes.

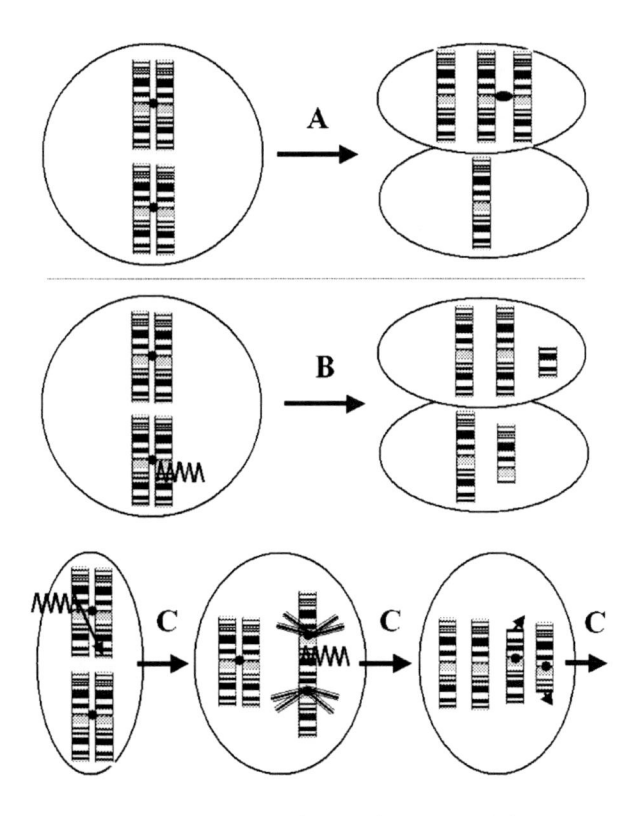

**Figure 2.7** *Mechanisms leading to chromosomal aberrations*
A: Mitotic non-disjunction typically yields trisomy and monosomy of a chromosome in the daughter cells. B: Double-strand breaks unrepaired until cell division can result in loss of genetic material. C: Alternatively, double strand breaks inititate breakage-fusion-bridge cycles. Note that the products resulting from disruption of the dicentric chromosome in the center figure are capable of starting further rounds of the cycle. Breakage-fusion-bridge cycles can also be set in motion by fusions of defective telomeres.

Structural chromosomal aberrations in tumor cells include the translocations and inversions discussed above, but also internal deletions of various sizes. Classical cytogenetic methods have identified 'marker' chromosomes in many cancer cells that are composed of fragments from several different chromosomes (cf. Figure 2.6). Modern cytogenetic analyses reveal that in addition many chromosomes appearing grossly normal in cancer cells also harbor deletions, inversions or are also composed from parts of several chromosomes.

Numerical chromosomal changes can be brought about by mitotic non-disjunction (Figure 2.7). If a chromosome does not attach properly to the mitotic spindle, one daughter cell may end up with an additional chromosome and the other with one less. The origin of structural chromosomal changes is probably more complicated. Some may result from double-strand breaks in the involved chromosomes which may lead to deletions or recombinations. Another mechanism are breakage-fusion-bridge cycles initiated by strand breaks or telomeric fusions (Figure 2.7).

The multitude of structural changes in chromosomes reveals a high degree of illegitimate recombination in tumor cells. While some of them are evident, others may not be detectable at the cytogenetic level, especially, if they result in the exchange of genetic material rather than net loss or gain or translocations. Indeed, molecular analyses of polymorphic DNA sequences reveal that tumor cells often contain identical copies of a DNA sequence that is heterozygous in normal cells of the patient. This is called loss of heterozygosity. It can be caused by deletions of one allele, but also by recombination (Figure 2.8).

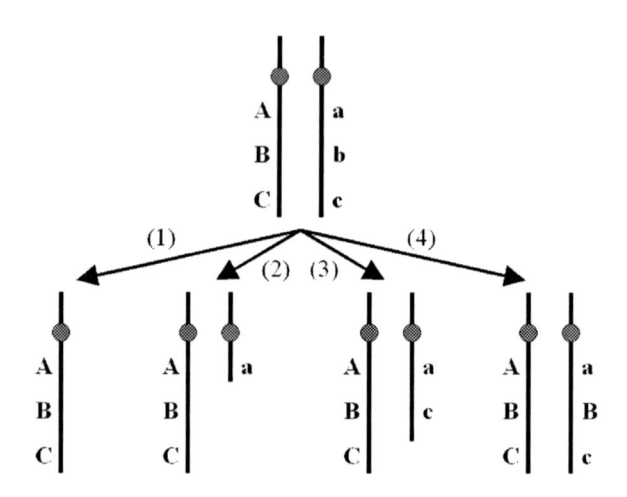

**Figure 2.8** *Mechanisms leading to loss of heterozygosity in cancer cells*
Two homologous chromosomes are shown with polymorphic markers Aa, Bb, Cc (microsatellites or single nucleotide polymorphisms). (1) Loss of one chromosome leading to monosomy; (2) Loss of part of a chromosome arm leading to partial monosomy; (3) Interstitial deletion with loss of marker b; (4) Recombination leading to substitution of b by B.

The various types of genetic alterations discussed in this section do not each occur to the same extent in every human cancer. Rather, in different tumors particular types of mutations tend to predominate. In some cancers point mutations are most prevalent, whereas chromosomal aberrations seem to be responsible for the majority of the genetic changes in others. In some cancers, distinct subtypes can be distinguished by this difference, e.g. in colon carcinomas (→13.5). Further differences may be more subtle. For instance, some cancers tend to lose or gain whole chromosomes, whereas others tend to delete, gain or rearrange chromosomal fragments. The reasons underlying such differences are the subject of a very active area of current research.

## 2.3 INHERITED PREDISPOSITION TO CANCER

Although most genetic alterations in tumor cells develop during the life-time of a patient, the predisposition to cancers can be inherited. Three different types of inheritance can be distinguished.

In some families cancers are very frequent and occur (essentially) in each generation. This is a general hallmark of autosomal-dominant inherited diseases with high penetrance. The families may be plagued by specific cancers, rarer ones such as retinoblastoma, or common cancers such as breast cancer, or by various types of cancer, such as in Li-Fraumeni-syndrome (Table 2.2). Typically, cancers manifest at a lower than average age of onset and also unusually often at multiple sites or bilaterally in paired organs, such as the eyes, kidney and breast. These are two further criteria pointing to inherited cancers. In some cases, cancer predisposition is associated with developmental defects, e.g. in the Gorlin and Cowden syndromes. This is a fourth, although not as strict criterion. The increased risk in cancer families with an autosomal-dominant mode of inheritance is caused by an inherited mutation in a single gene. The affected genes are usually tumor suppressor genes (→5) and more rarely oncogenes (→4). The inherited mutations are likely not sufficient to cause cancer, but they provide the first mutation of several that are required (→2.5).

In some families predisposition to cancer is inherited in a recessive mode. More often than in the dominantly inherited cases, cancer predisposition is found in the context of rare inherited syndromes (Table 2.3). So, the affected patients are initially afflicted by other symptoms and cancers appear later, but still at a relatively early age. Syndromes in this category include Xeroderma pigmentosum, Ataxia telengiectasia, Fanconi anemia, Nijmegen breakage-syndrome as well as the Bloom and Werner syndromes. These syndromes differ in the extent of the cancer risk and the predominant cancer types, but at least one type of cancer is substantially more prevalent than in the general population. In these syndromes, predisposition to cancer is evidently caused by mutations inactivating both copies of the same gene. The genes affected are usually involved in cell protection and DNA repair (→3.4). In general, the inherited defects in DNA repair favor genetic alterations in somatic cells that lead to cancer.

**Table 2.2.** *Tumor syndromes inherited in an autosomal dominant mode*

| Syndrome | Gene | Location | Cancer site | Chapter |
|---|---|---|---|---|
| Neurofibromatosis 1 | NF1 | 17q11.2 | peripheral nerves, eye, skin | 4.4 |
| Hereditary melanoma and pancreatic cancer | CDKN2A | 9p21 | skin, pancreas, others | 5.2 |
| Retinoblastoma | RB1 | 13q14 | eye, bone | 5.2 |
| Cowden | PTEN | 10q23.3 | many organs | 6.3 |
| Tuberous sclerosis | TSC1<br>TSC2 | 9q34<br>16p13 | soft tissues in several organs | 6.3 |
| Gorlin | PTCH | 9p22 | skin, brain | 12.3 |
| Familial adenomatous polyposis coli | APC | 5q21 | colon, rectum, others | 13.2 |
| Von Hippel-Lindau | VHL | 3p25 | kidney, adrenal glands, others | 15.4 |
| Hereditary leiomyoma renal cell carcinoma | FH | 1q24.1 | uterus, kidney | 15.3 |
| Burt-Hogg-Dubè | BHD | 17q11 | kidney, others | 15.3 |
| Hereditary gastric cancer | CDH1 | 16q22 | stomach | 17.2 |
| Multiple endocrine neoplasia type 1 | MEN1 | 11q13 | endocrine glands | -* |
| Neurofibromatosis 2 | NF2 | 22q | CNS, peripheral nerves | -* |
| Li-Fraumeni | TP53 | 17p13.1 | many organs | 5.1 |
| HNPCC | MLH1<br>MSH2<br>others | 3p21<br>2p15-16 | colon, endometrium, stomach, others | 13.4 |
| Hereditary breast and ovarian cancer | BRCA1<br>BRCA2 | 17q21<br>13q12 | breast, ovary | 18.3 |
| Multiple endocrine neoplasia type 2 | RET | 10q11.2 | thyroid and other endocrine glands | 5.4 |
| Hereditary papillary renal cancer | MET | 7q31 | kidney | 15.3 |

*see Vogelstein & Kinzler: 'The Genetic Basis of Human Cancer' or the online OMIM database for a detailed description of these tumor syndromes

Inheritance of mutated genes in autosomal-dominant or recessive cancer syndromes carries a greatly enhanced risk of developing cancer during a human's lifetime which may approach 100%, whereas the life-time risk of 'sporadic' cases in the general population may be minimal. Even in those cancers, like that of the skin, where the life-time risk is high in the general population, the risk to develop a cancer up to a certain age is strongly increased in persons with an inherited predisposition (Figure 2.9). Fortunately, cancer predisposition of this kind is infrequent. All high-risk mutations in dominantly and recessively inherited cancer syndromes together may account for less than 10% of all human cancers.

Nevertheless, an individual's cancer risk may be strongly influenced by the genotype. About one in a thousand base pairs differs between individual humans. Differences that occur in more than 0.5% of the population are called 'polymorphisms' and are thereby distinguished from rare changes considered as true mutations[1]. Polymorphisms are found in coding regions of genes as well as in regulatory sequences and in non-coding sequences throughout the genome. These differences comprise single nucleotide polymorphisms (SNPs) and differences in the size of micro- and minisatellite repeats, but also insertions or deletions of various sizes.

**Table 2.3.** *Tumor syndromes inherited in a recessive mode*

| Syndrome | Function affected | Gene(s) involved | Organ site | Chapter |
|---|---|---|---|---|
| Xeroderma Pigmentosum | Nucleotide excision repair | XP (ERCC) genes, others | skin | 3.2 |
| Ataxia Telangiectasia | Strand break repair signaling | ATM | multiple | 3.3 |
| Fanconi Anemia | Crosslink repair | FANC genes BRCA2 (homozygous) | hematopoetic system, others | 3.2 |
| Nijmegen breakage syndrome | Double-strand break repair | NBS | hematopoetic system | 3.3 |
| Bloom syndrome | Recombination | BLM | multiple | 3.3 |
| Werner syndrome | Recombination repair, telomere function | WRN | multiple | 7.4 |

---

[1] The distinction between polymorphism and mutation is not always made consistently, e.g., some set the cut-off for polymorphisms at 0.1%.

For instance, up to 50% individuals in some European populations lack the gene for GSTM1, a glutathione transferase enzyme metabolizing xenobiotics. This is called a 'null-allele' because no enzyme activity is present. Other polymorphisms have more subtle effects. Polymorphisms in genes involved in the metabolism of drugs and other exogenous compounds (Table 2.4) modulate the risk of cancer in

**Table 2.4.** *Some genetic polymorphisms important for cancer predisposition*

| Gene | Polymorphism | Gene function | Enviromental interaction |
|---|---|---|---|
| GSTM1 | Deletion | carcinogen metabolism | tobacco smoking and other sources of polyaromatic hydrocarbons |
| GSTT1 | Deletion | carcinogen metabolism | short aliphatic chlorinated alkenes |
| NQO1 | base/aa change with protein instability | carcinogen metabolism, prevention of redox cycling | benzene, quinones |
| NAT2 | multiple base/aa changes in coding region | carcinogen metabolism | arylamines |
| CYP1A1 | base/aa change | carcinogen metabolism | polyaromatic hydrocarbons |
| TS | repeat number in promoter | nucleotide biosynthesis | folate |
| MTHFR | base/aa change with reduced stability | nucleotide biosynthesis | folate, vitamin $B_{12}$, alcohol |
| TP53 | base/aa change | control of genome stability | radiation, clastogens |
| BRCA2 | base/aa change | DNA repair | radiation, clastogens |
| OGG1 | base/aa change | DNA repair | diet (?) |
| XRCC1 | base/aa change | DNA repair | UV (?) |
| ATM | base/aa change or truncation | DNA repair | ionizing radiation |
| MC1R | base/aa changes | melanocyte proliferation and differentiation | UV |
| IL1B | repeat number in promoter | immune response | *Helicobacter pylori* |
| HPC1 | base/aa change | response to dsRNA (?) | viral infection (?) |
| AR | repeat length in coding sequence | hormone response | diet (?) |

people exposed to them. For instance, detoxification of carcinogenic benzopyrene metabolites is in general more efficient in individuals with GSTM1 compared to those lacking the enzyme. Since benzopyrene is one of the carcinogens in tobacco smoke, the risk of GSTM1-/- smokers to develop cancer of the lung (and other organs) is increased. Even if the risk of lung cancer were only two-fold enhanced by the lack of the enzyme, this increase would apply to up to 50% of the smoking population in some European populations. Clearly, such polymorphisms may have profound effects at the population level, even if they modulate the risk for each individual only slightly.

It is also important to note that the effect of a polymorphism in a drug metabolism gene depends not only on the extent to which it alters the function of the gene, but also on the dose and the type of exposure. For instance, the product of another gene from the GST superfamily, GSTT1, also protects against carcinogens in general, but activates small chlorinated alkenes, such as trichloro-ethylene, to highly mutagenic compounds.

Other polymorphisms modulating cancer risks (Table 2.4) have been identified in genes involved in immunity, inflammation, hormone metabolism, and nucleotide metabolism. As with drug metabolism genes, the risk conferred by the polymorphic forms of these genes is contingent on non-genetic factors such as exposure or nutrition. Also, the differences in risk are always moderate, i.e. within one order of magnitude.

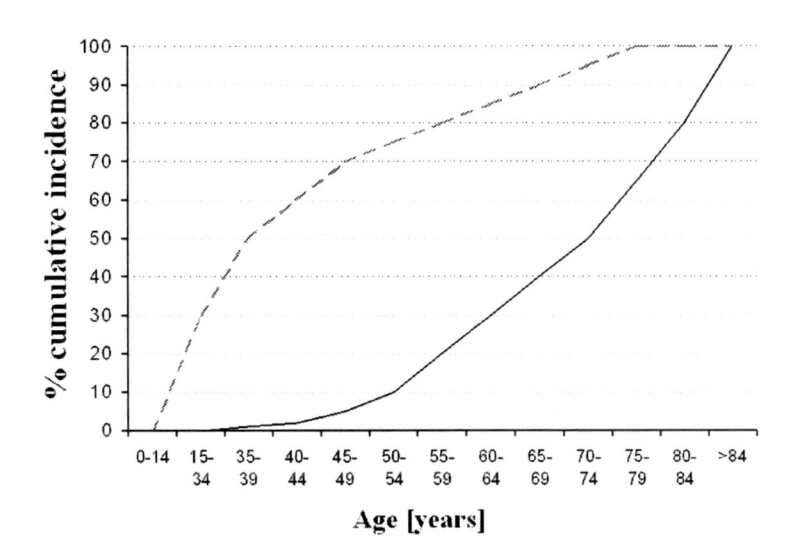

**Figure 2.9** *Age dependency of cancers in hereditary and sporadic cases*
In this hypothetical curve illustrating the typical experience from many cancers, the cumulative incidence curve of a cancer type differs strongly between hereditary vs. sporadic cases, because in the hereditary situation the cancers arise at a much earlier age.

A complicated situation is posed by polymorphisms in high-risk cancer genes. For instance, several polymorphisms are known in the gene mutated in Li-Fraumeni-syndrome, *TP53* (→5.3). At least one of these may be associated with only a small and more specific increase in cancer risk. An even more difficult problem concerns mutations in the *ATM* gene (→3.4), which lead to a high risk of cancer when present in a homozygous state (in fact, often different high-risk mutations combine, so precisely this is a 'compound' homozygosity). Homozygosity is fortunately rare, but heterozygosity for mutations like 7271T-->G could be as frequent as 1:200 in some populations (so, one would have to consider this as a polymorphism). Whether persons with polymorphisms are at increased risk, is a hotly debated issue (cf. 3.4).

Since one characteristic of inherited tumors is their precocious appearance compared to sporadic cases, the question arises to which extent childhood tumors are inherited. This is evidently so when tumors arise in the context of inherited syndromes, e.g. childhood brain tumors in Li-Fraumeni-syndrome families. Indeed, some childhood tumors occur as familial as well as sporadic cases, e.g. retinoblastoma (→5.1) and Wilms tumor (→11.2). Even though these are tumors of young children, the average onset is slightly earlier in familial cases.

Other cases of Wilms tumor, however, are caused by developmental defects in a parent's germ cell or during early development as a consequence of genetic or epigenetic changes (→11). A similar situation holds for testicular germ cell cancers, which are derived from germ cells that due to mutations acquired in the fetus have not completed their maturation properly. A few of these cancers become evident in young children, but most start to expand under the influence of rising androgen levels during puberty, leading to a peak in incidence in the third decade of life. In the same vein, studies on twin children coming down with leukemia indicate that the responsible chromosomal translocations have likely taken place very early during fetal development, although the disease manifests roughly a decade later. So, while these cancer may be suspected to be inherited, they are more precisely considered as 'congenital'.

## 2.4 CANCER GENES

In the cell of a cancer at a late stage of progression, several hundred genes may be mutated or rearranged and for many the dosage and accordingly as a rule their expression levels may be changed. Indeed, expression profiling of tumor cells by array techniques has revealed thousands of genes whose expression is increased or decreased. Moreover, expression changes cannot only be due to primary changes such as altered gene dosage or mutation, but also be secondary to mutations in regulatory genes or to epigenetic mechanisms. So, it is not immediately obvious which changes are essential for tumor development, which contribute to the phenotype of the cancer without being essential, and which are coincidental and irrelevant. Several different kinds of genes can be distinguished that are relevant in human cancers.

<u>Tumor suppressor genes</u>: The genes mutated in dominantly inherited cancers are obviously central to cancer development. Most of these belong to a class named

'tumor suppressor genes' (→5). Typically, their function is strongly diminished or obliterated in cancers by mutations or by epigenetic silencing. Not unexpectedly, these same genes are often found mutated in sporadic cases of the same cancer types and even in other types. Some genes of this kind are only found inactivated in sporadic cancers, but their inactivation can be shown to be essential for tumor development.

Oncogenes: A few responsible genes in dominantly inherited cancers, such as RET in hereditary endocrine cancers and MET in hereditary renal papillary carcinoma (→15.3) are activated by mutations instead of being inactivated like tumor suppressor genes. For a larger number of genes, activation by specific mutations or by overexpression as a consequence of gene amplification, or by other mechanisms, is found consistently in sporadic cancers. Some of these genes are related to those genes in oncogenic retroviruses known to cause cancers in animals. These genes are therefore called oncogenes (→4).

DNA repair and checkpoint genes: Most oncogenes and many tumor suppressor genes directly control cell proliferation, differentiation, and/or survival. Many genes inactivated by mutations in recessive cancer syndromes are not directly involved in this type of regulation (→3.4). Instead, defects in these genes increase the rate of mutations, of which some alter the function of genes directly involved in the development of tumors. In fact, mutations in genes of this type underlie certain cancer predispositions inherited in a dominant fashion. This type of tumor suppressor gene is sometimes called a 'caretaker' and those directly involved in growth regulation are called 'gatekeepers'. Mutations in some caretaker genes are also found in some sporadic human cancers.

Risk modulating genes: A related group of relevant genes is even less directly involved in cancer development. The products of genes in this group, exemplified by the GSTs, modulate the development of cancer, e.g., by influencing the level of active carcinogens, the reaction of the immune system to cancer cells, or the level of hormones and growth factors that stimulate tumor cell proliferation. Very often, their importance depends on environmental factors such as exposure to carcinogens. Once a cancer is established, they are not absolutely necessary.

Execution genes: In contrast, another group of genes only becomes important after the cancer is established. These genes may be activated by mutations, but are more often induced as a consequence of activation of oncogenes or inactivation of tumor suppressor genes. They may not be necessary for survival of the cancer cells, but for their sustained growth and specifically for invasion and metastasis (→9). These complex processes appear to require a relatively well coordinated program of gene expression which can be initiated by mutations in a limited number of genes, but requires the activity of a much larger number for its execution.

So, it is, in fact, anything but trivial to define a 'cancer gene'. There are some clear-cut cases, e.g. almost every retinoblastoma displays inactivation of the gene *RB1* (→5.1), and every Burkitt lymphoma shows activation of the gene *MYC* (→10.3), showing these are clearly tumor suppressor and oncogenes, respectively. However, such bona fide oncogenes and tumor suppressor genes apparently represent the extreme end of a continuum of more or less relevant cancer genes.

## 2.5 ACCUMULATION OF GENETIC AND EPIGENETIC CHANGES IN HUMAN CANCERS

Advanced human cancers usually contain a multitude of genetic changes, and often epigenetic alterations. In human cancers, it is rarely possible to determine, at which stage of tumor progression they were acquired. Most data, however, suggest that they gradually accumulate during tumor progression (→13.3). A gradual accumulation of multiple genetic changes would explain why most cancers appear at older age (Figure 2.10) and would fit with epidemiological data suggesting 4 to 5 essential 'hits' to be necessary for cancer development. The genes affected by such crucial mutations are likely oncogenes and tumor suppressor genes. Indeed, advanced human carcinomas typically contain mutations in several genes of both types. Genetic and epigenetic changes in further genes may modulate the tumor phenotype. Most genetic changes are the result of somatic mutations, but an inherited predisposition to cancer can be caused by one crucial mutation passed on in the germ-line, with further mutations occuring somatically.

Alternatively, inherited predisposition to cancer can be due to germ-line mutations in DNA repair genes or 'caretaker' genes. These defects appear to favor cancer development by increasing the probability of crucial mutations in oncogenes and tumor suppressor genes, but are associated with additional mutations that are more or less important for tumor progression. For instance, cancers with defects in DNA mismatch repair contain a large number of mutations, some in oncogenes and

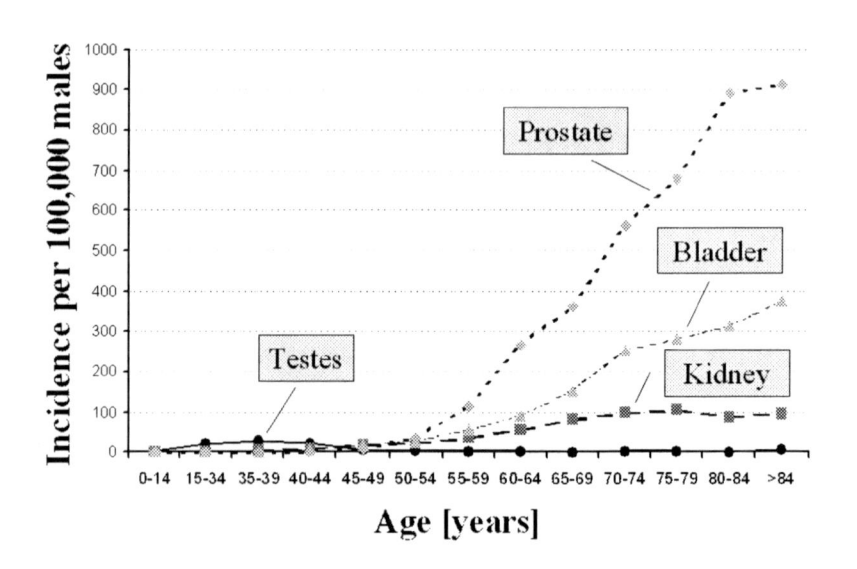

**Figure 2.10** *Age dependency of human cancers*
Incidence of selected cancers in males in relation to age in Germany 2000 (see www.rki.de). Note the early peak in testicular cancers and the plateau in the kidney cancer curve.

tumor suppressor genes, but others in irrelevant microsatellite repeats (→13.4).

So, tumors arising on that background exhibit a 'mutator phenotype'. Depending on the type of defect, a mutator phenotype can manifest as an increase in point mutations or as various kinds of chromosomal instability.

The multitude of genetic alterations observed in many advanced human cancers suggests that they have developed a sort of 'mutator phenotype', leading to an increased rate of point mutations, to chromosomal instability, or even to frequent epigenetic alterations. While the existence of genomic instability in advanced cancers seems evident, an interesting question is whether it is required for their development. In other words, can the accumulation of genetic alterations required for an advanced cancer be achieved by random mutations at a normal rate, or does it require the establishment of a 'mutator phenotype' at some stage of progression? This is a hotly debated question with ramifications for tumor therapy as well as tumor prevention.

## *Further reading*

Online Mendelian Inheritance in Man (OMIM database), at the ncbi.nih.gov website

Mitelman F (ed.) Catalog of chromosomal aberrations in Cancer. Wiley, 1998 (with regular updates)
Vogelstein & Kinzler (eds.) The genetic basis of human cancer 2$^{nd}$ ed. McGraw-Hill, 2002

Knudson AG (2000) Chasing the cancer demon. Annu. Rev. Genet. 34, 1-19
Ponder BAJ (2001) Cancer genetics. Nature 411, 336-341
Strausberg RL et al (2002) The cancer genome anatomy project: online resources to reveal the molecular signatures of cancer. Cancer Invest. 20, 1038-1050
Balmain A, Gray J, Ponder B (2003) The genetics and genomics of cancer. Nat. Genet. 33 Suppl. 238-244
Loeb LA, Loeb KR, Anderson JP (2003) Multiple mutations and cancer. PNAS USA 100, 776-781
Futreal PA (2004) A census of human cancer genes. Nat. Rev. Cancer 4, 177-183

## Box 2.1 Tumor viruses in human cancers

Retroviruses like the paradigmatic Rous Sarcoma Virus cause cancers in animals. They carry specific 'oncogenes' which are similar to cellular genes. Their investigation has led to the identification of important cellular oncogenes in human cancers (→4.1). However, only a rare retrovirus, **HTLV1**, appears to act in a comparable manner in humans (→Box 4.1).

Instead, several DNA viruses are implicated in human cancers, in a more or less causative manner. The strongest case can be made against human papilloma viruses (**HPV**) which are thought to initiate cervical and other carcinomas (→Box 5.1). Some strains of HPV express proteins which inhibit cellular proteins that control cell proliferation and genomic integrity, i.e. tumor suppressors.

Different proteins with the same function (→Fig. 5.10) are encoded by the DNA genomes of other viruses that can infect humans. This raises the question whether they too can cause human cancer. Adenoviruses encode the E1A and E1B proteins. They cause cancers in some animals and cold-like diseases in man. Nevertheless, it is generally agreed that they are not tumorigenic in man, likely because of efficient elimination by the immune system. The E1A and E1B proteins are thought to partially suppress cellular reactions to the viral infection, but are apparently not strong enough to cause transformation. Adenoviruses may, however, be exploited for cancer therapy (→22.6).

There is not as much agreement on the role of papovavirus in human cancers. The best characterized member of the family is a monkey virus, simian virus 40 (**SV40**), which is a potent tumor virus in rodents. It may have infected humans and has been found in brain tumors and mesotheliomas. Doubtless, its large T-antigen is capable of immortalizing cultured human cells (→7.4). The family members endemic in humans are the **JC** and **BK** viruses. They are present in many healthy humans and accordingly their genomes are found in some cancers, e.g. of the brain and the urinary tract. Their causative role is, however, debated.

In contrast, two members of the herpes virus family are accepted as co-carcinogens for human cancers. **HHV8** (or KHSV) is a crucial agent in the development of Kaposi sarcoma, which most often arises in the context of immunodeficiency caused by the retrovirus **HIV** (→Box 8.1). Epstein Barr Virus (**EBV**) is best known for its role as a co-carcinogen in Burkitt lymphoma (→10.3), but is very likely also involved in further malignancies including nasopharyngeal carcinoma and certain Hodgkin and non-Hodgkin lymphomas. It appears to act mainly by suppressing apoptosis (→7.2) of lymphoid cells.

Finally, the hepatitis virus B (**HBV**) can be understood as an intermediate between a DNA virus and a retrovirus. It is certainly a co-carcinogenic agent in a substantial fraction of human liver cancers (→16.3). It causes cancer by a mixture of direct effects of viral proteins and indirect effects resulting from chronic inflammation elicited by the viral infection. The same can be said of the RNA hepatitis C virus (**HCV**), which causes another substantial fraction of human liver cancers (→16.1).

# CHAPTER 3

# DNA DAMAGE AND DNA REPAIR

- DNA in human cells is continuously subject to damage. It is in most cases appropriately repaired, leaving relatively few permanent changes. The various kinds of damage comprise chemical modification or loss of DNA bases, single strand or double strand breaks as well as intra- and interstrand crosslinks. Each type of damage can lead to mutations. An important source of mutations are DNA replication and recombination. DNA replication is a particular critical phase, during which misincorporation of nucleotides, DNA polymerase slippage or stalling of replication forks may occur. A further source of mutations are physiological recombination processes that go astray, e.g. in germ cells or lymphocytes.
- In addition to endogenous processes such as oxidative stress and spontaneous reactions of DNA such as cytosine deamination, diverse exogenous physical and chemical carcinogens cause DNA damage. Some carcinogens cause specific point mutations while others induce strand-breaks or various types of alterations. Carcinogens that induce strand-breaks may act as 'clastogens', i.e. induce structural chromosomal aberrations. The involvement of specifically acting carcinogens is in some cases detectable by the kind of mutation found in a cancer.
- Tumor viruses can be mutagenic by insertion, by causing rearrangements or loss of chromosomes, or indirectly through viral proteins, which interfere with the cellular systems that control genomic integrity.
- The various DNA repair systems in human cells are tailored towards the different types of DNA damage. They share components such as DNA polymerases and DNA ligases, but each employ additional specific proteins.
- Specialized glycosylases remove damaged bases. AP (apurinic/apyrimidinic) endonucleases prepare sites lacking bases for short patch or long patch base excision repair. More problematic alterations such as carcinogen adducts and pyrimidine dimers caused by UV light are removed by nucleotide excision repair systems. Mismatched base pairs in DNA caused by mutagens or mistakes during DNA replication are the target of two interlinked mismatch repair systems. Double strand breaks pose a major challenge to cell survival and genomic integrity. They are recognized and handled by several different repair systems employing homologous or non-homologous recombination to avoid or minimize permanent damage. Still another repair system employs the FANC proteins to prepare crosslinked DNA for repair by recombination.
- Inborn errors in these DNA repair systems underlie syndromes associated with developmental defects, neurological disease and cancer. For instance, excision repair is defective in xeroderma pigmentosum, mismatch repair in HNPCC (hereditary non-poliposis carcinoma coli), double strand repair in ataxia

telangiectasia, and cross-link repair in Fanconi anemia, respectively. While each of these syndromes is rare, polymorphisms in DNA repair genes likely modulate cancer risk in the general population.
- ➤ A second layer of protective mechanisms helps to avoid DNA damage. Reactive mutagens are intercepted by low molecular protective compounds such as glutathione or by proteins such as metallothioneins or glutathione transferases. Specific mechanisms protect against reactive oxygen species and against radiation. Genetic polymorphisms again, but also diet and other environmental factors influence the efficiency of these mechanisms in individual humans.
- ➤ DNA damage can activate cellular checkpoints which prevent cell cycle progression and stop DNA synthesis and mitosis, or activate apoptotic cell death. Double-strand breaks elicit a particularly strong signal. Stress signals can also be activated by radiation or reactive oxygen species, through specialized signaling pathways. Infection by viruses also activates cellular checkpoints and stress signals.

## 3.1 DNA DAMAGE DURING REPLICATION: BASE EXCISION AND NUCLEOTIDE EXCISION REPAIR

The mutations and chromosomal alterations found in cancer cells (→2) represent only a small fraction of those that arise during the life-time of a human, because the great majority are removed by one of several repair mechanisms (Table 3.1). These are excellently tuned to the various types of DNA damage that might cause mutations (Figure 3.1). Moreover, cells with substantially damaged DNA or aneuploid genomes are normally eliminated or at least prevented from proliferation. Therefore, cancer cells displaying genomic instability need to inactivate the systems responsible for this surveillance. Accordingly, defects in DNA repair systems and cellular surveillance mechanisms are an important, if not even necessary factor in the development of human cancers. Such defects may be inherited or acquired.

Damage to DNA can result from endogenous as well as exogenous sources. DNA replication is a particularly critical process, with an increased potential for spontaneous mutations and an increased sensitivity towards induced damage. Proliferating cells are therefore more susceptible to neoplastic transformation. Problems that may arise during DNA replication comprise misincorporation of bases, slippage of the replisome in tandem repeat sequences, single-strand breaks being converted into double-strand breaks by replication, stalling of the replisome at 'difficult' sequences or at bases modified by chemical reactions with exogenous carcinogens or endogenous proteins.

Replication of nuclear DNA is extremely precise with a nucleotide misincorporation rate of $10^{-7}$- $10^{-6}$, since eukaryotic replication DNA polymerases discrimate well between the various nucleotides and the main replicase possesses a 3'-5' exonuclease proof-reading function. This level of precision is not always achieved by repair polymerases. In spite of the excellent fidelity of the replication proteins, in a genome of >3 x $10^9$ bp, several hundred mistakes are expected during each replication. Most misincorporations are corrected by base mismatch repair systems, leaving an estimated number of 1 x $10^{-10}$ base changes per cell division.

Base misincorporation is, however, only one of several problems that can occur during replication. Mismatch repair (MMR) systems also take care of single strand loops in replicated DNA that result from slippage of DNA in repeat sequences. Typically, slippage occurs in microsatellites which consist of tandem repeats of 1-4 bp repeats. Defects in mismatch repair therefore result in an increased frequency of base misincorporations, but also lead to microsatellite expansions or contractions. Even with fully functional mismatch repair, microsatellites are subject to a somewhat higher mutation rate than the average of the genome, which is one reason why they are normally polymorphic.

Another source of base misincorporation are mesomeric isoforms of the DNA bases that can mispair. The mesomeric isoforms of the standard four bases are shortlived, but some are stabilized by chemical modification. For instance, hydroxylation of guanine at the 8 position stabilizes a G:A mismatch. Most mismatches caused from frequent mispairing events such as OH-G:A are recognized by specific proteins that activate the mismatch repair system.

The precision of DNA replication also depends on the nucleotide precursor pools. Disparities in the relative levels of the deoxy-nucleotide triphosphates decrease the fidelity of base incorporation. As biochemistry textbooks discuss in detail, nucleotide biosynthetic pathways contain several crossregulatory and

**Table 3.1.** *An overview of DNA repair mechanisms in human cells*

| Type of DNA damage | Repair Mechanisms | Selected proteins involved |
|---|---|---|
| Base misincorporation | Mismatch Repair | MLH1, MSH2, MSH6, PMS1 |
| | Base excision + short patch repair | UGH |
| Chemical modification of bases | Base excision + short patch repair | UGH, OGG, MGMT, MBD4, MYH |
| Base loss | Short patch repair | AP endonucleases, DNA polymerase, DNA ligase |
| Formation of intrastrand base dimers or bulky adducts at bases | Nucleotide excision repair | XP (ERCC) proteins, CSA, CSB, XPG, DNA polymerases, DNA ligase I |
| | Bypass repair | DNA polymerase η or ι |
| Interstrand crosslinks | Crosslink repair | FANC proteins + HRR components |
| Double-strand breaks | Homologous recombination repair | RAD52, NBS1, MRE11, RAD50, RAD51, BRCA2, BLM(?) |
| | Nonhomologous DNA end joining | KU70, KU80, NBS1, MRE11, RAD50, FEN1, WRN(?), DNA ligase IV |

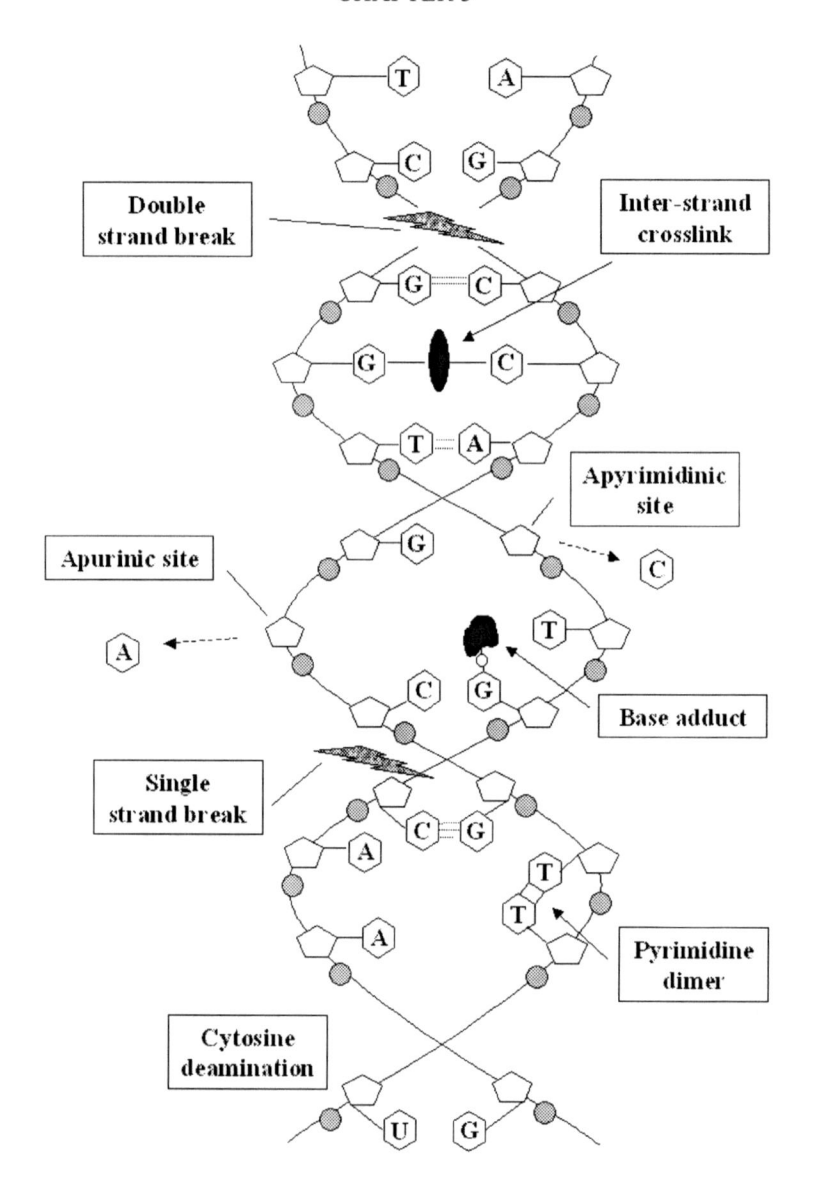

**Figure 3.1** *Types of DNA damage*
A schematic overview of several, but by far not all DNA damage types.

feedback mechanisms to minimize such disparities. In addition, deregulation of precursor pools, in particular of guanine nucleotides, activates cellular checkpoints through the TP53 protein (→5.3). Next to dGTP, dTTP may be most critical, because DNA polymerases also accept dUTP. In normal cells, dUTP levels are maintained low by enzymatic hydrolysis. In cells with suboptimal thymidine biosynthesis, e.g. as a consequence of low folate levels or of chemotherapy with methotrexate (→22.2), significant levels of uracil bases are incorporated. These, like the rarer ones in normal cells, are removed by a specialized hydrolyase, uracil-N-glycohydrolase (UNG). This removal, however, induces abasic sites and strand breaks in DNA (see below).

Spontaneous chemical reactions by DNA bases, outside of or during replication, represent a second type of problem. Hydrolytic cytosine deamination (Figure 3.2) yields uridine which is foreign to DNA. It is recognized as such and removed by the uracil glycohydrolyase. The capacity of this enzyme is more than sufficient to remove the estimated ≈1000 uracils that are spontaneously generated in each cell per day. The rate of cytosine deamination can be increased by exogenous and endogenous compounds. For instance, nitrosation at the amino group leads to deamination of cytosine. Similarly efficient, another specialized glycosylase eliminates hypoxanthin arising from purine base deamination. Hydrolysis of methylcytosine, which constitutes 3-5% of cytosines in human cells (→8), is more of a problem, since it yields thymidine upon deamination. A specialized enzyme, G-T mismatch glycosylase, removes the thymine from such mismatches. Still, the mutation rate at methylated CpG dinucleotides, at which methylcytosine is almost exclusively found in human cells, is higher than elsewhere in the genome, and mutations are usually C→T (or G→A). Deamination of methylcytosine is enhanced by oxidation of the methyl group towards hydroxy-methyl-cytosine. This yields hydroxymethyl-thymidine upon hydrolysis. This modified base can also be generated directly from thymidine by reactive oxygen species. In either case, it is removed by another specialized glycosylase.

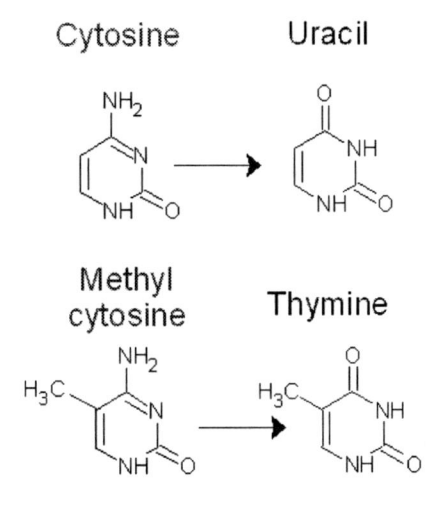

**Figure 3.2** *Effects of deamination of cytosine and methylcytosine*
See text for details

Next to the amino group of cytosine, guanine presents the most sensitive base target for chemical reactions on DNA. Many electrophiles react rather spontaneously at its N7 or O6 positions. An important endogenous electrophile is S-adenosylmethionine, the standard carrier for biological methylation reactions. Methylation of guanine can lead to mispairing and methyl groups are therefore removed by a specialized enzyme. The methyl-guanine methyltransferase MGMT has a broader specificity and also removes other alkyl groups by transfer to its own cysteine groups, inactivating itself in the course of the reaction. Guanine is also a major site for alkylation by exogenous compounds including several cytostatic drugs and major carcinogens. Again, MGMT acts protectively. Down-regulation of the enzyme by epigenetic mechanisms ($\rightarrow$8.3) is found in some cancers. It is an important factor in their responsiveness to chemotherapy, but likely also increases the rate of mutations in general.

Guanine is also the base most susceptible to reactive oxygen species. The most important product is 8-oxo-guanine (or 8-hydroxy-guanine, depending on which mesomeric form is considered). This base also mispairs and is removed by specialized glycosylases, prominently oxoguanine glycosylase 1 (OGG1). Since OGG1 is polymorphic in man, individuals may differ in their capacity of removing this type of damage.

As their designation indicates, base glycosylases in general hydrolyse the N-glycosidic bond between modified bases and deoxyribose, leaving abasic sites in DNA. Such sites also arise from spontaneous or induced hydrolysis of the N-glycosidic bonds of normal bases or of chemically modified bases. Purines are about 20-fold more susceptible to spontaneous loss from DNA than pyrimidines and >20,000 purine bases are estimated to be lost in a human cell each day.

Abasic sites having arisen spontaneously, been induced, or originated through enzyme action are filled in by short-patch repair (Figure 3.3). This is initiated by the action of one of several endonucleases, such as APE1 and APEXL2, which belong to a larger family of apurinic/apyrimidinic endonucleases (AP-endonucleases). They cleave the DNA strand with the missing base to provide a free 3'-hydroxyl group for DNA polymerase $\beta$. This enzyme removes the deoxyribose and replaces it with the correct nucleotide. DNA ligase III closes the strand break. The action of the polymerase and the ligase is coordinated by XRCC1. While only one base is replaced by short-patch repair, an alternative mode, 'long-patch repair' replaces several nucleotides (Figure 3.3). This repair system involves the FEN endonuclease, DNA polymerase $\delta$, PCNA, and DNA ligase I. It is one of several back-up systems to single-base repair.

The type of DNA repair resulting from the combined action of glycosylases, AP-endonucleases, DNA polymerases and DNA ligases is called 'base excision repair'. By comparison, the mismatch repair mechanism taking care of mismatched bases and enzyme slipping during DNA replication is one kind of 'nucleotide excision repair'. Similar to base excision repair, it involves the steps of damage recognition, incision, removal of a short stretch of nucleotides, resynthesis and ligation. Many, but not all components of this system are known in man (Figure 3.4). Damage recognition is achieved by different proteins depending on the type of damage.

Mismatched bases are recognized by the MSH2 and MSH6 proteins, whereas insertion or deletion loops resulting from slippage are recognized by MSH3 and MSH2. In either case, PMS2 and MLH1 are recruited. It is not clear whether any endonuclease is involved in DNA mismatch repair in humans. It seems that mismatch repair during DNA replication starts at existing single strand breaks and uses components of the DNA replisome plus the EXO1 exonuclease. All are coordinated by the PCNA subunit of the replisome.

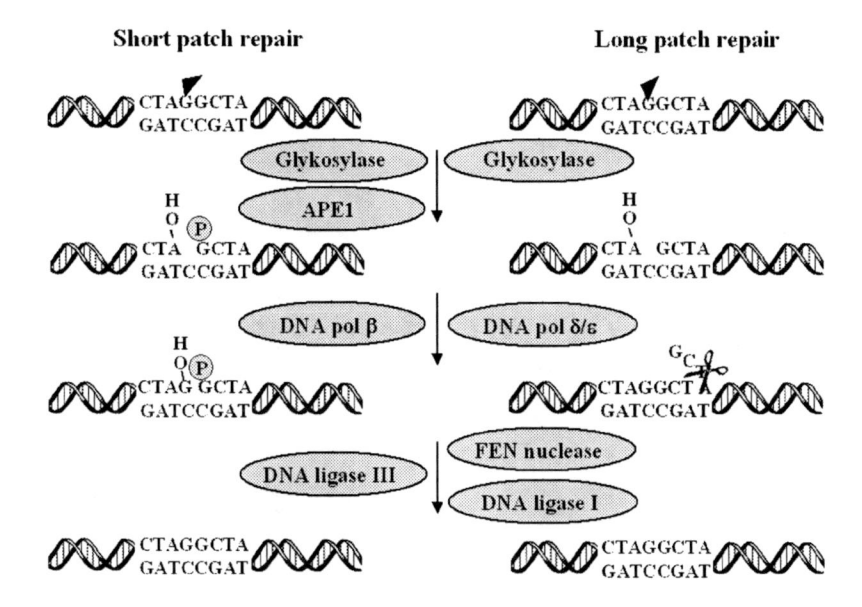

**Figure 3.3** *Short-patch vs. long-patch DNA repair*
Damaged DNA bases (symbolized by the G with the triangle) can be replaced by short patch (left) or long patch (right) repair employing the indicated enzymes (in the order indicated from top to bottom). Long patch repair is preferred or necessary if deoxyribose is as well damaged and/or a phosphate is lacking.

An evident dilemma during mismatch repair is how to decide which of the unmatched bases is to be excised or, likewise, whether a single-strand loop constitutes a deletion or an insertion. In some prokaryotes, this decision is facilitated, since the parental strand is methylated, but the daughter strand is methylated only later on. For instance, E. coli uses adenine (*dam*) methylation at GATC sites for this distinction. In humans DNA is postreplicatively methylated at cytosines (→8.3), but this modification does not seem to serve the same purpose. More likely, newly synthesized strands are distinguished by the presence of single-strand breaks that serve as the starting points for nucleotide excision.

Inherited mutations in components of the mismatch repair system carry a strong hereditary predisposition to certain cancers. Since in the affected individuals cancers in the colon and rectum are most conspicuous, with a life-time risk of up to 80%, these syndromes are summarized under the heading of HNPCC, for 'hereditary nonpolyposis colorectal cancer' (→13.4). They are genetically heterogeneous, since one or the other component of the mismatch repair can be defective. HNPCC is a dominant-autosomally inherited cancer syndrome. Cancers in families affected by HNPCC may also present in the endometrium, stomach, ovaries, hepatobiliary system and the upper urinary tract, in this approximate order of decreasing incidence. Mutations in at least 5 different genes can underlie the syndrome. Most frequently, one allele of *MSH2* and *MLH1* is mutated in the germ-line; mutations in *PMS2*, *MSH6*, and *PMS1* are less prevalent. Further candidates are *MBD4*, encoding a methylcytosine binding protein (→8.3) also involved in DNA repair, and *MYH* encoding a protein recognizing adenine-oxo-guanine mismatches.

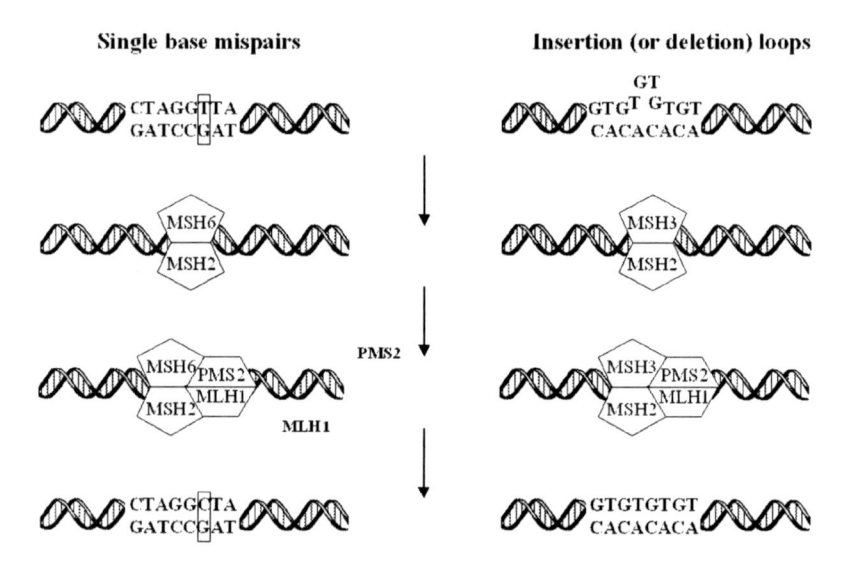

**Figure 3.4** *Two important pathways in DNA mismatch repair*
Left: repair of a T:G mismatch; Right: repair of an insertion/deletion at a microsatellite.

The dominant mode of inheritance in HNPCC is not due to a dominant effect of the mutated gene product. Instead, the one remaining functional allele is generally sufficient for mismatch repair. However, cancers arise when the second, intact allele of the affected mismatch repair gene is accidentially mutated or exchanged by recombination with the first mutated allele in somatic cells. During DNA replication, these cells then accumulate mutations at an increased rate.

Some of these mutations may be irrelevant, such as those in the length of microsatellite repeats which are not repaired after slippage. Others, however, lead to inactivation of genes crucial for the control of cell proliferation, because single base mutations arise from unrepaired mismatches or slippage in base repeats within coding regions are not amended (→13.4). For instance, the *MSH3* and *MSH6* genes each themselves contain cytidine and adenosine hexanucleotide stretches which tend to expand or contract in cells with defective mismatch repair. Since they occur regularly, microsatellite expansions and contractions, which are collectively called 'microsatellite instability' (MSI), can be used to diagnose cancers arising from defective mismatch repair. For this purpose, a standard set of five microsatellites that are most susceptible has been defined.

Cancers with microsatellite instability (MSI) are not restricted to HNPCC families, but also arise in sporadic (i.e. non-familial) cases. Overall, up to 15% of all colon cancers may belong to the MSI group, but only a few percent of all colon cancers arise in HNPCC families (→13.4). The cause of 'sporadic MSI' is mutation, deletion or epigenetic silencing of mismatch repair genes. The most frequent cause of MSI in sporadic cancers may be silencing of *MLH1* by promoter hypermethylation (→8.3).

## 3.2 NUCLEOTIDE EXCISION REPAIR AND CROSSLINK REPAIR

Processes like the hydrolytic deamination of cytosine or the oxidation of guanine lead to altered bases with an increased potential for mispairing. However, neither change interferes principally with DNA replication or transcription. This is different for some other types of damage inflicted on DNA. Ultraviolet radiation (UV) causes chemical reactions in DNA. UV radiation with wavelengths in the absorption maximum of DNA cannot penetrate into the body (→12.1), but the UVB range from 280-320 nm can and just reaches into the absorption spectrum of DNA. This type of UV induces mainly reactions between adjacent pyrimidine bases such as thymine-thymine cyclobutane dimers and thymine-cytosine (or cytosine-cytosine) 6-4 photoproducts (Figure 3.5). These intra-strand dimers present obstacles to transcription and replication of DNA.

Thymine-thymine cyclobutane dimer

Thymine-cytosine 6→4 product

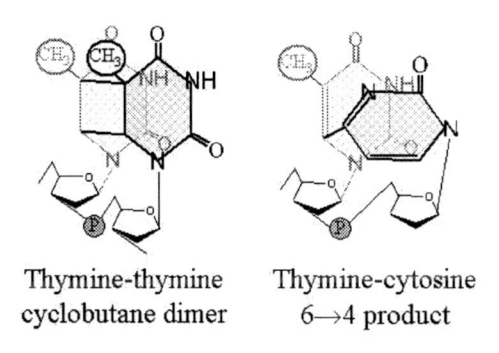

**Figure 3.5** *Pyrimidine base reactions induced by UV irradiation*

Likewise, chemical reactions of endogenous compounds and activated chemical carcinogens can lead to modified bases that are too bulky to fit into a double helix and cannot be recognized by polymerases. Adducts of aflatoxin or benzopyrene at guanines are important examples (Figure 3.6). Even proteins can become covalently linked to DNA bases. Transcription and replication are also prevented, when opposite DNA strands in the double helix are crosslinked. This is exploited in cancer therapy by compounds like cis-platinum and mitomycin C (→22.2).

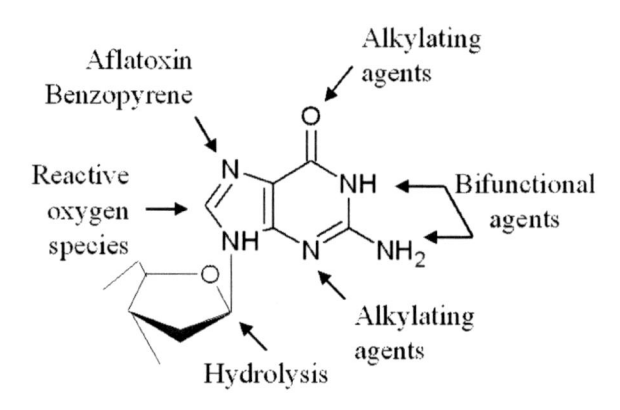

**Figure 3.6** *Reactions of chemical carcinogens with guanosine*

Photoproducts and bulky adducts are removed by nucleotide excision repair (NER). Two interlinked systems are known in man, called 'global-genome' and 'transcription-coupled' repair. Transcription-coupled repair is more rapid, but is restricted to regions of the genome transcribed by RNA polymerase II. When the transcription polymerase encounters a bulky adduct or a cyclobutane photoproduct that prevents further progress, it activates repair through its associated TFIIH complex. This complex contains ≈10 proteins, including Cyclin H. This cyclin regulates kinases that normally phosphorylate and activate PolII, but also the DNA helicases XPB and XPD (also known as ERCC2 and ERCC3) which are involved in NER. The complex successively binds the CSB and CSA proteins which start the actual repair sequence (Figure 3.7).

The actual repair mechanism appears to be identical in transcription-coupled and in global-genome repair. However, recognition of lesions in global-genome repair does not involve the RNA polymerase, but is performed by the XPC and HHR23 proteins. It does also not require the CSA and CSB proteins. Global-genome repair is slower than transcription-coupled repair and has a broader specificity. Following lesion recognition, however, both repair systems use TFIIH components such as XPB and XPD, as well as the single-strand binding protein RPA and the XPA protein to fully unwind and mark the lesion in an ATP-dependent manner. The damaged segment of DNA is excised as a 18-24 nt single strand through 5'-incision by the ERCC1/XPF endonuclease and 3'-incision by the XPG (also ERCC5) endonuclease. The DNA gap is filled by DNA polymerases δ or ε supported by PCNA and RFC and sealed by a DNA ligase, presumably DNA ligase I.

Independent of nucleotide excision repair, photoproducts and other lesions encountered by the replisome can be bypassed through 'translesional repair' which makes use of DNA polymerase η, a more robust enzyme: It it capable of replicating

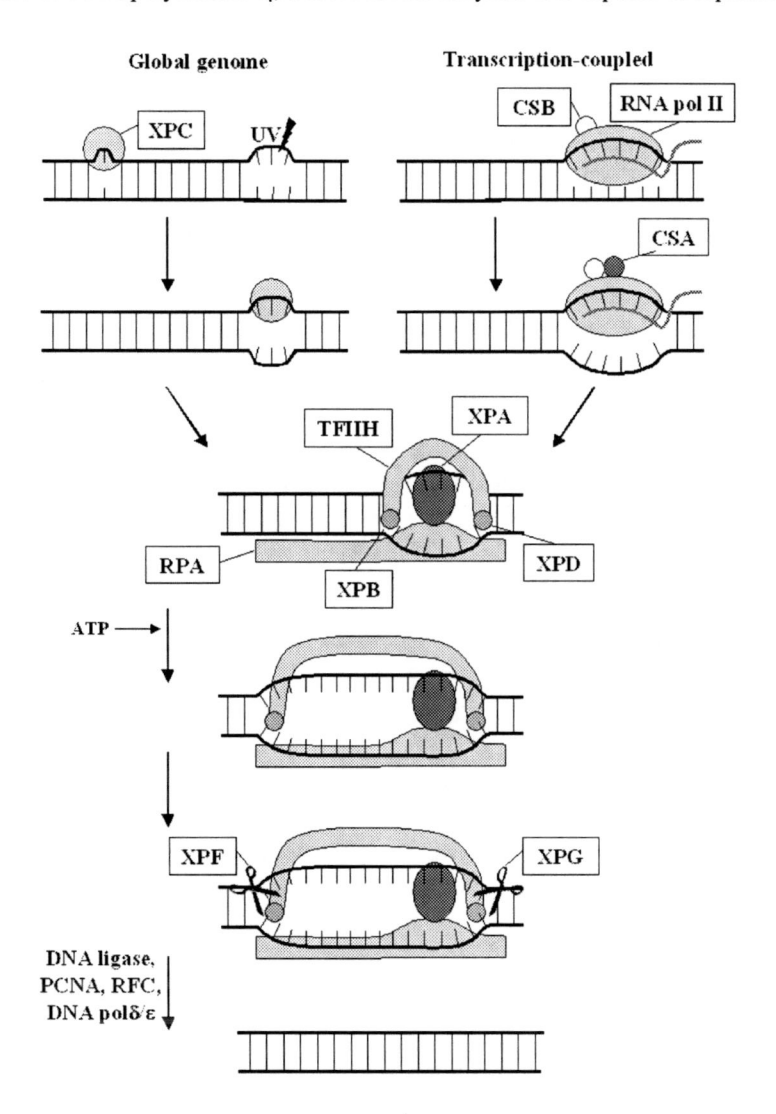

**Figure 3.7** *Nucleotide excision repair*
Nucleotide excision repair, e.g. of base dimers induced by UVB, can be performed by the convergent global genome (left) and transcription-coupled (right) pathways. See text for details.

DNA with very different types of damage, but at the price of a higher error rate than during replication by standard polymerases like Pol δ.

Mutations in genes involved in nucleotide excision repair underlie the diseases xeroderma pigmentosum, Cockayne syndrome, and trichothiodistrophy. These rare diseases are inherited in a recessive fashion. Patients with Cockayne syndrome and trichothiodistrophy suffer from growth defects and progressive mental retardation. Specific and less specific skin defects are apparent, in particular scaly skin (ichthyosis) and brittle hair and nails which are diagnostic for trichothiodistrophy. The patients do not seem to be particularly prone to cancers, but show some symptoms of premature aging (→7.4) and their life expectancy is diminished.

In contrast, patients with xeroderma pigmentosum usually do not display growth defects and mental retardation, except for those in a subgroup overlapping with Cockayne syndrome. Instead, they suffer from extreme photosensitivity of the skin, with abnormal pigmentation. UV-exposed parts of the eyes are also subject to damage. The main clinical problem in these patients is a huge increase in skin cancer risk, estimated as ≈2000-fold. Almost all xeroderma pigmentosum patients develop multiple skin cancers in sun-exposed areas before the age of 30, and often already during their first decade of life. All types of skin cancers are increased, basal cell carcinoma, squamous carcinoma, and melanoma (→12). Other cancer types may also occur at an increased frequency.

Typical Cockayne syndrome is caused by homozygous mutations in the *CSA* or *CSB* genes (hence the designation CS). Xeroderma pigmentosum is caused by mutations in *XPA – XPG* genes (now officially called *ERCC* genes), and likely in others. Some remain unidentified, but mutations in the gene encoding DNA polymerase η are responsible for a subgroup of the disease, XP-V. The very rare trichothiodistrophy syndrome is sometimes caused by mutations in a specific gene (*TTD-A*), but more often by certain mutations in *XPB* or *XPD*. Specific *XPD* mutations also account for most cases of combined xeroderma pigmentosum/Cockayne syndrome.

A simplified explanation for these relationships is illustrated in Figure 3.8. Mutations in Cockayne syndrome cause defects specifically in transcription-coupled repair which impair growth in general and the function of specific tissues such as the brain, since they diminish the efficiency of transcription. Apparently, other repair systems including global-genome repair are not fast enough to prevent this, but remove DNA damage eventually, at least preventing a large increase in cancer risk. In contrast, most defects in XP genes will compromise transcription-coupled as well as global-genome repair leading, in particular, to an increased sensitivity towards UV in exposed tissues. In the case of XPC, the defect is restricted to global-genome repair, with transcription-coupled repair apparently intact. It is not entirely clear, why defects in other XP genes do not regularly lead to impaired growth and neuronal degeneration. Neither is it obvious, why different mutations in the XPB and XPD genes lead to very different, in some respects even complementary phenotypes such as trichothiodistrophy and xeroderma pigmentosum.

Crosslinks between DNA strands are a still more severe impediment to transcription and DNA replication. Their repair is often only possible by sacrificing a fragment of DNA. The mechanisms involved in this type of repair are only partly understood. As in nucleotide excision repair, the actual mechanism may vary depending on when exactly the DNA modification is recognized, i.e. before, during or after DNA replication.

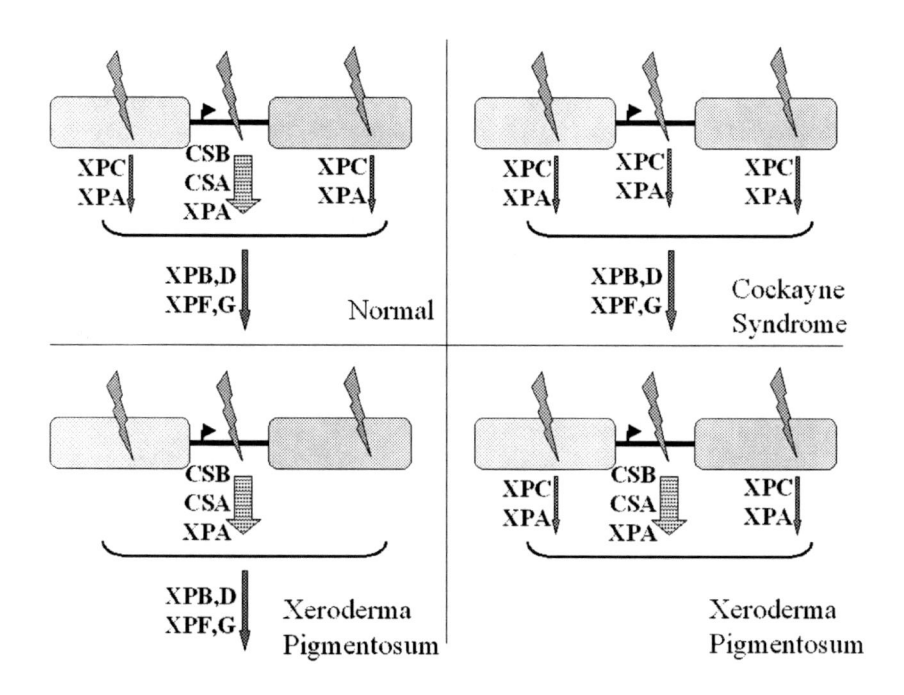

**Figure 3.8** *Relationship of xeroderma pigmentosum and Cockayne syndrome to transcription-coupled repair and global-genome repair*
In each panel, transcribed regions of the genome are indicated by the black line and non-transcribed regions by the large rectangles with rounded edges symbolizing denser chromatin. DNA damage is indicated by lightning symbols.

In non-replicating cells, several mechanisms may in principle be used. They range from outright deletion of the blocked double-stranded segment followed by non-homologous end-joining (→3.3) through error-prone excision/bypass-repair by components from the nucleotide excision repair arsenal to essentially error-free repair by homologous recombination with the homologous sister chromatid in G2 cells (→3.3). How they proceed exactly and how they are chosen, is still being investigated.

Similarly to the mechanism used in G2 cells, the presence of a second homologous sequence can be exploited in a still diploid cell, when crosslinks are encountered at a DNA replication fork (Figure 3.9). Here, an excision is made behind the lesion, likely by XPF/ERCC1. A gap is created by resection, one strand is filled in using the homologous sequence as a template and the resulting structure is resolved by recombination. The second strand is synthesized following excision of the cross-linked fragment. Most of these mechanisms use components of strand break repair systems discussed below and are crucially dependent on FANC proteins.

Mutations in either of >7 genes encoding FANC proteins cause the recessively inherited disease Fanconi anemia (hence: FANC proteins). Patients with Fanconi anemia are small of height and display an assortment of malformations in different organ systems. Most typical are malformations of the lower arm (radius) and thumb. Further parts of the skeleton may be affected as well as the genitourinary system, the gastrointestinal tract, the heart and the central nervous system. 'Café au lait' spots on the skin are an additional diagnostic sign.

The most problematic symptom in this pleiotropic disease is a diminished function of the hematopoetic system, often resulting in diminished production of all cell types (pancytopenia) which develops gradually during childhood. Malfunction

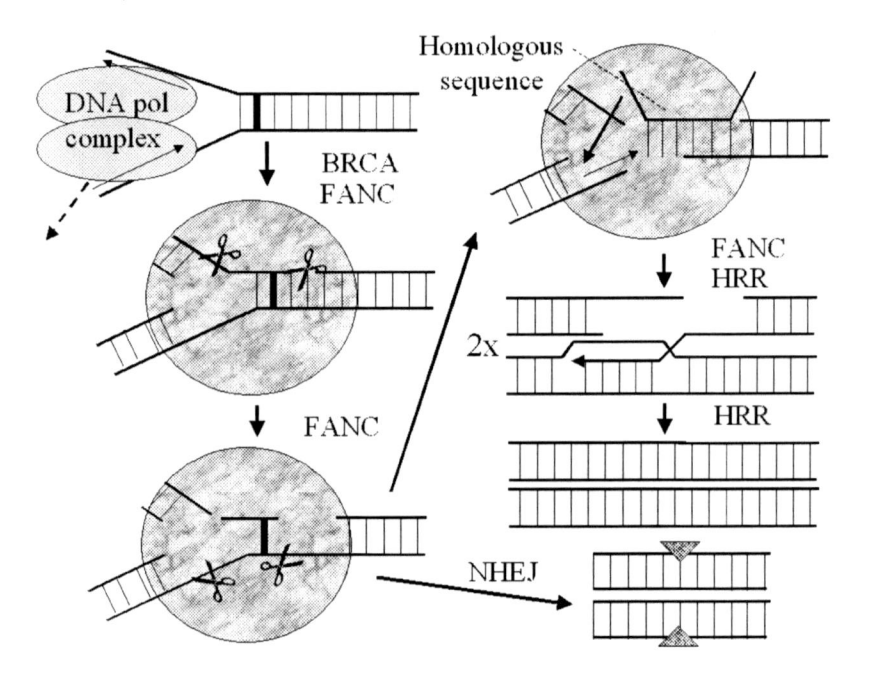

**Figure 3.9** *DNA crosslink repair during replication*
A largely hypothetical outline of DNA repair at replication forks stalled at an inter-strand crosslink (bold step in the DNA ladder). BRCA proteins are thought to regulate the activity of FANC proteins that coordinate excision and repair of the damaged DNA segment. The final stages of the process can be performed either by the HRR system (as in Figure 3.11) which can preserve the sequence or by NHEJ (as in Figure 3.10) which regularly causes a deletion.

of hematopoesis leads to bleeding, anemia, and susceptibility towards infections. Conversely, the patients often suffer from the preneoplastic 'myelodysplastic syndrome', which is prone to progression into outright leukemias, typically AML (acute myeloic leukemia). When challenged with DNA crosslinking compounds like mitomycin C or diepoxybutane, cells from Fanconi anemia patients prove hypersensitive and typically arrest in G2. This assay provides a much clearer margin towards other diseases than the increased rate of spontaneous chromosomal breakage per se, which is more variable and is also enhanced in other diseases.

The hypersensitivity towards DNA crosslinkers in Fanconi anemia underlines the importance of the FANC proteins in crosslink repair. However, it hardly accounts for the full phenotype of the patients. It is therefore thought that the FANC proteins have additional functions, e.g. in the regulation of cytokine synthesis and action or of apoptosis. FANCC, in particular, is implicated in the cellular defense against oxidative stress (cf. 3.5). The most clear-cut evidence points to a wider role of FANC proteins in DNA repair and signaling of DNA damage. In normal cells, several FANC proteins A, C, E, F, and G cooperate in the nucleus to mono-ubiquitinate FANCD2. The induction of this mono-ubiquitination by cross-linking agents is the basis of a biochemical assay for Fanconi anemia.

Ubiquitinated FANCD2 appears to activate 'repair foci' containing several proteins involved in homologous recombination repair of DNA. This type of recombination is not only used in crosslink repair, but also one of several alternatives for strand-break repair (→3.3). Among the components of the homologous recombination protein complex are the BRCA1 and BRCA2 proteins. Inherited mutations in these genes – even in heterozygotes - predispose to breast and ovarian cancers (→18.3). In fact, homozygous mutations in *BRCA2* lead to Fanconi anemia, and *BRCA2* is the *FANCD1* gene. Like BRCA2, BRCA1 also influences the FANC protein complex, regulating its interaction with BRCA2 and its ability to activate checkpoints (→18.3). Conversely, certain mutations in other *FANC* genes may predispose to breast and brain cancers.

The function of FANC proteins in signaling of DNA damage and activation of homologous recombination repair, and likely other systems, explains why cells from Fanconi anemia patients show a decreased ability to correctly repair double-strand DNA breaks in general, not only after crosslinking. This regulatory function may also relate to some of the defects in hematopoesis, since maturation of B- and T-cells involves gene rearrangements requiring joining of double-strand breaks introduced by the lymphocyte-specific recombinases. Indeed, these rearrangements have been found to be compromised and to be more imprecise in Fanconi anemia patients.

Finally, several commonly used cytostatic drugs are DNA crosslinkers. For instance, cis-platinum is an essential component in many cancer chemotherapy formulas and has proven something like a miracle drug in the treatment of testicular cancer. There is mounting evidence that whether individual cancers respond to such drugs may depend on their expression level of FANC proteins.

## 3.3 STRAND-BREAK REPAIR

Repair of damaged DNA bases is a permanent process in living cells. It is only one of several processes, including chemical reactions and enzymatic actions, that generate single-strand breaks in DNA. These are therefore common, and it is estimated that 100,000 DNA single-breaks occur per cell each day. In the most simple case, single-strand breaks are repaired by DNA ligase, but components from the short- and long-patch base excision repair or nucleotide excision repair systems may be needed, when severe base damage or chemical modification of the sugar are associated with the loss of a base.

Like mismatch repair during DNA replication, repair of DNA strand-breaks goes on rather 'quietly'. However, this changes dramatically, when DNA double-strand breaks are generated. These can arise by physiological and non-physiological mechanisms, endogenous processes and exogenous mutagens. During DNA replication, a double-strand break can result from a single-strand break, if this is not repaired, before it is encountered by the replisome. Sometimes two single-strand breaks may by chance occur closely together leading to a double-strand break. Other double-strand breaks are caused by exogenous agents. Some viruses encode enzymes that cut DNA in a similar fashion as restriction enzymes, e.g. retroviral integrases. Ionizing radiation can generate single- as well as double-strand DNA breaks as do several chemical carcinogens and some drugs used in chemotherapy. Bleomycin, e.g., cuts DNA directly, and topoisomerase inhibitors generate strand breaks by inhibition of these enzymes (→22.2). Repair of DNA crosslinks also involves the generation of double-strand breaks.

Double-strand breaks are also created, in a controlled fashion, during physiological recombinations. Important processes of this kind are meiotic recombination and generation of functional T-cell receptor (TCR) and immunoglobulin (IG) genes in lymphocytes, which yield occasionally errors. Unequal recombination in germ cells is an important cause of inherited disease including cancer. Aberrant joining of genes encoding the T-cell receptor or immunoglobulins to other genes such as *MYC* is a frequent source of chromosomal translocations in lymphomas (→10.2). Other translocations and deletions in cancers of the lymphoid lineage can result when the lymphocyte-specific recombination system acts accidentally at sites outside the TCR and IG gene clusters.

Independent of how they arise, double-strand breaks are dangerous as long as they exist, especially to a proliferating cell. They separate a fragment of DNA from the centromere, predisposing it to loss during mitosis. Moreover, the open ends can recombine with other parts of the genome, starting a chain reaction of recombinations and chromosome alterations that can lead to cell death or transformation. Double-strand repair therefore involves blocking of the open DNA ends in addition to actually mending the break. In addition, activation of double-strand DNA break repair is usually associated with the activation of cellular checkpoints that prevent the cell from entering or proceeding through S-phase and mitosis. Specifically, unrepaired DNA double-strand breaks in normal cells often elicit apoptosis. This mechanism provides another level of protection against carcinogenesis, in addition to DNA repair itself.

Several repair systems in human cells deal with double-strand breaks. They can be classified into non-homologous and homologous repair systems.

Non-homologous end-joining (NHEJ) is an imprecise mechanism which is nevertheless most often used in human cells (Figure 3.10). Double-strand breaks are protected by the KU70/KU80 protein heterodimer, and bound by the 'MRN' complex consisting of the MRE11, RAD50, and NBS1 (Nibrin) proteins. These proteins prevent them from illegitimate recombination and attempt to align them. Compatible ends may become ligated, but in many cases the ends are processed. Processing can involve filling in 5'-overhangs and degrading 3'-overhangs. MRE11 possesses nuclease activity. In addition the FEN1 nuclease may be involved as well as the WRN protein which may also supply helicase activity additional to that of the KU70/KU80 proteins. Apparently, processing, unwinding and alignment of the strands proceeds until short complementary base stretches are found which can be used to hybridize the two ends. Remaining overhangs are processed, gaps are filled in and the sugar-phosphate backbone is religated by DNA ligase IV/XRCC4. The end product of the repair process is a restored DNA double helix with a deletion, which is normally kept at a minimum. A distinct characteristic of sequences repaired by NHEJ are microhomologies, i.e. short stretches of 1-12 bp which were identical in the original sequences at both ends of the deletion. These stretches of homology

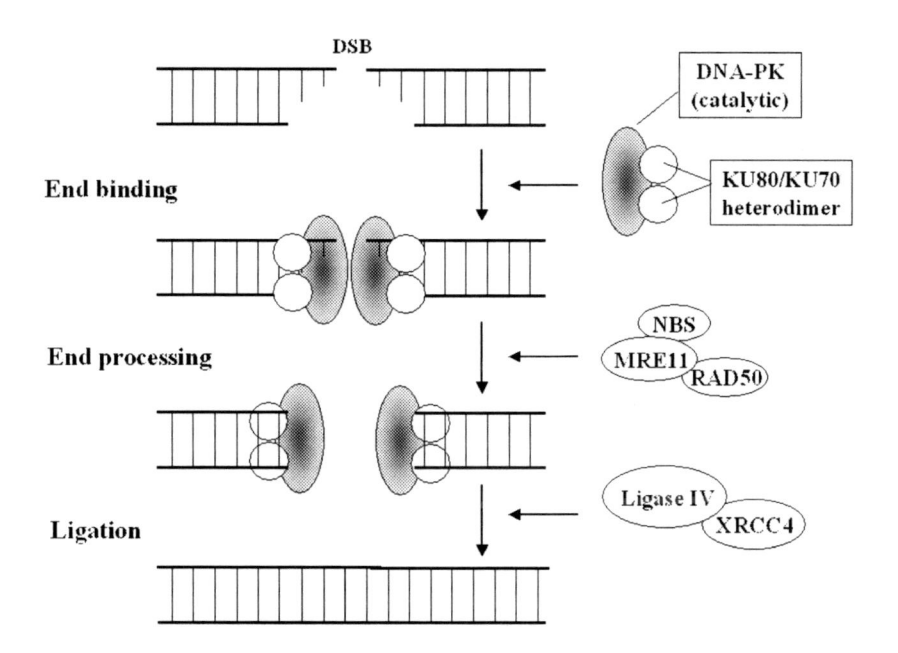

**Figure 3.10** *Repair of DNA double-strand breaks by non-homologous end joining*
See text for details.

are much longer when deletions arise by illegitimate homologous recombination. In some cases, NHEJ repair leads to the insertion of a few additional nucleotides, as during V(D)J joining in lymphocytes. This may help to anneal sequences.

When NHEJ begins, it elicits signals that activate cellular checkpoints. The KU proteins constitute the regulatory subunits of DNA-dependent protein kinase (DNA-PK), which is essential for proper DNA repair. Its catalytic subunit phosphorylates not only itself and other proteins directly involved in repair, but also activates the TP53 protein, which is one of the most important regulators of cellular checkpoints. The phosphorylation of TP53 by DNA-PK and/or further enyzmes such as the ATM and ATR kinases elicits cell cycle arrest or even apoptosis ($\rightarrow$5.3). The NHEJ protein complex itself is regulated by ATM and other proteins. Within the MRN complex, Nibrin seems to exert the major control. It is phosphorylated and activated by the ATM protein kinase and in turn interacts with BRCA proteins.

In contrast to NHEJ, homologous recombination repair (HRR) can be performed in an error-free fashion, at least in principle. In human cells, it is the mechanism of choice in the G2 phase of the cell cycle when a second sequence identical to the damaged one is available in the sister chromatid. NHEJ, in contrast, appears to be the predominant mechanism in G1 cells. HRR may also constitute the preferred method for the repair of double-strand breaks that arise when breaks in one DNA strand are extended into double-strand breaks during replication and the replisome has stalled. In this situation, the BLM helicase may be crucially involved. Demarcation of the double-strand lesion in all other cases is likely performed by the RAD52 protein (Figure 3.11). As a clear-cut difference towards NHEJ, the KU proteins are not involved. The double strand break is then processed to yield a 3'-overhanging single strand of several 100 bases. In this processing the MRE11/RAD50/NBS1 (MRN) complex is again involved together with additional, less well characterized components. With the help of the recombination protein RAD51, the single strands invade the intact homologous double-strand DNA forming D-loop structures ('D' for 'displacement'). The 3'-hydroxyls of the single-strands are then extended and a structure with two Holliday junctions forms. This is resolved by endonuclease action. There are several possible outcomes, depending on how the Holliday junctions are resolved. In one alternative, both original sequences are restored, in the other a crossover takes places. This does not result in a change of sequence when the sister chromatid is used. However, if a homologous sequence from a different chromosome was involved, gene conversion can happen.

Not all parts of the HRR mechanism are well understood, as for NHEJ repair. However, some components have been identified with certainty, because they are mutated in human inherited diseases. Homozygous mutations in the *NBS1* gene that compromise the function of Nibrin underlie the Nijmegen breakage syndrome. This very rare syndrome presents with mental retardation, immunodeficiency, and, tellingly, chromosomal instability and cancer susceptibility. Homozygous mutations in the *WRN* gene encoding a helicase/nuclease involved in double-strand break repair and telomere maintenance also increase the susceptibility to various types of cancer, particularly in soft tissues. However, the resulting Werner syndrome

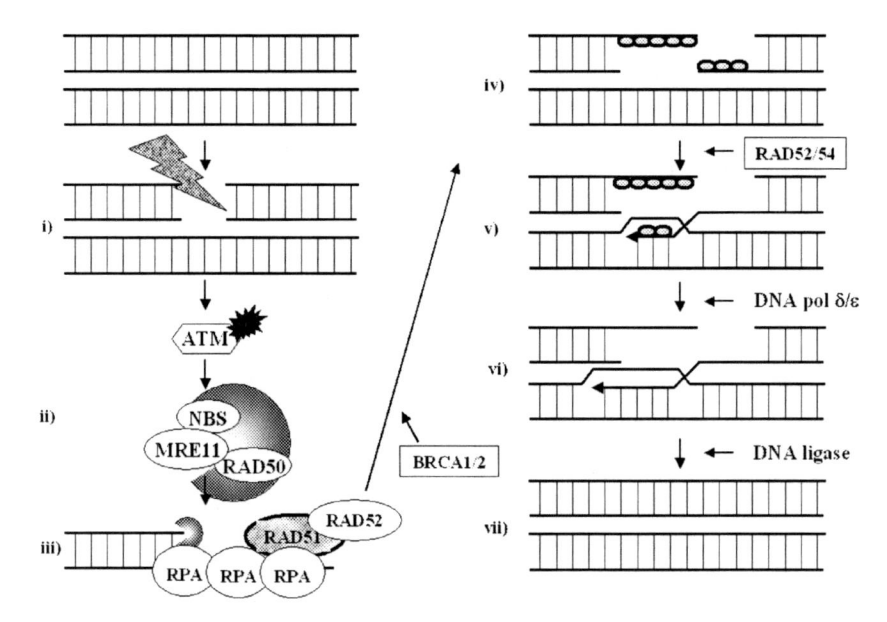

**Figure 3.11** *Mechanism of DNA double-strand break repair by homologous recombination*
See text for details

impresses primarily as a premature aging disease (→7.4) manifesting typically around puberty.

The most prevalent syndrome in this context is the recessively inherited ataxia telangiectasia (AT). It is caused by homozygous mutations in the gene encoding the ATM protein kinase that regulates DNA double-strand break repair. Like NBS patients, AT patients are prone to infections and chromosomal aberrations. They have a ≈100-fold increased risk of cancers, mostly of leukemias and lymphomas. Both syndromes share, in particular, a hypersensitivity towards ionizing radiation. However, AT patients are not usually mentally retarded. Instead, they develop a gradual decline of the function of the cerebellum, which progressively impedes movements, speech and sight. This very specific ataxia led to the name along with the diagnostic telangiectasias which are aggregates of small dilated blood vessel appearing in unusual places such as the conjunctiva of the eye. They are thought to be caused by inappropriate angiogenesis. The chain of events leading to these lesions may involve lack of ATM function leading to incomplete function of TP53 alleviating suppression of angiogenesis induced by hypoxia (→9.4). Other aspects of the pleiotropic ATM phenotype are less understood, including an elevation of the fetal albumin homologue α-fetoprotein that is useful for the diagnosis of the disease.

In contrast, the chromosomal instability and hypersensitivity towards ionizing radiation in the syndrome fit well with the known function of ATM as a central coordinator of double-strand break repair. DNA double-strand breaks caused by physiological recombination, by viral or retrotransposon enzymes, by ionizing radiation or chemicals, or by oxidative stress all appear to activate ATM. Likely, this occurs by different routes. The protein may itself sense damage to some extent, but the MRN complex through NBS1 certainly plays a part. A variant histone, H2AX, accumulates within 1 min at double-strand breaks to become phosphorylated by ATM; this could well be another sensor protein. H2AX can alternatively be phosphorylated by DNA-PK. Further candidates for damage sensors are RAD9 and RAD17 which are also ATM substrates.

Following its activation, ATM goes on to phosphorylate further proteins involved in DNA repair such as FANCD2, BRCA1, and RPA. Significantly, it also activates checkpoints that block the cell from further proliferation. Phosphorylation by ATM activates the TP53 protein, whereas it prevents the TP53 inhibitor protein HDM2 from binding to TP53. Together these actions lead to cell cycle arrest at the G1/S checkpoint via induction of the $p21^{CIP1}$ cell cycle inhibitor and at the G2/M checkpoint by other mediators ($\rightarrow$6.6). DNA replication can be arrested via phosphorylation of CHK2 (checkpoint kinase 2) and Nibrin, while phosphorylation of TP53 and BRCA1 also activates the G2/M checkpoint.

Some aspects of ATM function can also be provided by other protein kinases such as CHK2, ABL, and ATR. Severe damage by UV radiation, e.g., is signaled by the ATR protein kinase in an otherwise quite similar fashion, including phosphorylation of TP53 and CHK1 (instead of CHK2). The somewhat complementary functions of ATM and ATR are the likely explanation why AT patients and their cells are sensitive to ionizing radiation, but not to UV.

A public debate has developed on the issue of whether heterozygosity for ATM mutations leads to an increased cancer risk. This is a particular concern, since several methods commonly used in cancer screening and diagnosis employ ionizing radiation. The results of different investigations vary. It is possible that the cancer risk of heterozygous carriers of the disease may depend on which mutation is present. Some mutations may completely inactivate the affected allele. Others may show some degree of a dominant-negative phenotype, i.e. an altered protein product is formed which does not function in repair, but inhibits the function of the protein produced by the normal allele. In the case of the ATM protein, this is conceivable, since the protein normally exists as a dimer and its activation involves cross-phosphorylation between the subunits. So, dysfunctional subunits may inactivate some of the functional subunits as well.

## 3.4 DEFECTS IN DNA REPAIR AND CANCER SUSCEPTIBILITY

It is clear from the previous sections that inherited defects in DNA repair are an important source of susceptibility to cancer. A number of syndromes related to DNA repair carry an increased risk of cancers (Table 3.2). Homozygous mutations in the *ATM*, *NBS1*, *WRN*, *FANC*, and *XP/ERCC* genes underlie recessively inherited diseases, which confer an increased risk for cancers in the context of syndrome with a wider range of afflictions. Heterozygous mutations in MMR genes and in the *BRCAs* and perhaps certain ATM mutations lead to cancer predisposition in a dominantly inherited fashion. As a rule, no other consistent symptoms are associated with these mutations. With either type of predisposition, cancers develop at an increased rate as a consequence of an enhanced rate of mutations, either point mutations in MMR deficiency and XP, or chromosomal aberrations in the others. Also typically, in these diseases, cancers appear at an unusually early age.

Obviously, the question arises to what extent defects in DNA repair are involved in sporadic cancers (the great majority) which arise in people not carrying mutations in any of the above genes. In other words, since defective DNA repair is sufficient for cancer development, is it also necessary? Or can cancers arise in the course of the relatively long human lifetime just by accumulation of rare alterations that have occurred in spite of functional DNA repair? The answers to these questions are open.

One also has to consider in this regard that many genes involved in DNA repair are polymorphic. Several of these polymorphisms have been linked to an increased risk for one of the major cancers. Typically, the increases in cancer risk conferred by these polymorphisms to each individual are small compared to those resulting from

**Table 3.2.** *Inherited defects in DNA repair and predisposition to cancer*

| Syndrome | Mode of inheritance | Repair system affected | Gene(s) involved | Tissue with increased cancer risk |
| --- | --- | --- | --- | --- |
| Xeroderma pigmentosum | recessive | nucleotide excision repair | XPA-G (*ERCC1-7*) genes, others | skin |
| Cockayne | recessive | transcription-coupled nucleotide excision repair | *CSA*, *CSB* | no significant increase |
| Ataxia telangiectasia | recessive | strand-break repair | *ATM* | multiple |
| Fanconi anemia | recessive | crosslink repair | *FANC* genes, *BRCA2* | hematopoetic system, others |
| Nijmegen breakage | recessive | strand break repair | *NBS* | hematopoetic system |
| Bloom | recessive | strand-break repair (HRR?) | *BLM* | multiple |
| Werner | recessive | strand-break repair (NHEJ?) | *WRN* | multiple |
| Hereditary breast cancer | dominant | homologous recombination repair | *BRCA1*, *BRCA2*, others (?) | breast, ovary |
| Hereditary non-polyposis carcinoma coli | dominant | mismatch repair | *MSH2*, *MLH1*, *PMS2*, others | colon, endometrium, stomach, others |

mutations that lead to the full inactivation of a DNA repair system. However, frequent polymorphic forms of such genes could be important determinants of cancer frequency in the whole population (cf 2.3).

In addition to inherited mutations or polymorphisms, acquired mutations inactivate DNA repair genes in many cancers or they become silenced by epigenetic mechanisms. This is documented in the case of certain MSI cancers (→3.2) arising through inactivation of MMR genes. It is not known precisely, to what extent chromosomal instability in other cancers is caused by inactivation of *ATM*, *NBS1*, *WRN*, *FANC* or *BRCA* genes through somatic mutations. Overall, such cases appear to be rare.

In contrast, a failure of cell cycle checkpoints activated by DNA damage is detectable in many, if not most cancers (→5.4, →6.6). Therefore, defects in DNA repair as such may not be required for cancer development, but defective signaling to checkpoints, e.g. by loss of checkpoint kinase or of TP53 function, could be necessary. Defective checkpoint activation allows cell proliferation to continue in spite of DNA damage, with the consequence that some defects become permanent and are propagated by the following cell generations.

Finally, it is possible that through a human lifetime those occasional DNA defects that have escaped repair accumulate, and perhaps DNA repair becomes less efficient during aging. In addition, telomere dysfunction in aging cells (→7.4) may provide a new challenge to DNA repair systems which they cannot always master.

## 3.5 CELL PROTECTION MECHANISMS IN CANCER

Exogenous carcinogens and potential mutagens arising from endogenous processes are often prevented from encountering DNA by specific cellular protections mechanisms. These highly diverse mechanisms serve as a further tier of cancer prevention in addition to DNA repair and apoptosis. Often, they protect not only DNA, but cells in general from damage. Some of these mechanisms are very specific and some are very general. In the context of cancer, they are particularly important during two very different phases, viz. (1) during carcinogenesis and (2) during cancer therapy.

A number of low molecular weight compounds are employed in the cell to stabilize macromolecules and membranes, protect against altered osmolarity, buffer the redox state, and quench radicals and specifically reactive oxygen species. These chemically diverse compounds (Table 3.3) comprise polyamines, amino acids like taurine, the tripeptide glutathione, and the lipophilic and hydrophilic vitamins E and C, i.e. tocopherol and ascorbic acid. Glutathione, γ-glutamyl-cysteinyl-glycine (GSH), is part of a cellular redox buffer system and its thiol group also reacts readily with radicals. The normal oxidized form of glutathione is its disulfide GSSG, from which GSH can be recovered by glutathione reductase, which uses NADPH as the cosubstrate. GSH is also used in enzymatically catalyzed reactions for similar purposes. So, the selenium-containing enzyme glutathione peroxidase removes hydrogen peroxide generating GSSG. Lack of this enzyme may be one reason why selenium deficiency may increase the risk of cancer. Glutathione also reacts spontaneously with reactive electrophilic compounds, including activated carcinogens. These reactions are strongly accelerated by glutathione transferases (GSTs). The various isoenzymes in this family all catalyze the reaction of glutathione with several substrates, but each enzyme recognizes a different range of compounds. In most cases, carcinogens are inactivated by conjugation with glutathione. The conjugates are further metabolized and eventually excreted.

It is clear from this short description why polymorphisms in GST enzymes catalyzing reactions of glutathione modulate the risk of various cancers (→2.3). In addition, the level of glutathione itself and the ratio of GSSG:GSH are also relevant. However, these same reactions are also relevant in the context of cancer therapy. Cytotoxic cancer drugs also react with DNA and some act by inducing reactive oxygen species. So, both reaction with glutathione catalyzed by GSTs and the quenching of reactive oxygen species by GSH and other radical catchers diminish the efficacy of such drugs in cancer cells, while they protect normal cells. In fact, GSTP1, one isoenzyme of the family, is often overexpressed in cancers becoming resistant to therapy, in some cases as a consequence of gene amplification. Paradoxically, the same enzyme is down-regulated in a few selected cancers such as prostate carcinoma (→19.3).

A similar argument can be made for ionizing radiation. The effects of ionizing radiation on normal cells are mitigated by cellular protection mechanisms, in addition to DNA repair. For instance, polyamines stabilize cellular macromolecules such as DNA and structural RNAs. Tocopherol, ascorbate, carotenoids, and glutathione can all act to quench the effect of hydroxyl radicals and other reactive oxygen species. Therefore, it may in some cases be helpful to deplete such compounds prior to therapy.

GSTs and glutathione peroxidase (abbreviated GPx) are examples of cell-protective enzymes that modulate the effects of many different exogenous and endogenous agents. Others are more tailored towards specific compounds. For instance, metallothioneins are a group of small proteins protecting against toxic metal ions. They contain multiple thiol groups which are highly reactive towards

**Table 3.3.** *Some low molecular weight compounds and enzymes in cell protection*

| Low molecular weight compound | Function | Protein/enzyme | Function |
|---|---|---|---|
| Glutathione | protection against reactive oxygen species and electrophilic compounds | Glutathione peroxidase | removal of $H_2O_2$ |
| Ascorbic acid | protection against hydrophilic radicals | Catalase | removal of $H_2O_2$ |
| Tocopherol | protection against lipophilic radicals | Superoxide dismutases | removal of superoxide |
| Spermidine, Spermine (polyamines) | stabilization of ribonucleoprotein complexes | Metallothioneines | binding of toxic metal ions |
| Taurine, betaine | osmoprotection | | |

potentially carcinogenic metal ions including cadmium and nickel. However, while they may prevent carcinogenesis by these and other substances, they may also contribute to resistance against chemotherapy that uses metallo-organic compounds, and specifically against the widely employed platinum complexes. Moreover, these proteins as well react with radicals induced by cancer treatments.

It is important to realize that the efficiency of the cellular protection mechanisms discussed here and of others less well understood appears to be determined by an interaction of genetic and environmental factors. On the genetic side, polymorphisms in a large number of genes in this context are expected to modulate the risk of cancers, but also the responses to therapies. On the environmental side, the type of exposure is, of course, relevant, but also factors like diet and immune status which affect the levels of low molecular weight compounds as well as of proteins involved in cell protection. Genetic and environmental factors may, in particular, synergize with each other. There is evidence for this type of interaction in a wide variety of human cancers (see e.g. 21.3).

## Further reading

Zhou BBS, Elledge SJ (2000) The DNA damage response: putting checkpoints in perspective. Nature 408, 433-439

Hoeijmakers JHJ (2001) Genome maintenance mechanisms preventing cancer. Nature 411, 366-374

Elliott B, Jasin M (2002) Double-strand breaks and translocations in cancer. CMLS 59, 373-385

Goode EL, Ulrich CM, Potter JD (2002) Polymorphisms in DNA repair genes and associations with cancer risk. Cancer Epidemiol. Biomarkers Prevent. 11, 1513-1530

Obe G et al (2002) Chromosomal aberrations: formation, identification and distribution. Mutat. Res. 504, 17-36

Cherian MG, Jayasurya A, Bay BH (2003) Metallothioneins in human tumors and potential roles in carcinogenesis. Mutat. Res. 533, 201-209

Christmann M (2003) Mechanisms of human DNA repair: an update. Toxicology 193, 3-34

Cooke MS, Evans MD, Dizdaroglu M, Lunec J (2003) Oxidative DNA damage: mechanisms, mutation, and disease. FASEB J. 17, 1195-1214

D'Andrea AD, Grompe M (2003) The Fanconi anaemia/BRCA pathway. Nat. Rev. Cancer 3, 23-34

Dizdaroglu M (2003) Substrate specificities and excision kinetics of DNA glycosylases involved in base-excision repair of oxidative DNA damage. Mutat Res. 531, 109-126

Gisselson D (2003) Chromosome instability in cancer: how, when, and why? Adv. Cancer Res. 88, 1-29

Lucci-Cordisco E et al (2003) Hereditary nonpolyposis colorectal cancer and related conditions. Am. J. Med. Genet. 122A, 325-334

Powell SN, Kachnic LA (2003) Roles of BRCA1 and BRCA2 in homologous recombination, DNA replication fidelity and the cellular response to ionizing radiation. Oncogene 22, 5784-5791

Scharer OD (2003) Chemistry and biology of DNA repair. Angew. Chem. 42, 2946-2974

Shiloh Y (2003) ATM and related protein kinases: safeguarding genome integrity. Nat. Rev. Cancer 3, 155-165

Townsend D, Tew K (2003) Cancer drugs, genetic variation and the glutathione-S-transferase gene family. Am. J. Pharmacogenomics 3, 157-172

# CHAPTER 4

# ONCOGENES

- The first group of oncogenes to have been discovered form parts of the genomes of acutely transforming retroviruses, which cause hematological or soft tissue cancers in their avian or mammalian hosts. They act in a dominant manner and confer altered growth properties and morphology on specific target cells in mesenchymal tissues or the hematopoetic system.
- A second group of oncogenes consists of host proto-oncogenes that become activated when the insertion of slowly transforming retroviruses disrupts their regulation.
- Retroviral insertion is only one of several mechanisms that can activate cellular proto-oncogenes to dominantly acting oncogenes. Other mechanisms include chromosomal translocation, gene amplification and point mutations. These mechanisms alter the regulation and/or function of cellular genes, which are thereby activated from proto-oncogenes towards oncogenes. As a consequence of these alterations, their protein products become overexpressed or deregulated and/or become overactive or mislocalized in the cell.
- The oncogenes of acutely transforming retroviruses are in fact also derived from host genes, and have become deregulated and overactive by expression from the retroviral long terminal repeat and by mutations. Several genes such as NRAS, KRAS, ERBB1, and MYC orthologous to viral oncogenes have turned out to be overexpressed or mutated in human cancers. The cellular orthologs of other retroviral oncogenes are more subtly involved in human cancers.
- Many cellular proto-oncogenes regulate cell proliferation, differentiation and survival also in their normal state. Some act as extracellular growth factors, some as their receptors and some as juxtamembrane adaptors or transducers in signaling cascades emanating from growth factor receptors or other membrane receptors. A large class of proto-oncogene products are protein kinases. They include growth factor receptors with a crucial tyrosine kinase activity. Other kinases are located in the cytoplasm. A further large group of protooncogenes consists of transcription factors acting in the nucleus. So, oncogenes can be categorized according to their cellular localization and/or their biochemical function. Indeed, a surprisingly large number of proto-oncogenes functions within or interacts with a single signaling network. At its core is the mitogenic 'MAP kinase' cascade which links growth factor signaling to transcription in the nucleus and to the cell cycle, but also influences protein synthesis and the cytoskeleton.
- Growth factor signaling and the MAPK cascade are tightly regulated in normal cells by feedback regulation and by short half-lifes of activated states. Oncogenic mutations make oncogene proteins independent of input signals, disrupt feedback regulation or prolong their active state.

- There are very few cases in which a single oncogene is sufficient to fully transform a cell towards malignancy. Rather, a single oncogene confers some aspects of the malignant phenotype and cooperates with others or with defects in tumor suppressors for complete transformation. This relationship is illustrated in cellular assay systems such as rat embryo fibroblasts, in which two different types of oncogenes are required for transformation.
- Human cancers accumulate many genetic and epigenetic alterations during their progression. In a typical cancer many genes are overexpressed and many gene products are overactive. A fraction of these may indeed be necessary for the survival and sustained growth of the cancer. So, they might be regarded as oncogenes as well. A more strictly defined oncogene exhibits these same properties, but its overexpression or overactivity is caused by substantial changes in the gene, i.e. mutations or amplifications.

## 4.1 RETROVIRAL ONCOGENES

Several different types of viruses are involved in the development of cancers in humans and animals (→1.2). In humans, predominantly DNA viruses are implicated. Among the retroviruses, only HTLV1 (human T-lymphotropic virus) and HIV are involved in carcinogenesis, but in a very roundabout fashion (Box 4.1, Box 8.1). In animals, in contrast, several retroviruses have been documented to directly cause cancers, by two different mechanisms, which can be categorized as 'acute' and 'slow'. Alternatively, they can be designated as 'transducing' or 'cis-acting' (cf. 4.2).

Acute transforming retroviruses such as ALV (avian leukosis virus) or RSV[2] (Rous sarcoma virus) elicit leukemias, lymphomas or sarcomas rapidly after infection of their hosts. In fact, avian acute transforming retroviruses were often identified during epidemics that ravaged fowl farms. They were shown to be capable of transforming their target cells in a dominant fashion, without any apparent requirement for a co-carcinogen. Very early after their discovery at the begin of the 20$^{th}$ century, it was predicted that they carried genes which caused cell transformation and were accordingly termed oncogenes. The existence of these genes was formally and physically demonstrated in the second half of the 20$^{th}$ century and the first oncogene protein, v-src from RSV, was biochemically characterized in the late 1970's.

The v-src protein is a protein kinase located at the inside of the plasma membrane. While most protein kinases phosphorylate serine or threonine residues in their substrates, v-src phosphorylates tyrosine (which came as a surprise at the time of discovery). In the RSV genome, v-src is carried as an additional gene 3' to the standard gag, pol, and env gene complement. This is unusual for acutely transforming retroviruses. In most of them, one or two oncogenes replace parts of standard genes (Figure 4.1). Typically, oncogenes replace the pol and part of the gag gene and the oncogenic protein is expressed as a gag-fusion protein at very high

---

[2] This is a chicken virus unrelated to the human pathogen respiratory syncytial virus, also abbreviated RSV.

levels. This high expression level contributes to the dominant mode of action of retroviral oncogenes. Of course, the replacement of pol or other viral genes by an oncogene sequence obliterates the ability of the virus to replicate autonomously, rendering it 'defective'. For replication and propagation, defective retroviruses need intact replication-competent helper viruses that supply the proteins required for reverse transcription, integration, packaging and maturation. This requirement may explain why acutely transforming retroviruses are (fortunately) rare. Conversely, the unique ability of RSV to replicate autonomously may have contributed to its early isolation already one century ago.

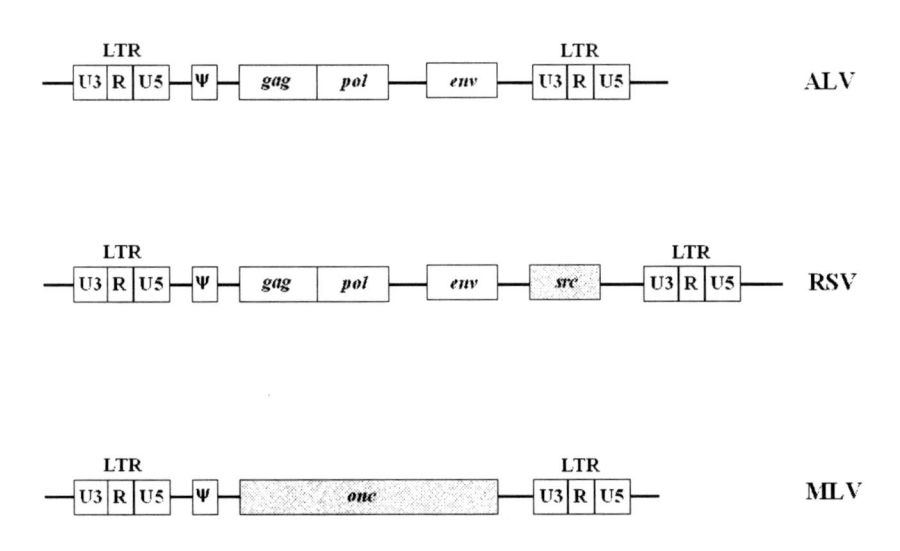

**Figure 4.1** *Prototypic retrovirus and onco-retrovirus genomes*
ALV (avian leukosis virus), RSV (Rous sarcoma virus) and MLV (murine leukemia virus)

Meanwhile, more than two dozen oncogenes have been isolated from retroviruses and have been biochemically characterized. For a selection of these, their origin, cellular localization and main biochemical function are listed in Table 4.1. Several points can be noted in this compilation. (1) Some oncogenes appear to be similar. In some cases this is due only to the nomenclature: v-myb and v-myc are not overtly similar beyond being transcription factors, but have both been discovered in retroviruses causing myeloid leukemias. In contrast, Ki-ras and Ha-ras are indeed highly similar proteins. This points to the existence of oncogene families. (2) The products of retroviral oncogenes appear to cover a relatively limited range of biochemical functions. Protein kinases like v-erbB, v-src and v-raf and transcriptional activators like v-fos, v-jun, v-myb and v-myc comprise the majority. Others are receptors, like v-erbB or v-fms, or proteins transducing signals from receptors, like the ras proteins. A limited number of other functions constitute the remainder. (3) Some oncogenic retroviruses carry two oncogenes which cooperate

**Table 4.1.** Some important retroviral oncogenes

| Oncogene | Virus | Species | Tumor type | Localization in the cell | Main biochemical function |
|---|---|---|---|---|---|
| *sis* | simian sarcoma virus | monkey | sarcoma | extracellular | growth factor |
| *erbB* | avian erythroblastosis virus | chicken | leukemia | cell membrane | tyrosine kinase |
| *fms* | feline sarcoma virus (SM strain) | cat | leukemia | cell membrane | tyrosine kinase |
| *kit* | feline sarcoma virus (HZ2 strain) | cat | sarcoma | cell membrane | tyrosine kinase |
| *src* | Rous sarcoma virus | chicken | sarcoma | inner cell membrane | tyrosine kinase |
| *abl* | Abelson murine leukemia virus | mouse | leukemia | cytoplasma | tyrosine kinase |
| *raf* | murine sarcoma virus (3611 strain) | mouse | sarcoma | inner cell membrane/ cytoplasm | tyrosine kinase |
| *Ha-ras* | Harvey sarcoma virus | rat | sarcoma | inner cell membrane | GTP-binding protein |
| *Ki-ras* | Kirsten sarcoma virus | rat | sarcoma | inner cell membrane | GTP-binding protein |
| *akt* | AKT8 virus | mouse | thymoma | inner cell membrane/ cytoplasma | serine protein kinase |
| *myc* | several avian myelocytomatosis viruses | chicken | leukemia | nucleus | transcription factor |
| *myb* | avian myeloblastosis virus | chicken | leukemia | nucleus | transcription factor |
| *rel* | avian reticuloendotheliosis virus | turkey | leukemia | nucleus | transcription factor |
| *fos* | murine osteosarcoma virus | mouse | osteosarcoma | nucleus | transcription factor |
| *jun* | avian sarcoma virus | chicken | sarcoma | nucleus | transcription factor |
| *erbA* | avian erythroblastosis virus | chicken | leukemia | nucleus | transcription factor |
| *tax* | HTLV1 | human | leukemia, lymphoma | nuclear | transcriptional regulator |

during transformation. For instance, the erythroblastosis virus contains two oncogenes, v-erbA and v-erbB. The first of these is a transcriptional repressor protein and the second a constitutively active cell membrane tyrosine protein kinase. Their cooperation likely results from v-erbA blocking differentiation of erythrocyte precursors and v-erbB stimulating proliferation and supporting survival of these cells to cause erythroblastosis.

## 4.2 SLOW-ACTING TRANSFORMING RETROVIRUSES

Slow-acting transforming retroviruses are as a rule replication-competent and do not transduce oncogenes. Instead, they cause transformation by integrating within or in the vicinity of cellular genes and altering their expression. In this fashion they convert cellular genes from proto-oncogenes into oncogenes in a cis-acting manner.

Integration of a slow-acting retrovirus disrupts negative regulatory elements of the targeted gene and/or activates its transcription by the transcriptional regulatory sequences contained in the retroviral LTR. Several mechanisms are conceivable by which the viral regulatory sequences could cause gene overexpression. Overall, the predominant mechanism seems activation of the cellular gene promoter by the enhancer in the retroviral LTR. This mechanism is most effective, if the retrovirus integrates in inverse orientation upstream from the promoter (Figure 4.2).

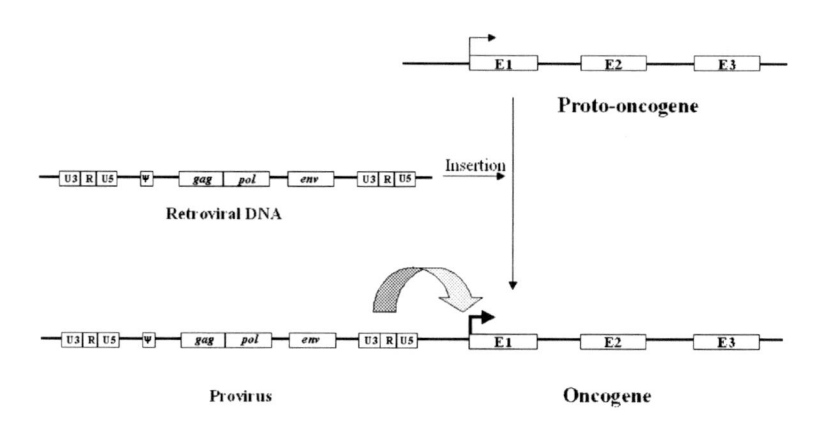

**Figure 4.2** *Oncogene activation by retroviral insertion*
The most frequent mode of oncogene activation is by enhancer activity of the 'idle' 3' provirus LTR on the proto-oncogene promoter shown in the figure. Often, provirus insertion additionally disrupts or separates negative regulatory elements of the host.

In the figure, this mode of activation is shown for the cellular myc gene (initially called c-myc). This gene consists of three exons and contains several negative regulatory elements upstream of its two transcriptional start sites and in the first intron. Transcription from a physiological start site proceeds into the first intron and pauses there until further signals arrive, similar as during transcriptional regulation by attenuation in bacteria or by the tat protein of HIV. A typical retroviral insertion of the myc gene disrupts this negative control mechanism (as well as others) and at the same time elicits strong transcription from an otherwise inactive ('cryptic') promoter near the end of intron 1. In this fashion, myc transcription becomes independent of extracellular signals. Specifically, the expression of the gene is not down-regulated in response to differentiation signals.

The c-myc gene activated by slowly transforming retroviruses is very similar to the oncogene carried by the myelocytomatosis virus (Figure 4.3). Apparently, the cellular gene is the precursor of the viral gene and has been picked up by an evolutionary precursor of the myelocytomatosis virus. This may have occured by recombination of the c-myc mRNA with the retroviral genomic mRNA followed by transduction. Even more likely, the recombination may have involved a retroviral genomic transcript and a transcript from a c-myc locus into which a retrovirus had inserted. In this fashion, slow-acting transforming retroviruses may give rise to the rarer acutely transforming types.

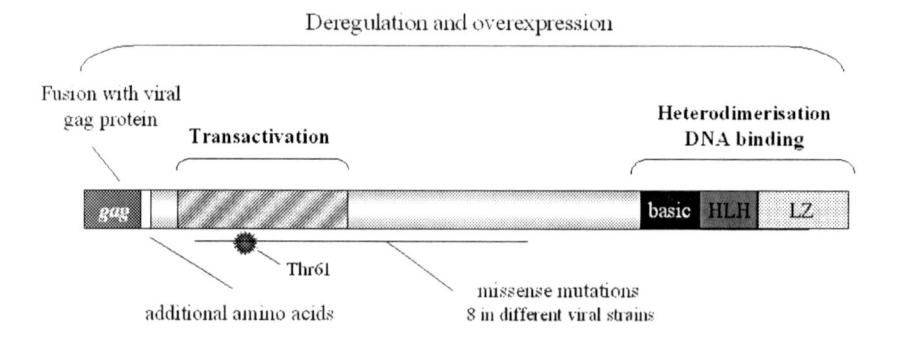

**Figure 4.3** *The v-myc protein*
In avian and murine transforming retroviruses, the v-myc protein retains all functional domains of its cellular ortholog. It is always overexpressed, usually as a fusion protein with viral gag sequences at the N-terminus plus linker amino acids. Thr61 is consistently mutated. Several other amino acids in the N-terminal part are mutated in individual viral strains.

When a slowly transforming retrovirus integrates into the c-myc gene, the target gene becomes deregulated and over-expressed (Figure 4.2). The v-myc gene contained in the myelocytomatosis retrovirus is likewise strongly expressed. In addition, there are several changes in the amino acid sequence of v-myc compared to the avian c-myc gene. These are not random. For instance, a change in most viral strains removes a threonine which is required for inactivation of the myc protein, further enhancing its oncogenicity (cf. 10.3). So, in summary, the acutely transforming retrovirus overexpresses an altered cellular protein, whereas a slowly transforming retrovirus deregulates the endogenous protein, which may remain unchanged, at least initially.

As in the example of c-myc/v-myc, retroviral oncogenes are almost always altered compared to their cellular orthologs, from which they were derived. These alterations are often more severe than in the case of myc. They comprise truncation, mutation or fusion to viral proteins which increase the activity, affect the regulation and alter the localization of the oncoproteins within the cell. For instance, the v-erbB product is derived from the cellular erb-B1 gene which encodes a growth factor receptor, EGFR (Figure 4.4). This receptor is basically composed of three domains: an extracellular domain binding the growth factor ligands, a transmembrane domain, and a cytoplasmic tyrosine kinase domain, which is controlled by an autoinhibitory loop. The viral product lacks most of the extracellular domain, but contains a small gag segment which causes aggregration of the protein, as would normally be induced by the ligand. Furthermore, a point mutation and a C-terminal truncation in the cytoplasmic domain relieve auto- and feedback inhibition. So, in summary, the virus encodes and overexpresses a constitutively active protein.

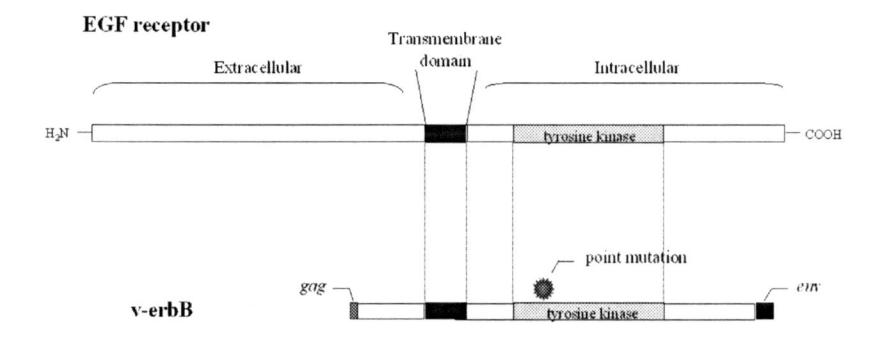

**Figure 4.4** *Activation of erbB1 to the v-erbB oncogene*
In the oncogenic receptor, truncation of the extracellular domain abolishes ligand binding, while fusion to a gag peptide leads to constitutive oligomerization. A truncation at the C-terminus and a point mutation in the autoinhibitory loop further enhance constitutive activity.

The different time-courses of transformation by acutely and slow-transforming viruses may be accounted for by these additional alterations in the transduced oncogene. Further differences may also contribute. Acutely transforming retroviruses transduce the activated oncogene into each cell they infect and thereby create a large pool of potentially transformed cells. In contrast, slowly transforming retroviruses integrate into many different sites in different infected cells and only rarely 'hit' a cellular proto-oncogene, yielding only a few potentially transformed cells. One reason for the time lag in tumor development is thus the time required for expansion of a tumor cell clone.

There are likely two more reasons. The first is that transduction of an activated oncogene into a large number of cells is bound to have a substantial effect on their interaction with each other and host cells. For instance, it could significantly alter the levels of autocrine or paracrine cytokines secreted by these cells or overwhelm an antitumor immune response. The second reason is that a larger pool of cells containing an oncogene increases the probability of a second mutation that causes complete transformation and subsequent tumor progression. There is indeed very good evidence that transformation by slowly transforming retroviruses often requires a second hit. Occasionally, this is provided by insertion of a second retrovirus elsewhere in the genome. Some acutely transforming retroviruses also carry two oncogenes, thereby achieving the 'two hits' at one stroke.

A selection of genes that have been found activated by retroviral insertion is shown in Table 4.2. A comparison with Table 4.1 reveals that several of these genes – like c-myc – possess viral homologs. This is expected, if one assumes that acutely transforming viruses arose from slowly-acting precursors. However, many genes activated by viral insertion have never been observed to be transduced. In some cases, this may be due to the size than can be accomodated in a retrovirus, but functional limitations are also conceivable.

Importantly, there is a cellular homologue for each and every retroviral oncogene found so far. So, all retroviral oncogenes are thought to have evolved from cellular precursors. For instance, the homologue of the v-src gene of RSV is c-src, a protein kinase located at focal adhesion points, at which the actin cytoskeleton is attached to the cell membrane. The c-src kinase relays signals from cell adhesion to the cytoskeleton and to other kinases that control cell proliferation. Such signals may be mimicked by the viral oncoprotein, which is altered towards the cellular protein by several point mutations and the replacement of the C-terminal 17 amino acids by an unrelated peptide.

## 4.3 APPROACHES TO THE IDENTIFICATION OF HUMAN ONCOGENES

No acutely transforming retroviruses have been observed in humans and even activation of cellular genes by retroviruses or related viruses such as HBV (→16.3) is exceptional. Two cases of iatrogenic oncogene activation may have happened when during attempted gene therapy of severely immunocompromised children retroviruses carrying a therapeutic gene integrated into the *MLO2* proto-oncogene locus.

While such cases are clearly exceptional, the elucidation of the mechanisms by which retroviruses cause cancers in animals has been very instructive for the understanding of human cancers. Even if genes in the human genome are almost never activated by retroviruses to become oncogenes, many genes orthologous to the viral and animal oncogenes can be activated by other mechanisms in humans. Following rapidly on the discovery of cellular oncogenes, many human cancers were also screened for oncogenic alterations in these genes. Indeed, a large number of genes are now implicated as oncogenes in human cancers (Table 4.3). These candidates were obtained through several lines of research.

<u>Analysis of human orthologs of retroviral oncogenes</u>: An obvious approach was to investigate the human orthologs of viral oncogenes for their level of expression and for mutations in human cancers. A huge amount of literature has resulted from

**Table 4.3.** *Some important oncogenes in human cancers*

| Oncogene | Tumor type | Activation Mechanism | Cellular localization | Main biochemical function |
|---|---|---|---|---|
| TGFA | many carcinomas | overexpression | extracellular | growth factor |
| FGF1 | many solid tumors | overexpression | extracellular | growth factor |
| WNT1 | selected carcinomas | overexpression | extracellular | growth factor |
| IGF2 | many cancers | overexpression | extracellular | growth factor |
| ERBB1 | many carcinomas | overexpression mutation | cell membrane | tyrosine kinase |
| ERBB2 | selected carcinomas | overexpression | cell membrane | tyrosine kinase |
| KIT | testicular cancers, gastrointestinal stromal tumors | mutation | cell membrane | tyrosine kinase |
| RET | thyroid and other endocrine gland cancers | mutation inversion | cell membrane | tyrosine kinase |
| MET | renal and other carcinomas | mutation overexpression | cell membrane | tyrosine kinase |
| IGFRI | liver and other carcinomas | overexpression mutation (?) | cell membrane | tyrosine kinase |
| SMO | skin and brain cancers | mutation | cell membrane | G-coupled receptor |
| HRAS | many cancers | mutation | inner cell membrane | GTP-binding protein |
| NRAS | many cancers | mutation | inner cell membrane | GTP-binding protein |
| KRAS | many carcinomas | mutation | inner cell membrane | GTP-binding protein |
| BRAF | melanoma, colon and other selected cancers | mutation | inner cell membrane, cytoplasm | tyrosine kinase |
| PI3K | selected cancers | overexpression | inner cell membrane, cytoplasm | phospholipid kinase |

**Table 4.3.** *continued*

| Oncogene | Tumor type | Activation Mechanism | Cellular localization | Main biochemical function |
|---|---|---|---|---|
| CTNNB1 | colon and liver carcinomas, and others | mutation | inner cell membrane, cytoplasm, nucleus | cytoskeleton transcriptional activation |
| MYC | many cancers | translocation overexpression mutation | nucleus | transcription factor |
| MYCN | selected cancers | overexpression | nucleus | transcription factor |
| MYCL | selected carcinomas | overexpression | nucleus | transcription factor |
| RELA | leukemias | translocation | nucleus | transcription factor |
| MDM2/HDM2 | sarcomas and other solid tumors | overexpression | nucleus, cytoplasm | transcriptional regulator/ubiquitin ligase |
| SKP2 | selected cancers | overexpression | nucleus, cytoplasm | ubiquitin ligase |
| CCND1 | many cancers | overexpression | nucleus | cell cycle regulation |
| CCND2 | selected cancers | overexpression | nucleus | cell cycle regulation |
| CDK4 | selected cancers | overexpression mutation | nucleus | cell cycle regulation |
| BCL2 | follicular lymphoma and many other cancers | translocation overexpression | mitochondria | apoptotic regulation |

this type of research. In summary, several orthologs of viral oncogenes are also over-expressed or mutated in human cancers and are clearly involved in their development. For instance, the human ortholog of v-erbB, *ERBB1*, and the human v-myc orthologue *MYC* are over-expressed, mutated and causally involved in many human cancers. Other genes are only directly involved in a more restricted range of cancers, such as rel in selected lymphomas (→6.9). On the other hand, many strong viral oncogenes have never turned up as dominant oncogenes in humans, prominently v-src and v-fos. Instead, the products of the corresponding human genes are more subtly involved in shaping the phenotype of human cancers. Of note, while *ERBB1* or *MYC* are overexpressed in a wide range of human cancers, this is not in each case caused by primary changes in their genes. In summary, human orthologs of almost all viral oncogenes contribute in some way to one or the other human cancer.

3T3 cell transformation assay: Cell culture assays such as the 3T3 fibroblast focus formation assay developed for the identification of viral oncogenes have also been used to discover human oncogenes. In the original 3T3 assay, the immortalized mouse fibroblast cell line 3T3 was infected with an oncogene-carrying retrovirus and yielded foci of transformed cells with altered morphology that grow out of the monolayer at confluence. Foci can also be obtained by transfection of DNA from human tumor cells. From such foci the gene responsible for the altered phenotype can be isolated. This assay has led to the identification of several human oncogenes. The most dramatic result was that three human orthologs of the v-ras genes can act as oncogenes. In humans orthologs of both the Ki-ras (KRAS) and Ha-ras (HRAS) genes exist as well as a third relative, NRAS, which possess similar structures and functions (→4.4). All three genes were recovered from 3T3 focus formation assays, in each case carrying point mutations at specific sites, i.e. codons 12, 13, or 61. Thus, some cellular genes like the RAS genes can be converted from proto-oncogenes to oncogenes by point mutation only. The 3T3 assay has also yielded several additional genes, as well as a number of biochemically interesting artifacts from genes rearranged during the transfection procedure.

Other transformation assays: The 3T3 assay is limited by a strong bias towards a certain type of oncogene that is 'ras-like', while others, like the potent myc oncogene, do not score. This limitation arises from the fact that the assay selects for the ability of oncogenes to keep fibroblasts growing beyond confluence and to alter their morphology. Mutated *RAS* genes confer these properties, but not all oncogenes do. Therefore, further cellular assays were developed to identify oncogenes acting on other properties. The most famous among these may be the REF assay, in which primary rat embryo fibroblasts are infected with retroviruses or transfected with purified oncogene or tumor cell DNA. In this assay, two oncogenes are required for focus formation, one 'ras-type' and one 'myc-type' oncogene. This is another instance of oncogene cooperativity, as found during retroviral transformation in animals. The precise molecular basis of this cooperation is still under investigation, almost twenty years after its discovery. In short, 'myc-type' genes immortalize rodent embryo fibroblasts and stimulate their proliferation, while 'ras-type' genes elicit overgrowth and altered morphology. Obviously, in 3T3 cells, the first step has already happened (Figure 4.5).[3]

Gene amplification: Since overexpression is often required for oncogenic function, genes that are strongly overexpressed in specific human cancers are obvious candidates for oncogenes. Specifically, some cancers contain recurrent amplifications of specific chromosomal regions. These can be detected by cytogenetic techniques. For instance, the overexpression of MYC and ERBB1 in human cancers often results from amplification of their genes located at 8q24 and 17p12. Investigation of other amplified regions has revealed further oncogenes. A

---

[3] It is now presumed that the relevant change is in fact a mutation of mouse p53 (→5.3). Indeed, mutated p53 acts as a 'myc-like' gene in the REF assay.

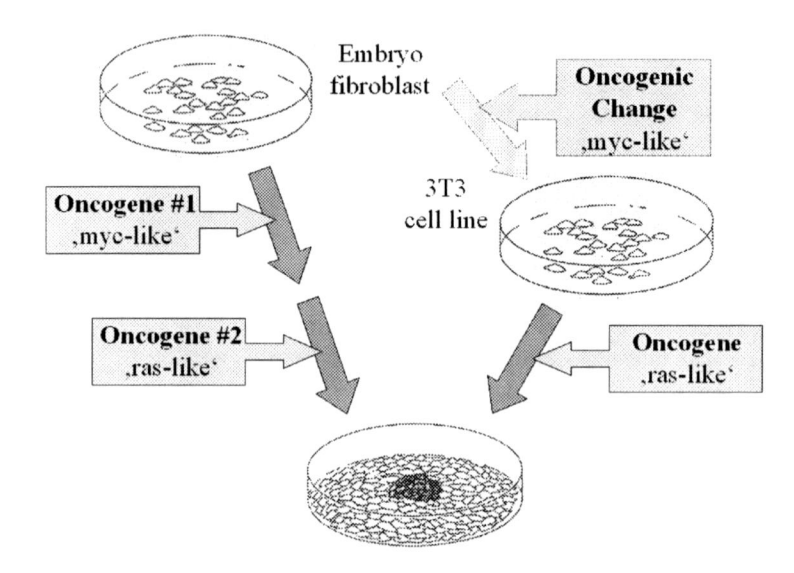

**Figure 4.5** *A comparison of the 3T3 and REF focus formation assays*
See text for explanation.

segment of chromosome 2p24 frequently amplified in neuroblastoma, but also in some carcinomas, contained a gene related to *MYC* that was named *MYCN* (N for neuroblastoma). Another related gene, *MYCL*, was found amplified and overexpressed in lung cancers. A different region originating from chromosome 17q11-12 amplified in breast cancer and other carcinomas yielded an overexpressed gene similar to ERBB1. This is now officially named ERBB2, but still doubles as HER-2 or NEU, the latter, because a mutated form was independently discovered in a chemically induced rat neuroblastoma.

Findings like these confirm that consistently amplified regions in the genome of cancer cells often contain oncogenes and that genes related to known oncogenes can be oncogenes (see below). However, the identification of oncogenes from amplified regions is not always straightforward, since amplified regions can encompass several Mbp. For instance, a region from 12q14 commonly amplified in human tumors contains the genes *GLI1*, *CDK4*, and *HDM2*[4]. As detailed in later chapters, each of these genes possesses properties which make it a good candidate for an oncogene. Perhaps, one or the other in different tumors or more than one could be relevant. A different kind of complication is that some amplifications are associated with gene silencing rather than overexpression.

Chromosomal translocations: Another type of chromosomal aberration in human cancers are translocations. In hematological cancers, in particular, they activate genes at the translocation sites to become oncogenes (→2.2, →10.2). So, systematic

---

[4] Both the acronyms HDM2 and MDM2 are in use for this gene. It is officially named HDM2, since MDM2 stands for mouse double minute

investigation of recurrent translocations in hematological cancers by cytogenetic and molecular cloning techniques has revealed several oncogenes. To a lesser extent, characterization of translocations in soft tissue tumors and carcinomas has been productive. One gene activated by several translocations is MYC, emphasizing what a potent oncogene it is. A second oncogene identified at a recurrent site on 11q13 in different cancers was initially named PRAD1 or BCL1, but has now been renamed CCND1 since it encodes Cyclin D1, a crucial regulator of cell cycle progression through the G1 phase (→5.2). Interestingly, the same gene is also overexpressed as a consequence of amplification in some carcinomas. Not untypical, several neighboring genes are often co-amplified with CCND1, including *GSTP1* (→3.5), and a growth factor gene. The typical translocation in follicular lymphoma activates the BCL2 gene (for breakpoint cluster 2). This gene belongs to an entirely different functional class: the BCL2 protein is a direct regulator of apoptosis at the mitochondrion (→7.2) and the first member of a larger family to be identified.

Oncogene families: Since oncogenes often seem to come in families (Table 4.3), it is tempting to speculate that genes related to oncogenes might also be oncogenes. This idea has inspired a lot of research activities. The gist of their many results is that the idea is in some instances correct, but overall more rarely than was originally expected. In some cases, the reasons are obvious (with hindsight). For instance, the MYC family includes members like *MXI1* and *MAD* that are actually antagonists of the proto-oncogenes *MYC*, *MYCN*, and *MYCL*, while *MAX* encodes a heterodimerization partner for all members. It is less straightforward to understand, why *ERBB1* and *ERBB2* are often oncogenically activated, but two further members of the family, *ERBB3* and *ERBB4*, rarely. Similarly, the closest human ortholog of the viral v-raf, *RAF1*, does not seem to be activated in human tumors, but its homologue *BRAF* certainly is, e.g., in melanomas (→12.4). So, homology to a proven oncogene is a good indication of importance, but does not allow firm conclusions on the oncogenic potential of a gene.

Another temptation is to regard every gene that is strongly over-expressed in a human tumor as an oncogene. As pointed out above, the significance of overexpression can be doubtful even for *bona fide* oncogenes like *MYC* and *ERBB1*. An observed overexpression of a gene in a human cancer is not always easy to interpret, since it can be difficult to test the functional implications of this overexpression. The distinction whether a gene is active as an oncogene as a consequence of overexpression or is overexpressed but not active as an oncogene is not only of intellectual interest, but also essential for the design of targeted therapy (→22.3).

Cancer pathways: The dilemma of how to test the functional importance of a gene in a human cancer has been partly alleviated by the recognition that oncogenes as well as tumor suppressor genes often interact in 'cancer pathways' (→6). If the product of a gene can be demonstrated to influence the activity of a 'cancer pathway' important in a certain tumor type, its overexpression or mutation can be understood as oncogenic activation. This line of investigation has yielded sufficient data to regard genes like *MDM2/HDM2* (→6.6), CTNNB1 (→6.10, →13.3, →16.2), or CDK4 (→6.4) as proto-oncogenes.

## 4.4 FUNCTIONS OF HUMAN ONCOGENES

The oncogenes of acutely transforming retroviruses (Table 4.1) as well as human oncogenes (Table 4.2) can be categorized according to their biochemical function or to the localization of their products. A sketch of these localizations (Figure 4.6) suggests that this spatial distribution may be more than incidential. Indeed, many proven or suspected human oncogenes belong to a functional network which transmits signals for proliferation and survival from the exterior of the cell to the nucleus.

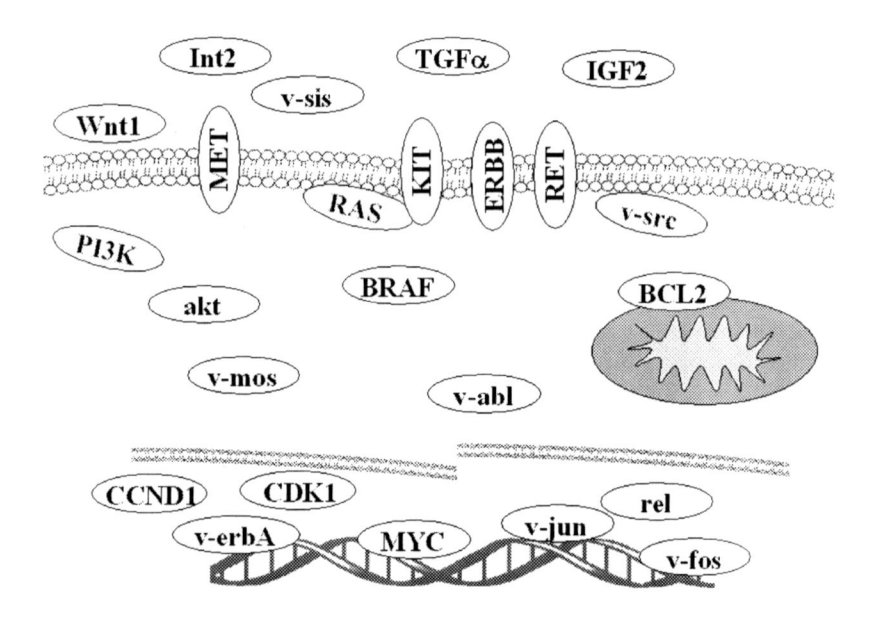

**Figure 4.6** *Cellular localisation of oncogene proteins*
See text for further discussion.

A normal cell proliferates in response to extracellular signals that are conferred by soluble growth factors and are modulated by signals elicited as a result of cell adhesion to the extracellular matrix and neighboring cells. The first group of presumed human oncogene products accordingly comprises peptide growth factors like TGFα (transforming growth factor alpha), FGF1 (fibroblast growth factor) or WNT1 (wingless/int-1) which are mitogens for epithelial and/or mesenchymal cells. These or related factors are overproduced by many carcinomas.

Peptide growth factors bind to and activate receptors at the cell membrane such as the EGFR (epidermal growth factor receptor, Fig. 4.4), the product of the *ERBB1* gene, or one of several FGF receptors. ERBB1 is overexpressed in different carcinomas, often as a consequence of gene amplification. FGFR expression is also altered in many human tumors and the FGFR3 in particular is activated by point mutations in cancers of the bladder (→14.3) and the cervix. Further growth factor receptors such as ERBB2 (→18.4), MET (→15.3), IGFR1 (→16.2), KIT (→22.4), and RET (→2.2) are also crucial oncogenes in human cancers. These receptors share structures and the mechanism of signaling and are summarized as receptor tyrosine kinases (RTKs, also: TRKs). RTKs constitute one of the biggest classes of oncogenes. However, some oncogene products are receptors belonging to different classes, e.g. cytokine receptors (→6.8).

Binding of a growth factor to the extracellular domain of a RTK leads to formation of receptor dimers or heterodimers, e.g. between ERBB1 and ERBB2. It also causes a conformational change by which a pseudosubstrate peptide loop in the intracellular domain of the receptor is removed from the active center of the intracellular tyrosine kinase domain. The tyrosine kinase becomes active and the receptor subunits phosphorylate each other in trans. Overexpression of RTKs in tumor cells favors dimer formation. Thereby, it sensitizes the cells to lower concentrations of ligand growth factors or even leading to growth factor-independent activation. Oncogenic point mutations typically occur in the inhibitory loop, thereby causing constitutive activation of the tyrosine kinase activity. Assembly of several receptor dimers to larger complexes in the cell membrane may occur followed by internalization into endosomes.

Cross- and auto-phosphorylation of RTKs provide phosphotyrosine for recognition by adaptor proteins containing SH2 domains which dock to the activated receptor. The SH2 domains in different proteins all bind phosphotyrosines, but recognize them in different peptide contexts. To various extents, RTKs also phosphorylate proteins other than themselves.

Usually, multiple proteins bind to one receptor by recognizing different phosphotyrosines. In this fashion, one receptor can activate several different signaling pathways. For instance, activated EGFR binds the adaptor proteins GRB2 and SHC (which again binds GRB2), PLCγ (phospholipase Cγ), the regulatory subunit of PI3K (phosphatidylinositol-3'-kinase), and GAP (GTPase activator protein).

Phosphotyrosines are substrates for tyrosine phosphatases and also serve as recognition sites for proteins which lead to internalization and degradation of the receptor, such as CBL proteins. Together, these help to limit the strength of growth factor signals in normal cells and ensure that the signals are transient.

The adaptor protein GRB2 in turn assembles a further protein named SOS to the complex, which interacts with and activates RAS proteins (Figure 4.7). All RAS proteins, HRAS, KRAS, or NRAS, are ≈21 kDa proteins and are linked to the inside of the cell membrane through their C-terminus which is posttranslationally modified by myristylation, farnesylation and methylation (→Figure 22.8). They belong to a larger superfamily of small monomeric G proteins that can bind alternatively GTP or GDP. In the active state, GTP is bound. Hydrolysis of GTP to GDP by the combined action of RAS and a GTPase activator protein (i.e., GAP) restores the basal inactive state. Normally, activation of RAS depends on interaction with SOS which acts as a guanine nucleotide exchange factor (GEF), loading the RAS protein with GTP instead of GDP. The activated state of normal RAS proteins is short-lived, since

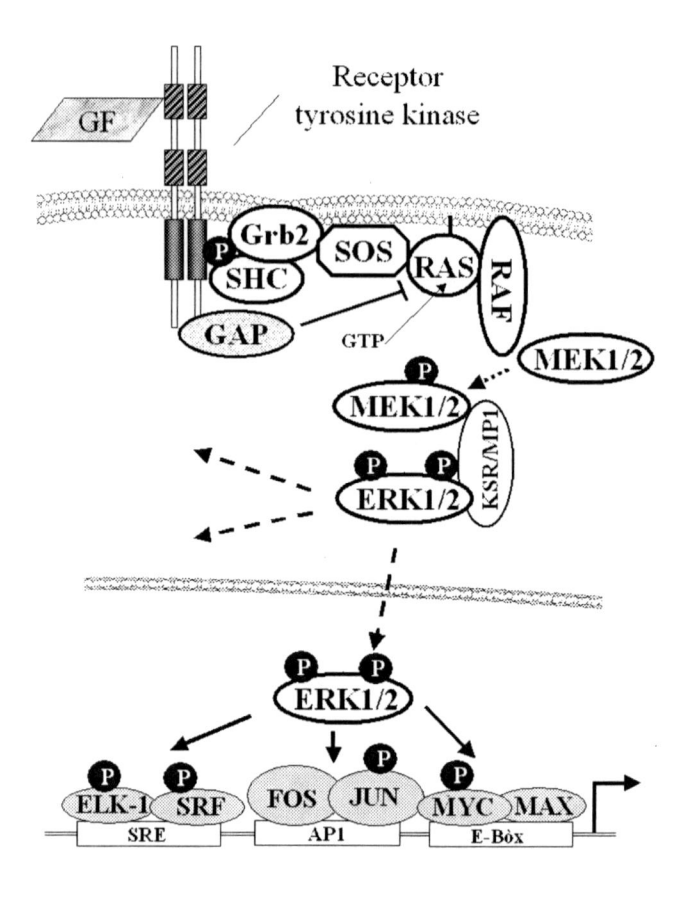

**Figure 4.7** *The main MAPK pathway*
See text for further details

RTKs in parallel to the SOS GEF activate GAPs that stimulate GTP hydrolysis. However, mutations in RAS amino acids 12, 13, or 61, which surround the GTP binding site, obstruct the access for the GAP protein and prolong the active state.

RAS is the next branching point in the signaling network described here, since activated RAS acts on several pathways (→Fig. 6.2) which affect protein synthesis, the cytoskeleton and cell survival, notably the PI3K pathway (→6.3).

The main route by which RAS relays a proliferation signal is via interaction with RAF proteins. The three RAF proteins in human cells are Ser/Thr protein kinases. Like many protein kinases, they contain a regulatory domain and a C-terminal catalytic domain. In the inactive state, the protein resides in the cytosol and the kinase activity is blocked by the regulatory domain. Activated RAS translocates RAF to the cell membrane and relieves the inhibition of the regulatory domain. Interestingly, RAF activity is also modulated by phosphorylation of the CR2 segment in its regulatory domain. A tyrosine in this domain is phosphorylated by SRC family kinases, and serine and threonine residues are phosphorylated by protein kinase C isozymes which can be activated indirectly by PLC (→6.5). So, here are further links in the network. In some human cancers, BRAF is altered by specific point mutations to become overactive (→12.4). Mutations in BRAF occur in the same tumor types as mutations in RAS genes, but alternately to them. Since RAS proteins act on several pathways, this finding is important in showing that signaling via RAF is indeed significant for transformation. Tellingly, the main modification in the retroviral v-raf oncogene is the inactivation of the regulatory domain of the cellular protein.

Activated RAF is the first one in a cascade of protein kinases that further comprises MEK and ERK proteins. Alternative names for these are MAPK (mitogen-activated protein kinase) for ERK and MAPKK for MEK; RAF proteins can therefore be counted as MAPKKK (or MEKK). There are several additional MEKKs in parallel pathways in human cells (→6.2). Most human cells contain two MEK and two ERK protein kinases. MEK are highly specific and phosphorylate predominantly ERK proteins at a tyrosine and threonine each in a TEY sequence. Activated ERKs phosphorylate a variety of substrates to activate protein synthesis, alter the structure of the cytoskeleton, and induce gene expression. Interaction of MEKs, ERKs and MAPKKs is supported and its specificity is enhanced by scaffold proteins. Phosphorylation of MAPKKs, MEKs, and ERKs is removed by several specific and less specific protein phosphatases in order to terminate the signal. Some of these phospatases, like MKP1 (MAP kinase phosphatase 1), are themselves activated by MAPK signaling, while others may be constitutively active.

Increased protein synthesis and cytoskeletal changes elicited by MAPK signalling in the cytoplasm are important for cell growth and also necessary for cell migration. Stimulation of cell proliferation in addition requires an altered pattern of gene expression and activation of the cell cycle in the nucleus. The mitogenic signal from growth factors for altered transcription is to a large extent relayed by the MAPK signaling cascade. Activated ERKs phosphorylate several transcription factors (Figure 4.7) directly or stimulate other protein kinases like p90$^{RSK1}$ to do so. This induces the transcription of a larger set of genes, which again are organized as a cascaded network in the nucleus.

The first set of genes induced upon activation of the MAPK pathway by growth factors in previously resting cells are the 'early-response' genes. They are, e.g. induced by treatment of quiescent cells in cell culture with growth-factors or serum. Among them are several from the oncogene lists, such as FOS, JUN, MYB, and MYC. The most important one of these in the context of human cancers is MYC which is frequently activated by overexpression or deregulation and thereby becomes active independent of growth factor signaling. As a rule, FOS, JUN and MYB are necessary for the growth of human cancers, but do not usually seem to act as oncogenes, strictly spoken. It is not entirely clear, why this is so. Certainly, one obvious reason is that the viral counterparts of these genes are severely altered and over-expressed. Moreover, cells from long-lived humans may have better checks against tumor formation than cells of many animals. Indeed, overexpression of proteins like FOS or JUN can induce apoptosis in some human cells.

Following their induction, the products of the 'early response' genes induce the expression of genes required for cell cycle progression, directly or indirectly. In normal cells, one of the most important proteins that links the mitogenic signal to cell cycle regulation is Cyclin D1 (→6.4), the product of the *CCND1* gene (→4.3). Overexpression of Cyclin D1 caused by gene amplification or translocation of *CCDN1* is important in several different human cancers. Moreover, Cyclin D1 is often overexpressed without alterations in the gene itself, likely as a consequence of 'upstream' mutations in the MAPK pathway. Its overexpression may (partly) mediate the oncogenic effect of these alterations. The related gene *CCDN2* is expressed in a more restricted range of tissues than *CCDN1*. Accordingly, overexpression and amplification of this gene occur in a more limited range of human tumors, e.g. prominently in testicular cancers. D-Cyclins activate the cyclin-dependent kinase 4 to promote progression through the cell cycle (→5.2, →6.4). Amplification and overexpression of CDK4 are also observed in human cancers, e.g. in gliomas and hepatomas.

Finally, physiological signaling for cell proliferation requires a parallel signal for cell survival, by the same growth factor and its receptor or by a complementing pathway (→6.4). For instance, insulin-like growth factors also stimulate the MAPK cascade, but more strongly confer a survival signal, which is mediated through PI3K (→6.3). Therefore, the overexpression of the insulin-like growth factors IGF1 and IGF2 as well as PI3K and its downstream kinase AKT must be considered as an oncogenic event in many human cancers. Alternatively, cell death by apoptosis as a consequence of inappropriate stimulation of cell proliferation can be prevented by oncogenic overexpression of proteins like BCL2 (→7.3). This is the crucial event in follicular lymphoma, but this change or equivalent ones are also found in other hematological cancers as well as many carcinomas. The synergism between oncogenes that stimulate cell proliferation and oncogenes that prevent apoptosis induced by this stimulation is evident in many human cancers, most transparently in certain lymphomas (→10.2). This is another example of oncogene cooperation.

Since human cancers accumulate many genetic and epigenetic alterations during their progression, in a typical cancer many genes are overexpressed and many of their products are overactive, and some may show mutations. Many of these genes may well be functionally important and promote tumor growth. It is tempting to regard every gene of this kind as an oncogene, particularly, if it shows a fitting biochemical property such as a protein kinase activity or DNA binding. In the context of human cancers, however, caution is warranted. This is because many cancers have developed over a long period, with many displaying genomic instability and/or an increased rate of point mutations (→3.4). As a consequence, there may be a larger number of genes with alterations in their sequence and dosage in the definitive tumor clone than absolutely required for its growth. To prove that a gene of this sort is an oncogene by a strict definition, one would have to show that the altered or overexpressed gene product indeed dominantly confers an essential property for the survival and sustained growth of the cancer, and that its overexpression and/or overactivity are caused by substantial changes in the gene itself, i.e. mutations or amplifications.

## Further reading

Varmus HM, Weinberg RA (1993) Genes and the biology of cancer. Scientific American Publishing
Hesketh R (1997) The oncogene and tumor suppressor factbook. Academic Press.
Knipe DM, Howley PM (eds.) Fields Virology 4$^{th}$ ed. 2 vols. Lippincott Williams & Wilkins 2001

Lewis TS, Shapiro PS, Ahn NG (1998) Signal transduction through MAP kinase cascades. Adv. Cancer Res. 74, 49-139
Bar-Sagi D, Hall A (2000) Ras and Rho GTPases: a family reunion. Cell 103, 227-238
Schlessinger J (2000) Cell signaling by receptor tyrosine kinases. Cell 103, 211-225
Wilkinson MG, Millar JBA (2000) Control of the eukaryotic cell cycle by MAP kinase signaling pathways. FASEB J. 14, 2147-2157
Blume-Jensen P, Hunter T (2001) Oncogenic kinase signalling. Nature 411, 355-365
Kerkhoff E, Rapp UR (2001) The Ras-Raf relationship: an unfinished puzzle. Adv. Enzyme Regul. 41, 261-267
Savelyeva L, Schwab M (2001) Amplification of oncogenes revisited: from expression profiling to clinical application. Cancer Lett. 167, 115-123
Klein G (2002) Perspectives in studies of human tumor viruses. Front Biosci. 7, d268-274
Pelengaris S, Khan M, Evan G (2002) c-MYC: more that just a matter of life and death. Nat. Rev. Cancer 2, 764-776
Mikkers H, Berns A (2003) Retroviral insertional mutagenesis: tagging cancer pathways. Adv. Cancer Res. 88, 53-99
Nilson JA, Cleveland JL (2003) Myc pathways provoking cell suicide and cancer. Oncogene 22, 9007-9021
Mercer KE, Pritchard CA (2003) Raf proteins and cancer: B-Raf is identified as a mutational target. BBA 1653, 25-40

## Box 4.1 Carcinogenesis by HTLV-I

About 20 million people worldwide are estimated to be infected with HTLV-I. Up to 2% of them eventually develop a slow-growing, but obstinate and usually fatal malignancy of clonal CD3+ CD4+ cells, termed adult T-cell lymphoma/leukemia (ATLL). This cancer is clearly initiated by the retrovirus, but as it progresses, it develops chromosomal aberrations and may become completely independent of proteins expressed from the proviral genome it harbors. Through much of its development, it appears to also require stimulation by the antigen recognized by the particular T-cell clone.

In addition to the standard gag, pol, and env proteins of retroviruses(cf. Fig. 4.1), HTLV-I expresses several accessory proteins thought to be involved in the initial immortalization and clonal expansion of a T-cell infected by the virus. The most important one is tax, a transactivator protein with gross similarities to HIV tat.

Tax acts in a pleiotropic fashion, i.e. it influences the expression of many proviral and cellular genes. It binds to certain transcriptional activators and augments their interaction with transcriptional co-activators, in particular with CBP/p300. In this fashion, gene activation by $NF_\kappa B$ (→6.9), AP1 and other factors activated by MAPK signaling (→6.2), as well as CREB (→12.4) is enhanced. In T-cells, this leads to the increased production of cytokines, e.g. IL2, of cytokine receptors, e.g. IL2R, and of anti-apoptotic proteins such as $BCL-X_L$ and the IAP survivin (→6.9).

In addition to proliferation and apoptosis, Tax and other viral proteins also influence cell cycle regulation directly and the control of genomic stability by the TP53 protein (→5.3). Tax may sequester CBP/p300 from TP53, prohibiting gene activation by TP53 in response to chromosomal defects. Moreover, the provirus integration site can act as a hotspot for chromosomal breaks.

HTLV-I action in carcinogenesis might be schematically illustrated as follows:

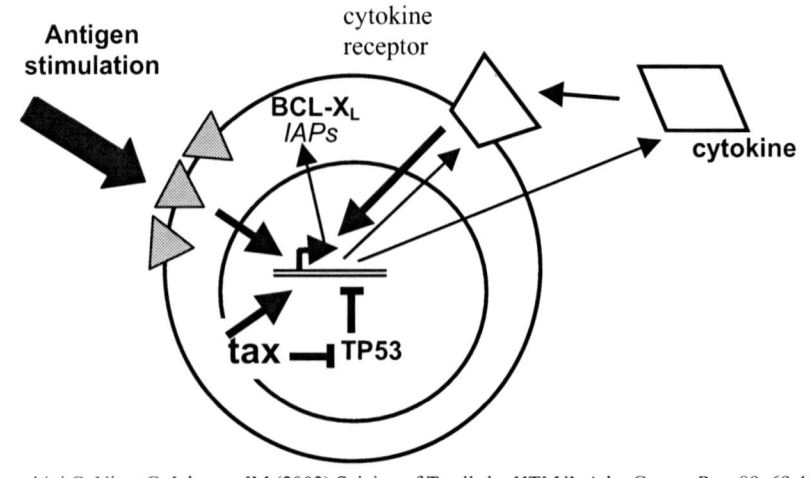

Franchini G, Nicot C, Johnson JM (2003) Seizing of T cells by HTLVI. Adv. Cancer Res. 89, 69-107.

# CHAPTER 5

# TUMOR SUPPRESSOR GENES

- While oncogenes promote tumor development by increased activity or deregulation, tumor suppressor genes have to undergo loss of function for tumor development.
- Many hereditary cancers result from germ-line mutations in tumor suppressor genes that are passed on in families. In most cancers with a dominant mode of inheritance, one mutant allele of a tumor suppressor gene is inherited. When the second allele becomes as well inactivated by mutation, deletion, recombination with the mutated allele, or by epigenetic mechanisms, cancer development is initiated. In sporadic cancers, two mutations in the two alleles of the gene must occur within one cell line. Therefore, inherited cancers typically occur at earlier ages and are more often multifocal than sporadic cancers of the same type. Of note, while inheritance is dominant, tumor suppressor genes behave as recessive genes at the cellular level.
- Loss of a tumor suppressor allele by deletion or recombination often becomes apparent as loss of heterozygosity (LOH) of polymorphic markers, e.g., microsatellites, in its vicinity. LOH analysis is a useful method to detect deletions or recombinations in tumors and can be employed to discover tumor suppressor genes in regions of the genome which consistently show LOH in one type of tumor.
- Retinoblastoma has provided the paradigm for tumor suppressor genes, that also fits for several other inherited cancers. The tumor suppressor gene inactivated in retinoblastomas, *RB1*, encodes a central regulator of the cell cycle, prominently of the G1→S transition, but also aides to ensure correct mitotic segregation, affects chromatin structure and regulates apoptosis. RB1 acts by binding and controlling other proteins, prominently E2F transcriptional factors. These can activate genes required for entry into the S phase and DNA synthesis, but in certain circumstances also induce apoptosis. RB1 itself is regulated via phosphorylation by cyclin-dependent kinases (CDKs) which in turn are dependent on their regulatory cyclin subunits.
- Multiple mechanisms are employed in human cells to regulate the cell cycle and RB1 phosphorylation, including induction and proteolysis of cyclins, phosphorylation and dephosphorylation of CDKs and induction, stabilization and proteolysis of CDK protein inhibitors. The CDK inhibitors comprise the INK4 family which specifically inhibits CDK4 and the CIP/KIP family inhibiting several CDKs. CDK inhibitors can be tumor suppressors. The most important one in this regard is CDKN2A/p16$^{INK4A}$, but p27$^{KIP1}$, p21$^{CIP1}$, and p57$^{KIP2}$ are also relevant.
- Alterations in cell cycle regulation in different tumors are alternatively brought about by loss of tumor suppressor gene (*RB1* or *CDKN2A*) function or by

oncogene (*CCND1* or *CDK4*) activation. There are further cases, in which tumor suppressors essentially act as antagonists of oncogenes, e.g. the tumor suppressor NF1 is a GAP that limits the action of RAS proteins. Loss of tumor suppressors and activation of oncogenes therefore often have similar, albeit not fully equivalent consequences.

➤ Whereas loss of RB1 function directly impinges on cell proliferation and differentiation, loss of tumor suppressor genes such as TP53 promotes tumor formation in a different fashion. Inactivation of TP53 compromises the ability of a cell to react to genomic damage, e.g. by ionizing radiation, as well as to hyperproliferation induced by oncogenes, or to viral infection. Therefore, loss of TP53 function permits survival and proliferation of cells accumulating mutations and thereby the emergence of a cancer. Several other tumor suppressors, e.g. BRCA1 and BRCA2, which are defective in familial breast cancers, also act primarily by protecting against genomic instability. Such 'caretakers' can be considered as a different class of tumor suppressors from 'gatekeepers' like RB1.

➤ *TP53*, as a central regulator gene of genomic stability, may be the most frequently altered gene in human cancers. Several upstream pathways responding to different kinds of genomic instability lead to activation of TP53, mainly by post-translational regulation. In turn, TP53 acts on several downstream pathways as a transcriptional activator or repressor, and may even directly participate in DNA repair. Activated TP53 can arrest the cell cycle via induction of cell cycle inhibitors such as $p21^{CIP1}$, or elicit apoptosis by induction of proteins like BAX. TP53 action is limited by its induction of MDM2/HDM2 which inhibits TP53 and initiates its proteolytic degradation.

➤ In different human cancers, TP53 is inactivated by different mechanisms. Most widely, one gene copy is inactivated by point mutations in the central DNA binding domain of the protein, whereas the other one is lost by deletion or recombination. In some tumors, HDM2 is overexpressed. The CDKN2A gene encodes not only $p16^{INK4A}$, but also an activator of p53, $p14^{ARF1}$ in a different reading frame. Homozygous deletions, therefore, and certain point mutations in this gene impede the function of both TP53 and RB1.

➤ DNA tumor viruses, e.g. the papovavirus SV40 and tumorigenic strains of human papilloma viruses, contain proteins inactivating both TP53 and RB1. This underlines the crucial role of these two proteins in the prevention of human cancer.

## 5.1 TUMOR SUPPRESSOR GENES IN HEREDITARY CANCERS

As described in Chapter 3, several recessively inherited syndromes caused by defects in DNA repair genes are associated with an increased cancer risk, often for leukemias and lymphomas (→Table 3.2). Other 'cancer syndromes' are inherited in a dominant mode. Most of these predispose to a restricted range of cancers or even to a single tumor type, while a few are rather unselective, e.g., the rarer Li-Fraumeni and Cowden syndromes (Table 5.1).

**Table 5.1.** *Some dominantly inherited cancer syndromes in man*

| Syndrome | Gene | Location | Cancer site | Function |
|---|---|---|---|---|
| Retinoblastoma | *RB1* | 13q14 | eye, bone | gatekeeper tumor suppressor |
| Li-Fraumeni | *TP53* | 17p13.1 | many organs | caretaker tumor suppressor |
| Hereditary melanoma and pancreatic cancer | *CDKN2A* | 9p21 | skin, pancreas, others | gatekeeper tumor suppressor |
| Familial adenomatous polyposis coli | *APC* | 5q21 | colon, rectum, others | gatekeeper tumor suppressor |
| Cowden | *PTEN* | 10q23.3 | many organs | gatekeeper tumor suppressor |
| Gorlin | *PTCH* | 9p22 | skin, brain | gatekeeper tumor suppressor |
| Von Hippel-Lindau | *VHL* | 3p25 | kidney, adrenal glands, others | gatekeeper tumor suppressor |
| Hereditary breast and ovarian cancer | *BRCA1* *BRCA2* | 17q21 13q12 | breast, ovary | caretaker tumor suppressor |
| HNPCC | *MLH1* *MSH2* others | 3p21 2p15-16 | colon, endometrium, stomach, others | caretaker tumor suppressor |
| Multiple endocrine neoplasia type 2 | *RET* | 10q11.2 | thyroid and other endocrine glands | oncogene |
| Hereditary renal papillary cancer | *MET* | 7q31 | kidney | oncogene |

Typically, these syndromes show high penetrance and the life-time risk of cancer may approach 100%. As a further important characteristic, patients with familial cancers often develop cancers at a significantly lower age than in other 'sporadic' cases. Thus, most sporadic colon or breast cancers present in patients in their sixties or seventies, but familial cases can appear already in the second or third decade of life. In addition, patients with inherited cancers may develop more than one cancer of the same type or cancers of different types. Multifocality or bilaterality is obvious in cancers of paired organs such as breast, kidney or the eyes. Patients with FAP (→13.2) or HPRCC (→15.3) can have literally thousands of individual tumors in their bowel or kidneys, respectively. A theoretical explanation for these properties of hereditary cancers was developed and later experimentally confirmed for hereditary retinoblastoma.

Retinoblastoma is a rare tumor which occurs in young children. It consists of undifferentiated or incompletely differentiated retinocyte precursors ('retinoblasts') forming an expanding cell mass in the back of the eye. The incidence in general is around 1:20000 live births, but in some families approximately every second child is affected on average, as expected for an autosomal dominantly inherited disease with high penetrance. Bilateral cases are extremely rare outside retinoblastoma families. Even though all patients suffering from this disease are young, on average the cancers manifest at lower age in familial cases. Retinoblastoma was recognized as such and could be treated by surgery already in the 19$^{th}$ century. Attentive surgeons of the time noted that patients cured of retinoblastoma tended to develop other cancers later in life, notably of the bone (osteosarcoma), and that their children also tended to develop the disease. Based on the assembled statistical data, in the early 1970's, a model was developed by Knudson that accounted for these observations.

The Knudson model (Figure 5.1) assumes that development of retinoblastoma requires two mutations ('hits') within one cell to form the initial tumor cell clone. In

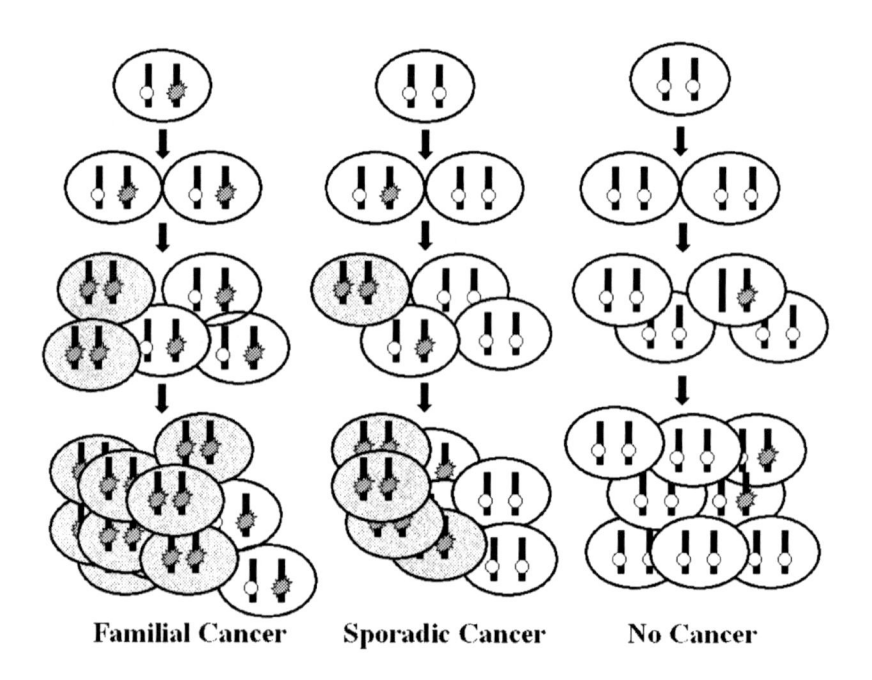

**Familial Cancer**     **Sporadic Cancer**     **No Cancer**

**Figure 5.1** *The Knudson model for retinoblastoma*
Right: a single mutation (denoted by the exploding star) does not suffice to cause cancer. This requires the rare event of a second mutation in the same cell line (center). Since in hereditary cases the first mutation is already present in the germ line (left), cancers are frequent and may even occur in both eyes.

hereditary cases, one of these hits has already occurred and is passed along in the affected families. The development of a retinoblastoma then depends on a single mutation to take place in any cell during the critical period when retinoblasts proliferate during fetal development. The probability of a mutation in a specific human gene is in the order of $10^{-6}$ - $10^{-7}$ per cell generation. So it is not unlikely for one or several mutations to occur leading to one or more retinoblastomas. If no mutation is inherited, two mutations in two cells within the same line are required. The probability for this double accident is very low and it is even more unlikely that it occurs more than once within the same person. So, in familial cases tumors are much more likely to occur at all, tend to develop more rapidly (because their expansion begins already after one hit), and they can be bilateral, whereas in children without an inherited defective allele tumors form much less frequently, on average later, and almost never in more than one place.

This theoretical model so far does not make assumptions about which genes are affected by hit 1 and hit 2. They could be two alleles of the same gene or one allele each of different genes. However, the model required that the genes behave recessively at the cellular level. Since oncogenes act in a dominant fashion, the model indicated a different sort of tumor gene undergoing inactivation in cancers. Indeed, some retinoblastomas occur in patients lacking parts of chromosome 13, with a common region of deletion within band 13q14.1. Deletions of this region were also seen in sporadic cases. Therefore, at least one of the hits in retinoblastoma formation involved loss of a gene and its function. In fact, in almost all cases of retinoblastoma the second hit also concerns the same gene and inactivates the function of its second allele. Tumor formation therefore requires inactivation of the function of this gene, which is designated a tumor suppressor gene.

The gene that is defective in the overwhelming majority of retinoblastomas is designated *RB1*. It encompasses ≈180 kb of genomic sequence, in which 27 exons code for a 4.7 kb mRNA and a 110 kDa phosphoprotein. Mutations of this gene are observed in familial as well as in sporadic retinoblastoma and osteosarcoma, but also in sporadic cases of several other cancers such as glioma, breast and bladder cancer (→14.2).

In familial cases of retinoblastoma, mutations in *RB1* are passed on in the germ-line. Mutations in different families comprise deletions of various sizes, ranging from cytogenetically detectable (i.e. several Mbp) to single bases, small insertions, nonsense and splice mutations and specific missense mutations which usually alter amino acids in the central portion of the protein, the 'pocket' domain.

The second allele can be inactivated independently by deletions, insertion, or various kinds of point mutation (Figure 5.2). In a few cases, an entire chromosome 13 is lost, e.g. by mitotic nondisjunction, more frequently, large deletions obliterate sequences including 13q14.1. In some cases (only in childhood cancers), the otherwise intact second allele is not transcribed as a consequence of hypermethylation of the *RB1* promoter (→8.3).

Another important mechanism involves the replacement of the intact allele by a defective copy. This can again occur by several means. In some cases, the cell contains two identical chromosomes 13, either by duplication of one remaining after loss of the other or by misdistribution of chromosomes during mitosis. In other cases, recombination between the two different chromosomes leads to replacement

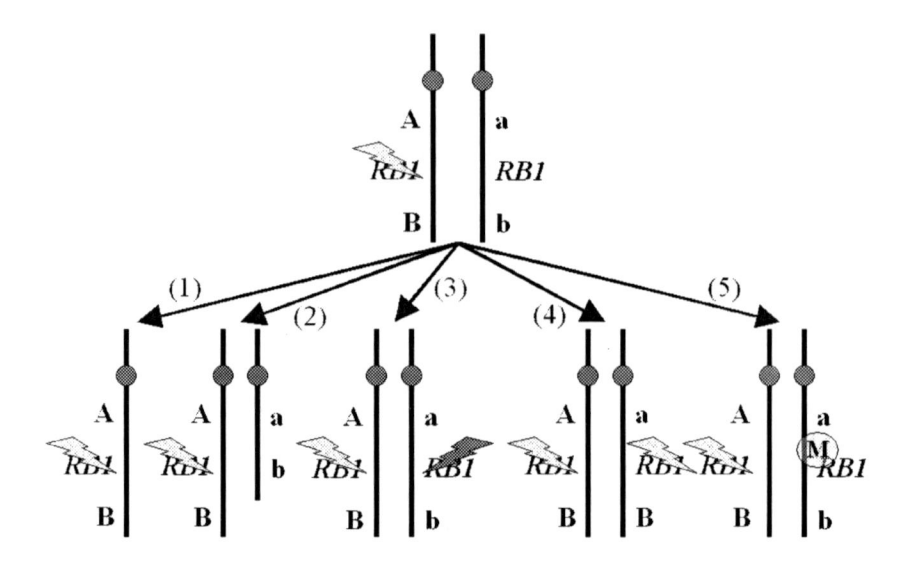

**Figure 5.2** *Mechanisms causing inactivation of the second* RB1 *allele*
(1) Loss of chromosome 13 carrying the intact allele; (2) Deletion; (3) Independent mutation; (4) Recombination; (5) Promoter hypermethylation. Mechanisms (1), (2), and (4) are associated with LOH. cf. also Figure 2.8. The mutated first allele is marked by a lightning.

of larger or smaller regions of one chromosome by sequences from the other and specifically to conversion of the intact RB1 to a defective copy.

Several of the above mechanisms, specifically larger deletions and unequal recombinations or recombination followed by chromosomal loss, leave the cell not only without a functional allele of the *RB1* tumor suppressor gene, but also abolish other differences between the two chromosomal copies. So, polymorphisms in single nucleotides (SNPs) or in microsatellite sequences disappear, as heterozygous sequences become homozygous (Figure 5.2). This process is called loss of heterozygosity (LOH). So, LOH marks regions in a tumor genome where deletions or illegitimate recombination have taken place. Since such processes are often involved in the inactivation of tumor suppressor genes, the observation of consistent LOH in one region in a particular tumor type can be used as an indication of the presence of a tumor suppressor gene.

Many further tumor suppressors also fit the Knudson model well (Table 5.1). Among them are *APC* in colon carcinoma (→13.2), *VHL* in renal cell carcinoma (→15.4), and *CDH1* in gastric carcinoma (→17.2). Others follow the model at least partly. So, *BRCA1* and *BRCA2* behave according to the model in hereditary breast and ovarian cancer (→18.3), but are rarely involved in sporadic cases. The converse case is more common, i.e. several tumor suppressor genes are never mutated in the germ-line, but both copies are inactivated in cancers. In other words, not all common human cancers also occur in an inherited form. It is also conceivable that mutation of one copy of a tumor suppressor gene could be sufficient to initiate the development of a cancer, albeit not as efficiently as the inactivation of both alleles. This situation is called 'haploinsufficiency'. It is discussed for several tumor suppressors, including *PTCH* in basal cell carcinomas of the skin (→12.3), *PTEN* in a variety of cancers (→6.3), and *HPC1* in prostate cancer (→19.3). In the case of inherited mutation in *PTCH* and *PTEN*, the case for haploinsufficency is supported by the fact that the patients tend to have developmental abnormalities unrelated to cancer.

An entirely different situation is provided by germ-line mutations that activate oncogenes and thereby result in dominantly inherited cancers. This situation has been experimentally achieved with very high efficiency in transgenic mice, e.g. by introduction of a mutated *Ras* gene. In humans, very few inherited cancers are caused by activated oncogenes. Multiple endocrine neoplasia type 2 is caused by mutations in the *RET* proto-oncogene predisposing to cancer of several endocrine glands, notably the thyroid and adrenal glands. RET is a tyrosine receptor kinase and the inherited mutations lead to its constitutive activation. Comparable mutations in the *MET* gene, which also encodes a receptor tyrosine kinase, cause one type of papillary renal cell carcinoma (→15.3). The reason why tumor suppressor gene mutations prevail in inherited human cancer is not really understood. It may partly reflect a better protection of human cells against oncogenic transformation compared to rodent cells (→7.4) or a pronounced sensitivity of human development to disturbances by mutated oncogenes.

## 5.2 RB1 AND THE CELL CYCLE

The product of the *RB1* gene, $pp110^{RB1}$, is most of all a central regulator of the cell cycle (Figure 5.3). The RB1 protein controls the transition from the G1 to the S phase by binding to E2F1, E2F2, or E2F3 proteins and thereby repressing the promoters of genes needed for the entrance into S phase. This repression is relieved and binding to E2F alleviated when $pp110^{RB1}$ becomes hyperphosphorylated[5] towards the end of G1. At least two successive phosphorylations are needed to inactivate RB1. Normally, the first phosphorylation is performed by a CDK4/Cyclin D holoenzyme, and is followed by further phosphorylations by the CDK2/Cyclin E holoenzyme. Hyperphosphorylated RB1 is inactive as far as G1/S cell cycle regulation is concerned. The protein likely has also functions in the S phase, where it may be involved in chromatin organization, and during mitosis, where it may help to organize proteins for chromosome segregation. Following mitosis, RB1 is partly dephosphorylated and its hypophosphorylated state restored. In this fashion, RB1 switches between hypophosphorylated and hyperphosphorylated states during the cell cycle.

---

[5] The RB1 protein is always phosphorylated at some sites; therefore its phosphorylation varies between hypo- and hyperphosphorylation rather than between non-phosphorylated and phosphorylated. Up to 14 phosphates can be incorporated; their individual functions are not fully elucidated (understandably...).

Clearly, therefore, loss of RB1 function, at a minimum, upsets cell cycle regulation and may lead to unrestrained cell proliferation. Specifically, in the absence of RB1 immature cells such as retinoblasts may not spend sufficient time in G1 to enter a differentiated state or establish stable quiescence, i.e. G0. Worse, since RB1 may also be required for proper chromatin structure and chromosome segregation, cells may tend to become genomically instable and acquire additional alterations that favor tumor progression. At least one mechanism may protect against loss of RB1 function: Over-activity of E2F factors, particularly of E2F1, can induce apoptosis.

The mechanism of cell cycle regulation sketched so far is in fact much more complex and consists of several layers of control, even when only the G1 to S transition is considered (Figure 5.4). The activity of the CDK4 and CDK2 protein kinases depends strictly on the presence of their regulatory subunits, i.e., D-Cyclins and Cyclin E, respectively. Both fluctuate in a coordinate fashion in the course of the cell cycle. Cyclin D expression is directly dependent on stimulation by exogenous signals such as growth factors ($\rightarrow$6.4). Moreover, CDKs are only active, if they are phosphorylated in a certain pattern. Phosphorylation at one specific threonine residue by the CDK activating kinase (CAK, identical to CDK7) in conjunction with its regulatory subunit Cyclin H is activating, while phosphorylation at two different threonines is inhibitory. The phosphates at these sites are removed by CDC25 phosphatases, which also respond to external signals ($\rightarrow$6.4).

A further layer of control is provided by protein inhibitors of CDKs (Table 5.2). There are two classes of such inhibitors, the CIP/KIP and the INK proteins. The first comprises the $p21^{CIP1}$, $p27^{KIP1}$, and $p57^{KIP2}$ proteins, the second $p15^{INK4B}$, $p16^{INK4A}$, $p18^{INK4C}$, and $p19^{INK4D}$. Their genes are now systematically designated *CDKN1A – CDKN1C* and *CDKN2A - CDKN2D*. All proteins are named according to their molecular weights. The function of the INK4 (inhibitor of kinase 4) proteins is straightforward. They compete with D-Cyclins for binding to CDK4 and block its kinase activity.

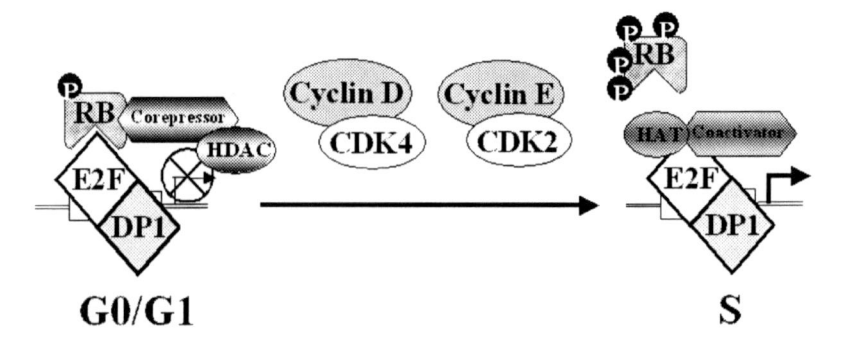

**Figure 5.3** *Function of RB1 in cell cycle regulation*
DP1 is a heterodimer partner of E2F proteins. HDAC: histone deacetylase; HAT: histone acetyl transferase. The functions of HDACs and HATs are explained in 8.4.

The function of the CIP/KIP proteins (CDK/kinase inhibitory proteins) is more complicated. At high concentrations, they inhibit the activity of CDKs in general, but different from INK4s they block the CDK-Cyclin holoenzymes. At moderate concentrations, they rather stimulate the assembly of CDK-Cyclin complexes. In proliferating cells, CIP/KIP proteins, in particular p27$^{KIP1}$, help to coordinate the cell cycle. Until late G1, p27$^{KIP1}$ is bound to the CDK2/Cyclin E complex delaying its activity until late G1 when CDK2 activity is required to inactivate RB1. At this point, the p27$^{KIP1}$ inhibitor is phosphorylated and rapidly degraded (cf. 6.4). On the other hand, in cells that are not supposed to proliferate, high levels of inhibitor proteins arrest the cell cycle.

The different CDK inhibitors respond to different signals that lead to cell arrest, allowing fine-tuned cellular responses to different signals. For instance, in some cell

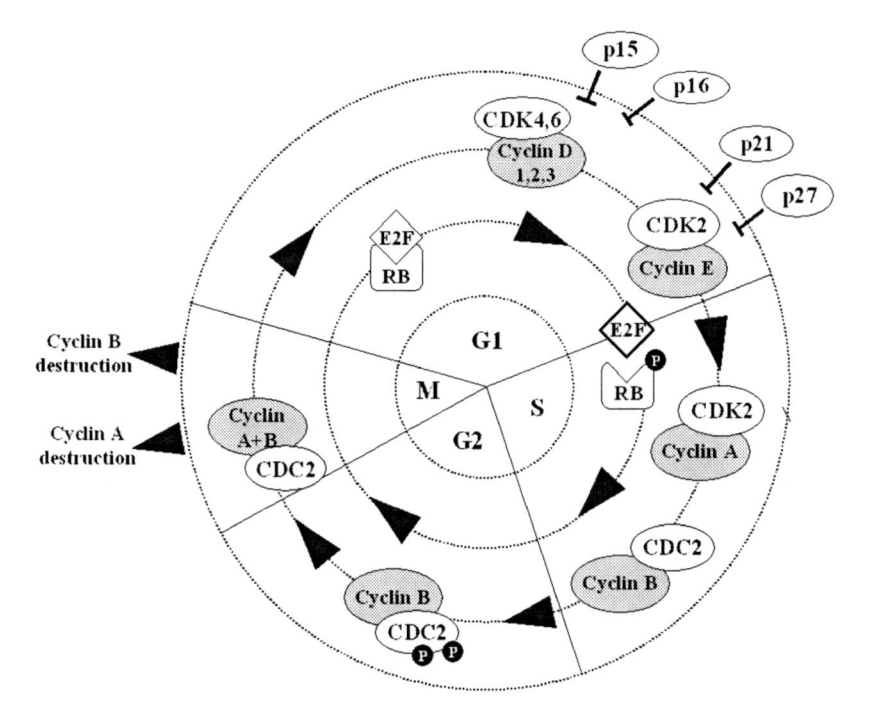

**Figure 5.4** *Three layers of cell cycle regulation*
The figure focuses on the G1→S transition, which is the usual point of regulation in human cells (mechanisms regulating G2→S transition are used in specific cells). The inner layer consists of the RB1 phosphorylation cycle which determines E2F activity. It is dependent on the second layer formed by the CDK/cyclin cycle (CDC2 is also named CDK1). This layer is additionally regulated by a third one involving phosphorylation and dephosphorylation of the CDKs (only one example is shown) and CDK inhibitor proteins. This layer interacts mutually with the inner layers, and is influenced by various mitogenic and anti-proliferative signals.

types $p15^{INK4B}$ is induced by inhibitory growth factors like TGFβ. High levels of the inhibitor protein dissociate Cyclin D/CDK4 complexes and block CDK4. They also redirect any $p21^{CIP1}$ or $p27^{KIP1}$ present towards CDK2/Cyclin E, blocking this kinase as well. By comparison, $p16^{INK4A}$ has a very slow turnover and accumulates gradually in continuously proliferating cells until its concentration becomes high enough to slow down the cell cycle or arrest it irreversibly (→7.4). $p21^{CIP1}$ accumulates in a similar fashion during frequent replication, but also in response to various growth factors and to DNA damage (→5.3). $p57^{KIP2}$ is expressed in a restricted range of cell types, most prominently during embryonic development where the protein appears to help establish a terminally differentiated state in specific tissues. The gene encoding $p57^{KIP2}$, *CDKN1C*, is as a rule only expressed from the maternally inherited allele, i.e. it is imprinted (→8.2).

In summary, therefore, RB1 forms a node in the cell cycle regulation network which ensures that cell proliferation occurs only in response to proper sets of signals, e.g. following stimulation by growth factors via the MAPK and further signaling pathways (→6.4). It is easy to imagine how this regulation network becomes disrupted by loss of function of the central RB1 protein. Alternatively to loss of RB1 function itself, other components of the regulatory network may be affected in human cancers. So, overexpression of D-Cyclins or CDK4 as a consequence of gene amplification, or mutations of CDK4 that make it unresponsive to CDK inhibitors may exert similar effects.

In addition, alterations in CDK inhibitors are found in a variety of human cancers. In a wide range of human cancers, the *CDKN2A* gene is inactivated by point mutation, promoter hypermethylation, or homozygous deletion. In fact, *CDKN2A* must also be regarded as a 'classical' tumor suppressor gene, since both alleles are affected in such cancers, and germ-line mutations of the gene have been found in families prone to pancreatic cancer and melanoma (→12.4). Of note, not all these changes may be equivalent. There is some evidence that loss of RB1 function may be the most severe defect, as one might guess intuitively (→14.2).

**Table 5.2.** *Inhibitor proteins of cyclin-dependent kinases*

| CDK Inhibitor | Gene | Location | CDK inhibited | Inducers |
|---|---|---|---|---|
| $p21^{CIP1}$ | *CDKN1A* | 6p21.2 | several | TP53, growth factors, senescence |
| $p27^{KIP1}$ | *CDKN1B* | 12p13.1-p12 | CDK2 | TGFβ, GSK(?) |
| $p57^{KIP2}$ | *CDKN1C* | 11p15.5 | several | cell differentiation, senescence |
| $p16^{INK4A}$ | *CDKN2A* | 9p21 | CDK4 | growth factors, senescence |
| $p15^{INK4B}$ | *CDKN2B* | 9p21 | CDK4 | TGFβ |
| $p18^{INK4C}$ | *CDKN2C* | 1p32 | CDK4/6 | ? |
| $p19^{INK4D}$ | *CDKN2D* | 19p13 | CDK4/6 | ? |

Surprisingly, mutations or deletions of the genes encoding CDK inhibitors other than p16$^{INK4A}$ are by far not as frequent in human cancers. The *CDKN2B* gene encoding p15$^{INK4B}$ is located within 40 kb of *CDKN2A* (cf. Figure 5.9) and is often deleted together with its neighbor. However, inactivation of the gene may only be crucial in certain leukemias. Inactivation of *CDKN1C*, in accord with its more circumscribed expression, may be relevant in a tighter range of cancers. Mutations in the genes encoding p21$^{CIP1}$ and p27$^{KIP1}$ seem very rare in human cancers, but down-regulation of their expression is highly prevalent in many different cancers. It is often a good indication of tumor progression and a marker for cancers taking a more aggressive clinical course. It is neither well understood, why these genes are so rarely mutated nor which mechanisms exactly lead to down-regulation of their expression during tumor progression. In some cases, loss of p27$^{KIP1}$ may be caused by over-expression of proteins involved in its degradation at the end of G1 (→6.4).

## 5.3 TP53 AS A DIFFERENT KIND OF TUMOR SUPPRESSOR

*RB1* is a tumor suppressor gene mainly because its inactivation removes an essential point of regulation of the cell cycle. Thus, its loss takes a tumor cell directly towards unrestricted proliferation and diminished dependence on extracellular signals. Of note, while RB1 inactivation favors tumor development, its actual function is not so much to suppress tumors, but to allow coordinated proliferation and differentiation during the development and maintenance of normal tissues. Accordingly, mice lacking *Rb1* are not viable and exhibit defects in the development of several tissues.

In contrast, animals can in principle live without another tumor suppressor, Tp53. However, mice lacking Tp53 succumb to tumors after a few months of life. In a rare human disease, Li-Fraumeni-syndrome (Table 5.1), one mutant copy of the *TP53* gene is inherited. The patients develop various types of tumors in different tissues, including blood, lymphoid organs, soft tissues, the nervous system and epithelia, often early in life. In accord with the Knudson model, the tumors contain mutations inactivating the second *TP53* allele or exhibit LOH at chromosome 17p where the gene resides. The *TP53* gene is also inactivated in many different types of sporadic cancers in man. It is arguably the most frequently mutated gene in human cancer. Indeed, the main function of the TP53 protein appears the prevention of damage to the genome and of cancer, making it literally a 'tumor suppressor'.

The approximately 20 kb *TP53* gene on chromosome 17p13.1 encodes in eleven exons a 2.2 kb mRNA, from which, as the name indicates, a 53 kDa phosphoprotein is translated. The TP53 (or simply p53) protein has a structure typical for transcriptional activators (Figure 5.5). It is composed of a central DNA binding domain, an N-terminal transactivation domain, and a C-terminal domain which contains sequences necessary for oligomerization and regulation of DNA-binding.

Indeed, TP53 can function as a transcriptional activator at several hundred genes. In many of these genes, TP53 binds as a tetramer to a symmetric specific binding sequence in the promoter region or an (often intronic) enhancer. In other genes, TP53 acts as a repressor of transcription by PolII and PolIII by binding and blocking the basal transcription factor TBP or by interacting with transcriptional repressors,

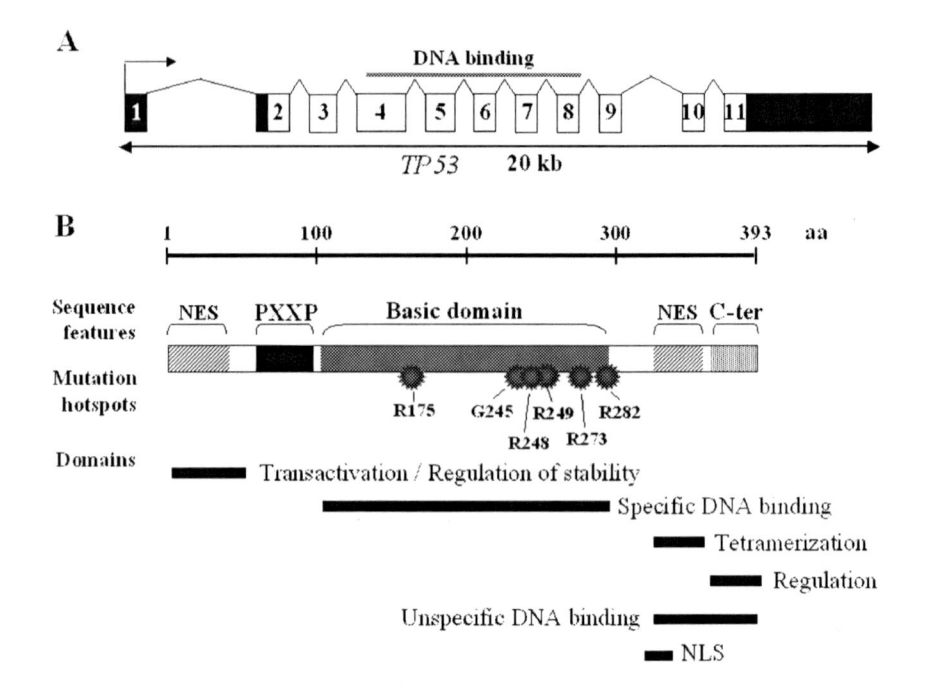

**Figure 5.5** *Structure and functional domains of TP53*
A: Organization of the gene. The introns are not fully to scale. The grey bar marks the exons encoding the basic DNA-binding domain. B: Features of the protein. C-ter: C-terminal interaction domain; NES/NLS nuclear export/localization signal; PXXP: proline-rich domain.

e.g. SIN3A. Moreover, TP53 has several other functions that are less well characterized, such as the ability to bind to damaged DNA directly, to function as an exonuclease, to anneal nucleic acids to each other, and perhaps even to regulate protein biosynthesis at the ribosome.

Through these various functions TP53 coordinates the cellular response towards many kinds of damage to the genome, in particular, DNA double-strand breaks induced by chemical mutagens and ionizing radiation, as well as to certain kinds of cellular stress, such as guanine nucleotide imbalance, viral infection, and oncogene-induced hyperproliferation (Figure 5.6). TP53 is also involved in the regulation of replicative senescence (→7.4).

TP53 action is mostly regulated posttranslationally by phosphorylation and other modifications (Figure 5.7) which alter the stability and activity of the protein. Under normal circumstances, TP53 undergoes rapid turnover in the cell, with a half-life in the 10-20 min range. For this short half-life, the MDM2/HDM2 protein is mostly responsible. It binds to the N-terminal domain of TP53 blocking its transcriptional activity and initiating its transport out of the nucleus. Furthermore, MDM2/HDM2 acts as a specific E3 ubiquitin ligase for TP53, which following oligo-ubiquitination is rapidly degraded by the proteasome.

Different pathways signal various types of damage and stress to TP53 leading to a variety of post-translational modifications (Figure 5.7). Double-strand breaks induced by ionizing radiation activate ATM and/or DNA-dependent protein kinase and extensive UV damage activates ATR (→3.3) which phosphorylate TP53 at Ser15 and Ser37. The checkpoint protein kinases CHK1 and CHK2, which respond to DNA damage and specifically to mitotic disturbances phosphorylate Ser20. Phosphorylation at Ser15 and Ser 20 in particular block the interaction of TP53 with MDM2/HDM2 and increase the half-life of the protein. Moreover, most phosphorylations in the N-terminal transactivation domain enhance the strength of TP53 as a transcriptional activator. In the C-terminal regulatory domain, sumolation at Lys386 helps to guide TP53 within the nucleus. Acetylation at Lys320, Lys373, and Lys382 may modulate DNA-binding, transcriptional activation, and nuclear localization. Phosphorylation at Ser392 by the double-strand RNA-dependent protein kinase PKR may activate TP53 in response to viral infections. Conversely, other phosphorylations may restrain TP53. For instance, TP53 is phosphorylated at several serines (371, 376, and 378) by PKC and at Ser315 by CDK2. Phosphorylations by p38$^{MAPK}$ and JNK (→6.2) at Ser46 and Thr81 appear to modulate the pro-apoptotic function of TP53.

Inappropriate cell proliferation, e.g. induced by oncogenic RAS, also activates TP53, mainly by an indirect mechanism (Figure 5.8). Increased proliferation is associated with increased activity of E2F factors such as E2F1. In addition to genes required for cell cycle progression, E2F1 activates the transcription of p16$^{INK4A}$

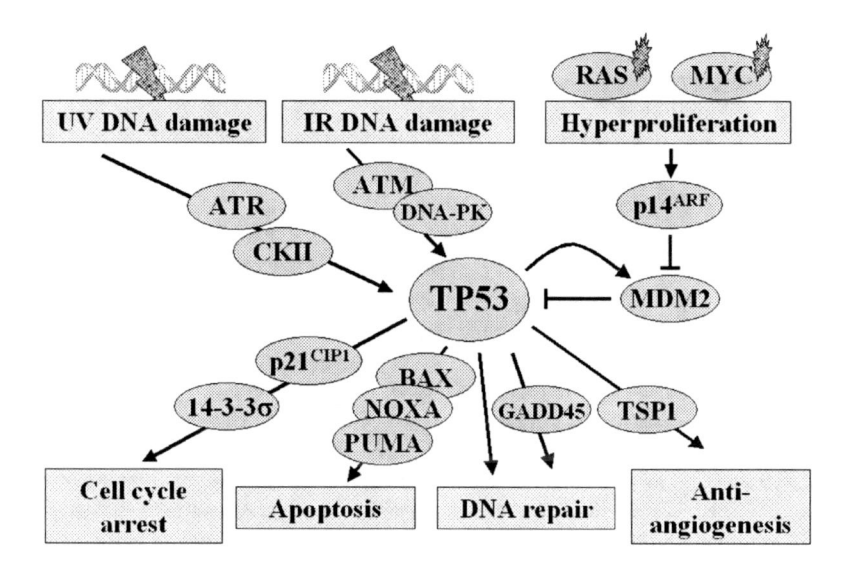

**Figure 5.6** *A sketch of the TP53 network*
Only a central part of the network is shown. See text for details

# CHAPTER 5

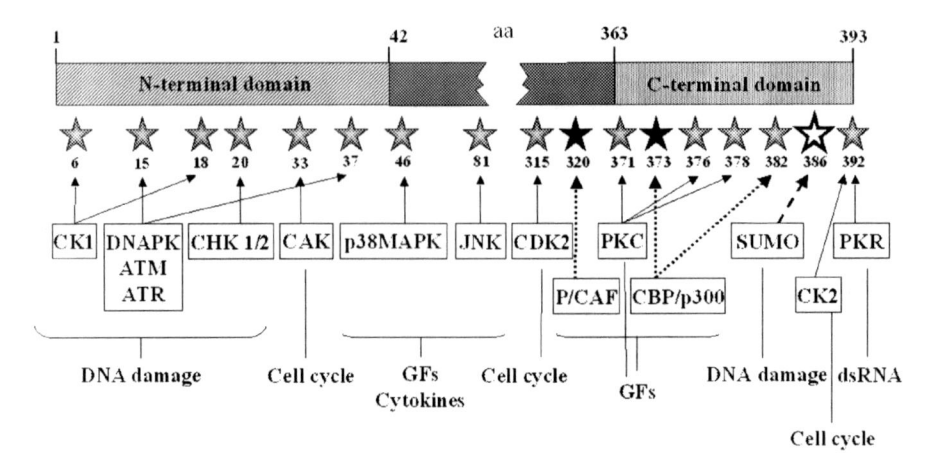

**Figure 5.7** *Posttranslational regulation of TP53*
Note that only the N-terminal and C-terminal domains of the protein are shown. Most modifications are phosphorylations except at Lys320 and 373 (acetylation) and 386 (sumolation). The modifying enzymes are boxed.

mRNA from the *CDKN2A* gene, but also a second promoter in the gene, leading to expression of a 14 kDa protein in an <u>a</u>lternative <u>r</u>eading <u>f</u>rame. This protein is therefore designated p14$^{ARF}$ (p19$^{ARF}$ in mice). It is an inhibitor of MDM2/HDM2. Therefore, 'inappropriate' proliferation signals induce p14$^{ARF}$ which blocks MDM2/HDM2 leading to stabilization and activation of TP53. The *CDKN2A* gene is thus, in fact, a double locus with two common exons, of which exon 2 codes for both p16$^{INK4A}$ and p14$^{ARF1}$ in different reading frames (Figure 5.9). So, many mutations in this exon inactivate both proteins, as do homozygous deletions of *CDKN2A*. Therefore, alterations in this locus normally compromise the functions of both TP53 and RB1 indirectly (Figure 5.8).

Following its activation by one of the above pathways, TP53 induces multiple cellular responses (Figure 5.6), i.e. cell cycle arrest, DNA repair, altered secretion of growth factors (particularly of angiogenic factors), apoptosis, and, finally, its own inactivation by MDM2/HDM2. Which responses are induced in particular, appears to depend on several factors, such as the cell type, the extent of DNA damage or cellular stress, the basal and induced pattern of phosphorylation of TP53 by multiple kinases (Figure 5.7), and on competing signals. Cell cycle arrest and apoptosis occur alternatively to each other, and the fate of a cell may sometimes simply depend on which response is induced faster. Overall, several hundred genes can be activated or repressed by TP53. Therefore, the proteins shown in Figure 5.6 to act 'downstream' of TP53 are a selection of the more established and more general mediators of TP53 action in human cells.

Arrest of the cell cycle by TP53 is mediated through rapid and strong induction of two inhibitory proteins: p21$^{CIP1}$ and 14-3-3σ, which block the cell cycle in G1 or G2. This response appears to be common to all cell types and to follow the activation of TP53 by a variety of signals. Another common response to TP53 activation is induction of GADD45, a protein involved in DNA repair.

Induction of apoptosis by TP53 appears to proceed through different factors in different cells. Most widespread may be induction of BAX, an antagonistic homolog of BCL2 (→4.4), and an effector of apoptosis at mitochondria (→7.2). Conversely, *BCL2* is one of the genes repressed by TP53. Several further proteins in the 'mitochondrial' or 'intrinsic' apoptotic pathway (→7.2) can be induced by activated TP53, such as p53AIP, NOXA, PUMA, and APAF1. In addition or in parallel, TP53 increases the sensitivity towards exogenous apoptotic signals, e.g. by increasing expression of the 'death receptor' FAS (→7.2). One could interpret this variety of pro-apoptotic signals as a series of back-up mechanisms evolved to ensure that apoptosis cannot be circumvented. Alternatively, this variety may allow a better choice between apoptosis and survival (with cell cycle arrest) depending on further intracellular and exogenous signals and on the cell type.

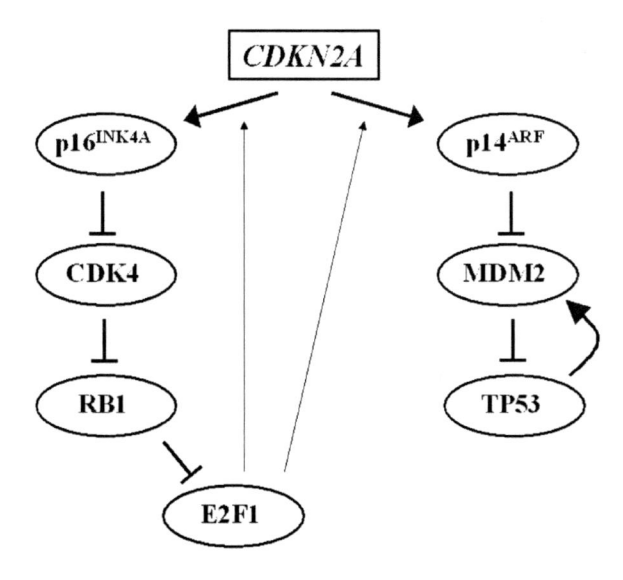

**Figure 5.8** *Regulation of TP53 and RB1 by proteins from the* CDKN2A *locus*

A further group of genes induced or repressed by TP53 functions in the communication with neighboring tissue, particularly with endothelial cells. For instance, TP53 induces thrombospondin (TSP1) which inhibits the proliferation of endothelial cells and thereby blocks angiogenesis (→9.4).

Finally, TP53 induces a feedback mechanism to limit its own action. The *MDM2/HDM2* gene is a direct transcriptional target of TP53 and becomes relatively rapidly induced following TP53 activation. Accumulating MDM2/HDM2 blocks transcriptional activation by TP53 and causes its degradation. The efficacy of this mechanism is most dramatically illustrated by an experiment in mice. Mice lacking

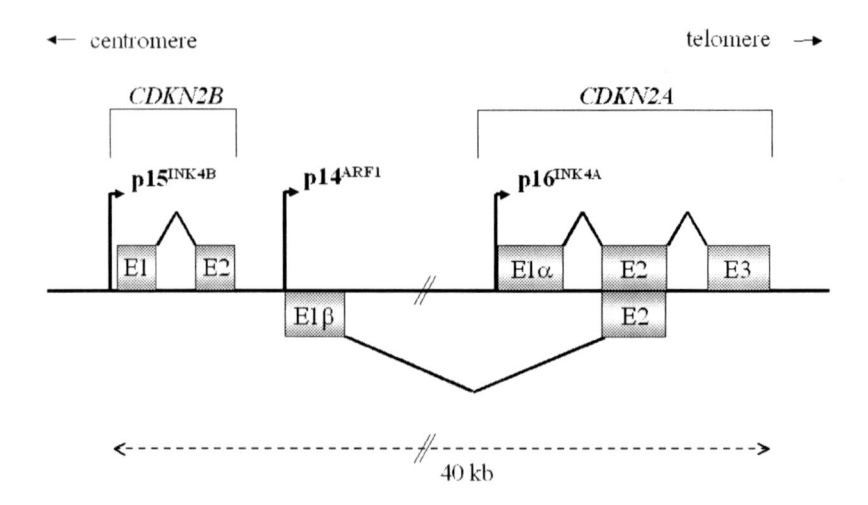

**Figure 5.9** *Organization of the* CDKN2A *locus at 9p21*

Tp53 are usually viable, although they die of tumors at an early age. In contrast, mice lacking Mdm2 die in utero from widespread apoptosis. Knockout of the Tp53 gene as well largely corrects this defect.

In summary, TP53 can be considered as a central node in an important network which regulates the cellular response to most kinds of genomic damage and many types of cellular stress. This explains why defects in TP53 are so widespread in human cancers. Cancers lacking functional TP53 will tolerate more DNA damage, including DNA strand breaks and aneuploidy, will condone inappropriate proliferation signals and nucleotide stress, and will less easily enter replicative senescence. Compared to cancers of the same type with wild-type TP53, cancers with mutant TP53 therefore tend to accumulate more genomic alterations and to run a higher risk of progression.

The inactivation of TP53 function is brought about by several distinct mechanisms in different human cancers (Table 5.3).

Missense mutations plus LOH: The most common mechanism in many different kinds of human tumors consists of missense mutations in one allele and loss of the second functional allele by deletion or recombination, which is detectable as LOH on 17p. It is likely favorable for the two changes to occur in this order, since some missense mutation in the first allele may already impede the function of TP53, with LOH completing its inactivation. In experimental models, certain mutant TP53 clearly compromise the function of unaltered ('wild-type') protein. They act as dominant negatives, probably by sequestering wild-type protein monomers in inactive (tetrameric) complexes. It is less clear, how important this effect actually is during the development of human cancers. Conceivably, the initial missense mutation might increase the probability of the loss of the second allele by compromising the ability of the cell to react to genomic damage such as illegitimate recombinations or deletions of chromosome 17p.

Most missense mutations in TP53 affect the central domain of the protein necessary for specific binding to DNA (Figure 5.5). This domain consists of a large structure made up from ß-sheets stabilized by zinc ions which supports the α-helix that contacts DNA through arginine and further residues. These arginines are mutational hot-spots, but many other amino acids in the central domain can also be mutated. The exact site of the mutation also appears to depend on the carcinogen involved (→12.1, →16.1). Mutations in the N-terminal and C-terminal domains are found with lower frequencies and may compromise the activation and oligomerization of TP53.

Nonsense and splice mutations: Nonsense and splice mutations in TP53 do occur, but are less frequent. Apparently, missense mutations provide some sort of advantage, either by acting as dominant-negatives as described above, or by retaining some functions of TP53 which are advantageous to tumor cells. This issue is still under investigation.

MDM2 overexpression: Some tumors harbor amplifications of the HDM2/MDM2 gene. In others this gene is overexpressed by unknown mechanisms. Such overexpression likely diminishes the function of TP53. Indeed, in some cancers like sarcomas and gliomas amplifications of HDM2 and mutations of TP53 occur in a mutually exclusive fashion. This is a very good indication that they are indeed complementary. It should be mentioned, though, that HDM2 has functions beyond the regulation of TP53 which could be important in specific cancers (e.g. sarcomas).

Conversely, in cancers with mutated TP53, HDM2 cannot be induced by TP53. If the mutation does not otherwise destabilize the TP53 protein, this will lead to an accumulation of the mutated protein, since HDM2 is the rate-limiting enzyme for its degradation. This is one reason, why in many cancers TP53 protein levels paradoxically are higher than in the corresponding normal tissues. Thus, in some cases, detection of increased TP53 protein levels (i.e. accumulated mutant protein) can be used for the detection of tumor cells.

Table 5.3. *Mechanisms of TP53 inactivation in human cancers*

| Mechanism of inactivation |
| --- |
| Missense mutations |
| Nonsense and splice mutations |
| Deletion of one allele |
| Overexpression of HDM2 by gene amplification or deregulation |
| Loss of p14$^{ARF}$ function by gene deletion, mutation or promoter hypermethylation |
| Loss of function of upstream activators, e.g. ATM or CHK1 |
| Loss of function of downstream effectors, e.g. BAX |
| Inactivation by altered post-translational modification |
| Inactivation by viral oncoproteins, e.g. HPV E6 |

Upstream activator inactivation: The function of TP53 is expected to be impeded by mutations in upstream pathways that provide signals for its activation (Figure 5.6). This is likely the case in ataxia telengiectasia (→3.3). In this disease, activation of TP53 in response to ionizing radiation is diminished as a consequence of defects in the ATM kinase. More generally in human cancers, the loss of p14$^{ARF}$ may be important. The *CDKN2A* gene is inactivated in many different types of human cancers by deletions or point mutations, in specific cancers also by hypermethylation of its promoters. The loss of p14$^{ARF}$ would be predicted to specifically disrupt the activation of TP53 in response to inappropriate cell proliferation, e.g. as induced by oncogene activation. Of note, loss of p14$^{ARF}$ also occurs in cancers with TP53 inactivation by other mechanisms, but then, p16$^{INK4A}$ is a mutation target in the same gene. Intriguingly, loss of p16$^{INK4A}$ and RB1 appear to occur in a mutually exclusive fashion.

Cell-type specific down-regulation: There are a few cancers in which mutations of TP53 are extremely rare. In testicular germ cell tumors, e.g. seminomas, mutations may be very rare, because the gene is normally not very active at the stage of development from which the tumor cells arise. However, TP53 can be activated by strong signals and the presence of an intact gene is one factor responsible for the high efficacy of radiotherapy and chemotherapy in these cancers.

Inactivation by viral proteins: Still another mechanism is found regularly in cervical cancers and occasionally in some other cancer types (Box 5.1). These cancers are caused by specific oncogenic strains of human papilloma virus (HPV). These express a protein, E6, which binds to TP53 and promotes its degradation, quite similar as HDM2. This mechanism inhibits the cellular response to the infection and active replication of the virus. If E6 expression is sustained, e.g. by activation of the viral gene through cellular enhancers (Box 5.1), it contributes to cell immortalization (→7.4) and favors the accumulation of genetic alterations promoting cancer by impeding the response to DNA damage and chromosomal defects.

Indeed, other DNA viruses have developed similar mechanisms to cope with TP53 during infection. Adenoviruses, which are tumorigenic in rodents, employ the E1B protein to sequester and inhibit TP53, although the protein is not degraded. In fact, TP53 was originally discovered as a protein regularly associated with the major transforming protein of the DNA tumor virus SV40, a papovavirus infecting primates including man. This virus encodes two tumor antigens, named large-T and small-T. Large-T antigen is a multifunctional protein that regulates SV40 transcription and replication. It also binds and inactivates several host proteins, including TP53.

A second important protein sequestered and inactivated by SV40 large-T is RB1 (Figure 5.10). Other DNA tumor viruses as well inactivate RB1, HPV by its E7 protein and adenoviruses by their E1A proteins. Nevertheless, adenoviruses are probably not transforming in man. However, SV40 and related papovaviruses such as BK and JC may play a role in specific human cancers. The best evidence is available for mesothelioma, where SV40 infection may synergize with exposure towards asbestos.

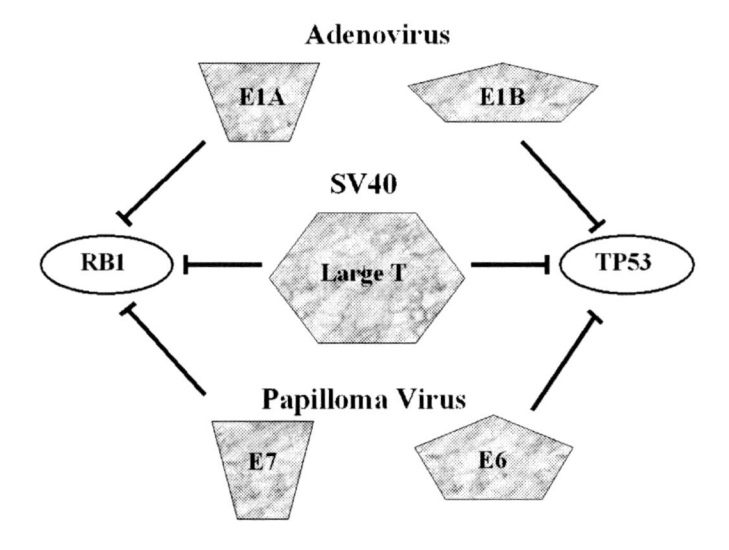

**Figure 5.10** *Inactivation of TP53 and RB1 by DNA tumor viruses*

## 5.4 CLASSIFICATION OF TUMOR SUPPRESSOR GENES

The fact that DNA tumor viruses so specifically target TP53 and RB1 is another indication that these proteins are of central importance in normal cell growth as well as in cancer. They reside at nodes of networks which control cell proliferation and genomic integrity. These must be subverted or overcome for tumor growth to proceed. As discussed in the previous chapter (→4.4), many oncogenes can be arranged around one central signaling pathway (which could with some right too be called a network) and can be usefully classified by their biochemical function. Considered overall, tumor suppressors may exhibit a wider array of biochemical functions and act in a wider range of regulatory networks in addition to cell cycle regulation and DNA damage response.

In some of these networks, both tumor suppressors and oncogenes can be spotted and tend to be antagonists in the normal functioning of the pathway. For instance, PTEN is a negative regulator of the PI3K pathway (→6.2), PTCH1 is a negative regulator of SMO in the Hedgehog pathway (→6.10, →12.3), APC promotes the inactivation of CTNNB1 in the WNT pathway (→6.10, →13.2), and NF1 mutated in familial neurofibromatosis is a tissue-specific GAP regulating RAS function (→4.4). When tumor suppressors were first discovered, it was suggested to designate them as 'anti-oncogenes'. Clearly, there is some justification for this name. However, it is misleading in so far as there is no pairwise complementarity between oncogenes and tumor suppressors. Moreover, activation of proto-oncogenes and inactivation of tumor suppressors from the same pathway typically have similar, but not identical consequences. In any case, the presence of alternatively occuring alterations in tumor suppressor genes and oncogenes suggests that it is the function of 'cancer pathways' (or networks) that is crucial for cancer development, rather than that of individual genes (→6.1).

Even a crude comparison between *TP53* and *RB1* suggests that there are at least two classes of tumor suppressors. Most importantly, loss of TP53 function does not directly lead to altered cell growth. Rather, it permits alterations in the cell to take place that are then directly responsible for altered growth at the cell and tissue level. One such alteration is loss of RB1 which causes altered growth and differentiation directly. Vogelstein and Kinzler have proposed to call these two kinds of tumor suppressors 'caretakers' and 'gatekeepers', respectively (→13.3). The designation 'gatekeeper' implies in addition that a cancer of a certain type can only arise, if the function of this particular tumor suppressor is abolished. This may indeed be true in some cases, e.g. in colorectal cancer, where the concept was developed (→13.3).

Of course, further classes of tumor suppressor genes may exist. For instance, it has been proposed that some genes may be irrelevant for the growth of a primary tumor, but their loss of function would be essential for metastasis. These would then be considered 'metastasis suppressor genes'. There is some evidence for such genes (→9.1), but the concept is not as well developed as that of 'caretakers & gatekeepers'.

So, finally, what is the precise definition of a tumor suppressor gene? The strictest definition, according to the Knudson model, would encompass all genes which lead to cancers inherited in a autosomal-dominant fashion, but behave in a recessive mode at the cellular level. Thus, one mutant, functionally inactive allele is inherited and the second one is inactivated in the ensuing tumors by point mutation, deletion, insertion, recombination, or promoter hypermethylation. Often, both alleles of these same genes are also functionally inactivated in sporadic cases of the same type. No conceptual problem arises from extending this definition to all genes that show inactivation of both alleles in many sporadic cases of a specific tumor type, if this occurs consistently or in a sizeable subgroup, and if the functional relevance for tumor development or progression can be shown.

The concept is more difficult to apply when inactivation of both alleles does not occur regularly or only one allele of a gene is consistently lost. The first case could be a random event and the second case could, e.g., be a consequence of proximity to a real tumor suppressor gene. However, haploinsufficiency is a real possibility in such cases. A gene exhibiting consistent loss of only one allele could represent a tumor suppressor, whose expression at half the normal dose is insufficient to inhibit tumor formation. Unfortunately, haploinsufficiency is very difficult to ascertain in the context of human cancers, as demonstrated by the case of NKX3.1 in prostate cancer (→19.3) or by the elusive tumor suppressor gene at chromosome 9q in bladder cancer (→14.2). A particularly difficult case is raised by genes whose expression is consistently down-regulated in one type of cancer, but which do not show genetic alterations in structure or dosage, and not even a clear-cut epigenetic mark of stable silencing such as promoter hypermethylation (→8-3). This is similar to the dilemma whether overexpressed genes are generally oncogenes (→4.4). As in that case, defining the precise function of the gene and protein in question is an essential requirement for approaching this problem.

## Further reading

Stein GS, Pardee AB (eds.) The molecular basis of cell cycle and growth control $2^{nd}$ ed. Wiley & Sons, 2004

Weinberg RA (1996) How cancer arises. Sci. Am. 275, 62-70
Kinzler KW, Vogelstein B (1997) Gatekeepers and caretakers. Nature 386, 761-762
Ruas M, Peters G (1998) The p16INK4a/CDKN2A tumor suppressor and its relatives. BBA 1378, F115-177
Knudson AG (2000) Chasing the cancer demon. Annu. Rev. Genet. 34, 1-19
Vogelstein B, Lane D, Levine AJ (2000) Surfing the p53 network. Nature 408, 307-310
Evan GI, Vousden KH (2001) Proliferation, cell cycle and apoptosis. Nature 411, 342-348
Irwin MS, Kaelin WG (2001) p53 family update: p73 and p63 develop their own identities. Cell Growth Different. 12, 337-349
Classon M, Harlow E (2002) The retinoblastoma tumour suppressor in development and cancer. Nat. Rev. Cancer 2, 910-917
Gazdar AF, Butel JS, Carbone M (2002) SV40 and human tumours: myth, association or causality? Nat. Rev. Cancer 2, 957-964
Leblanc V, May P (2002) Activation et modifications post-traductionnelles de p53 après dommage de l'ADN. Med. Sci. 18, 577-584
Sherr CJ, McCormick F (2002) The RB and p53 pathways in cancer. Cancer Cell 2, 103-112
Zheng L, Lee WH (2002) Retinoblastoma tumor suppressor and genome stability. Adv. Cancer Res. 85, 13-50
Fridman JS, Lowe SW (2003) Control of apoptosis by p53. Oncogene 22, 9030-9040
Hofseth LJ, Hussain SP, Harris CC (2004) p53: 25 years after its discovery. Trends Pharmacol. Sci. 25, 177-181
Iwakuma T, Lozano G (2003) MDM2, an introduction. Mol Cancer Res. 1, 993-1000
Lowe SW, Sherr CJ (2003) Tumor suppression by Ink4a-Arf: progress and puzzles. Curr. Opin. Genet. Devel. 13, 77-83
Vargas DA, Takahashi S, Ronai Z (2003) Mdm2: a regulator of cell growth and death. Adv. Cancer Res. 89, 1-33
Santarosa M, Ashworth A (2004) Haploinsufficiency for tumour suppressor genes: when you don't need to go all the way. BBA 1654, 105-122
Sherr CJ (2004) Principles of tumor suppression. Cell 116, 235-246

## Box 5.1 Human papilloma viruses

Multiple strains of human papilloma viruses (HPV) occur and are transmitted by sexual and other close contacts. Some strains cause benign lesions of the epidermis and other epithelia, while others cause malignancies, particular of the cervix, the penis, the anogenital region, and likely also of the epidermis and squamous epithelia of the head and neck. The differences between the strains can be traced to differences in the potency of their encoded E6 and E7 proteins. The most prevalent oncogenic strain is HPV16.

Many women and men are infected by HPV, but only some develop cancers. This is partly due to strain differences, and partly to differences in the ability of an individual's immune system to clear the virus and repair the lesions in the infected tissues. It appears that genetic differences in this ability modulate the predisposition towards HPV-induced cancers, in addition to the frequency and kind of exposure.

The viral E6 and E7 proteins inhibit and destabilize the tumor suppressors TP53 and RB1, respectively. Some cancers caused by HPV16 express E6 only, and show disruption of *CDKN2A* encoding p16$^{INK4A}$. Others express E6 and E7 and in these p16$^{INK4A}$ is regularly overexpressed, since it is under feedback control by RB1 and further induced by continuous cell proliferation

HPVs contain circular DNA genomes of ≈7.9 kb encoding 7 early and 2 late proteins with one control region (CR). While they replicate as circles, carcinogenicity requires integration of the viral genome into the cellular genome. Typically, the circle is opened within the E2 gene and several viral genes are lost. E6 and E7 expression is activated by cellular promoters or enhancers (this is pretty exactly the opposite of activation of cellular oncogenes by retroviral insertion):

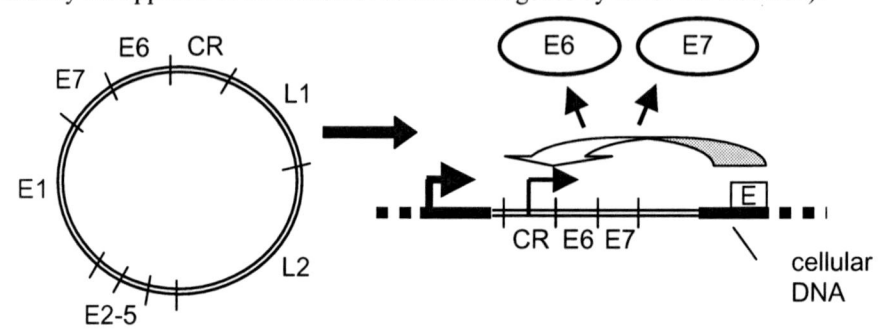

Because HPV infection is the initial step in the development of most cervical cancers, vaccination against HPV seems particularly promising for primary prevention of this cancer type (→20.2) and is under development. Immunotherapy of established cervical cancers has been found to be of little avail, likely because only the E6 and/or E7 viral proteins are consistently expressed in established carcinomas. Conceivably, drug inhibitors of these proteins will be found.

zur Hausen H (2002) Papillomaviruses and cancer: from basic studies to clinical applications. Nat. Rev. Cancer 2, 342-350.

# CHAPTER 6

# CANCER PATHWAYS

- Proliferation, differentiation, and survival of normal cells are regulated by a limited number of pathways, which are partly interlinked. They transmit and integrate signals from growth factors, hormones, cell-cell- and cell-matrix-interactions. The pathways turn into 'cancer pathways' by deregulation. Inappropriate activation of 'cancer pathways' in some cases and inactivation in others are crucial for the development and progression of human cancers. In particular, many proto-oncogenes and tumor suppressors act within or upon 'cancer pathways'.
- MAP kinase pathways transduce signals from the cell membrane to the nucleus, but also to the translational machinery and to the cytoskeleton. The canonical MAPK pathway activated by receptor tyrosine kinases proceeds through RAS proteins, RAF, MEK, and ERK protein kinases. It typically promotes proliferation in response to growth factors. Its activity is modulated by other signals emanating, e.g., from adhesion molecules. Other MAPK pathways mediate stress responses and may alternatively elicit apoptosis. While the functions of the various MAPK pathways are overlapping, the ERK pathway is the most important one in the control of cell proliferation.
- Another frequent target of membrane receptors is phospholipase C, which activates PKC isoenzymes that can stimulate MAPK signaling and other cellular processes. The diverse PKC enzymes are also targets of other factors, including the tumor-promoting phorbol esters.
- Many growth factors, through receptor tyrosine kinases and RAS proteins, activate PI3K, thereby triggering the PDK/PKB/AKT protein kinase cascade that increases cell survival, protein synthesis, and cell proliferation. The PTEN and TSC tumor suppressors are antagonists of this cascade. In many instances, the PI3K pathway acts synergistically with the canonical MAPK pathway.
- MAPK signaling promotes progression of the cell cycle at several distinct steps leading ultimately to phosphorylation and temporary inactivation of RB1 by CDKs. CDKs, Cyclins, CDK phosphatases, and CDK inhibitor proteins form a multi-layered network which ensures orderly cell proliferation, integrates proliferative and inhibitory signals received by the cell, and responds to cellular checkpoints monitoring cell cycle progress and genomic integrity. In human cancers, inappropriate stimulation of the cell cycle by alterations in 'upstream' pathways and/or mutations in immediate regulators of the cell cycle are essential for uncontrolled proliferation.
- Many signals from checkpoints of genomic integrity are channeled through TP53 which elicit cell cycle arrest, apoptosis, or replicative senescence. Inactivation or curtailment of TP53 function by one of various mechanisms is therefore a prerequisite for genomic instability in human cancers.

- Another limit to the expansion of carcinomas is provided by TGFβ signaling. As a rule, TGFβ factors stimulate the proliferation of mesenchymal cells, but inhibit the proliferation of epithelial cells. In response to TGFβ receptor activation, SMAD transcription factors become phosphorylated and translocate to the nucleus where they stimulate the expression of genes encoding e.g. CDK inhibitors or proteins involved in formation and remodeling of the extracellular matrix. Other SMAD factors act as feedback inhibitors. Disruption of this pathway is important during the progression of many carcinomas. TGFβ synthesized by non-responsive carcinoma cells activates tumor stroma and inhibits immune cells.
- Stimulation of cell proliferation by cytokines often proceeds via the JAK/STAT pathway in hematopoetic cells and lymphocytes, and to a lesser degree in epithelial cells. Overactivity of this pathway is responsible for increased proliferation in hematological cancers. Its role in carcinomas is more ambiguous. In some, it is even down-regulated. In this short pathway, JAK protein kinases recruited by cytokine receptors phosphorylate STAT transcription factors which translocate into the nucleus to activate transcription. The JAK/STAT pathway is autoregulated by SOCS proteins.
- Signaling through NFκB also regulates activation and proliferation of immune cells. In epithelial cells, this pathway can protect against apoptosis during cellular stress. Its activation may therefore contribute to prevention of apoptosis in carcinoma cells.
- The patterning of tissues during fetal development involves several specialized regulatory systems, in particular WNT, Hedgehog (SHH), and NOTCH ligands, receptors, and pathways. These pathways remain important in adult tissues for maintenance of stem cell compartments and the regulation of cell fate. Disturbances in these pathways are fundamental for the development of specific human cancers. Thus, colon cancers invariably display constitutive activation of the WNT pathway, basal cell carcinoma of the skin of the SHH pathway, and certain T-cell leukemias of the NOTCH pathway. The responsible alterations are alternatively mutations in proto-oncogenes or tumor suppressor genes.

## 6.1 CANCER PATHWAYS

Since >250 different genes have been demonstrated to be causally involved in human cancers and many more are implicated, it would seem that a bewildering variety of mechanisms may cause human cancer. While this could still turn out to be so, the great majority of currently known oncoproteins can be assigned to a limited number of pathways or at least be demonstrated to act on these. These 'cancer pathways' are, of course, systems that regulate fate, survival, proliferation, differentiation, and function of normal cells as well. In cancer cells, they are accordingly overactive or inactivated to cause the characteristic properties of cancer cells such as uncontrolled proliferation, blocked differentiation, decreased apoptosis, altered tissue structure, etc. (→1.3). Some 'cancer pathways' are rather specifically involved in particular cancers, whereas others play crucial roles in a broad range of malignant tumors. To avoid considering each and every cellular regulatory system as a cancer pathway, a working definition might be formulated as: A 'cancer pathway' is a cellular regulatory system whose activation or inactivation by a genetic or epigenetic mutation is essential for the development of at least one human cancer. Typically, cancer pathways become evident by alterations in different components of the same regulatory system in individual cases of one cancer type or in distinct cancers.

By this latter criterion, several regulatory systems treated in the previous two chapters can be regarded as prototypic cancer pathways (Table 6.1). These comprise the MAPK pathway (→4.4), the TP53 regulatory system (→5.3), and the cell cycle regulatory network centered around RB1 (→5.2). These pathways all interact with each other. Further pathways and proteins are also connected to them, such as the PI3K pathway, the PKC kinases, the STAT pathway, the $NF_\kappa B$ pathway, and the TGFß response pathway. A third group of cancer pathways comprises the WNT and Hedgehog response pathways and the NOTCH regulatory system. They are essential in regulating the shaping and differentiation of tissues during fetal development and remain important even in grown-up humans for the maintenance of tissue homeostasis, particularly in tissues that undergo rapid turnover or frequent regeneration. They are particularly important in specific cancers (→13.2; →12.3).

## 6.2 MAPK SIGNALING AS A CANCER PATHWAY

Signaling through the MAPK pathway composed of RAF, MEK, and ERK proteins (→4.4) is required for the proliferation of normal cells as well as for many cancer cells. In normal cells, it typically relays signals receptor tyrosine kinases (RTKs) activated by growth factors to the nucleus, resulting in the activation of gene expression. It can also be activated by a range of other extra- and intracellular stimuli and it elicits effects besides altered gene expression. In tumor cells, MAPK signaling is often enhanced, e.g. as a consequence of oncogenic activation of RTKs or RAS (→4.4). This MAPK pathway is therefore considered as the 'classical mitogenic cascade' or as the 'canonical' pathway

In reality, this pathway is one of about six similar modules[6] which each consist of a MAP kinase (like ERK1/2), a MAPK kinase (like MEK1/2), and a MAP kinase kinase (like RAF1, ARAF, or BRAF). These other MAPK modules respond likewise to extracellular signals from growth factors, but also to signals from hormones and cytokines acting through G-proteins, to cell adhesion molecules, and stress signals generated in the cell or at the cell membrane. Like the canonical pathway, these modules can stimulate cell proliferation, but also apoptosis, cell differentiation, or specific cell functions. None of them, however, has been as straightforwardly implicated in the control of cell proliferation and in cancer as the canonical pathway.

The best studied of the parallel pathways (Figure 6.1) involves MEKK1, MEK4, and JNK (JUN N-terminal kinase) leading to phosphorylation of the transcriptional

---

[6] The number depends on how one counts incomplete modules.

Table 6.1. *An overview of cancer pathways*

| Pathway or network | Cancers involved in | Oncogene proteins in the pathway | Tumor suppressors in the pathway | Remarks |
|---|---|---|---|---|
| MAPK pathway (canonical) | many | RAS, BRAF, (MYC) | | also effector of oncogenic receptor tyrosine kinases |
| PI3K pathway | many | PI3K, AKT | PTEN, CTMP | also effector of oncogenic receptor tyrosine kinases |
| TP53 network | many | MDM2/HDM2 | | |
| RB1 network | many | Cyclin D1, CDK4, (MYC) | RB1, p16$^{INK4A}$, p15$^{INK4B}$, p57$^{KIP2}$ | |
| TGFβ pathway | carcinomas, selected soft tissue cancers, selected leukemias | | TGFβRII, SMAD2, SMAD4, RUNX | typically inactivated in carcinomas, but activated in soft tissue cancers |
| JAK/STAT pathway | selected carcinomas, many leukemias and lymphomas | STAT3, STAT5(?) | STAT1(?), SOCS1 | also effector of oncogenic cytokine receptors and fusion proteins |
| NFκB pathway | selected leukemias, many carcinomas | REL proteins | CYLD | effect strongly dependent on cell type and context |
| WNT pathway | Carcinomas of colon, liver, breast, stomach and others | WNT1, β-Catenin | APC, AXIN | modulated by E-Cadherin and SFRPs |
| SHH pathway | Specific skin, brain, and lung cancers | SHH(?), SMO, GLI1(?) | PTCH1, PTCH2, SUFU | |
| NOTCH pathway | T-cell lymphomas, carcinomas | NOTCH1 | NOTCH1 | effect strongly dependent on cell type, context, and gene dosage |

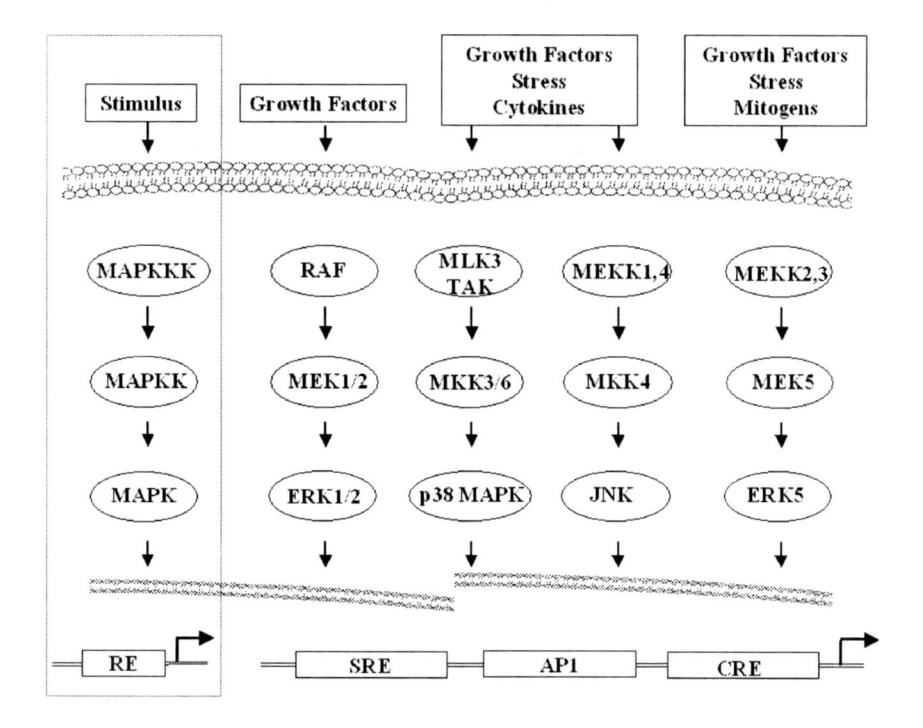

**Figure 6.1** *MAPK modules*
The box on the left illustrated the general setup of a MAPK module. Several actual modules are shown on the right. These are in reality more diverse than shown (e.g., there are five known JNK isozymes) and may 'cross-talk' to each other. See text for further details.

activator JUN. This pathway is activated by RAS proteins, specifically in response to stress resulting from UV irradiation and heat, or to certain cytokines. Alternatively, other GTP-binding proteins such as RAC or CDC42, which respond to cell adhesion signals, activate the pathway through MEKK1,2, or 3. The outcome of activation of this MAPK pathway is more usually a stop to cell proliferation than its stimulation. In some circumstances, it can induce apoptosis. Accordingly, the JNK pathway appears to be important in the progression of some cancers, but more often it is down-regulated than overactive in cancers. Specifically, its components do not seem to constitute a primary target of mutations in human cancers. Interestingly, there are several points of cross-talk between different MAPK pathways, e.g. at the level of MAPKKs. Moreover, JUN is phosphorylated as a consequence of activation of the JNK MAPK pathway, while the protein is transcriptionally induced by the ERK MAPK pathway. So, during normal cell proliferation, these two MAPK pathway synergize.

Both the ERK and the JNK MAPK pathways can be activated through RAS proteins (Figure 6.2). RAS-GTP acts directly at the level of the MEKKs by recruiting RAF proteins or MEKK1 to the cell membrane and relieving their

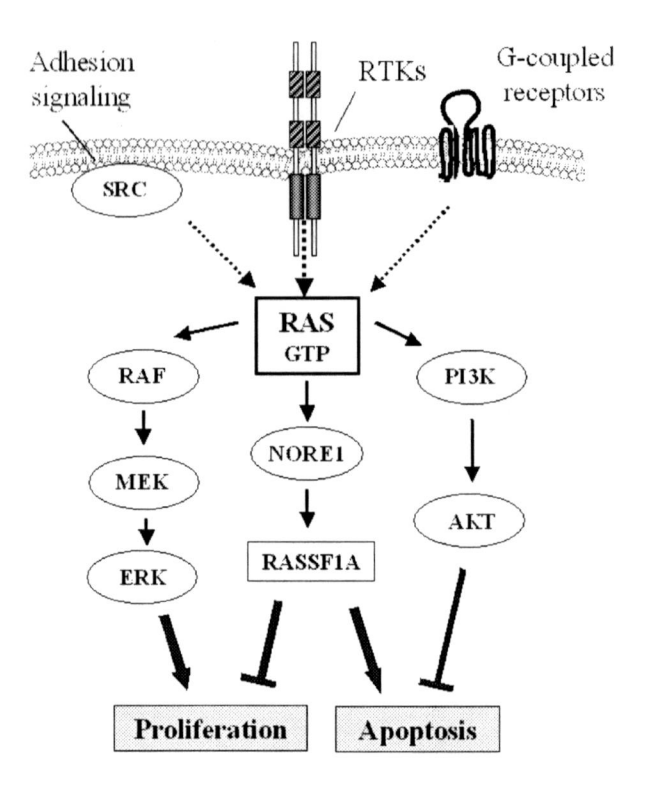

**Figure 6.2** *RAS as a node in cellular signaling*
RAS activation can occur as a consequence of several upstream signals and in turn multiple pathways emerge from activated RAS proteins. Only some are shown. See text for further details.

autoinhibition. RAS proteins also act in an indirect manner on the G-protein RAC through activation of PI3K, a crucial component of a further cancer pathway (→6.3). In addition to RAC, RAS proteins regulate further similar small G-proteins, such as RAL and RHO, by activating their GTP-exchange factors (GEF) or GTPase activating proteins (GAP). Through these routes, RAS is thought to elicit the changes in the cytoskeleton and consequently cell shape, adhesion, and migration which are commonly seen in cells with oncogenic RAS, e.g. in the 3T3 assay system (→4.3).

Clearly, such changes are expected to also influence the behavior of a tumor cell in vivo (→9.2). The action of RAS on the PI3K pathway may be particularly significant in those cancers in which that pathway is not autonomously activated (→6.3). The high prevalence of RAS mutations in human cancers is thus well explained by its function as a nodal regulator of several pathways relevant to essential properties of cancer cells. Among these various effects, the activation of the ERK MAPK pathway is likely most crucial for stimulation of cell proliferation in cancers with RAS mutations, as well as in those with oncogenic activation of receptor tyrosine kinases upstream of RAS. One of several arguments for this tenet are mutually exclusive mutations in *RAS* genes and *BRAF* in several cancers, prominently in melanoma (→12.4).

The activation of the JNK MAPK pathway by RAS points to another facet in the pleitropic function of these proteins. Overactivity of RAS proteins in normal cells can induce growth arrest, apoptosis or replicative senescence. In some cell types, RAS activation may also induce differentiation. So, in addition to their effect on cell proliferation, RAS proteins in their active state also stimulate responses that eventually terminate their effects or under some circumstances evoke fail-safe reactions against hyperproliferation. Depending on the circumstances, the JNK MAPK pathway may act in this fashion. Another mechanism of this kind is induction of *CDKN2A* with increased transcription of p14$^{ARF1}$ and p16$^{INK4A}$ that leads to stabilization of TP53 and slows phosphorylation of RB1 (→5.3). Still another fail-safe mechanism may be exerted through the RASSF1A protein. This protein is encoded by a complex locus on chromosome 3p21.3, binds to microtubuli and may block mitosis. Loss of 3p is a common finding in clear-cell renal carcinoma and many other carcinomas (→15.2). Furthermore, *RASSF1A* is inactivated by promoter hypermethylation in a wide range of human cancers.

Since RAS proteins are almost invariably activated when signals emanate from active growth factor receptors, they mediate many effects of the physiological or oncogenic activation of receptor tyrosine kinases, specifically an activation of the MAPK pathway downstream of RAS proteins. In fact, RAS proteins mediate not only signals by receptor tyrosine kinases, but are also involved in cellular responses to cytokines and hormones acting through G-protein coupled serpentine receptors. Conversely, receptor tyrosine kinases and others modulate the activities of the MAPK, PI3K, and further cancer pathways not only through RAS, but also by diverse other mechanisms (cf. 6.3 and 6.5).

## 6.3 THE PI3K PATHWAY

The PI3K pathway functions in the control of cellular metabolism, particularly that of glucose transport and utilization, in the regulation of cell growth, particularly of protein biosynthesis, and it prevents apoptosis. The pathway (Figure 6.3) is stimulated by RAS proteins and by active receptor tyrosine kinases, but it also feeds back into pathways downstream of RAS. It is distinguished as a cancer pathway, since its major inhibitory component, PTEN, is a classical tumor suppressor. Two other proteins affected in a human tumor syndrome, TSC1 and TSC2, modulate intermediate steps of the pathway. Moreover, in some instances PI3K is implicated as an oncogenic protein. For instance, its gene is amplified in ovarian cancers.

The abbreviation PI3K stands for phosphatidyl inositol phosphate kinase, since the enzyme phosphorylates the membrane phospholipids phosphatidyl inositol (PI), PI(4)P, and PI(4,5)P to yield PI3P, PI(3,4)P, and PI(3,4,5)P, respectively (Figure 6.4). Several different isoenzymes of PI3K exist, of which PI3Kα is the one most relevant in the context of cancer. PI3Kα also has protein kinase activity, but the

120  CHAPTER 6

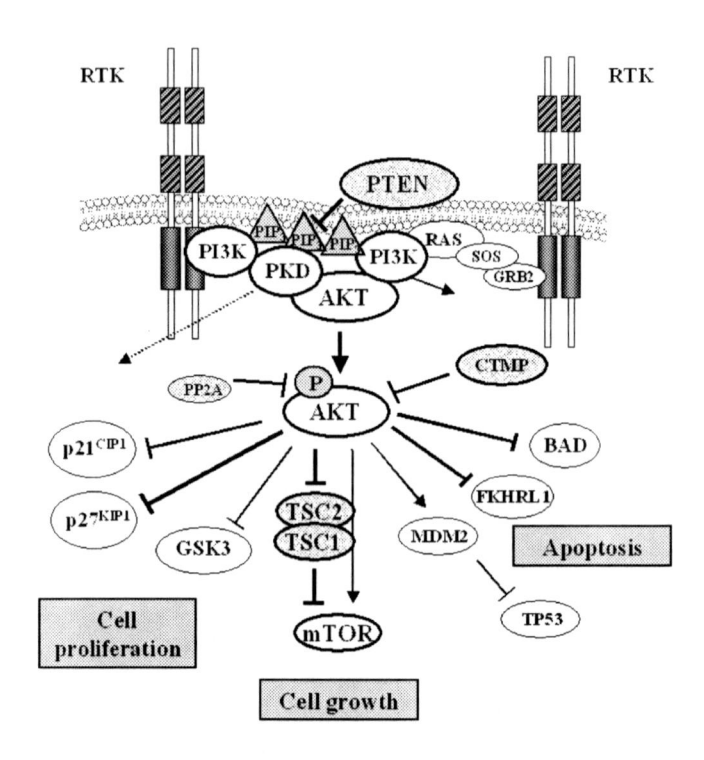

**Figure 6.3** *The PI3K pathway*
Central events in the PI3K pathway promoting cell proliferation and growth or blocking apoptosis; note that PKD and PI3K effects beyond AKT phosphorylation are not detailed. Effects of AKT occur mostly through phosphorylation. Main PI3K pathway components are ringed by bold lines; inhibitory proteins are marked by a grey background.

substrates have not been well defined. It is a heterodimer consisting of a p110 catalytic subunit and a p85 regulatory subunit which in the basal state restricts the activity of the catalytic subunit.

Following receptor tyrosine kinase activation and autophosphorylation, the PI3K heterodimer binds to tyrosine phosphate via the SH2 domain of p85 leading to activation of the catalytic subunit. This is also – independently or synergistically - stimulated by direct interaction with RAS-GTP. Lipid phosphorylation generates PI(3,4,5)P at the inner face of the plasma membrane. This creates binding sites for proteins containing a pleckstrin homology (PH) domain, which in this way become relocated to the membrane. There are several such proteins, among them GEFs for RAC and other RAS-like proteins involved in the organization of the cytoskeleton.

The actual PI3K pathway (Figure 6.3) proceeds through the protein kinases PDK1 and AKT, which is also known as PKB. PDK1 becomes directly activated by binding to PI(3,4,5)P, whereas AKT additionally needs to be phosphorylated by PDK1 to become active. PDK1 phosphorylates further protein kinases, including the

PKC isoenzyme $\xi$ (→6.5) and the p70$^{S6K}$, which is also a target of ERK MAP kinases. The main effector of the pathway, however, is the AKT protein kinase. Following its own activation by PDK1, it can phosphorylate a variety of proteins. One major substrate is mTOR, a protein kinase controlling the initiation of translation at the EIF4 step. So, the combined actions of PDK1 and AKT lead to an increase in protein synthesis. This effect is important in the anabolic action of insulin and insulin-like growth factors (IGFs) in many tissues, but it is also a prerequisite to cell proliferation and therefore central to tumor growth. AKT activity may be antagonized by CTMP (C-terminal membrane protein), a potential tumor suppressor.

The acronym mTOR stands for 'mammalian target of rapamycin', since rapamycin inhibits this kinase specifically. Physiologically, mTOR is inhibited by the TSC1/TSC2 protein complex. TSC1 and TSC2 are also called Tuberin and Hamartin. Either protein is mutated in tuberous sclerosis, an autosomal dominant disorder affecting ≈1/6000 persons. The *TSC* genes, which are located at chromosomes 9q34 and 16p13, respectively, behave as classical tumor suppressors. Thus, in affected persons, one defective copy of a *TSC* gene is inherited. Mutation of the second allele leads to hamartomas, usually small benign tumors consisting of different connective tissue components which occur in several organs including brain, heart, lung and kidneys. Although carcinomas are not usually found in this

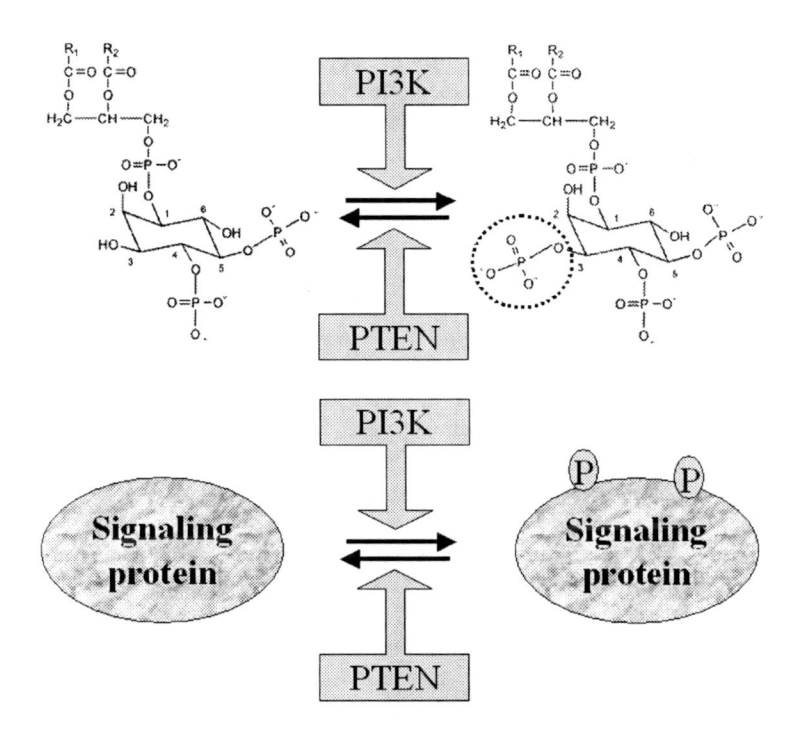

**Figure 6.4** *Reactions catalyzed by PI3K and PTEN*

syndrome, the *TSC1* gene may be involved in bladder and perhaps renal carcinomas. During growth, inhibition of mTOR by the TSC proteins is relieved by phosphorylation of the large TSC2 protein at Ser939 and Thr1462 by the AKT protein kinase. As loss of TSC function leads to overactivity of mTOR, rapamycin is now investigated for the treatment of this condition.

The tumor syndrome caused by mutation of *TSC* genes underlines the importance of activation of protein synthesis and stimulation of cell growth by the PI3K pathway. This pathway has, however, further functions. This is dramatically demonstrated by the hereditary Cowden tumor syndrome caused by mutations in the *PTEN* gene, located at 10q23.3. Its product is a lipid phosphatase and protein phosphatase, as PI3K is a lipid kinase and protein kinase. PTEN antagonizes PI3K by hydrolytically removing phosphates at position 3 of phosphatidyl inositols (Figure 6.4). Like mutations in the TSC genes, PTEN mutations in the germ-line predispose to multiple hamartomas, but unlike tuberous sclerosis, the Cowden syndrome carries a risk for major common cancers, such as colon and breast cancer. Deletions, mutations and inactivation of the *PTEN* gene are as well observed in many different sporadic cancers. The wide range of cancers in which its inactivation is prevalent points to PTEN as one of the most important tumor suppressors. Typically, in sporadic cancer loss of PTEN is associated with tumor progression and a more aggressive clinical course.

The AKT protein kinase not only stimulates protein synthesis but also cell survival and cell proliferation. Cell survival is promoted by several routes. Phosphorylation of forkhead transcription factors like FKHR-L1 prevents their transcriptional activation of proapoptotic and growth-inhibitory genes. Phosphorylation by AKT also inhibits the important apoptosis activator BAD (→7.2). In either case, the proteins phosphorylated by AKT are recognized by 14-3-3 proteins. The protein complex is retained in the cytosol, preventing FKHR-L1 from entering the nucleus and BAD from interacting with mitochondrial proteins, respectively. Through the same mechanism, phosphorylation by AKT limits the activity of the p27$^{KIP1}$ cell cycle inhibitor. TP53 activity may also be diminished, since AKT appears to activate MDM2. Further effects on cell proliferation result from phosphorylation of glycogen synthase kinase 3 (GSK3) by AKT. As implied by its name, this enzyme is an important regulator of glucose metabolism, and a major target of insulin action. In addition, it regulates cell proliferation by phosphorylating and inhibiting Cyclin D1 and MYC. Perhaps its most important function is in the regulation of β-Catenin levels within the WNT pathway (→6.10), which is particularly important in colon cancer (→13.2).

In summary, during normal cell proliferation the PI3K pathway may be essential to complement the MAPK pathway (Figure 6.5). Somewhat simplified, the MAPK pathway mainly stimulates cell proliferation (i.e. DNA synthesis and mitosis), but requires additional signals through the PI3K pathway which provides the necessary stimulus for cell growth (including protein biosynthesis) and, importantly, counteracts the pro-apoptotic effects caused by isolated MAPK stimulation. Appropriate stimulation of cell proliferation by growth factors therefore involves a coordinated activation of these two pathways. Alternatively, certain growth factors preferentially stimulate the PI3K pathway, prominently members of the insulin family, such as IGF1 and IGF2, and act primarily as 'survival growth factors'. They can in this fashion complement growth factors which act predominantly through the MAPK pathway. Accordingly, in tumor cells, mutations leading to the constitutive activation of the MAPK pathway can be complemented by increased activity of survival growth factors or by mutations in the PI3K pathway, e.g. those disabling PTEN function.

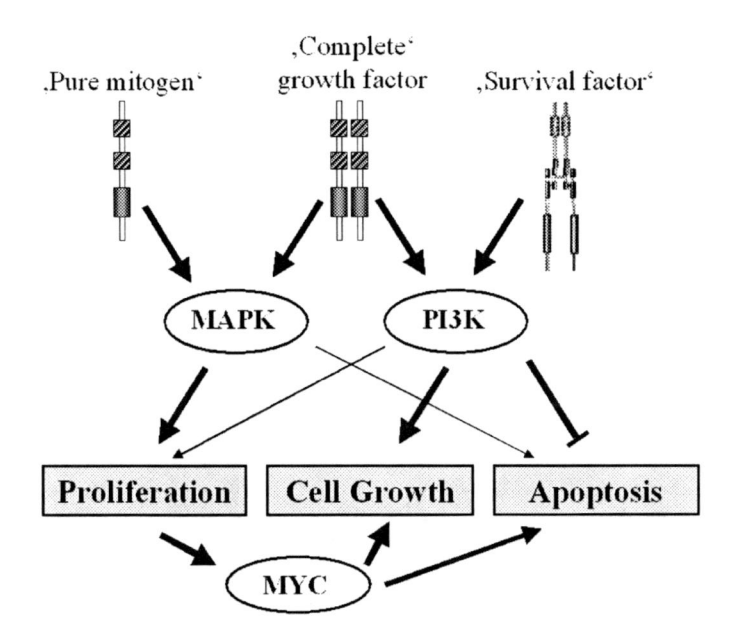

**Figure 6.5** *Synergism between the PI3K and MAPK pathways*
A simplified interpretation of the interaction of the two pathways activated by growth and survival factors acting mostly through receptor tyrosine kinases. See text for details.

## 6.4 REGULATION OF THE CELL CYCLE BY THE MAPK AND PI3K PATHWAYS

Normal cell proliferation thus often involves the coordinate activation of MAPK and PI3K pathways. Together, these pathways elicit a panoply of changes, such as increased metabolism, enhanced protein synthesis, reorganization of the cytoskeleton, inhibition of proapoptotic signals, and, of course, stimulation of cell cycle progression from G1/G0 into S-phase (cf. 5.2). Several mechanisms synergize to achieve this last effect (Figure 6.6).

Following their translocation into the nucleus as a consequence of MAPK activation, activated ERK protein kinases phosphorylate several transcriptional activators involved in cell cycle progression. In addition, ERK kinases

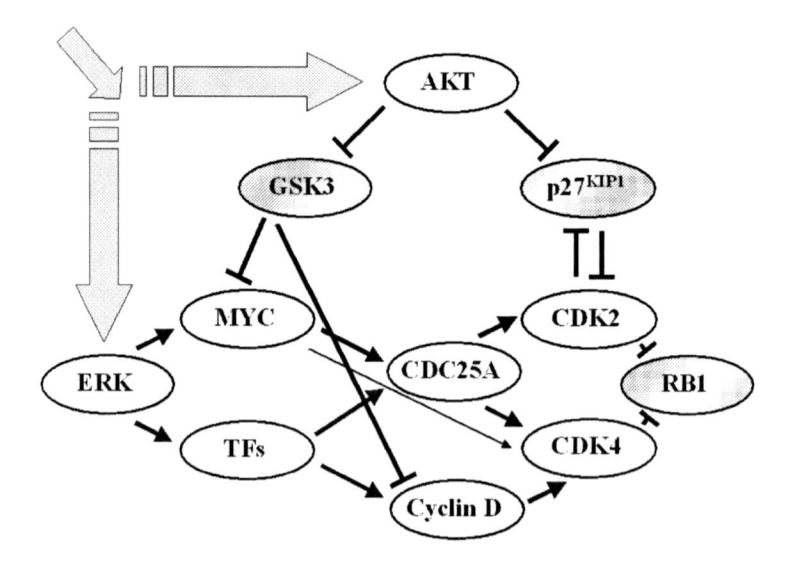

**Figure 6.6** *Cell cycle activation by coordinated action of the MAPK and PI3K pathways*
Following activation by mitogenic signals (top left corner), the pathways interact to stimulate cell cycle progression towards S phase. Negative regulators of the cell cycle are shaded.

phosphorylate the pp90$^{RSK}$ kinases which then as well translocate to the nucleus to phosphorylate a further set of transcriptional activators. Among the more important targets are the SRF, ETS1, ELK1 and MYC factors, which are all stimulated by this type of phosphorylation, and, of course, the AP1 components FOS and JUN (→4.3), which are also induced at the transcriptional level.

In many cells, an important consequence of these events is a stimulation of Cyclin D transcription[7]. Accumulation of Cyclin D drives cell cycle progression during most of the G1 period. It is counteracted by degradation of the protein which is increased if Cyclin D is phosphorylated by GSK3. This inhibitory phosphorylation is blocked by phosphorylation of GSK3 by AKT. The coordinate activation of Cyclin D is thus an important point of synergy between the MAPK and PI3K pathways.

Further genes activated as a consequence of MAPK pathway stimulation are CDC25A and – to a lesser extent - CDK4. Induction of the CDC25A phosphatase synergizes with Cyclin D accumulation to increase CDK4 activity, since the phosphatase removes the inhibitory threonine phosphate from CDK4, as well as from CDK2.

As the CDK4/Cyclin D holoenzyme accumulates, it binds more and more p27$^{KIP1}$, diverting it from CDK2. Since Cyclin E begins to accumulate at this stage of the cell cycle, the activity of the CDK2/Cyclin E enzyme increases

---

[7] In most cell types, this would be Cyclin D1, Cyclin D2 and D3 being involved in a more restricted range of cell types.

consequentially. The Cyclin E gene promoter is activated by E2F1. So, as RB1 starts to become hyperphosphorylated and less capable of inhibiting E2F, more Cyclin E is produced leading to still higher activity of CDK2. This autocatalytic loop eventually leads to irreversible commitment to S-phase, but is limited by $p27^{KIP1}$.

For CDK2 to become fully active, the $p27^{KIP1}$ inhibitor must be removed. This occurs by successive phosphorylation, oligo-ubiquitination, and proteolytic degradation. In normal cells, the down-regulation of $p27^{KIP1}$ is a gradual process which accelerates when CDK2 becomes sufficiently active to phosphorylate some of the inhibitor protein, thereby liberating further CDK2 molecules to phosphorylate more $p27^{KIP1}$, in another autocatalytic loop. A crucial component in this process is MYC. MYC is induced by many growth factors, but its half-life and activity are also regulated by phosphorylation. ERK phosphorylation at Ser62 leads to increased stability and activity, whereas GSK phosphorylation at Thr58 has the opposite effect. So, again, the inhibitory phosphorylation of GSK by AKT synergizes with activation of the MAPK to activate MYC. Activation of MYC contributes to the increased transcription of Cyclin D1, CDK4, and CDC25A, and leads to a small decrease in transcription of $p27^{KIP1}$. Moreover, MYC induces CUL1, a component of the protein complex that ubiquitinates $p27^{KIP1}$. Still another mechanism involved is decreased translation of $p27^{KIP1}$, likely as a consequence of PI3K activation. So, regulation of MYC and $p27^{KIP1}$ are further points of synergy between the pathways.

The synergisms exerted on the regulation of the cell cycle by the concomitant activation of the MAPK and AKT pathways during normal cell proliferation are reflected in the alterations observed in cancers (Figure 6.7). Overactive receptor tyrosine kinases may be such potent oncogenes, since they are capable of activating both pathways. Oncogenic RAS may likewise activate both pathways, albeit less efficiently. Conversely, oncogenic MYC itself, while very efficient in promoting cell growth (also in a broader sense) and stimulating cell cycle progression, also tends to drive cells into apoptosis, which is again counteracted by factors that activate the PI3K pathway, such as IGF peptides or RAS mutations. This explains the cooperation of RAS-like and MYC-like oncogenes in rodent cell transformation assays (→4.3). This cooperativity is also at work in many human cancers.

However, these combinations of alterations are probably not sufficient to cause human cancers, since they are counteracted by further fail-safe mechanisms. The most important one may be activation of TP53 by hyperproliferation signals and through accumulation of CDK inhibitors like $p16^{INK4A}$ and $p21^{CIP1}$ during continuous rapid growth of human cells. These fail-safe mechanisms lead to apoptosis in some cases, but more typically to replicative senescence (→7). Thus, at some point in the development of human cancers, latest during progression, these mechanisms need to be inactivated. The most radical way for their inactivation is loss of both RB1 and TP53. Another frequent mechanism is deletion of the *CDKN2A* locus encoding $p16^{INK4A}$ and $p14^{ARF}$. Of note, while the prime significance of these alterations may be the inactivation of fail-safe mechanisms against hyperproliferation, they also decrease the dependence of tumor cells on external growth factors and increase their tolerance of genomic instability.

126                                   CHAPTER 6

So, alterations that lead to an increased or constitutive activity of the MAPK and PI3K pathways may in some cases be self-limiting. At the least, they exert a strong selection pressure favoring further changes that lead to progression towards a more aggressive phenotype. This sort of progression is observed in many human cancers, in carcinomas (e.g. 13, 14, 19) as well as hematological cancers (→10).

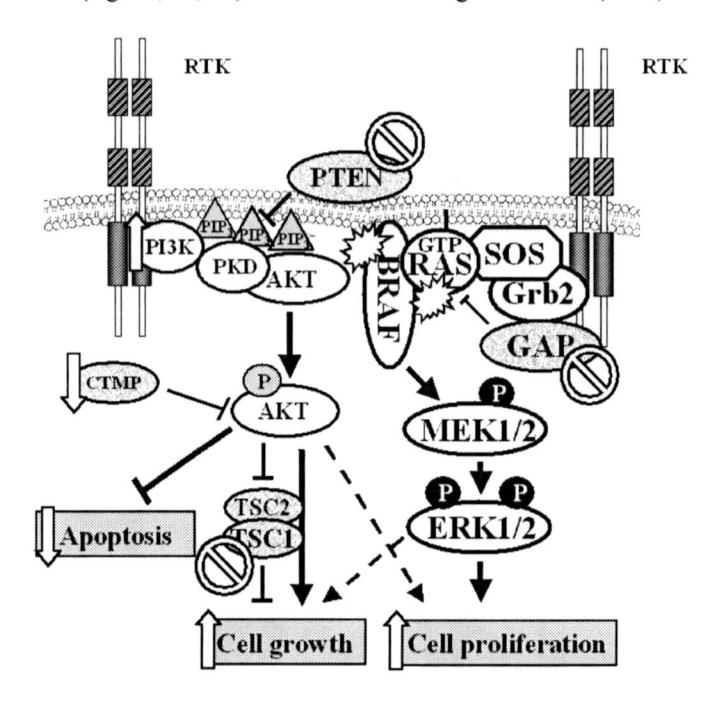

**Figure 6.7** *An overview of alterations in MAPK and PI3K pathways in human cancers*
Alterations in the MAPK (right) and PI3K (left) pathways synergize to increase cell proliferation and growth and to decrease apoptosis. Prohibition sign: inactivation by deletion, mutation and/or promotor hypermethylation. Exploding star: Activation by mutation. Arrow down: decrease; arrow up: increase. Activities of both pathways are also induced by receptor tyrosine kinase overactivity or can partly be mimiced by oncogenic activation of transcription factors, especially of the MYC family (not shown).

## 6.5 MODULATORS OF THE MAPK AND PI3K PATHWAYS

The MAPK and PI3K pathways have multiple effects in the cell, influencing a variety of protein kinases, transcription factors, cytoskeletal proteins, proteins acting at the ribosome, and apoptotic regulators. Conversely, they are activated not only by signals from receptor tyrosine kinases, but also modulated by many others. These emerge, e.g., from other types of receptors, such as cytokine receptors and G-protein coupled serpentine receptors, as well as from various cell adhesion molecules.

This part of intracellular signaling involves a number of additional protein kinases and other enzymes. Several have in fact been discovered in substantially altered form as retroviral oncogenes, but in human cancers they rarely appear as oncogenes or tumor suppressors fitting the strict definitions proposed at the ends of the chapters 4 and 5. However, these proteins are clearly important in human cancers for the establishment of typical properties of tumor cells and as modulators of the MAPK and PI3K 'cancer pathways'.

Most receptor tyrosine kinases (RTK) and some G-protein coupled receptors (GPCR) activate the MAPK cascade through RAS activation. Many GPCRs and some RTKs employ phospholipases in this activation, typically PLCβ in the case of GPCRs and PLCγ in the case of RTKs. These phospholipases cleave phosphatidyl inositol bisphosphate in the membrane to yield diacylglyerol and inositol triphosphate (IP$_3$). IP$_3$ is a second messenger molecule that regulates cytosolic Ca$^{2+}$ levels, which again affect many different cellular functions, including metabolism, transcription, and the state of the cytoskeleton.

The diacylglycerol liberated by the PLC reaction is also a signal molecule which binds and activates protein kinases of a class named PKC. It comprises >10 members in humans, with different tissue distributions and specificities. The most widely distributed member is PKCα. This enzyme, like the 'classical' PKCβ and PKCγ isoenzymes is activated synergistically by diacylglycerol and Ca$^{2+}$, other isoenzymes such as the likewise widely distributed PKCδ are independent of Ca$^{2+}$ levels. Still another class represented by PKCξ are independent of both activators. The structure of the PKCs is modular like that of many other protein kinases. A distal catalytic domain containing the ATP and substrate binding site is normally inhibited by a pseudosubstrate in the proximal regulatory domain. In the classical PKC isoenzymes, the regulatory region contains binding sites for diacylglycerol and for calcium ions. Binding of both activators relieves the autoinhibition. Like RAF, PDK, and AKT, many PKCs are activated at the inner face of the plasma membrane.

The diacylglycerol binding site in PKCs also recognizes phorbol esters like tetradecanoyl phorbol acetate (TPA), which is also called phorbol myristyl acetate (PMA). This compound from plants is a strong irritant and stimulates the proliferation of some cells. In animal experiments, it promotes tumor growth in the skin, although it is not mutagenic. It is thus the prototype of a 'tumor promoter'. However, the effect is dependent on the tissue and the species. This may relate to the type of PKC present and how important it is for the regulation of proliferation in a particular tissue. PKCs may also bind and react to other ligands.

PKCs are serine/threonine protein kinases with a (wide) range of their substrates, differing between the individual isoenzymes. Since PKCs such as PKCα and PKCε can phosphorylate RAF, they can mediate a stimulation of the MAPK cascade independent of or synergistically with RAS. By phosphorylating cytoskeletal proteins like MARCKS, they affect the structure of the actin cytoskeleton. This influences cell motility and transport processes which are also directly regulated by PKCs. PKCs also phosphorylate nuclear receptors like the VDR. Last, not least PKCs can phosphorylate and regulate receptor tyrosine kinases, including the EGFR.

In effect, PKCs are typically used in the cell to relay signals from one activated pathway to others, i.e. in 'cross-talk'. This may be more of a coordinating than a determining task, which could explain why they have not appeared as 'true' oncogenes in spite of their close connection to the control of cell proliferation and motility. However, inhibition of PKCs does block the proliferation of some cancer cells and clearly, activation by phorbol esters can principally stimulate the proliferation of normal and particularly transformed cells. Moreover, PLCs and PKCs are apparently critical mediators of the effect of several oncogenes, including retroviral oncogenes like yes and cellular oncogenes like TRK that act in lymphoid cells. They may also be necessary in those (rarer) cases, where the proliferation of cancers is caused by G-coupled receptors.

PKCs are also participants in the cross-talk in the cell between cell adhesion molecules and growth-regulatory pathways. Signals emerge from cell-cell-adhesion as well as from cell adhesion to the substrate. Cell-cell-adhesion is often mediated by cadherins which are connected to the cytoskeleton and influence its activity (→9.2). Different cadherins, to different extents, activate or inhibit cancer pathways. E-Cadherin, the major cadherin in many epithelia, binds β-Catenin and may thereby limit the activity of the WNT/β-Catenin pathway (→6.10) in some cell types. Likewise, E-Cadherin activates the PI3K pathway to yield a robust survival signal. Both signals are reasonable for a cell firmly integrated into a relatively quiescent epithelium. In some (but not all) invasive carcinomas, E-Cadherin is down-regulated or becomes replaced by other cadherins that activate different pathways.

Integrins (→9.2) mediate cell-matrix interactions by binding to extracellular fibronectin. At the inner surface of the cell membrane, they provide a focus to organize the attachment of the actin cytoskeleton at focal adhesion contacts. Focal adhesion contacts are the hub of a signaling network that influences the cytoskeleton, cell proliferation and survival (Figure 6.8). The integrin-linked kinase ILK regulates and signals the connection between integrins and the cytoskeleton. The focal adhesion kinase FAK then signals the establishment of the focal adhesions. It is autophosphorylated and can accordingly be recognized by SH2 domains in SRC protein kinases.

SRC kinases are the human orthologs of the v-src gene product of RSV (→4.1). They are located to the inner face of the plasma membrane by a myristyl anchor. Upon binding to FAK they become activated and further phosphorylate this kinase in turn. Activated SRCs behave very similar to activated receptor tyrosine kinases, binding adaptor proteins like SHC/GRB2, phosphorylating and activating a number of proteins, including RAS as well as PLCs and consequentially PKCs, and leading to stimulation of the MAPK and PI3K pathways. Integrins are also capable of activating PKCs directly and of interacting with other small GTPases like RAC that regulate the structure of the cytoskeleton.

SRC kinases are often overexpressed and overactive in human cancers, although not mutated like their viral counterpart. It appears that the activation of these enzymes is often associated with advanced stages of carcinomas, where cells become highly invasive and migratory. Increased activity of FAK and ILK is also observed in such cancers. It is not entirely clear whether the increased activity of these enzymes is cause or effect of increased motility and altered attachment. However, in either case, the activation is likely to be necessary, as cell proliferation requires an attachment signal. Stimulation of proliferation without such a signal may lead to anoikis, a specific type of programmed cell death. In a cancer cell with altered attachment, activation of the signaling pathways from adhesion molecules may be needed to compensate. Like in their subversion of the TP53/RB1 checkpoint mechanisms (→6.4), during their progression cancer cells appear to acquire the ability to circumvent this additional barrier against inappropriate proliferation.

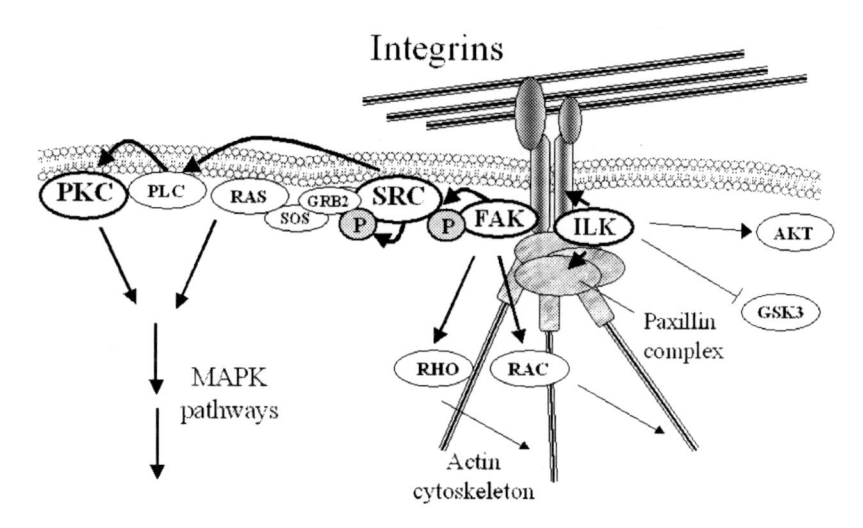

**Figure 6.8** *Signaling from integrins*
See text for explanation.

## 6.6 THE TP53 NETWORK

The TP53 protein is the central node in a network that is linked to several others both upstream and downstream of TP53 (→Figure 5.6). Upstream events that modulate TP53 function include prominently activation of protein kinases that sense DNA damage such as ATM and DNA-dependent protein kinase (→3.3). They phosphorylate TP53 at its N-terminus, increasing its transcriptional activity and decreasing its sensitivity to inhibition by HDM2, in the process enhancing also its half-life. Other types of cellular stress, such as hypoxia or imbalances in nucleotide metabolism, e.g. following treatment with cytostatic antimetabolites, also activate TP53.

A further set of mechanisms controls the activation of TP53 upon inappropriate cell proliferation. The relationship is in fact intricate. In many cells that enter the cell cycle, the TP53 gene is more strongly transcribed. This may be regarded as a kind of precautionary measure providing sufficient TP53 for proper checkpoint function.

The activity of TP53 is limited during normal cell cycles through phosphorylation of its C-terminal domain by G1 and G2 CDKs and by the cell-cycle dependent casein kinase II (→Figure 5.7). Moreover, TP53 accumulation is prevented by transport out of the nucleus, poly-ubiquitination and degradation mediated by HDM2. Phosphorylation of TP53 as a consequence of DNA damage overcomes these restraints. Similarly, checkpoint failures, e.g. incomplete replication noticed at the G2→M boundary, activate checkpoint-dependent kinases such as CHK1 and CHK2 which also phosphorylate the TP53 N-terminus blocking HDM2 interaction and alleviating the effects of inhibitory phosphorylation. Defects in the RB1-dependent restriction point at the G1→S boundary, caused e.g. by oncogenic MYC or RAS, lead to increased transcription of p14$^{ARF1}$ which blocks HDM2 and thereby increases TP53 activity and concentration.

There are also several different pathways downstream of the TP53 node (→Figure 5.6). TP53 induces cell cycle inhibitors, prominently p21$^{CIP1}$ in G1 and 14-3-3σ in G2 which block the cell cycle. The extent of TP53 activation is, of course, limited by its induction of HDM2 (→5.3). It seems that induction of apoptosis by TP53 through the intrinsic and extrinsic pathway (→7.2) is induced in a competitive fashion to cell cycle inhibition. So, the outcome of TP53 activation may not only depend on its extent, but also on the circumstances of a cell that determine whether cell cycle arrest or apoptosis are first in place, each preventing the other. One can imagine, e.g., that a cell with strong expression of apoptotic inhibitors like BCL2 may undergo cell cycle arrest, while a cell that is exposed to ligands of TNFRSF receptors (→7.2) which are induced by TP53 may undergo apoptosis. Importantly, cell cycle arrest induced by TP53 may often be irreversible and take the form of 'replicative senescence' (→7.4).

The fate of a cell in response to TP53 is also determined by the activity of several other 'cancer pathways'. Many of these interactions are known, but not all are understood. So, increased activity of ERK and JNK MAPK pathways tends to activate TP53. Similar, increased activity of the WNT pathway in colon cancer (→13.3) and particularly in liver cancer (→16.2) appears to be incompatible with intact TP53 over a longer period. Conversely, activation of the PI3K and NF$_κ$B pathways in cancers tends to counteract the effects of TP53. In the case of the PI3K pathway, this may be due to the activation of HDM2/MDM2 by AKT and to increased 'survival signaling' at large (→6.3) which may compensate for decreased signaling by survival factors caused by TP53. Among others, TP53 strongly induces IGFBP3, a secreted protein that binds and sequesters IGFs. As IGFs act largely through the PI3K pathway, autonomous activation of this pathway, e.g. by loss of PTEN function, will obliterate this branch of TP53 action. While this change will diminish the proapoptotic effect of TP53, AKT phosphorylation of p21$^{CIP1}$ moderates the influence of TP53 on growth arrest. So, overactivity of the PI3K pathway will limit the effects of TP53 activation on cell growth and survival. In comparison, overactivity of the NF$_κ$B pathway found in many cancers (→6.9) is expected to more selectively impede apoptosis induced by TP53. The induction of antiapoptotic proteins like BCL-X$_L$ and IAPs (→7.2) may be a key element in this protective effect. AKT also phosphorylates and activates NF$_κ$B factors.

Finally, the targets of activated TP53 include genes involved in cellular metabolism, in the shaping of the extracellular matrix and of angiogenesis. This facet of TP53 action would be expected to counteract the effects that tumor cells exert on their environment.

The diverse effector functions of TP53 thus relate to typical aspects of the tumor phenotype, in a surprisingly consistent one-to-one fashion. Conversely, TP53 is activated as a consequence of a diverse set of aberrations which have the common denominator of potentially leading to cancer. These relationships explain why the TP53 network has such a strongly nodal character. As a consequence, loss of TP53 as such is the most common alteration in the network observed in human cancers and perhaps the most common alteration overall. Alterations in its immediate regulators like HDM2 and p14$^{ARF1}$ are still relatively frequent. Loss of function of the protein kinase activators which activate TP53 such as ATM or CHK1 is found in hereditary diseases associated with an increased predisposition to cancer (→3.4), but more rarely in sporadic cancers. This may mean that the most crucial property of TP53 in established cancers is its ability to respond to hyperproliferation. Alterations in the effectors downstream of TP53 may not have the same impact as TP53 loss itself. Where they occur, they appear to target primarily effectors of TP53 induced apoptosis. So, mutations of BAX and promoter hypermethylation of APAF1 and other presumed apoptotic mediators of TP53 are found in some cancers. Hypermethylation also inactivates 14-3-3σ which is involved in G2 arrest induced by TP53, but this protein has additional functions.

## 6.7 SIGNALING BY TGFβ FACTORS

The TGFβ superfamily of growth factors comprises ≈30 members with diverse functions. For instance, members of the family such as mullerian inhibitory factor or bone morphogenetic proteins (BMPs) shape tissues during development.

In the context of cancer, TGFβ1 has drawn the main interest, as it is a potent growth inhibitory factor for epithelial cells and because mutations in the pathway that relays TGFβ signals in the cell are frequent during the progression of many carcinomas. TGFβ1 is also an inhibitory factor for many cells of the immune system. In contrast, it stimulates the proliferation of many mesenchymal cell types. This last property had originally led to the designation 'transforming growth factor', because the EGF-like TGFα and TGFβ cooperate to stimulate anchorage-independent growth of mesenchymal cells in culture, which is an indication of a transformed phenotype.

In carcinoma tissues, not only the loss of response to TGFβ in the tumor cell, but also the effects on immune and mesenchymal cells are significant. TGFβ produced by carcinoma cells or liberated from preprotein forms bound to the extracellular matrix inhibits the immune response against the cancer, while stimulating stromal cells to proliferate and to lay down extracellular matrix on which the carcinoma can extend its growth. This relationship may explain why defective responses to TGFβ often portent the onset of invasion and metastasis in the course of carcinoma development. In particular, the ability of carcinoma cells to secrete or activate TGFβ for other cells while being themselves unresponsive to its growth-inhibitory effect helps to establish an environment favorable for the establishment and expansion of metastases, e.g. in prostate cancers (→19.4).

The main intracellular pathway for TGFβ signaling is surprisingly straightforward (Figure 6.9). The growth factor associates with receptor type II at the cell membrane to form a trimeric complex with receptor type I. Receptor type I becomes phosphorylated and is itself activated as a serine/threonine kinase. It phosphorylates SMAD2 or SMAD3 proteins which are presented by the SARA protein at the inner face of the cell membrane. The phosphorylated SMADs are set free and heterodimerize with SMAD4. Together they are transported into the nucleus where they activate or repress various genes. SMAD factors are not very potent transactivators and their action is strongly dependent on the interaction with other transcription factors binding in their vicinity. This may partly explain why their action is context-dependent and different among various cell types.

Activation of the pathway is counteracted by the inhibitory SMADs, SMAD6 and SMAD7. These are encoded by genes induced by TGFβ or BMP signaling, leading to a feedback inhibition. SMAD2 and SMAD3 are categorized as R-SMADs, 'R' denoting receptor-regulated. Further R-SMADS, SMAD1,5, and 8 are activated by distinct receptors for BMPs, which otherwise function in the same manner. Signaling through the pathway can be down-regulated by ERK and JNK kinases which phosphorylate R-SMADs. Conversely, MAPK pathways are also activated by some members of the TGFβ receptor family. Further cross-talk takes place between TGFβ signaling and the canonical WNT pathway (→6.10). In general, these pathways tend to inhibit each other.

Loss of responsiveness to TGFβ in carcinomas can be brought about by alterations at several steps of the pathway. They include receptor mutations, loss of R-SMADs or Co-SMADs by deletions or their inactivation by mutation. Such alterations are, e.g., observed in invasive colon carcinomas (→13.3) or in metastatic prostate carcinomas (→19.4).

## 6.8 SIGNALING THROUGH STAT FACTORS

Signaling through STAT factors in human cancers can also have quite different consequences. As in SMAD factor signaling, this is partly due to varying and 'context-dependent' effects of the >7 different STAT factors in different cell types. In cells of the hematopoetic system, STAT signaling pathway promotes cell proliferation in response to interleukins and other cell-type specific growth factors and cytokines, e.g. GM-CSF and erythropoetin. STAT signaling is also activated in epithelial cells as one of several pathway in the action of growth factors such as EGF and the cytokine IL-6 as wells as by protein kinases relaying adhesion signals, such as SRC, and of activated ABL kinase. STAT factors also mediate interferon effects. So, in hematological cancers, constitutive activation of STAT signaling is a common finding, e.g. in chronic myelogenic leukemia (→10.4). Here, STAT5 is the major factor involved. In some carcinomas as well, activation of STAT pathways is found, e.g. in squamous cell carcinomas of the head and neck as a consequence of

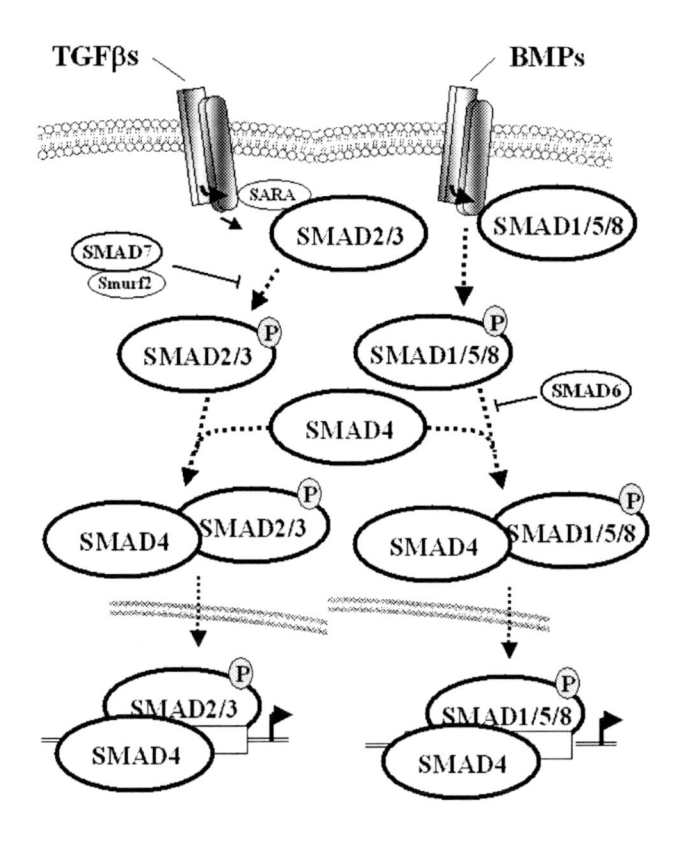

**Figure 6.9** *Intracellular signalling by TGFβ and BMPs*
See text for explanation.

overexpression of the EGFR prevalent in this cancer type. Here, STAT3 may be the most relevant member of the family. In other carcinomas, STAT signaling appears to be actively repressed, likely in conjunction with the inhibition of the cellular response to interferons. The down-regulation mainly concerns STAT1.

The STAT signaling pathway is similarly straightforward as the SMAD pathway (Figure 6.10). Like R-SMADs (→6.7), STATs are phosphorylated at the cell membrane and travel to the nucleus to act as transcriptional activators. Upon binding of their ligand, many cytokine receptors, e.g. the IL-6 receptor gp130, begin to recruit janus kinases (JAK1 or JAK2) which phosphorylate the receptor protein at tyrosine residues. These phosphotyrosines serve as the binding sites for the SH2-domain of the various STAT proteins, predominantly STAT1, STAT3, or STAT5. Upon binding to the receptor/JAK complex, the STAT factors also become activated by phosphorylation. The phosphorylated factors homo- or heterodimerize and travel to the nucleus to bind to specific recognition sites in gene promoters. A typical binding sequence for STAT factors is the GAS element mediating γ-interferon

(IFNγ) responses; the slightly different ISRE elements respond to IFNα/β. Like SMAD factors, STAT proteins are not strong transcriptional activators and usually combine with other factors binding to neighboring sequences for efficient transcriptional activation. Among the STAT target genes are those encoding SOCS factors which act as JAK inhibitors and contribute to termination of STAT signals. This termination is supported by dephosphorylation through SHP protein phosphatases. The activity of STAT factors is moreover cross-regulated by protein kinases of the MAPK and PI3K pathways.

Constitutive activation of STAT pathways in hematological cancers is achieved by several mechanisms. In some leukemias STAT factors are part of fusion proteins (→10.2), in others they are indirectly activated by other fusions events such as the formation of the BCR-ABL protein (→10.4). Inactivation of SOCS genes, e.g. by promoter hypermethylation, may compound such changes. This change is accordingly also found in carcinomas with constitutive STAT activation resulting from receptor tyrosine kinase activation. In contrast, in some other carcinomas, expression of STAT factors, specifically STAT1, is down-regulated, perhaps as cells become unresponsive to the growth-inhibitory effects of interferons.

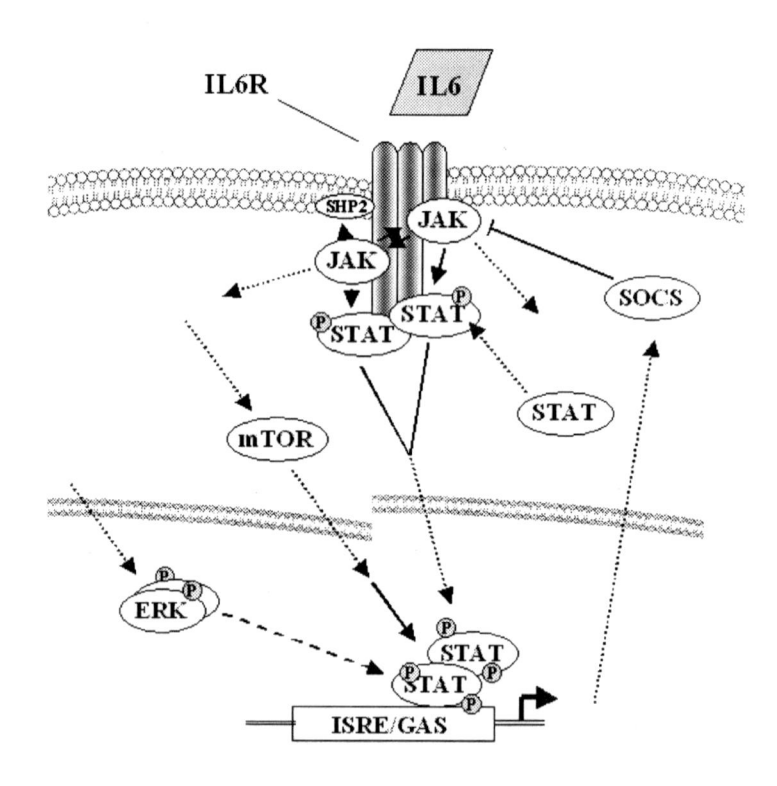

**Figure 6.10** *Activation of the JAK/STAT pathway by IL6*
See text for details.

## 6.9 THE NFκB PATHWAY

Similar to the STAT pathway, the NFκB pathway is not unambiguously a pathway whose activation promotes cancer, and like the STAT pathway, it has different functions in hematopoetic compared to other cell types. The primary functions of the NFκB pathway are the regulation of activation of lymphoid cells, of inflammation and of apoptosis. However, in some cell types, it is also involved in the regulation of cell proliferation, specifically in response to cytokines.

The designation NFκB means nuclear factor regulating the expression of the Ig kappa chain in B cells. NFκB factors are heterodimeric transcription composed of a larger REL subunit and a smaller p50 or p52 subunit. The REL subunits comprise RELA, RELB and c-REL. RELA is the most abundant and widespread of these and therefore commonly known as p65. It is also the strongest transcriptional activator. RELA and c-REL form heterodimers with p50, which is derived from a larger p105 precursor. These heterodimers are normally retained in the cytoplasm by a protein inhibitor IκB (Figure 6.11). RELB associates with the precursor of the p52 subunit, p100, which also contains an inhibitory domain similar to IκBs. In this fashion, the RELB-pre-p52 complex is also retained in the cytoplasm.

Activation of the transcription can be elicited by different signals, including reactive oxygen species like singlet oxygen (→Box 1.1) and other types of cellular stress, and in a more controlled fashion by cytokine receptors. In lymphoid cells, these would be receptors for interleukins or co-receptors for activation. In macrophages, they could be receptors for cytokines or bacterial lipopolysaccharides acting through toll-like-receptors. Activation of NFκB factors in many cell types is also typically elicited by receptors of the TNFRSF family (→7.2) which regulate apoptosis, and this activation modulates the efficiency of the response. Cytokine actions on epithelial cells also involve NFκB factors in addition to STAT factors and cross-talk with MAPK pathways.

The two different types of cytosolic complexes are activated by two similar mechanisms. A multiprotein complex which contains the IKK protein kinase phosphorylates the inhibitory subunit, which in the case of the p65/p50 complex permits immediately transport into the nucleus. In the case of RELB/p52, additional phosphorylation by the NIK protein kinase contained in the complex initiates the proteolytic removal of the inhibitory domain of the p52 pre-protein, and the complex can enter into the nucleus.

Four kinds of target genes of NFκB factors can be distinguished. (1) As in STAT and TGFβ responses, feedback inhibitors are induced, in this case IκB proteins. (2) A limited but illustrous set of regulators of cell proliferation are induced, prominently MYC and Cyclin D1. (3) A much larger set of proteins is induced that modulate apoptosis. Most, like BCL-$X_L$ and FLIP are anti-apoptotic, while others like the FAS receptor and its ligand (CD95 and CD95L) are pro-apoptotic (→7.2). (4) An even wider set of proteins is induced in immune cells, but also non-immune cells that relate to immune function and particularly inflammation. These comprise proteins for cell-cell communication, e.g. adhesion molecules, chemokines, and cytokines, but also factors involved in specific or unspecific immune responses.

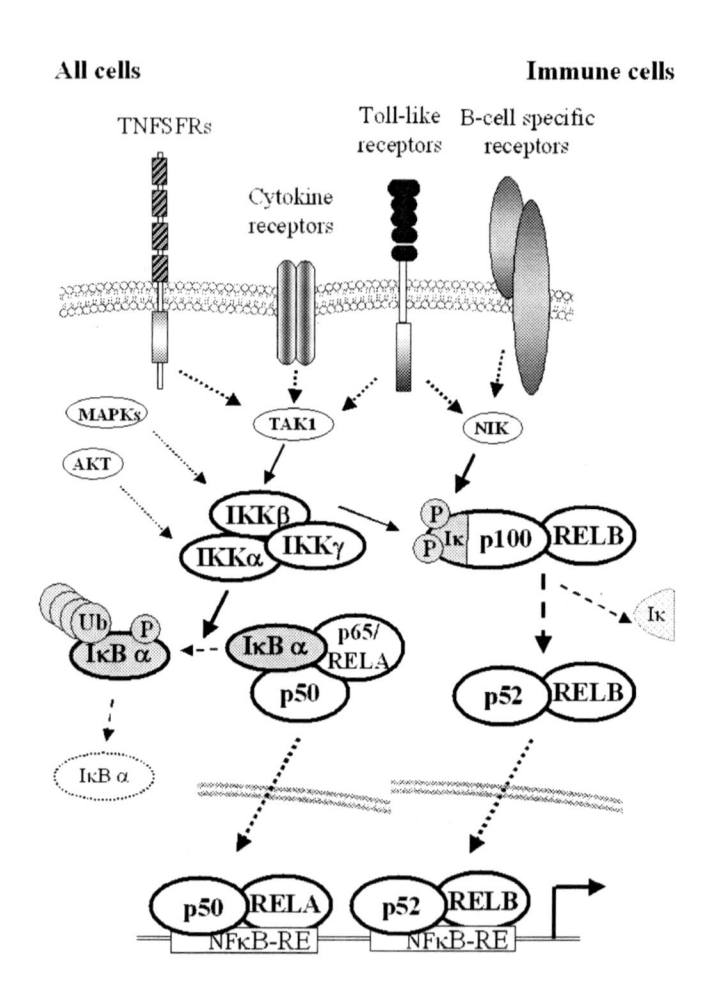

**Figure 6.11 An outline of NF$_\kappa$B pathways**
The scheme emphasizes the activation of cytoplasmic NF$_\kappa$B proteins to nuclear transcriptional activators. The events leading to their activation by a variety of signals are considerably more complicated than shown here. Moreover, they differ between cell types.

Specifically, NFκB factors are the main mediators of the induction of the inducible nitrogen oxide synthase (iNOS) and of cyclooxygenase 2 (COX2), which contribute to the burst in oxidative radicals in immune defense, but also in inflammation.

This last function of the NFκB pathway may perhaps be the most problematic during carcinogenesis. Activity of the pathway is a requirement for inflammation and specifically for chronic inflammation. So, overactivity of the pathway is crucial in the development of cancers associated with chronic inflammation (→13.6, →16.3), and in some cases, is relatively specifically elicited by the pathogenetic agent, such as *Helicobacter pylori* in the stomach (→17.3). Consequentially, agents that block activation of the pathway can diminish the extent of inflammation and prevent such cancers. Some commonly used non-steroidal anti-inflammatory drugs (NSAIDs) like acetylsalicylic acid (aspirin) inhibit the IKK.

Obviously, the antiapoptotic effect of NFκB factors is also problematic, if the pathway becomes constitutively activated in cancer cells. Normally, this antiapoptotic effect appears to be employed to select and protect lymphoid cells during immune defenses. So, translocations and amplifications of REL genes in various human lymphomas may act through this mechanism, but in these cells the pathway also tends to stimulate proliferation. The retroviral orthologue of the REL genes, v-rel (→4.1) is a potent oncogene causing leukemias and lymphomas.

The effect of NFκB activation in carcinomas is less well understood. Certainly, its effect on apoptosis may be relevant. Even the induction of the FAS ligand by these factors may be employed by the carcinoma cells for a counterattack against the immune system (→7.3). However, NFκB could also influence proliferation, particularly in metastases. Some cancers may establish metastases in certain organs by becoming responsive to local growth factors, including chemokines and cytokines. For instance, breast and prostate cancers metastasize preferentially to bone by a sort of mimicry in which they assume some properties of osteoblasts or osteoclasts, including responsiveness to growth factors for these cells that act partly through NFκB. This is, however, only one part of a much more complex interaction (→9.6, →19.4).

Finally, the pathway is implicated in a human dominant cancer syndrome. The rare disease cylindromatosis is caused by inherited mutations in the *CYLD* gene at chromosome 16q. It behaves like a classical tumor suppressor gene (→5.1). Mutations in the second allele lead to tumors of the hair follicles and sweat glands. The protein encoded by the *CYLD* gene is a component of the IKK protein complex. It appears to regulate the duration of the NFκB pathway activation by cytokines like TNFα by de-ubiquitination of a regulatory subunit of the TNF receptor complex. Treatment of the disease and tumor prevention is now attempted by inhibitors of the NFκB pathway.

## 6.10 DEVELOPMENTAL REGULATORY SYSTEMS AS CANCER PATHWAYS

As in adult tissues, signaling through MAPK and other pathways is involved in the regulation of cell proliferation, cell differentiation and apoptosis during ontogenetic development. However, the development of an embryo is, of course, neither a homeostatic replacement nor a simple expansion, but involves many decisions on the developmental fates of individual cells and of cell populations.

From the pluripotent cells in the epiblast, some are developed into primordial germ cells, while others form the germ layers which interact further to form tissues and organs. This requires again cell fate decisions as well as an excellent coordination between the expansion of more or less committed precursor populations by cell proliferation and their differentiation into temporally quiescent or terminally differentiated cells.

In many developing tissues, cells are set aside to form stem cell or precursor populations that allow a continuous supply of differentiated cells exerting specific tissue functions or at least a replenishment of the differentiated cell compartments after tissue damage. Many growth factors and signaling pathways discussed in the preceding part of the chapter are also involved in these processes, but several additional pathways function specifically in cell fate decisions and in the establishment of tissue stem cells and their regulation in adult tissues. Three of them have also become notorious in the context of human cancers. They are the WNT (Figure 6.12), the Hedgehog/SHH (Figure 6.13) and the NOTCH (Figure 6.14) pathways.

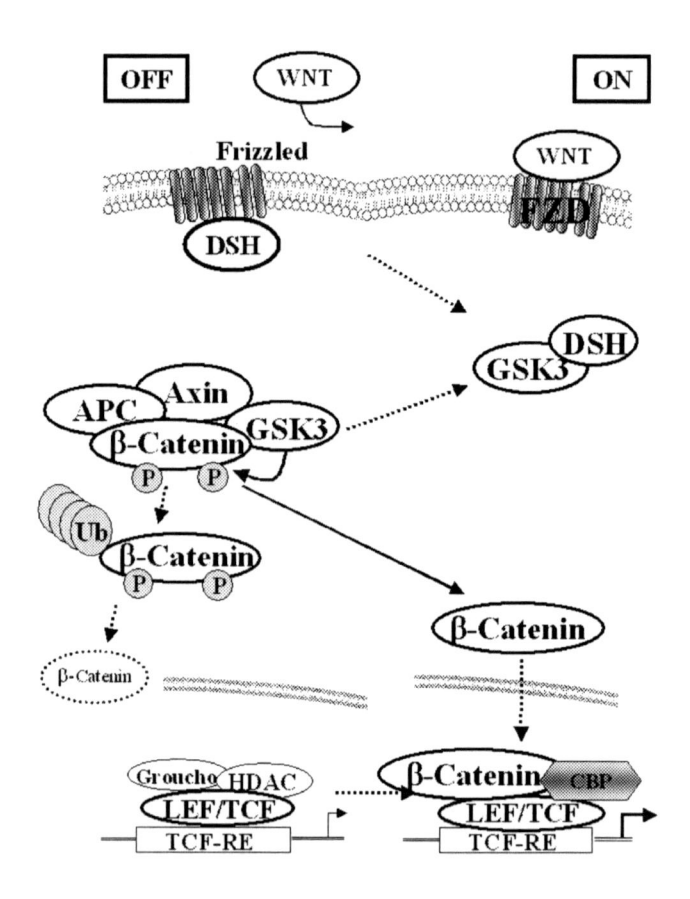

**Figure 6.12** *The WNT/β-Catenin pathway*
The inactive state (in the absence of WNT factor binding) is shown on the left and the active state on the right. This simplified scheme does not incorporate cross-talk with other pathways (see e.g. Figure 6.3) and pathway modulators discussed in chapters 13.2 and 16.2.

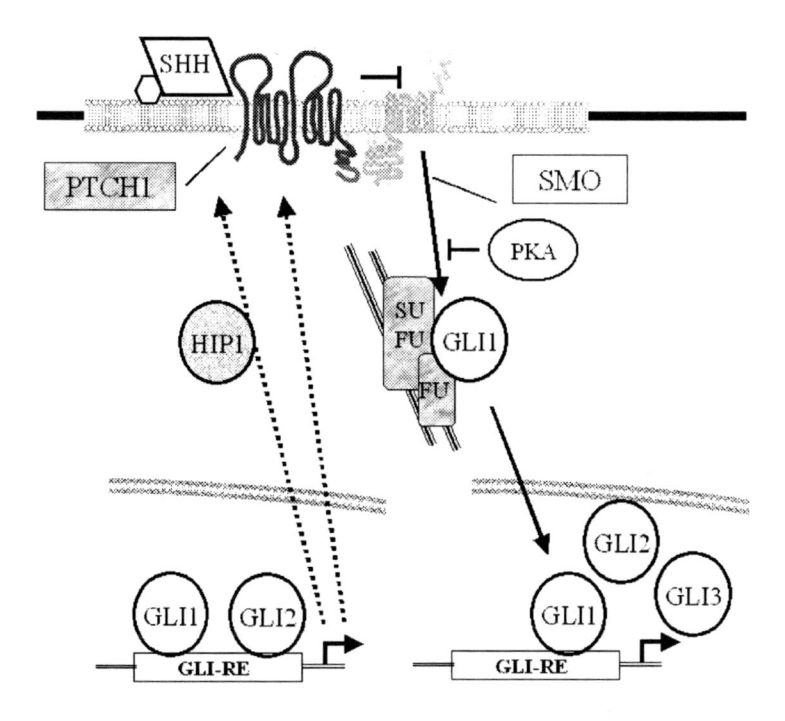

**Figure 6.13** *The Sonic Hedgehog (SHH) pathway*
SHH binding to the PTCH1 membrane receptor alleviates inhibition of SMO. By unknown mechanisms GLI1 is released from a protein complex at the microtubules containing SUFU and likely a FU-like protein. GLI1 migrates to the nucleus activating transcription of target genes, which include GLI factors itself as well as PTCH1 and another feedback inhibitor, HIP1. The pathway is also inhibited by active PKA. Note that the GLI factor cascade may differ between cell types; the version shown here is the most likely one in keratinocytes.

More than 20 factors belong to the WNT growth factor family in humans. They are ≈40 kDa proteins with >30% homology towards each other. The proteins are secreted after being being glycosylated and covalent linkage of a lipophilic moiety. Accordingly, the factors bind well to the extracellular matrix and stick to cell membranes, which restricts their diffusion. Therefore, they are limited to acting as paracrine or autocrine factors. During development, WNT proteins drive the expansion and morphogenesis of many different tissues, in particular of tubular and ductular epithelial structures in the kidney and the gut. Of course, WNT factors and pathways interact with others, e.g. with Hedgehog-dependent pathways in the development of the limbs. Individual WNT factors remain involved in the homeostasis of adult tissue and specifically seem to control stem cell compartments, e.g. in the gastrointestinal tract.

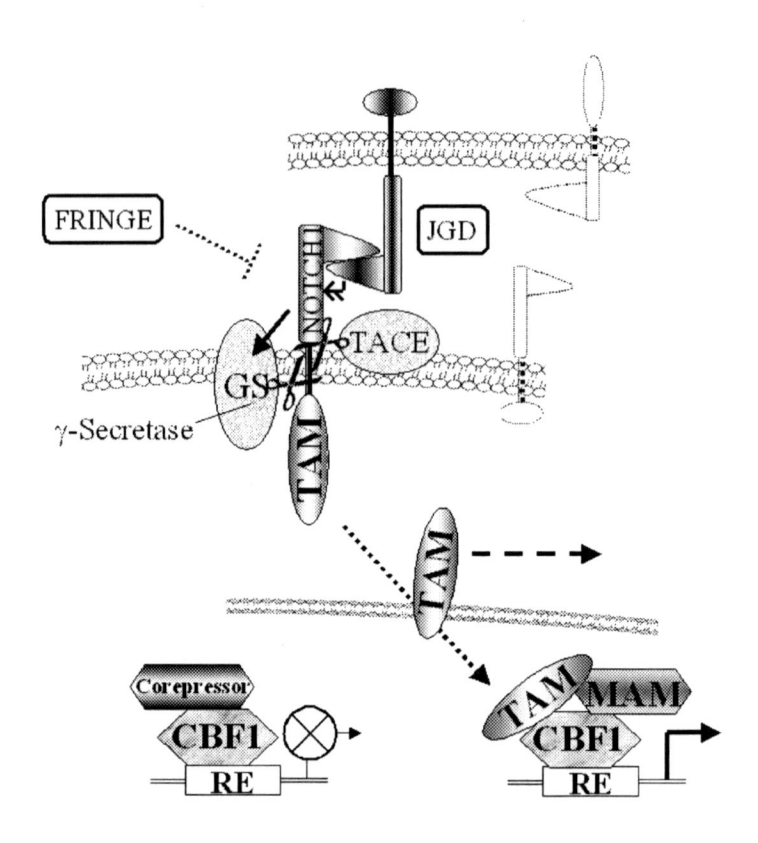

**Figure 6.14** *The NOTCH pathway*
See text for explanation.

WNT factors are normally recognized by Frizzled receptor proteins (FZD) at the cell surface. There are again several of these that respond to different WNT factors. They are supported by specific LRP proteins (LRP5 and LRP6) which seem to recognize the lipid part of the WNTs and are counteracted by secreted Frizzled related proteins (SFRP). The FZD receptors may not only differ with respect to which WNT they bind, but also with respect to which intracellular pathways they stimulate. At least three different pathways are known, but only one is really well characterized. The decisive step in this pathway is the accumulation and activation of β-Catenin which is prevented by the coordinated action of several proteins including APC, Axin and GSK3β (Figure 6.12, →13.2). This particular pathway is therefore called the WNT/β-Catenin or 'canonical' WNT pathway. Its constitutive activation is crucial in the development of colon cancer (→13.2). It is also important in other cancers, e.g. in hepatocellular carcinoma (→16.2). Its deregulation is alternatively brought about by loss of function of negative regulators, such as APC or Axin or by oncogenic mutations activating β-Catenin. *APC* and *CTNNB1* (encoding β-Catenin) therefore behave as tumor suppressor and oncogene, respectively (Table 6.1).

The WNT/Ca$^{2+}$ pathway involves activation of classical PKCs and other Ca$^{2+}$-dependent protein kinases. Its function is largely undefined, but it may promote cellular differentiation to a greater extent than the 'canonical' pathway. The WNT/polarity pathway involves the small GTP-binding protein RHO and acts on the polarization of the cytoskeleton. It also leads to phosphorylation of JNK1 kinase and thus to cross-talk with this MAPK pathway. These two other pathways are little studied so far in the context of human cancers. Likewise, the significance of altered expression of WNT factors, of FZD receptors and of SFRP1 in several different types of human cancers is still under study. However, at least one instance of WNT factor activation by a retroviral insertion (→4.2) has been documented.

Somewhat comfortingly, there are only three Hedgehog proteins in man. They are Sonic Hedgehog (SHH), Desert hedgehog (DHH), and Indian Hedgehog (IHH). Like WNT factors, they act in a paracrine fashion; and as for WNTs, their diffusion is limited by a lipophilic modification, in this case covalent attachement of cholesterol. Hedgehog factors are perhaps best studied for their role in limb development, but they are certainly also crucial for the development of many other organs, as diverse as the brain and the prostate. Also like WNTs, they contribute to tissue homeostasis in fully grown humans and, in fact, are also implicated in the maintenance of stem cell populations. The most important factor in adult tissues is SHH. Hedgehog factors bind to smoothened receptors such as SMO1 and initiate their own intracellular signaling pathway which ultimately leads to the activation of GLI transcriptional activators. Activation of SMO1 is prevented by PTCH1 (Figure 6.13, →12.3). Perhaps unsurprisngly, there is accumulating evidence of cross-talk between the SHH and WNT pathways inside the cell.

Constitutive activation of SHH signaling appears to lie at the core of several cancers with a precursor cell phenotype such as small cell lung cancer and basal cell carcinoma of the skin (→12.3). This activation can alternatively be brought about by mutations in activating components, such as point mutations in SMO1 or – perhaps - overexpression of GLI transcription factors, or by inactivation of a negative regulatory component, usually PTCH1. So, these have to be regarded as oncogenes and tumor suppressors, respectively (Table 6.1).

Like the WNT and SHH pathways, the NOTCH pathway was first encountered in Drosophila and has been more extensively studied for its function in development in model organisms than in humans. It is now firmly established as a cancer pathway in man, and is also considered to factor in the neurological Alzheimer degenerative diseases. The emerging consensus is that both disruption and overactivity of the NOTCH pathway can promote cancer development. This ambiguity is not entirely unprecedented for cancer pathways, since the TGFβ and STAT pathways, e.g., act differently in different cell types. In the case of NOTCH signaling, its ambivalence is related to its pronounced dosage sensitivity which derives from its biological function.

Like the WNT and SHH pathways, the NOTCH pathway also controls stem and precursor cell compartments. More precisely, its characteristic function is the regulation of binary cell fate decisions. This includes the decision of whether a cell remains a tissue precursor or goes on to differentiate, such as in the basal layer of the epidermis. NOTCH signaling is also involved in decisions like whether a differentiating intestinal cell becomes an enterocyte or a secretory Paneth or goblet cell or whether a cell in the lymphocyte lineage enters the T-cell or B-cell sublineage. In system theory, this kind of decision is called a bifurcation and its control requires a metastable equilibrium which develops towards either of two opposite states upon slight perturbations, like a coin standing on its edge. This comparison describes the kind of function provided by the NOTCH signaling system.

Four different NOTCH receptors, NOTCH1-4, are known, and are expressed on the cell surface. They are activated by ligands expressed on the cell surface of neighboring cells. Two different kinds of ligands are known. In humans, they comprise the three 'delta-like' DLL1, DLL3, and DLL4 and the 'jagged-like' JAG1 and JAG2 ligands. These ligands differ somewhat in the response they elicit and more substantially in their sensitivity towards modification of NOTCH receptors by FRINGE glycosylases. These elongate glycosyl chains on NOTCH receptors that prevent binding of JAG, but not of DLL proteins.

Importantly, expression of NOTCH receptors and their ligands is each self-reinforcing and cross-inhibitory and therefore tends to become mutually exclusive. Thus within an organized tissue, different types of cells express either receptors or specific ligands. A precursor cell population, e.g., may express a receptor, while cells that have taken a step towards differentiation express a ligand, or the other way round. The latter situation is found, e.g., in human epidermis, where basal cells express NOTCH ligands and cells in upper, differentiated layers express NOTCH1.

NOTCH receptors are heterodimers formed by proteolytic cleavage from a single precursor (Figure 6.14). The protease involved is the γ-secretase presenilin whose dysfunction is one cause of Alzheimer disease. The extracellular domain of NOTCH binding the ligands on neighboring cells contains EGF-like repeats and a cystein-rich domain termed LN. The second subunit remains bound to it by a small extracellular domain, continues through the membrane into its larger intracellular segment. This segment contains ankyrin repeats, which mediate protein-protein interactions, a nuclear localization signal, a transcriptional transactivation domain, and a PEST sequence likely responsible for regulated proteolytic degradation.

Following activation by ligands, the intracellular NOTCH domain (sometimes called TAM) becomes free to move into the nucleus, where it replaces repressor proteins from the CBF1 transcription factor (also known by multiple other names such as CSL or RBJ1) to activate its target genes. Activation of NOTCH receptors additionally elicits CSL-independent events which are not as well elucidated. NOTCH target genes appear to be different in different cell types. Thus, in neuronal precursors NOTCH signaling inhibits expression of neuron-specific genes. In keratinocyte precursors, it induces differentiation markers and the $p21^{CIP1}$ CDK inhibitor causing cell cycle arrest. NOTCH signaling often inhibits WNT/β-Catenin, SHH and AP1 signaling, and, conversely, supports $NF_\kappa B$ activation.

As would be expected from its function in normal epidermis, NOTCH receptors appear to function as tumor suppressors in this organ. Down-regulation of NOTCH1 and NOTCH2 is a regular finding in basal cell carcinoma, although the extent to which it contributes to this tumor is debated. As NOTCH signaling inhibits the SHH pathway, its loss may exacerbate the overactivity of this pathway that causes this type of cancer (→12.3). Similarly, loss of NOTCH function may be a prerequisite for formation of small cell lung cancers, which also show activation of the SHH pathway, likely by an autocrine mechanism. In this case, NOTCH activity appears to prevent the precursor cells from adopting the neuroendocrine, 'stem-cell-like' phenotype displayed by these cancers.

In other cancers, NOTCH proteins undoubtedly function as oncogenes. One type of T-cell acute leukemias (T-ALL) is characterized by a translocation between chromosome 7 and chromosome 9, t(7;9) (q34;q34.3), which leads to the overexpression of the cytoplasmic domain of NOTCH1 under the influence of the T-cell receptor $\beta$ enhancer. Constitutively active NOTCH signaling appears to direct an inappropriately large fraction of lymphocyte precursors towards a T-cell fate where they become malignant by further mutations. In a similar fashion, NOTCH overactivity appears to cooperate with viral oncoproteins such as the SV40 T-antigen in mesothelioma and HPV E6 and E7 in genital cancers. Here, the inhibitory effect of NOTCH signaling on the cell cycle is abrogated by the viral oncoproteins whereas its precursor cell maintenance function remains active and contributes to expansion of the tumor.

## *Further reading*

Krauss G (2003) Biochemistry of signal transduction and regulation 3$^{rd}$ ed. Wiley-VCH

Polakis P (2000) Wnt signaling and cancer. Genes Devel. 14, 1837-1851

Taipale J, Beachy PA (2001) The Hedgehog and Wnt signalling pathways in cancer. Nature 411, 349-354

Attisano L, Wrana JL (2002) Signal transduction by the TGFβ superfamily. Science 296, 1646-1647

Cantley LC (2002) The Phosphoinositide 3-Kinase pathway. Science 296, 1655-1657

Hood JD, Cheresh DA (2002) Role of integrins in cell invasion and migration. Nat. Rev. Cancer 2, 91-100

Martin KH et al (2002) Integrin connections map: to infinity and beyond. Science 296, 1652-1653

Sears RC, Nevins JR (2002) Signaling networks that link cell proliferation and cell fate. JBC 277, 11617-11620

Vivanco I, Sawyers CL (2002) The phosphatidylinositol 3-kinase/AKT pathway in human cancer. Nat. Rev. Cancer 2, 489-501

Derynck R, Zhang YE (2003) Smad-dependent and Smad-independent pathways in TGF-β family signalling. Nature 425, 577-584

Heinrich PC et al (2003) Principles of interleukin (IL)-6-type cytokine signalling and its regulation. Biochem. J. 374, 1-20

Knowles MA, Hornigold N, Pitt E (2003) Tuberous sclerosis complex (TSC) gene involvement in sporadic tumours. Biochem. Soc. Transact. 31, 597-602

Maillard I, Pear WS (2003) Notch and cancer: best to avoid the ups and downs. Cancer Cell 1, 203-205

Nickoloff BJ, Osborne BA, Miele L (2003) Notch signaling as a therapeutic target in cancer: a new approach to the development of cell fate modifying agents. Oncogene 22, 6598-6608

Parsons JT (2003) Focal adhesion kinase: the first ten years. J. Cell Sci. 116, 1409-1416

Pouysségur J, Lenormand P (2003) Fidelity and spatio-temporal control in MAP kinase (ERKs) signalling. Eur. J. Biochem. 270, 3291-3299

Radtke F, Raj K (2003) The role of Notch in tumorigenesis: oncogene or tumour suppressor. Nat. Rev. Cancer 3, 756-767

Siegel PM, Massagué J (2003) Cytostatic and apoptotic actions of TGF-β in homeostasis and cancer. Nat Rev Cancer 3: 807-820

Van Es, Barker N, Clevers H (2003) You Wnt some, you lose some: oncogenes in the Wnt signaling pathway. Curr. Opin. Genet. Devel. 13, 28-33

Greten FR, Karin M (2004) The IKK/NF-kappaB activation pathway-a target for prevention and treatment of cancer. Cancer Lett. 206, 193-199

Jin H, Varner J (2004) Integrins: roles in cancer development and as treatment targets. Brit. J. Cancer 90, 561-565

Lum L, Beachy PA (2004) The hedgehog response network: sensors, switches and routers. Science 304, 1755-1759

Slee EA, O'Connor DJ, Lu X (2004) To die or not to die: how does p53 decide? Oncogene 23, 2809-2818

Yu H, Jove R (2004) The STATs of cancer – new molecular targets come of age. Nat. Rev. Cancer 4, 97-105

# CHAPTER 7

# APOPTOSIS AND REPLICATIVE SENESCENCE IN CANCER

- In normal tissues, cell proliferation is limited by terminal differentiation and counterbalanced by cell loss, often via apoptosis. Cell damage also elicits apoptosis or necrosis. Moroever, apoptosis can be induced in response to inappropriate proliferation or by cytotoxic immune cells.
- Replicative senescence provides a different kind of limit to cell proliferation. Cells survive, but irreversibly exit from the cell cycle, as during terminal differentiation. Replicative senescence is established after cells have undergone a large number of replicative cycles, or more rapidly as a response to 'inappropriate' proliferation signals.
- Apoptosis can be induced by two largely separate signaling pathways which converge into a common executing cascade. Signaling and execution use proteases called caspases. The intrinsic pathway elicited, e.g., by strong DNA damage signals or by activated oncogenes, generates a mitochondrial permeability transition ultimately leading to the establishment of the 'apoptosome' protein complex that activates execution caspases. The extrinsic pathway is initiated by cell surface receptors, labelled 'death receptors'. These are activated by cytokine ligands and surface proteins of cytotoxic immune cells. The intracellular 'death domains' of the activated receptors associate with FADD adaptor proteins in a 'DISC' complex that puts distinct initiator caspases into action. These initiate the execution cascade, with or without support by the intrinsic pathway.
- Apoptosis is regulated at several steps in both pathways. BCL2 prohibits the related pro-apoptotic proteins BAX or BAK from acting at the mitochondria. They are regulated by further members of the BCL2 family, which mediate pro- or anti-apoptotic signals. FLIP inhibits the extrinsic pathway at the receptors, whereas IAPs like survivin act at the apoptosome inhibiting caspases. They are in turn antagonized by SMAC/Diablo liberated during the mitochondrial permeability transition.
- Replicative senescence was first observed in cultured human fibroblasts which undergo >50 population doublings but then stop to proliferate, with distinctive changes in morphology and gene expression. Since some evidence of replicative senescence is found in aging tissues, senescence may be related to human aging. More unequivocally, replicative senescence acts as a fail-safe mechanism against inappropriate proliferation, alternatively to apoptosis.
- Cancer cells must overcome the barriers presented by apoptosis and replicative senescence. The apoptotic response in cancers is inadequate or blocked.

Likewise, cancer cells are in general 'immortalized' and proliferate beyond the limits of replicative senescence.
- Diminished apoptosis in tumor cells can be caused by several means. Down-regulation of death receptors or secretion of decoy receptors mutes the extrinsic pathway. Some cancers may even use death receptor ligands to counterattack the immune system. Mutational inactivation of TP53 as well as silencing or mutation of its downstream pro-apoptotic mediators diminishes the response to signals activating the intrinsic pathway. Moreover, the PI3K or NFκB pathways conferring survival signals are often overactive. Overexpression of inhibitory proteins, such as BCL2, survivin and other IAPs, occurs in almost all cancers. More rarely, inactivation of genes acting in the executing cascade is observed.
- Replicative senescence is induced after multiple cell doublings when telomeres lengths have reached a critical minimum size. It is established by mechanisms similar to those that activate cell cycle checkpoints as a consequence of DNA double-strand breaks.
- Telomeres in humans consist of several hundred repeats of the hexanucleotide TTAGGG. They form a specialized T-loop structure which is protected by binding of several different proteins. In most somatic cells, telomeres shorten during each division. In contrast, germ cells and likely tissue precursor cells express the specialized reverse transcriptase hTERT and its RNA subunit hTR which serves as a template during telomere elongation by hTERT. Many cancers express telomerase and thereby stabilize telomere length. An alternative (hence: ALT) mechanism for the stabilization of telomere length is rarer and at present obscure. It may use DNA recombination.
- Induction of replicative senescence can also be mediated through activators of RB1 or TP53. Proliferating cells gradually accumulate the CDK inhibitors $p21^{CIP1}$, $p57^{KIP2}$, and particularly $p16^{INK4A}$, which induce cell cycle arrest through RB1. A more acute induction of replicative senescence is mediated by induction of $p14^{ARF1}$ and $p16^{INK4A}$ by oncogenes or viral infection and activation of TP53 as well as RB1. Therefore, defects in RB1, TP53 and CDK inhibitors in cancer cells also prevent replicative senescence.
- Shortened telomeres possess an increased potential for recombination and fusion with each other. If replicative senescence cannot be established, reactive telomeres contribute to chromosomal instability in cancer.

## 7.1 LIMITS TO CELL PROLIFERATION

The number of cells in normal tissues as well as in tumors is determined by the number of cells newly produced by proliferation and division and by the number of cells that die or are otherwise lost (e.g. by abrasion). Two principal modes of cell death exist, apoptosis and necrosis, with variations and intermediate forms (Figure 7.1). The overall growth rate of a tissue is additionally dependent on the proportion of cells with an active cell cycle, i.e. the proliferative fraction. Terminal differentiation and replicative senescence are two mechanisms that irreversibly

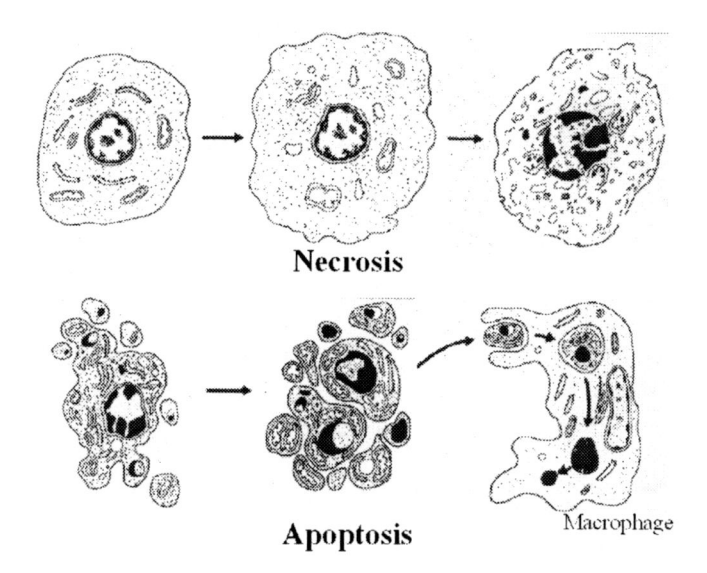

**Figure 7.1** *A morphological comparison of apoptosis and necrosis*
A standard representation of the description of necrosis and apoptosis according to Wyllie, Kerr and coworkers. Courtesy: Prof. K. Schulze-Osthoff

remove cells from the proliferative fraction, although the cells are not destroyed, at least not in the short run.

Apoptosis is a rapid process by which cells are destroyed in a thoroughly controlled fashion within hours[8]. Cells develop multiple blebs on their surface, lose their connections to other cells and the extracellular matrix, and round off while the nucleus and later the entire cell are fragmented into small membrane-enclosed particles. DNA is first cut into large >50 kb fragments, and later into smaller fragments corresponding to multiples of the nucleosomal units. So, free DNA ends can be labeled in apoptotic cells and cytologically detected, while isolated DNA from apoptotic cells often appears as a 'nucleosomal ladder' on agarose gels. During uncomplicated apoptosis, all cell remnants are phagocytosed or lost from the tissue, e.g. into the lumen of an organ, without evoking an inflammatory reaction.

Apoptosis can be elicited by external or internal signals. It is employed to shape tissues during development, to eliminate superfluous or autoreactive immune cells, to maintain homeostasis of tissues with rapid or cyclic cell turnover, and to destroy cells infected by viruses. An absolute or relative decrease in the apoptotic rate or a failure to respond properly to apoptotic signals is an almost universal property of human cancers. The mechanisms of apoptosis are therefore under intense scrutiny to better understand the development and progression of cancer and to find new angles for therapy.

---

[8] Depending on the cell type and stimulus, apoptosis lasts between 1 h and 1 d. This short duration and its variablity make the estimation of the apoptotic rate in a tissue difficult.

Necrosis is a different form of cell death elicited, e.g., by mechanical, chemical, or thermal damage as well as by some infectious agents. Necrotic cells burst, usually after swelling, releasing their content into the surrounding tissue in an uncontrolled fashion (Figure 7.1). Frequently, inflammation follows. Unlike apoptosis, necrosis may proceed in the absence of cellular energy. Thus, in a large solid tumor, hypoxic areas will typically show an enhanced rate of apoptosis, but the central core, which is almost anoxic (→9.1) and devoid of nutrients, will be necrotic. Theoretically, the inflammatory reaction ensuing from necrotic tumor cells could be beneficial, since it attracts immune cells. In reality, it might be a sort of double-edged sword, because inflammation may not only eliminate tumor cells but also contributes to the destruction of normal tissue structures and thereby may facilitate invasion and metastasis.

The specific functions of many tissues are exerted by cells that have once and for all exited from the cell cycle. Their formation is meant by the more precise usage of 'terminal differentiation'. Diverse strategies of terminal differentiation are used in various tissues. In many tissues, terminally differentiated cells are polyploid and/or multinuclear, e.g. the syncytia of the skeletal muscle or, less spectacularly, polyploid (tetraploid or octoploid) hepatocytes of the liver or umbrella cells of the urothelium. Robust terminal differentiation can also be efficiently achieved in diploid cells with normal, very active nuclei, as impressively demonstrated by neurons. In some tissues, however, the nuclei of terminally differentiated cells shrink and/or are expelled or dissolved. The epidermis, the lens of the eye, and the erythrocyte lineage provide examples of this strategy. In some instances, the destruction of the nucleus resembles an apoptotic process and is indeed implemented by similar mechanisms. In fact, in some tissues terminal differentiation is a prelude to actual apoptosis, e.g. in the gut. During differentiation, enterocytes move from the crypts towards the tips of the villi, where they undergo apoptosis, their remnants being lost into the lumen.

A decreased rate of terminal differentiation is a fundamental requirement for tumor growth. Many cancers are characterized by its complete lack, while some produce terminally differentiated cells at a diminished rate. Accordingly, in many cancers, proteins that are only expressed in terminally differentiated cells are not detectable. In others, some such proteins are expressed in a fraction of the tumor cells that still differentiate, or the cancer cells express some proteins of their differentiated tissue counterparts, but do not exit from the cell cycle.

Replicative senescence is also defined by an irreversible exit from the cell cycle, after which cells survive for an extended period. This distinguishes replicative senescence from apoptosis, but the borderline towards terminal differentiation cannot always be drawn as easily. Cultured cells undergoing replicative senescence often take on a characteristic morphology (Figure 7.2) with a flattened appearance, large nuclei, and many small granuli. They express some characteristic proteins, such as SAβ-GAL, a β-galactosidase with a comparatively acidic pH optimum, and high levels of CDK inhibitors like $p21^{CIP1}$ or $p16^{INK4A}$. Conversely, senescent cells express few or none of the markers that are diagnostic for terminally differentiated cells. There are overlaps, however, and senescent fibroblasts have been as well considered as terminally differentiated. In human tissues, replicative senescence is

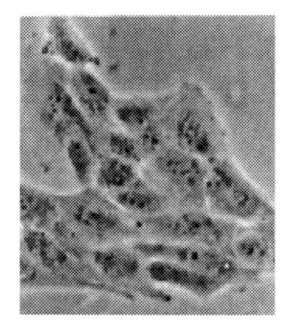

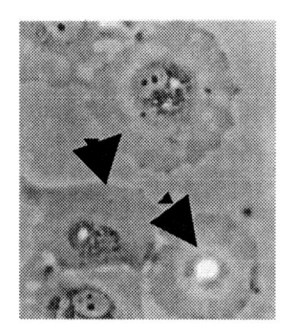

**Figure 7.2** *Morphological characteristics of cells with a senescent phenotype*
A tumor cell line containing cells with a typical senescent phenotype (right, arrowheads).

not easily ascertained, although cells with conspicuous morphologies that express SAβ-GAL have been observed.

In cultured cells, replicative senescence can be more clearly defined and is elicited in two very different instances. The 'classic' mode of induction occurs after propagation of normal human cells over many passages. It sets in gradually. In fibroblasts, where the phenomenon was first described, it may occur after as many as 50-80 cell doublings. In cultured epithelial cells, it appears much earlier. It can be prevented by infection with certain DNA viruses, typically the SV40 papovavirus or its large T-antigen (→5.3). Thus, replicative senescence presents a limit to the life-span of normal human somatic cells.

Replicative senescence can also be induced in a rapid mode, long before cells have exhausted their normal life-span, by inappropriate proliferation signals, specifically by overexpressed mutant RAS proteins. Unlike apoptosis and terminal differentiation, replicative senescence does not seem to be employed in the human body for tissue homeostasis. Rather, it appears to act as a fail-safe mechanism. Its only normal function may be setting a maximum to the human life-span (Box 7.1).

Evidently, cells in the germ-line must be exempt from replicative senescence. Moreover, it is plausible that tissue stem and/or early precursor cells must be subject to replicative senescence to a lower degree than more differentiated somatic cells, since during a human life-time in a continuously replicating tissue they will have to undergo many more than 100 divisions.

Replicative senescence is circumvented in many cancers. Cancer cells grown in culture or as xenografts in experimental animals can often be propagated for many more than 100 cell doublings and apparently indefinitely, without introducing T-antigen or its like. They are therefore considered 'immortalized'. It is, of course, difficult to ascertain immortalization as such in a human cancer tissue. It is easier in vitro, but not all human cancers can be grown in culture or as xenografts. Therefore, it is not certain whether all human cancers are really immortalized. Indefinite growth is not a necessary condition for a cancer to kill its host, because 50 cells doublings can theoretically produce more cells than an entire human body holds.

However, mechanisms that allow cancer cells in vitro to circumvent replicative senescence can be shown to be also active in many cancer tissues (→7.4). In other cases, tumor cells may evade replicative senescence by acquiring a kind of 'stem cell' character. This is evident in germ-cell cancers, e.g. in the testes or ovary. In addition, some cancers originating in somatic tissues may acquire (or maintain) properties of the respective tissue stem cell or early precursor, e.g. basal cell carcinoma of the skin (→12.3) and colon carcinoma (→13.2).

## 7.2 MECHANISMS OF APOPTOSIS

Apoptosis can be divided into several stages, i.e. initiation, execution and removal of the cell remnants ('burial'). Initiation can be performed by two separate pathways, often designated 'intrinsic' and 'extrinsic', which converge towards a common execution pathway. The intrinsic pathway responds to internal signals, e.g. from DNA damage, whereas the extrinsic pathway responds to external signals, e.g. by cytotoxic T-cells. In some cells, the extrinsic pathway can proceed towards execution on its own while in other cells it needs some contribution from the intrinsic pathway. All steps in these pathways are well defined and controlled. Considering apoptosis as a 'programmed cell death' is certainly justified.

The decisive step of the intrinsic pathway (Figure 7.3) takes place at the mitochondria. Its most important regulators are members of the BCL2 family. Around 20 members are known, some of which are pro-apoptotic, whereas others are antiapoptotic (Table 7.1). They share common domains, termed BH1 to BH4. The founding member, BCL2, was discovered as the oncogene at the characteristic translocation site t(14;18) in follicular B-cell lymphoma. This translocation places the *BCL2* gene under the control of the immunoglobulin heavy chain enhancer. Overexpression of BCL2 prevents apoptosis of follicular B-cells and is the initiating event in this relatively slow-growing cancer.

Induction of apoptosis by the intrinsic pathway requires inactivation of BCL2 and other anti-apoptotic proteins such as BCL-$X_L$. These proteins are located at the mitochondrial membrane, probably forming heterodimers with BAX and BAK, which are pro-apoptotic members of the BCL2 family. Other members of the family are distinguished by containing only a BH3 domain, and none of the other motifs. They also lack a transmembrane domain. These relay different pro-apoptotic signals to the mitochondria (Table 7.1).

Cellular stress of various kinds can induce apoptosis. In many cases, activation of TP53 is involved, e.g. when DNA double-strand breaks are created by radiation. Activated TP53 induces transcription of one or several pro-apoptotic BCL2 signaling proteins such as NOXA or PUMA, also increasing BAX, while down-regulating BCL2. The precise mechanisms may differ according to cell type and type of cellular stress. The pro-apoptotic proteins induced by TP53 override anti-apoptotic signals to initiate the next step in the intrinsic apoptotic pathway, i.e. formation of pores in the mitochondria.

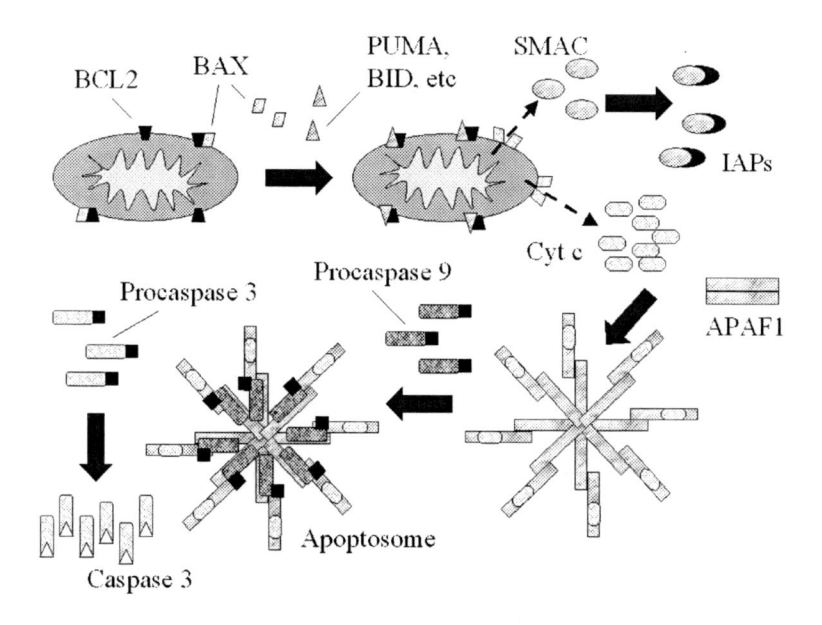

**Figure 7.3** *The intrinsic pathway of apoptosis*
The pathway starts in the upper left corner, leading to the processing of Procaspase 3 to Caspase 3 in the lower left corner. See text for more explanation.

BID mediates the 'cooptation' of the intrinsic pathway in cells in which it is needed for amplification of the extrinsic pathway. In that case, two shorter forms of BID, p15 or p13, are produced by proteolytic cleavage of an inactive precursor protein by Caspase 8 or Caspase 10. These then activate BAX or BAK, respectively.

The next step in the intrinsic pathway, mediated a.o. by BAX and BAK, consists of a change in the mitochondrial structure and function designated 'mitochondrial permeability transition' (MPT). At the contact sites between the outer and the inner mitochondrial membrane, pores are formed by a multiprotein complex to which proteins from all mitochondrial compartments contribute. One crucial component is the adenine nucleotide translocator which otherwise exchanges ADP + $P_i$ against ATP across the inner mitochondrial membrane. The pores let molecules <1500 D pass. This leads to a break-down of the mitochondrial transmembrane potential, since protons and other ions, including $Ca^{2+}$, can now move freely across the membrane. Mitochondria are thus the first organelles functionally inactivated in the intrinsic apoptotic pathway.

In conjunction with the mitochondrial permeability transition, mitochondria release several proteins, mostly from the intermembrane compartment, such as cytochrome c and the SMAC/Diablo protein. Apoptosis-inducing factor (AIF), a flavoprotein is released from the mitochondrial matrix. In the cytoplasm, 8 molecules of cytochrome c associate with an equal number of APAF1 proteins to form a large structure resembling the spokes of a wheel. It is therefore called 'wheel of death'[9] or 'apoptosome'. The apoptosome binds a stochiometric number of the pro-protease, pro-caspase 9 and supports its autocatalytic activation in an ATP-dependent process.

Caspases are cysteine proteases which cleave the peptide bond following an Asp in the consensus sequence QAD↓RG. In man, 14 caspases are known, which are categorized in three groups, initiator caspases (including caspase 9), executor caspases (including caspase 3), and inflammatory caspases, which are not directly involved in apoptosis, but process cytokines. The prototypic enzyme of that group is caspase 1, or interleukin-converting enzyme (ICE). Pro-caspase 9 is a homodimer containing two 'CARD' domains by which it binds to the APAF1 adaptor proteins in the apoptosome. Following autocatalytic cleavage, the active caspase 9 is a heterotetramer of two smaller and two larger subunits each. Active Caspase 9 goes on to process and activate the executioner caspase 3 to initiate the execution phase of apoptosis.

Activation of caspases is supported by the AIF protein released from the mitochondrial matrix. The other protein liberated from this compartment, SMAC/Diablo, has a distinct function (Figure 7.3). Activation of initiator - and sometimes even executioner caspases - is not always sufficient to actually elicit apoptosis, because a number of small proteins in the cell are capable of inhibiting these proteases. They belong to a group of proteins called IAPs (inhibitors of apoptosis). IAPs are mostly small proteins characterized by one or several 'BIR' domains. XIAP and survivin inhibit the intrinsic pathway at the step of activated caspase 9 and even caspase 3. They bind caspases through their BIR domains and inhibit their protease activity. SMAC/Diablo binds and sequesters IAPs like survivin

Table 7.1. *The BCL2 family of apoptotic regulators*

| Subfamily | Structure elements* | Representative members | Regulators** | Other members |
|---|---|---|---|---|
| Anti-apoptotic | BH4-BH3-BH1-BH2-TM | BCL2 BCL-XL | TP53↓ NFκB↑ | BCL-W, MCL-1, A1, NRF3 |
| Pro-apoptotic*** | BH3-BH1-BH2-TM | ***BAX*** | TP53↑, AKT↓ | **BAK**, BOK |
| BH3-only pro-apoptotic | BH3-TM or BH3 | BID BAD, NOXA, PUMA NIX | Caspase 8 (↑ by cleavage) TP53↑ Hypoxia | BAD, BIK, BLK, BMF |

\* BH: BCL2 homology domains; TM: transmembrane domain,
\*\* ↑ activation/induction, ↑ inactivation/downregulation
\*\*\* BAX and BAK indicated in bold print are likely effectors at mitochondria

[9] This is not the only case where the nomenclature in apoptosis tends towards the morbid; these terms are however actually in use.

and XIAP, thereby removing a further obstacle to apoptosis. Interestingly, several viruses - oncogenic or not - express their own IAPs to prevent a cell from apoptosis while they replicate.

The extrinsic pathway (Figure 7.4) is initiated when specific cell surface receptors are engaged by their specific ligands. As a rule, these 'death receptors' belong to the TNF receptor superfamily (Figure 7.5). Tumor necrosis factor α (TNFα) is one of several cytokine ligands of this receptor superfamily. This peptide is secreted by monocytes, macrophages and other cells of the immune system during inflammatory reactions and in response to cellular stress. It elicits various reactions, including apoptosis in some cells containing the TNFRI receptor.

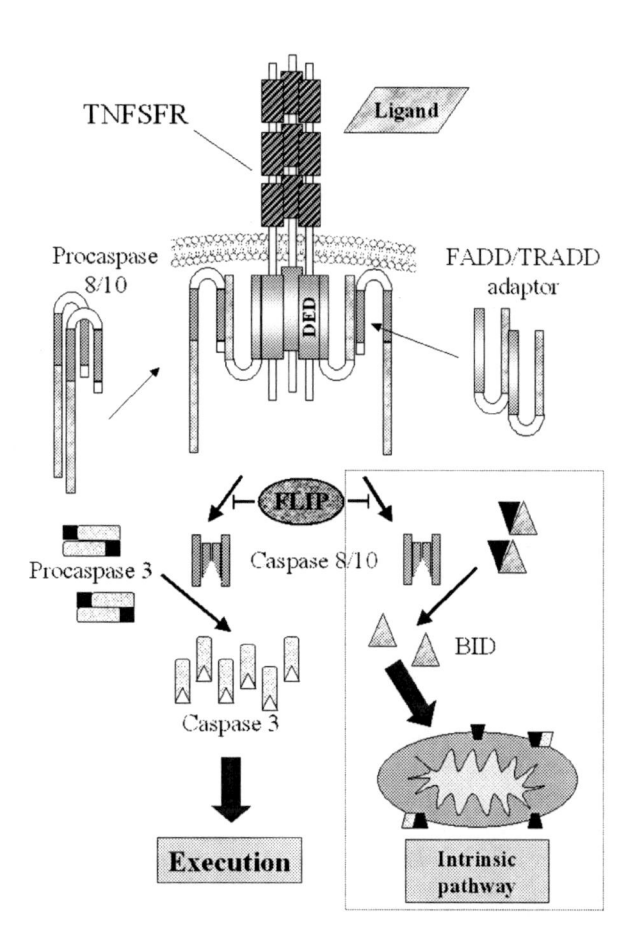

**Figure 7.4** *The extrinsic pathway of apoptosis*
Caspase 8 and/or Caspase 10 may be activated depending on cell type and receptor. Activation of BID to coopt the intrinsic pathway (dotted box) is not obligatory in all cell types.

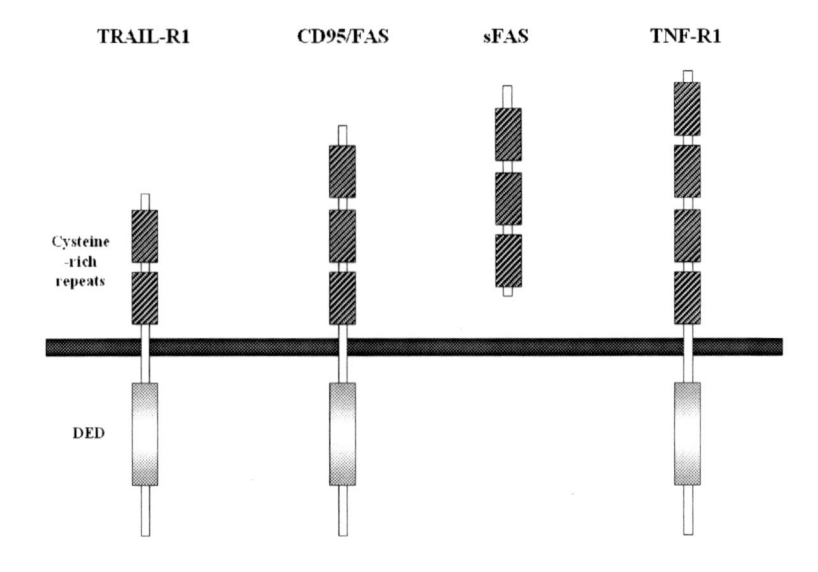

**Figure 7.5** *Some members of the TNFRSF family*
All members of the family share similar cysteine-rich domains in their extracellular domains, whose numbers vary. In the intracellular domain, they consistently contain a DED domain. Some members have additional signaling functions. The sFAS protein is otherwise identical to CD95/FAS, but lacks transmembrane and intracellular domains and acts as a decoy receptor. The ligands for TRAIL-R1 and TNFR1 are TRAIL and TNFα, respectively.

Other ligands of TNFRSFs are present mainly on the surface of immune cells, e.g. CD95L, and the ligand-receptor interaction is part of a cell-to-cell-interaction (→9.5). CD95L is also called FAS ligand and activates TNFRSF6, alias CD95, FAS, or APO-1. The CD95/CD95L system is considered one of the most important components in killing of infected and tumor cells by cytotoxic T-cells. It is also employed in the selective elimination of auto-reactive immune cells. Defects in CD95 function occur in autoimmune diseases as well as in cancers (→7.3).

In fact, the borderline between membrane-bound and soluble ligands is blurred. Some cytokines, including TNFα, are also present as a active membrane-bound form on the cell surface and CD95L is also secreted. The regulation of receptor-ligand interactions in this system is in fact very complex. For instance, at the receiving end, the response is modulated by the presence of modulating and decoy receptors. The response to TNFα is modulated by the TNFRII receptor. When present at the cell surface, the TNFRII appears to bind the cytokine and pass it on to the TNFRI which mediates the actual response. When TNFRII is sheared off the cell, it acts as a 'decoy receptor', sequestering the ligand and preventing it from acting on TNFRI.

Similarly, in addition to the membrane-bound form of CD95 (also tmFAS), a soluble form is generated by alternative splicing (sFAS), which also acts as a decoy receptor to decrease the responsiveness to CD95L. Apoptotic responses to the TNFα-related cytokine TRAIL are also dependent on the relative expression of four different TNFRSF members, TRAIL-R1 through TRAIL-R4, two of which are true receptors and two are decoys.

Members of the TNFRSF family act through several pathways, notably the NF$_\kappa$B pathway (→6.9). Family members that can activate the extrinsic apoptosis pathway differ from their homologues by the presence of an additional intracellular domain, called the 'death domain' (Figure 7.4). This part of the protein is required for the activation of the extrinsic apoptotic pathway. Since the NF$_\kappa$B pathway as a rule counteracts apoptosis, the actual cellular response will often depend on the relative strengths of the two pathways activated in parallel. In specific cell types, cytokine receptors like TNFRI also stimulate cell proliferation.

Following ligand binding to an active TNFR, such as TNFRI or TNFRSF6, the ligand/receptor complexes trimerize and the receptor death domains bind FADD proteins by interaction with the homologous domains in this adaptor. In addition, FADD contains a death effector domain homologous to that in initiator caspases. By binding to the death receptor, this domain is exposed and binds an initiator pro-caspase, usually pro-caspase 8 or pro-caspase 10. The resulting complex appears sometimes in the literature as 'death inducing signaling complex' (DISC). Its function in apoptosis is to bring pro-caspase molecules into close proximity to dimerize and activate one another. The activated initiator caspases 8 or 10 then activate executioner caspases like caspase 3 setting the execution phase into motion. The FLIP protein acts as an inhibitor of the extrinsic pathway by interfering with initiator caspase dimerization.

In some cells, activation of the extrinsic pathway by certain death receptor ligands is sufficient to elicit apoptosis. In such cases, the expression levels of BCL2 and BCLX$_L$ are quite irrelevant. In others, induction of apoptosis requires the participation of the intrinsic pathway. In response to external signals, this is typically stimulated via the BID protein cleaved by caspase 8. Conversely, the intrinsic pathway also influences the extrinsic pathway. For instance, TP53 induces activators of the extrinsic pathway, but also increases the expression of CD95, thereby sensitizing cells to pro-apoptotic external signals.

The multiple biochemical and morphological changes that take place during the execution phase of apoptosis are caused by proteolytic cleavage of >300 cellular proteins by caspase 3 and other executioner caspases like caspase 6 (Table 7.2). The substrates comprise regulators of the cell cycle such as RB1, DNA repair proteins such as DNA-PK and poly-ADP-ribosyl polymerase (PARP), and cytosketelal proteins such as actin, lamins, and keratin 18. The characteristic 'nucleosomal ladder' DNA fragmentation is caused by several DNases, prominently CAD (caspase activated DNase) that are liberated by cleavage of inhibitory proteins to which they are normally bound. Cleavage of FAK (focal adhesion kinase), PAK2 (p21-associated kinase), and Gelsolin contributes to the loss of adhesion and the characteristic membrane changes such as blebbing and redistribution of membrane

**Table 7.2.** *Protein substrates of caspases during execution of apoptosis*

| Category | Examples |
| --- | --- |
| Cytoskeleton and structural | Fodrin, β-Catenin, Plakoglobin, actins, Gelsolin, cytokeratins, lamins |
| Cell cycle and DNA replication | MCM3, MDM2/HDM2, RB1, p21$^{CIP1}$, p27$^{KIP1}$, WEE1, CDC27, Cyclin A |
| DNA repair and metabolism | Topoisomerase I, Poly-ADP-ribosyl-polymerase (PARP), DNA-dependent protein kinase (DNA-PK), Inhibitor of caspase-activated DNases (ICAD) |
| Transcription and splicing | sterol regulatory element binding proteins (SREBPs), transcription factors STAT1, NFκB (p65, p50), and SP1, IκB, various SNRNPs |
| Signal transduction | protein kinases PKCδ, PKCθ, MEKK1, FAK, and others, PP2A, RAS-GAP, PLA$_2$ |
| Proteases, protease inhibitors and apoptotic regulators | po-caspases, Calpastatin, Huntingtin, presenilins, ataxins, BCL2, BCL-X$_L$ |

proteins and phospholipids. These redistributions create signals for the subsequent burial phase. Importantly, phosphatidylserine which is normally restricted strictly to the inner layer of the membrane phospholipid bilayer, is flipped to the outer layer and recognized by receptors on macrophages that are attracted by further chemotactic signals diffusing out from the dying cell.

## 7.3 MECHANISMS OF DIMINISHED APOPTOSIS IN CANCER

Diminished apoptosis of cancer cells is important for a number of reasons.

(1) In some cancers decreased apoptosis is the primary cause of tumorous growth, e.g. in follicular B-cell lymphoma and perhaps in early prostate cancer (→19.1). In these tumors, cells that ought to undergo apoptosis in the course of normal tissue homeostasis survive, which leads to an oversized and progressively disorganized tissue mass.

(2) A diminished rate of apoptosis exacerbates hyperproliferation in many different cancers.

(3) Apoptosis is a fail-safe mechanism in response to 'inappropriate' proliferation signals (as is replicative senescence) and following pronounced DNA damage, e.g. unrepaired double-strand breaks. Therefore, a decreased response to 'internal' pro-apoptotic signals allows cells to proliferate in spite of proliferation signals being inappropriate or in spite of persisting severe DNA damage. This occurs in many cancer types, at the latest during progression.

(4) Cytotoxic T-cells from the immune system which protect against cancer and infections employ induction of apoptosis as a mechanism of cell killing. Decreased responsiveness to 'external' pro-apoptotic signals therefore is one of the

mechanisms by which tumor cells evade the immune response (→9.5). This aspect becomes particularly relevant during invasion and metastasis. At an earlier stage of cancer development, some viruses, e.g. Epstein-Barr virus (→10.3) or HHV8 (Box 8.1), express anti-apoptotic factors which diminish apoptosis in response to both internal and external signals, thereby creating a population of cells more susceptible to carcinogenesis.

(5) Many cytotoxic drugs employed in chemotherapy as well as radiotherapy act by inducing apoptosis (→22.2). Decreased apoptotic responsiveness therefore contributes to primary and secondary resistance (→22.2) to chemo- and radiotherapy.

In human cancers, diminished apoptosis can originate from alterations in many different steps of apoptosis by a variety of mechanisms (Table 7.3). Proteins that relay internal or external pro-apoptotic signals can be inactivated, the extrinsic or the intrinsic pathway become deactivated or desensitized, and even the execution stage can be impeded. In one and the same cancer, several different steps can be affected. Moreover, decreased apoptosis in cancer cells is often caused by overactivity of survival signal pathways rather than by primary alterations in apoptotic pathways. Nevertheless, some degree of apoptosis does take place in human cancers, but not at the same rate as would be elicited in normal cells by comparable external and internal signals. Typical changes that diminish the apoptotic rate in human cancer cells include the following.

Desensitization of death receptors: The CD95/CD95L system is inactivated or desensitized in many different human cancers, in hematological cancers as well as in carcinomas. While mutation of the *TNFRSF6* gene encoding CD95 is occasionally observed, in most cases down-regulation of receptor expression is the major mechanism responsible. In some cancers, a shift in expression from the transmembrane towards the soluble (decoy) receptor takes place. Altered expression of FADD proteins, decreased expression of Caspase 8, and overexpression of FLIP which inhibits the activation of Caspases at the DISC have been identified as causes of post-receptor defects in some cancers. In each case, the overall consequence is a decreased response to cytotoxic immune cells, but also to chemotherapeutic agents, which induce apoptosis partly through increased expression of both CD95 and its ligand. Other members of the TNFRSF family like the TRAIL receptors are also often inactivated by similar mechanisms.

**Table 7.3.** *Mechanisms causing diminished apoptosis in human cancers*

| Mechanism |
| --- |
| Desensitization of 'death receptor' (initiation and signaling of extrinsic pathway) |
| Counterattack (avoidance of death receptor signaling) |
| Loss of TP53 function |
| Desensitization or inactivation of the intrinsic pathway |
| Overexpression of IAPs |
| Activation of anti-apoptotic pathways |

Counterattack: Additionally, decreased expression of the CD95/FAS receptor in some cancers is accompanied by increased expression of soluble CD95 ligand. It is thought that secretion of CD95L normally contributes to the establishment of 'immune-privileged' sites in the human body. Immune-privileged sites are established in organs such as the anterior eye chamber or the testes that could not function properly in the presence of lymphocytes and therefore have to keep them out. In consequence, increased production of CD95L may help cancers to prevent immune responses and may even destroy T-cells and other cells that express CD95. This 'counterattack' may account for some tissue damage caused by invasive cancers locally and perhaps even in distant organs like the liver.

Loss of TP53: While downregulation or mutation of CD95 inactivate external pro-apoptotic signaling, inactivation of TP53 may be the most common alteration that compromises internal pro-apoptotic signaling. TP53 mediates induction of apoptosis in response to DNA damage as well as to hyperproliferation (→6.6). Some think that loss of its pro-apoptotic function may be the most important consequence of TP53 inactivation.

Intrinsic pathway inactivation: The most varied assortment of alterations affect the intrinsic apoptotic pathway. BCL2 was discovered as an oncogene protein activated by the most characteristic translocation in follicular lymphoma (→4.3). It is also over-expressed in a wide range of other cancers, including different types of carcinoma, prominently breast and prostate cancer (→18.4, →19.2). Alternatively to BCL2, cancers contain high levels of BCL-$X_L$, which is induced a.o. by the NFκB pathway (→6.9). Conversely, pro-apoptotic members of the BCL2 family such as BAD or NOXA (Table 7.2) are down-regulated in a variety of cancers, in some cases by promoter hypermethylation, and cannot be induced by activated TP53 or other signals. The most generally down-regulated member of the family may be BAX, perhaps due to its effector function at the mitochondria. An even stronger block to apoptosis may ensue when APAF1 expression is down-regulated, typically by promoter hypermethylation (→8.3). In summary, in almost all cancers the balance between pro-apoptotic and anti-apoptotic members of the family is tilted. The overall result of this imbalance is a decreased sensitivity towards apoptotic signals, particularly those elicited by hyperproliferation and aneuploidy, but also by chemotherapy.

IAP overexpression: Both the intrinsic and extrinsic pathway are affected by overexpression of IAP proteins which block signaling through caspases as well as the actual execution caspases, e.g. caspase 3. Survivin, e.g., is expressed at high levels during fetal development, but almost undetectable in normal resting tissues. Some expression is found associated with normal proliferation and regenerative processes such as wound repair. However, in many human cancers the levels of this IAP protein are so strongly and consistently increased that it is being developed as a tumor marker. IAPs may also be expressed or be induced by viruses present in a tumor cell such as EBV (→10.3) or HBV (→16.3). The overexpression of IAPs that block the execution phase can result in chaotic situations within a cancer cell, viz. partial activation of caspases which is not sufficient for execution of cell death. Consequences of such partial activation could be altered cell morphology and adhesion as well as genomic instability. Overexpression of other IAPs or of c-FLIP impede primarily the signaling phase of the extrinsic pathway by inhibiting the signal from FADD to caspase 8 or 10.

Activation of antiapoptotic pathways: The defects in the actual apoptotic signaling and execution cascades occurring in cancer cells are almost regularly complemented by increased activity of pathways that convey survival signals. Perhaps the most important ones in this regard are the PI3K (→6.3) and the $NF_κB$ (→6.9) pathways. The PI3K pathway is activated in many cancers, indirectly by growth factors, oncogenic mutation of receptor tyrosine kinases, or RAS mutations, or directly by inactivation of negative regulators in the pathway like PTEN or by oncogenic overexpression of PI3Kα. The pathway does stimulate proliferation and particularly the growth of cells, but in many cancers, the main importance of its activation may lie in the ensuing inhibition of apoptosis. This is mediated through activity of AKT/PKB which phosphorylates BAX preventing it from activating the intrinsic apoptotic pathway. The kinase also phosphorylates and activates the forkhead transcription factor FKHR-L1, which counteracts apoptosis at the level of transcription. Activation of the PI3K pathway also diminishes the effect of cancer chemotherapy. Compared to the PI3K pathway, the $NF_κB$ pathway is less frequently subject to direct activation by mutation in human cancers. However, in many cell types, it limits the extent of induction of apoptosis resulting from activation of TNFRSF death receptors. Therefore, its indirect activation in tumor cells, which can be achieved by a variety of cytokines and stress signals, contributes to the resistance towards induction of apoptosis by external signals such as TNFα or CD95L.

## 7.4 REPLICATIVE SENESCENCE AND ITS DISTURBANCES IN HUMAN CANCERS

Apoptosis is normally a rapid process, occurring within a period between one hour and one day. It is elicited likewise by rapid signals, such as the binding of cytokines to death receptors at the cell membrane or the activation of TP53 by ATM following a DNA double-strand break. Moreover, cells subjected to apoptosis vanish quite rapidly by phagocytosis. In all these respects, replicative senescence differs. It sets in slowly, it is usually elicited by signals that accumulate gradually, and cells persist, at least in the short run.

Replicative senescence can be evoked by two different signals which use overlapping pathways for execution. One type of signal emanates from short telomeres, and the second type from CDK inhibitors.

Telomeres in human cells are 5-30 kb long and made up of 1000-5000 repeats of TTAGGG hexamers. The bulk of each telomere consists of double-stranded DNA, but 75-150 nt at the ends are single-stranded. Normally, these single strands are folded back into the double strand, forming a T-loop (Figure 7.6). This is a structure similar to the D-loops occuring during DNA repair by homologous recombination (→3.3). In humans, telomeric DNA is wrapped around nucleosomes. Therefore, core histones are present, but in addition an unusual assembly of further proteins. The TRF2 (telomeric repeat binding factor 2) protein induces and seals T-loops. It also serves as an anchor for a number of further proteins that are located to the telomere under normal circumstances, in particular the RAD50/NBS/MRE11 complex. This complex processes double-strand breaks during DNA repair (→3.3). The KU70 and KU80 proteins which mark and protect DNA double-strand breaks during repair are also present at telomeres. Thus, telomeres appear to serve as reservoirs for these proteins on one hand, but on the other the repair proteins are strategically placed for dealing with damage to the telomeres themselves. A further protein, TRF1, which is homologous to TRF2, limits telomere length, being regulated itself by tankyrase, a poly-adenosine diphosphate ribosylase and TRF1-interacting nuclear protein 2. TRF1 also helps to maintain the RAD50/NBS/MRE11 complex at the telomere.

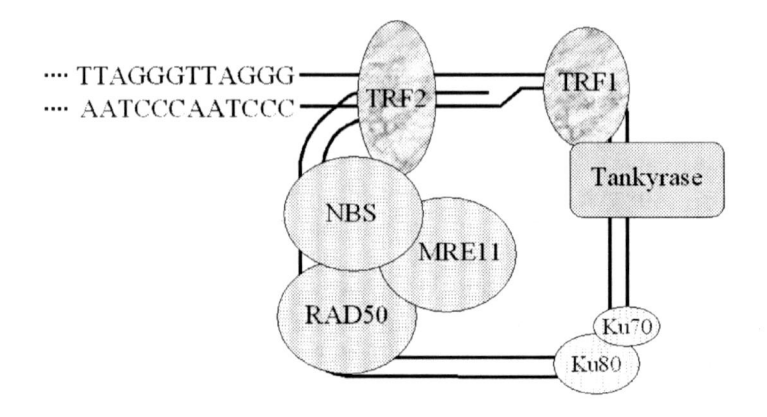

**Figure 7.6** *Structure of human telomeres*
The T-loop structure of human telomeres with some proteins located there. Their arrangement is largely hypothetical.

With each DNA replication in somatic cells, telomeres shorten. This is caused in principle by the end-replication problem. The top strand (with a 5'-end at the telomere) is replicated by elongation of an RNA primer at or near its end. When it is removed by RNase H after DNA synthesis has proceeded, the resulting gap cannot be filled, since DNA polymerases work invariably in the 5'→3' direction. This end-replication dilemma predicts a theoretical minimum loss of telomere sequences during each replication. In reality, its extent can be larger and is regulated by TRF1.

In germ-line cells, the decrease in telomere length is prevented by a specialized enzyme, telomerase (Figure 7.7). Accordingly, telomeres in germ line cells are approximately twice as long as in somatic cells. Telomerase is a specialized reverse transcriptase that uses an RNA template (AAUCCC) provided by its hTERC subunit to elongate telomeres. While the hTERC subunit is expressed in almost all human cells, the catalytic subunit hTERT is restricted to a small set of cells with high replicative potential, like germ-line cells, tissue stem cells and memory immune cells. Expression of the *TERT* gene is induced by a number of proliferation-

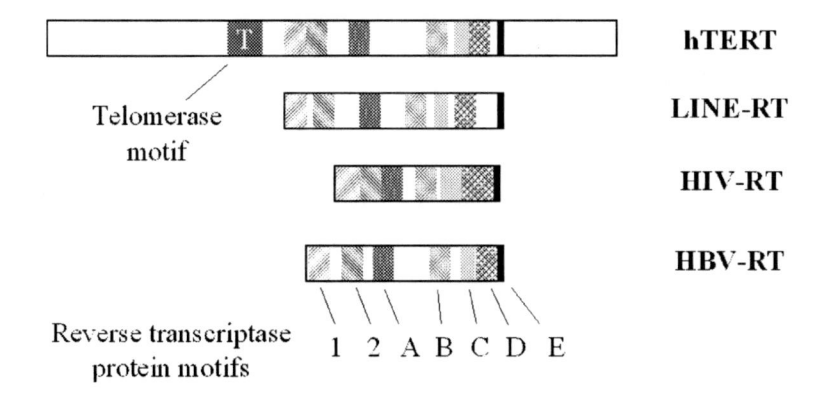

**Figure 7.7** *Structure of the human telomerase catalytic subunit*
The 127 kDa human hTERT enzyme contains a motif (T) shared by all telomerases and several motifs characteristic of reverse transcriptases (RT), including those of endogenous human LINE-1 retrotransposons, HIV and HBV.

stimulating and stem-cell maintaining factors. In particular, its promoter is a target of MYC proteins.

Shortening of telomeres to below a certain length causes replicative senescence. In cells cultured over longer periods, actually two successive steps can be distinguished, which are called M1 and M2. They are operationally defined: M1 can be bypassed by obliteration of RB1 and TP53 function. In the laboratory, this can be achieved by introduction of viral proteins such as SV40 large T antigen (→5.3). After 40 - 50 further doublings, senescence sets in irreversibly at the M2 point. Circumventing the M2 point requires activation of telomerase. It is not precisely known what happens at M1 and M2. Human telomeres are very variable. So, one idea is that M1 is triggered by the first telomere reaching a critical length. This would then activate a checkpoint response through the RB1 and TP53 pathways. At M2, further telomere shrinking has taken place. Some telomeres may have become so short that they can no longer form a T-loop. In addition, they may not be capable of storing DNA repair proteins any more. So, some sort of DNA double-strand repair response may be initiated, likely through ATM, which induces replicative senescence once and for all. While some of these ideas are not fully proven, DNA damage signaling is certainly involved in replicative senescence.

Telomere shortening leads to chromosomal instability. Of course, shortened unsealed telomeres are expected to become substrates for exonucleases which would gradually degrade a chromosome. In fact, a greater danger to genomic integrity may be recombination between different telomeres that are not protected by proteins. Recombination between the telomeres of two chromosomes can generate a dicentric chromosome. During mitosis, this may become missegregated or be pulled to opposite sides of the spindle and disrupted. Disruption would cause two open chromosome ends which could again fuse to other chromosomes and form further dicentrics to continue the cycle. Of note, in this classical breakage-fusion-bridge sequence (→Figure 2.7), the breakpoints tend to move from the telomeres towards the centromere.

In human cancers, establishment of replicative senescence as a consequence of telomere shortening is impeded, often at M1 as well as M2. Many human cancers contain defects in the RB1 and TP53 pathway. While these have many other consequences as well (→6.4, →6.6), loss of RB1 and TP53 function would be expected to permit the bypassing of the M1 limit. This would establish a population of cells continuing to proliferate with at least some critically shortened telomeres and therefore an enhanced potential for genomic instability. Dicentric chromosomes and a movement of chromosome breaks towards the centromere are quite common observations in carcinoma cells. Moreover, telomere instability due to telomerase dysfuntion is the cause of a human disease, dyskeratosis congenita. Patients with this rare inherited affliction do not only present with defects in skin, hair, and the hematopoetic system, but are also prone to cancer.

In addition to the defects in the RB1 and TP53 pathways, many human cancers express hTERT which can be shown to be enzymatically active in tissue extracts. In some tissues, hTERT expression or activity could therefore serve as a cancer biomarker. In cancers with hTERT expression, telomere lengths are at least stabilized at a low level, albeit they do not always rebound.

There is evidence for a different, alternative mechanism of telomere stabilization, named ALT, in some cancers and even in normal tissues, where telomeres are stabilized or even expanded in the absence of detectable telomerase activity. The unspecific designation ALT reveals that the mechanism is presently mostly based on conjecture, with hints from alternative mechanisms employed in organisms that lack telomerase. There, telomere expansion can be achieved by a kind of homologous recombination double-strand repair (→3.3). Indeed, there is some evidence for such a mechanism in humans and, specifically, that the WRN helicase might be involved.

Telomere erosion is certainly to a large degree responsible for the limited lifespan of cultured human cells. It can be regarded as a mechanism counting the number of cycles a cell has undergone. A second mechanism appears to rely on CDK inhibitor proteins, in particular $p16^{INK4A}$, $p21^{CIP1}$, and $p57^{KIP2}$ (Figure 7.8).

Among the CDK inhibitors, $p21^{CIP1}$ is strongly induced by TP53 and may be largely reponsible for the arrest of the cell cycle after telomere shortening. However, it is thought that $p21^{CIP1}$ also accumulates in cells that proliferate continuously, independently of TP53, since it is induced by many proliferative stimuli. This is certainly so for $p16^{INK4A}$ which is not regulated by TP53. In somatic human cells $p16^{INK4A}$ is induced by E2F and other transcription factors activated during cell cycle progression. Because the protein has a relatively long half-life, it accumulates when successive cell cycles follow rapidly upon each other. In some cell types that express $p57^{KIP2}$, this inhibitor behaves in a similar fashion. So, the level of certain CDK inhibitors - like telomere length - depends on the number of successive cell cycles. This may provide a second counting mechanism.

However, in this mechanism counting not only depends on the actual number of cell cycles, but more critically on how quickly they follow each other and on which

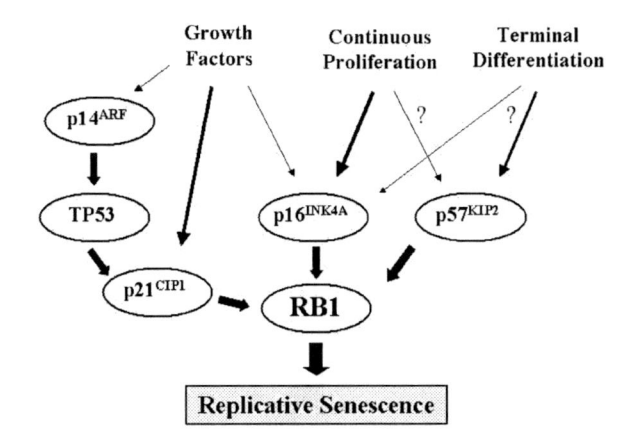

**Figure 7.8** CDK inhibitors as regulators of replicative senescence
The width of the arrows indicates the presumed strength of the influences.

signals elicit proliferation. An extreme case is hyperproliferation induced by oncogenes such as *RAS* and *MYC*. In human cells, such hyperproliferation induces not only p14$^{ARF1}$ to sensitize TP53, but also p16$^{INK4A}$. Together, these proteins lead to a rather quick arrest of the cell cycle, certainly more rapidly than the telomere shortening mechanism would. This mechanism could account for the different lifespans of different human cell types in culture, because it may be more sensitive in epithelial cells that become relatively soon senescent in culture. More generally, the involvement of both p14$^{ARF1}$ and p16$^{INK4A}$ in the response to hyperproliferation in human cells may explain why the *CDKN2A* locus is such a frequent target for inactivation in such a wide variety of human cancers (→5.3). Specifically, it may solve the enigma why p16$^{INK4A}$ of all INK4 proteins is the most important tumor suppressor.

The mechanisms involved in the regulation of replicative senescence constitute one of the more important differences between humans and rodents with regard to cancer. Since these mechanisms may be related to organism aging (Box 7.1), this is plausible. A two year old mouse is approaching old age, whereas a two year old human is a toddler and a long way from maturity. Moreover, 70 kg humans living for 70 years or so may require additional mechanisms for protection against cancer than 50 g mice living for 30 months. On a less intuitive argument, it has been observed for a long time that human cells are much more difficult to transform in vitro than rodent cells. It had been a long-standing speculation that there might be (at least) one additional mechanism that protects them from becoming cancerous. It is now established that somatic cells in rodents more generally express telomerase and telomeres in rodents are longer than in humans. Moreover, the regulation of

CDK inhibitors is different, particularly that of p16$^{INK4A}$. There is good reason to believe that the long-sought difference may reside here.

---

## *Further reading*

Johnson FB, Sinclair DA, Guarante L (1999) Molecular biology of aging. Cell 96, 291-302

Los M et al (2001) Caspases: more than just killers? Trends Immunol. 22, 31-34

Chen G, Goeddel DV (2002) TNF-R1 signaling: a beautiful pathway. Science 296, 1634-1635

Wajant H (2002) The Fas signaling pathway: more than a paradigm. Science 296, 1635-1636

Brenner C, Le Bras M, Kroemer G (2003) Insights into the mitochondrial signaling pathway: what lessons for chemotherapy? J. Clin. Immunol. 23, 73-80

Cory S, Huang DC, Adams JM (2003) The Bcl-2 family: roles in cell survival and oncogenesis. Oncogene 22, 8590-8607

Feldser DM, Hackett JA, Greider CW (2003) Telomere dysfunction and the initiation of genomic instability. Nat. Rev. Cancer 3, 623-627

Franke TF et al (2003) PI3K/Akt and apoptosis: size matters. Oncogene 22, 8983-8998

Hahn WC (2003) Role of telomeres and telomerase in the pathogenesis of human cancer. J. Clin. Oncol. 21, 2034-2043

Kucharczak J et al (2003) To be, or not to be: NFκB is the answer – role of Rel/ NFκB in the regulation of apoptosis. Oncogene 22, 8961-8982

Schwerk C, Schulze-Osthoff K (2003) Non-apoptotic functions of caspases in cellular proliferation and differentiation. Biochem Pharmacol. 66, 1453-1458

Ben-Porath I, Weinberg RA (2004) When cells get stressed: an integrative view of cellular senescence. J. Clin. Invest. 113, 8-13

Castedo M et al (2004) Cell death by mitotic catastrophe: a molecular definition. Oncogene 23, 2825-2837

Debatin KM, Krammer PH (2004) Death receptors in chemotherapy and cancer. Oncogene 23, 2950-2966

Sharpless NE, DePinho RA (2004) Telomeres, stem cells, senescence, and cancer. J. Clin. Invest. 113, 160-168

## Box 7.1: Human aging and cancer

**Theories on the causes of human aging** basically fall into two groups. One group assumes that the phenotypic changes associated with aging are caused by the **accumulation of unrepaired damage** to tissues, cells and macromolecules. One variant emphasizes oxidative damage by reactive oxygen species. Indeed, changes progressing with age can be found in extracellular tissue and in cells, including base mutations and epigenetic changes in the nuclear and mitochondrial genomes.

A second group of theories stresses that the regularity of the changes occurring with age reminds one of a **genetic program** – just like that controlling development and maturation. A minimum version of this sort of theory suggests that humans are 'build' to last for a certain period sufficient for reproduction, protective mechanisms holding out only so long. This version is easily reconciled with damage theories.

**Replicative senescence** is defined at the cell level. Although 'senescent' cells can be observed **in aging humans**, it is not clear to what extent this phenomenon contributes to human aging at the tissue level and the entire organism. Replicative senescence can be straightforwardly integrated into theories of programmed aging, but is neither incompatible with damage accumulation theories.

The fact that **the majority of human cancers arise in older people** and the incidence, prevalence and mortality of many cancers increase with age (cf Fig. 2.9) is compatible with both theories, perhaps better with the damage theory. In fact, there are indications that at very old age (>85 years), the incidence and aggressiveness of cancers also diminish. Again, both theories hold explanations for this (uncertain) effect, but the explanation by program theories is more elegant, i.e. cancer cells, too, are affected by the programmed loss of 'vigour'.

One might have thought that the elucidation of the **genetic basis of human premature aging syndromes** would have decided the debate. Their very existence has traditionally been used as an argument in favor of program theories. While premature aging is observed in several syndromes, including some resulting from defects in DNA repair and cell protection (→3.4), the prototypic diseases are the **Hutchison-Gilford** and **Werner syndromes**, which differ in the age of onset and the range of symptoms. Hutchison-Gilford syndrome is caused by mutations in the lamin A gene. This is puzzling, since it is everything but clear why defects in the nuclear membrane should be associated with prepubertal aging. The Werner syndrome is caused by muations inactivating the WRN helicase-exonuclease. The protein is certainly involved in DNA repair, making a good case for damage accumulation theories. Still, it may be particularly important for the maintenance of telomeres (→7.4), as might be expected for a protein involved in programmed aging. Moreover, the syndrome sets in at puberty, apparently dependent on hormonal changes, thereby fulfilling another postulate of program theories.

Hisama FM, Weissman SM, Martin GM (eds) Chromosomal instability and aging, Marcel Dekker, 2003.
Hayflick L (1994) How and why we age. Ballantine.

# CHAPTER 8

# CANCER EPIGENETICS

- In humans, cell differentiation does not involve changes in the base sequence or in the amount of DNA, with few exceptions. Rather, 'epigenetic' mechanisms are employed to establish stable patterns of gene expression. In this case, 'epigenetic' mechanisms are those which establish stably inherited patterns of gene expression in somatic cells without changes in the content or sequence of genomic DNA.
- Specific epigenetic mechanisms are involved in X-inactivation in female cells and for genomic imprinting, i.e. selective expression of alleles inherited from mother or father. Aberrant genomic imprinting is a cause of certain pediatric tumors, e.g. Wilms tumors. Loss of imprinting is observed in many carcinomas also of older people.
- An important component of epigenetic mechanisms is DNA methylation at cytosine residues in the palindromic CpG dinucleotide sequence. In normal somatic human cells, CpG dinucleotides are mainly methylated in repetitive sequences, in the body of genes and in the regulatory regions of non-expressed genes. In contrast, relatively CpG-rich sequences overlapping the transcriptional start site of many human genes, called 'CpG-islands', are usually devoid of methylation.
- In many human tumors, some CpG-islands become aberrantly methylated. This 'hypermethylation' as a rule is associated with silencing of the hypermethylated gene. In spite of such increases in methylation at specific sites, the overall methylcytosine content is decreased in many tumor cells, owing to partial demethylation of repetitive sequences and gene coding regions. This phenomenon is designated 'global hypomethylation'. It may be related to chromosomal instability. Both changes are relatively straightforwardly detected and monitored, and can be used for tumor diagnostics.
- DNA methylation is one of several interacting mechanisms that down-regulate gene expression in an increasingly stable fashion. The most dynamical of these mechanisms is deacetylation of histones in the nucleosomes of gene regulatory regions. Acetylation is enhanced by transcriptional activators binding to DNA and by co-activators with histone acetylase (HAT) activity. Conversely, deacetylation is catalyzed by histone deacetylases (HDACs) recruited by repressors or co-repressors. Methylation of histones at specific sites, prominently at the K9 of H3, by histone methyltransferases (HMTs) is a further step towards inactivation, while methylation at other sites, e.g. K4 of H3 stabilizes gene activation. Modification at K9 attracts repressor proteins, e.g. HP1, but also DNA methyltransferases (DNMTs), which 'lock in' gene silencing. DNA methylation directly inhibits the binding of some transcriptional activators and promotes the binding of repressory protein complexes which recognize

methylcytosine via MBD proteins. DNMTs also interact with HDACs and HMTs, thereby reinforcing silencing. Gene activation as well as gene inactivation employ chromatin remodeling complexes which mutually interact with activators and repressors.
➢ Aberrant gene silencing by epigenetic mechanisms in tumor cells is often, but not always accompanied by DNA hypermethylation. The underlying rules are not understood. A variety of HATs, HDACs, HMTs, and chromatin remodeling factors are implicated as oncogenes or tumor suppressors in human cancers.
➢ Activated gene states are also propagated by epigenetic mechanisms, including specific chromatin modifications. While epigenetic mechanisms leading to inappropriate gene over-expression in human cancers are overall less well understood than those leading to gene silencing, it is clear that epigenetic mechanism contribute to the inappropriate expression of oncogenic proteins.
➢ The concept of 'epigenetics' can be extended to include phenomena beyond the nucleus and even beyond a single cell. It is likely that such mechanisms contribute to the establishment of stably inherited patterns of gene expression in normal tissues and in tumors. They could encompass autoregulatory loops in transcription factor networks or growth factor signal transducing pathways acting within one cell, but also stable interaction loops between different cell types, particularly mesenchymal and epithelial cells or stromal and carcinoma cells.

## 8.1 MECHANISMS OF EPIGENETIC INHERITANCE

It may seem trivial to say that cancers are caused by genetic alterations in their constituent cells (→2.1), but it is not. Many properties of tumor cells are determined by their pattern of gene expression and do neither necessarily require structural alterations of gene products by mutations nor alterations in the structure of gene regulatory elements nor in gene dosage. Evidently, very different patterns of gene expression are established in normal cells of the human body and can in many cases be stably maintained during proliferation.

For instance, tissue stem cells retain their phenotype through several thousand divisions in a human life-time. Likewise, cell differentiation in humans is in general achieved without alterations in the sequence and amount of DNA. There are a few exceptions. Differentiation of B- and T-lymphocytes involves gene rearrangements with loss of small DNA segments from the immunoglobin and T-cell receptor genes, respectively. In some tissues, terminally differentiated cells are polyploid.

So, theoretically, a tumor cell phenotype could be achieved by mechanisms similar to those that determine normal differentiated states. In general, mechanisms leading to a stably inherited phenotype without changes in the DNA sequence and content of a cell are designated as 'epigenetic'. In reality, no malignant tumors in humans appear to be caused exclusively by epigenetic mechanisms. Instead, in most cancers, epigenetic alterations complement genetic alterations and in many, they appear to be essential.

The definition of what is considered as epigenetic has undergone fluctuations over the last decades (Table 8.1). It is generally agreed that genomic imprinting and

X-chromosome inactivation are prime examples. In both cases, identical DNA sequences are differentially expressed in a stably inherited fashion. One mechanism involved in fixing this differential expression is DNA methylation at cytosine residues, which is thus another exemplary epigenetic mechanism. DNA methylation is also instrumental in other instances of gene silencing and of facultative heterochromatin formation. Further mechanisms contribute, notably posttranslational modifications of histones, in particular methylation at specific lysine residues. In comparison, histone acetylation certainly regulates gene activity, but it is questionable whether this modification should be considered an epigenetic mechanism, because it is readily reversible, even without a cell division.

As DNA methylation and related epigenetic mechanisms are important for stably inherited gene silencing, other mechanisms must be responsible for stably inherited gene activation. Gene activation requires modification of chromatin in the regulatory regions of the gene and the assembly of a protein complex consisting of transcription factors binding to DNA at specific sites and co-activators. This complex interacts with basal transcription factors and RNA polymerases to initiate transcription, but also further modifies regional chromatin. It is clear that, but not entirely how active gene states are propagated through DNA replication and mitosis. Histone phosphorylation it thought to play a role. Another factor in this propagation is that cell differentiation is often achieved through transcription factor cascades which include an autoregulatory amplification step that make the process essentially irreversible. This could certainly be considered an epigenetic mechanism.

While the above mechanisms all occur esentially within the nucleus of a single cell, one could extend the concept of epigenetics to phenomena outside the nucleus and even to certain stable interactions between different cells. For instance, signals from one cell may elicit a response from another one which re-acts on the first and so on, until a stable steady-state is reached to which the system returns even after perturbations. Such signals are indeed exchanged in a homotypic or heterotypic fashion during normal tissue function and during tissue repair and adaptation. Intercellular loops are important in the regulation of tissue proliferation and differentiation and can be stably maintained throughout life. It does stretch the

**Table 8.1.** *Some examples of epigenetic processes in humans*

| Accepted | Considered |
|---|---|
| Genomic imprinting | Posttranscriptional histone modification, specifically histone acetylation |
| X-chromosome inactivation | Regulation by polycomb and trithorax proteins |
| Gene regulation by DNA methylation | Chromatin remodeling |
| Posttranscriptional histone modification, specifically histone methylation | Autoregulatory transcription factor networks |
|  | Mutual paracrine cell-to-cell interaction networks |
|  | Stem cell specification and maintenance |

concept, but one could regard human embryonic development with some right as a sequence of epigenetic events.

Disturbances of each of the above mechanisms contribute to human cancers.

## 8.2 IMPRINTING AND X-INACTIVATION

Most genes in humans are expressed equally strong from both alleles. About 50 genes, however, are genomically imprinted. They often occur in clusters, i.e. several imprinted genes are located within one chromosomal region (see Figure 11.5 for an example). The expression of imprinted genes differs between the alleles inherited from the mother ('maternal') and father ('paternal'). Depending on the gene, expression differences between the maternal and paternal alleles may be found in every or in selected tissue, and they may be qualitative or quantitative. The most pronounced differences are found in fetal tissues and in the placenta. This observation underlies the 'battle of the sexes' hypothesis. According to this interpretation, genes preferentially expressed from the paternal allele promote growth of the fetus and the placenta, thereby straining the mother's ressources. In contrast, genes expressed from the maternal allele tend to limit growth. This interpretation fits amazingly well with all observations. At the least, it is helpful to memorize which genes are preferentially expressed from which allele.

The best studied example of imprinted genes involves the mini-cluster consisting of *IGF2* and *H19* located near the tip of chromosome 11 at 11p15.5 (Figure 8.1).

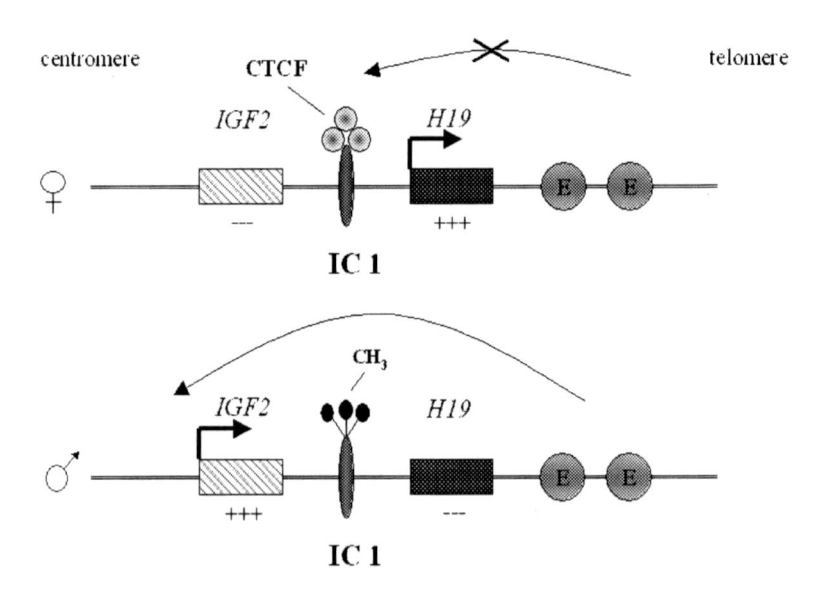

**Figure 8.1** *Regulation of the imprinted loci* IGF2 *and* H19
A simplified illustration of the mechanism by which alternate activation of *IGF2* and *H19* is achieved at maternal (top) and paternal (bottom) alleles. E: enhancer

They are imprinted in opposite ways. *IGF2* encodes a growth factor from the insulin family and is expressed from the paternally inherited allele. *H19* is located telomeric of *IGF2* and encodes a non-coding RNA which is expressed only from the maternal allele. It is not clear whether the H19 RNA has a function. Each gene has its own promoter, but both share the same enhancer which is located telomeric of the *H19* gene. On the paternal allele, the enhancer interacts with the *IGF2* promoter; on the maternal allele it interacts with that of *H19*. The choice between them is imposed by a boundary element located in an intron of the *IGF2* gene. This is the 'imprinting center' of the gene. This DNA sequence can bind the chromatin protein CTCF which prohibits the interaction between enhancer and promoter across the boundary. Binding of CTCF is sensitive to methylation of cytosines within its recognition sequence. Methylation of the boundary element on the maternal allele therefore directs the enhancer towards the *H19* promoter, diminishing expression of the *IGF2* gene. Conversely, the CTCF binding region is unmethylated on the paternal allele, allowing expression of *IGF2*.

This elegant regulatory system is disturbed in many human cancers. Most frequently, overexpression of the growth factor IGF2 is found, due to expression from every allele in the cancer cells. This corresponds to a loss of imprinting (sometimes abbreviated as 'LOI'). LOI can have several causes. In some pediatric cancers, notably in Wilms tumors (→11.3) and in germ cell cancers, imprinting may be lacking because it has never been properly set up during development. In some cancers of adults, the maternal allele has been lost by deletion or recombination. Alternatively, imprinting may be disturbed by loss of DNA methylation at the boundary site or by altered expression of chromatin proteins involved in maintaining the boundary. In some cases, the regulation of the twin locus is so disturbed that both IGF2 and H19 become overexpressed. The issue is further complicated by differential use of promoters in the *IGF2* locus. In cancers, the P3 and P4 promoters are used preferentially, other than in normal tissues.

The *IGF2/H19* locus is certainly not the only imprinted locus deregulated in human cancers. Rather, it is best studied and due to the potency of IGF2 as a growth factor highly relevant. It is likely that LOI occurs at other imprinted loci, too, where the responsible mechanisms are at present incompletely understood.

A case in point is *CDKN1C*, which encodes the CDK inhibitor p57$^{KIP2}$ (→5.2). As would be expected, this growth inhibitor is expressed from the maternal allele, albeit in a tissue-specific fashion. The *CDKN1C* gene (Figure 8.2) is also located on chromosome 11p15.5, centromeric of *IGF2*/H19, at a distance. Its imprinting is regulated by a different mechanism, although it again involves an 'imprinting center', which is in this case located within an intron of the neighboring gene *KCNQ1*. Like the boundary element in the *IGF2/H19* locus, this imprinting center is differentially methylated on maternal and paternal alleles. Intriguingly, it harbors a promoter from which another non-coding RNA is transcribed in opposite orientation to *KCNQ1*. Presently, it is not known how this leads to control of *CDKN1C* expression. However, disruption of the physical proximity between the imprinting center and *CDKN1C* by translocations leads to LOI. Such translocations are one cause of the human Beckwith-Wiedemann syndrome (→11.2), which is

172                               CHAPTER 8

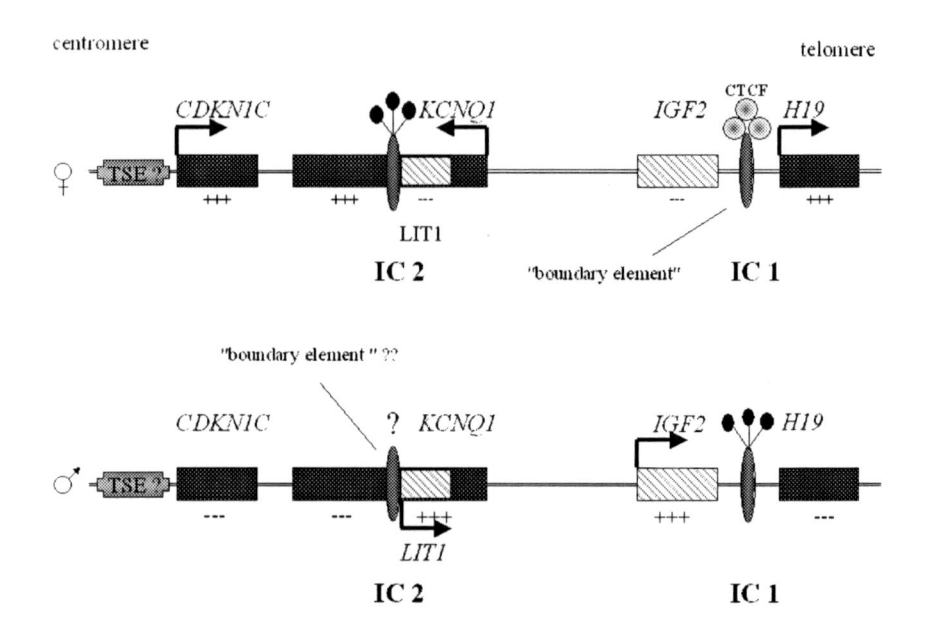

**Figure 8.2** *Regulation of the imprinted gene* CDKN1C
Mechanisms considered for the regulation of *CDKN1C* imprinting, which are largely modeled on the better understood mechanisms at the *IGF2/H19* loci (cf. Figure 8.1). TSE: tissue-specific regulatory element (hypothetical).

characterized by fetal overgrowth and a propensity to childhood tumors such as nephroblastoma and hepatoblastoma. One variant of the syndrome is caused by mutations in *CDKN1C*. Moreover, p57$^{KIP2}$ is down-regulated in several carcinomas.

Mechanisms very similar to those responsible for genomic imprinting are employed in X-chromosome inactivation. As in other mammals, the second X-chromosome in human females is largely inactivated and heterochromatized, except for the small 'pseudo-autosomal' region which is homologous to a stretch of the Y chromosome. In humans, the choice of the X-chromosome subject to inactivation is entirely random, even in extrafetal tissues. Inactivation sets in during gastrulation due to increased expression of the non-coding (!) XIST RNA from the X-inactivation center on the chromosome destined for inactivation. This increase is initially achieved predominantly by posttranscriptional stabilization and leads to coating of the chromosome to become inactivated with XIST RNA. Chromatin is remodeled and histones are hypoacetylated and hypermethylated. Heterochromatin proteins such as HP1 bind and DNA methyltransferases methylate the promoter regions of genes, thereby fixing the inactive state. Conversely, the XIST gene becomes inactivated and methylated on the active X-chromosome of males and females. The activity states of the chromosomes are then faithfully maintained through many cell generations.

In cancer research, the present of X-inactivation has been used to investigate whether cancers are monoclonal or polyclonal based on the argument that if a cancer contains cells from only one clone, then the same X-chromosome should always be inactivated (Figure 8.3). If more than one clone was involved, expression from both should be found. Formerly, this approach used enzyme variants, e.g. the isozymes A and B of glucose-6-phosphate dehydrogenase. More recently, polymorphisms in the androgen receptor mRNA (the *AR* gene is located at Xq12) were used. In most cases, tumors were found to be monoclonal. This conclusion is a bit problematic. X-inactivation is established during gastrulation and mostly finished, when organogenesis sets in. Thus, e.g, of ≈8 cells in the foregut committed to form the liver, four each (on average) will have one or the other X-chromosome inactivated. The probability to find a liver cancer with inactivation of the same X throughout is rather high because of this fact alone. The test is also limited to cancers in females. Modern methods that follow the pattern of chromosomal alterations by microsatellite analysis (Figure 8.3) yield a more reliable and more detailed picture of clonal development in human cancers.

An interesting and somewhat neglected issue is what happens to supernumerary chromosomes in aneuploid cancer cells. During development, supernumerary X-chromosomes are also inactivated. However, in some cancers they may remain active and contribute to the tumor phenotype. Conversely, germ cell cancers in males can be detected by the presence of transcriptionally active, unmethylated XIST sequences.

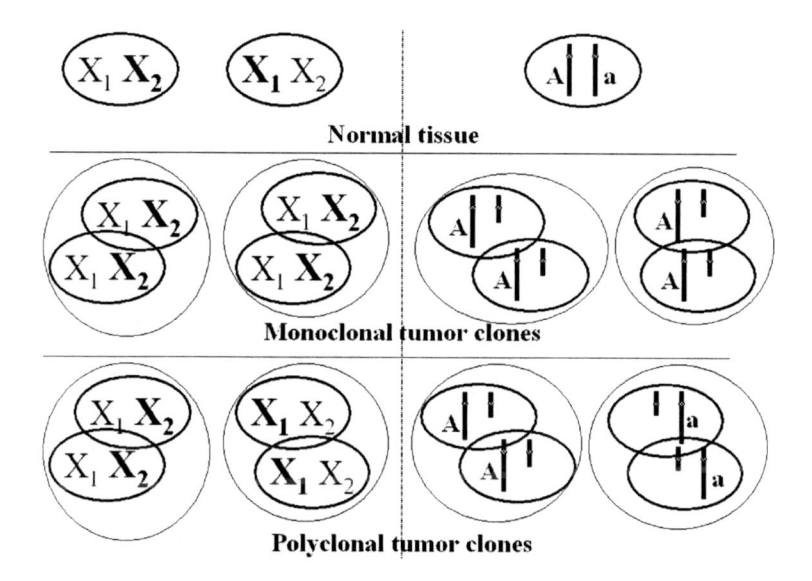

**Figure 8.3** *Clonality assays for cancer*
Left: traditional assay based on X-inactivation; Right: assay based on LOH analyses. See text for detailed explanation.

## 8.3 DNA METHYLATION

DNA methylation is instrumental for both imprinting and X-chromosome inactivation. In mammals, physiological methylation of DNA is restricted to the 5-position of cytosine residues, and again only to those in CpG dinucleotides. Since CG is a palindromic sequence, a CpG site can be non-methylated, hemi-methylated, i.e. in one strand only, or fully methylated, i.e. symmetrically in both strands (Figure 8.4). Except during replication, the usual state of methylated sites in human DNA is symmetrical methylation. After replication, which creates a hemimethylated site, symmetrical methylation is re-established by a maintenance DNA methyltransferase. If no re-methylation occurs, the site remains hemi-methylated and can become unmethylated in one daughter strand during the next round of replication. So, normally, removal of DNA methylation requires at least two cell cycles.

Methylation levels vary somewhat among normal tissues, with more pronounced changes during germ cell and embryonic development. In typical somatic cells, 3.5-4% of all cytosines are methylated. This is an average value, since DNA methylation is unequally distributed across the genome. Most methylcytosines are contained in repetitive sequences such as LINE and SINE retrotransposons interspersed in the genome and in CpG-rich satellites concentrated in peri- and juxtacentromeric regions. Genes and intergenic regions, too, are mostly methylated.

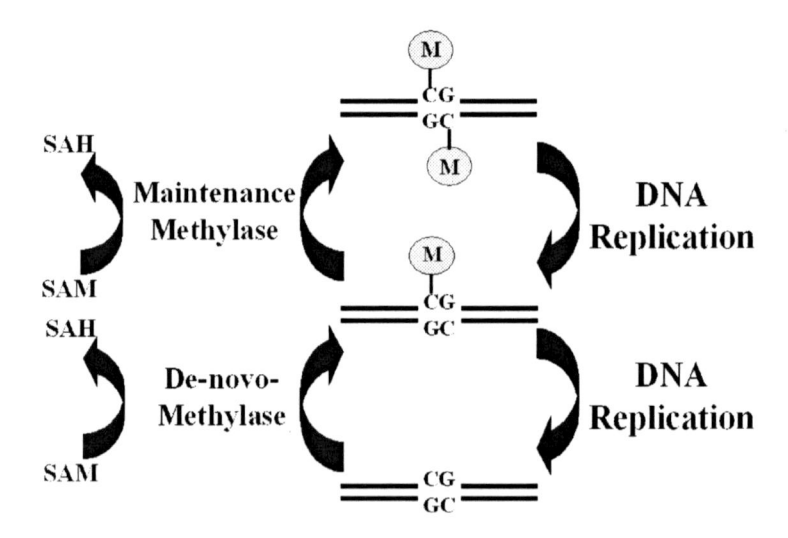

**Figure 8.4** *Establishment and changes of methylation status at CpG sites*
Hemimethylated DNA formed during DNA replication from symmetrically methylated DNA is reconverted by a maintenance methylase (normally DNMT1). Insufficient activity of the enzyme can lead to loss of methylation after two rounds of replication. Unmethylated sites can be methylated by the successive action of de-novo-methylases (normally DNMT3A or DNMT3B) and the maintenance enzyme. All enzymes use S-adenosylmethionine (SAM) as the methyl group donor, converting it to S-adenosylhomocysteine (SAH).

In contrast, regulatory regions of active genes are generally undermethylated. Specifically, in <50% of all human genes, 0.5 – 2 kb stretches around the transcriptional start site, including the basal promoter, are richer in CpG-dinucleotides than the rest of the genome, with a frequency of >0.6 found/expected in a random sequence and a higher GC content than the rest of the genome (Figure 8.5). These sequences are called 'CpG-islands'. As a rule, they remain unmethylated throughout development and in all tissues. A prominent exception are the CpG-islands on the inactive X-chromosome which are methylated and the respective genes are silenced.

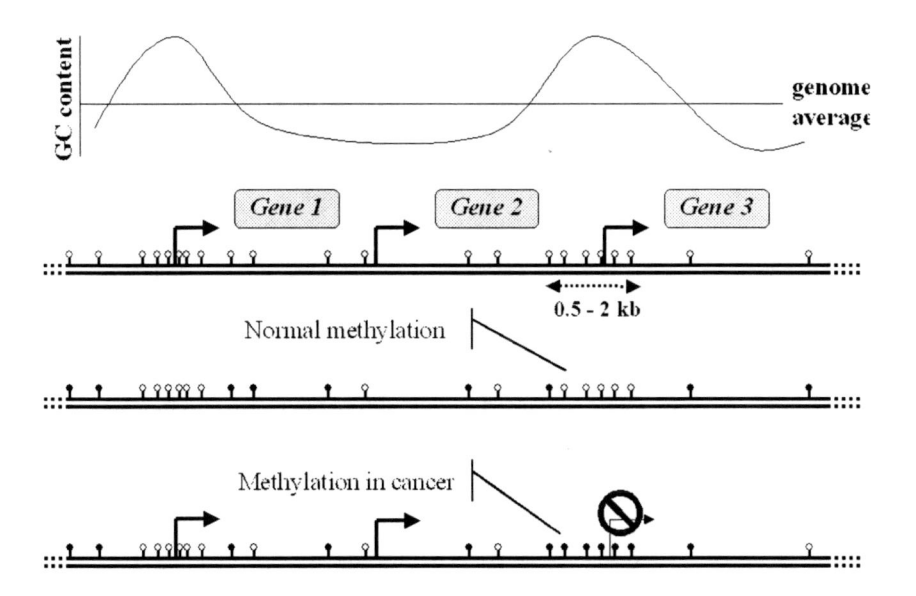

**Figure 8.5** *CpG islands in the human genome*
Top: Schematic illustration of variations in GC content and distribution of CpG sites (stick and circle symbols) in the human genome. Note that gene 2 does not possess a CpG island, like ≈40% of human genes. Bottom: methylation patterns in normal and cancer cells. As customary, methylated sites are indicated by filled and unmethylated sites by open circles.

Apparently, the lack of methylation helps to mark CpG-islands as regions of potential transcription in the genome. Typically, genes with CpG-island type promoters can be transcribed in several different cell types. DNA methylation also regulates the transcription of some other genes without CpG-islands, including some with cell-type specific expression.

CpG-islands stand out from the rest of the genome, because they contain more CpG-dinucleotides. More precisely, the rest of the genome contains less, as a consequence of cytosine loss during evolution. Hydrolytic deamination of cytosine occurs frequently, spontaneously or induced by chemicals, and yields uracil. This base is very efficiently recognized as incongruous and accordingly repaired (→3.1).

In contrast, methylcytosine yields methyluracil, i.e. thymine, albeit in a G-T mismatch. Such mismatches are accordingly repaired preferentially towards G-C, with the help of the protein MBD4 which recognizes the methylcytosine in the opposite strand of the CpG palindrome (→3.1).

In spite of such precautionary mechanisms, over evolutionary periods, CpGs have become depleted from heavily methylated sequences by mutating to TpG (or CpA). This depletion has not affected sequences exempt from methylation and in this fashion has sculpted CpG-islands out of the genome background. In fact, the mutation rate at methylated cytosines remains higher in the present. Therefore, methylated CpGs are preferential sites of mutations not only in the human germline, but also in cancers.

DNA methylation patterns change substantially during development. During germ cell development, DNA is first widely demethylated and then remethylated to yield distinctive patterns in oocytes and sperm. Differential methylation at imprinted genes is also established during this period. Following fertilization, methylation again decreases across the genome, although some specific sites, e.g. in imprinted genes, are exempt from these changes. Extraembryonal tissues remain strongly demethylated, whereas in the cells of the fetus proper the genome is subjected to a wave of de-novo-methylation during gastrulation. This process largely establishes the overall level of methylation found in the DNA of somatic cells. Demethylation of genes expressed in a cell-type specific fashion then leads to the DNA methylation patterns of the various cell types. Of note, CpG-islands are in general exempt from these changes and remain unmethylated throughout development. Likewise, methylation patterns of imprinted genes follow their own rules.

These wide swings in overall methylation levels likely reflect the necessity to completely reprogram the expression of the genome, once during the development of germ cells and then again during embryonic development. A major reason why cloned embryos are often defective appears to be a failure to achieve this reprogramming properly. Very low levels of DNA methylation in germ cells may moreover signify a state of chromatin that facilitates recombination during meiosis.

Given this background, it is not unexpected that the altered state of cancer cells is often also accompanied by alterations in DNA methylation. Basically, two types of alterations can be distinguished (Figure 8.6). Both can occur in the same cell. In many cancers, overall DNA methylation levels are diminished by up to 70% compared to the corresponding normal cell type. This decrease affects mostly methylcytosine contained in repetitive sequences and is therefore termed 'global' or 'genome-wide' 'hypomethylation'. In contrast, specific sites can become 'hypermethylated'. Hypermethylation occurs, in particular, at CpG-islands which are never methylated otherwise. Like the extent of hypomethylation, that of hypermethylation differs widely between cancers, even of the same histological type. In some cancers, only individual CpG-islands become hypermethylated, whereas several hundreds are afflicted in others. Moreover, hypermethylation affects different genes in different cancers, although some genes are prone to hypermethylation in many cancer types (Table 8.2). Alterations of DNA methylation may also affect the expression of imprinted genes (→8.2).

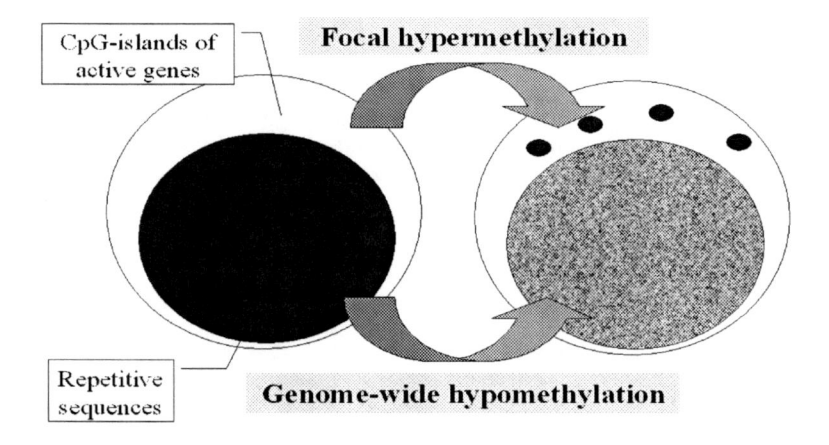

**Figure 8.6** *Alterations of DNA methylation in human cancers*
In normal somatic cells, most of the genome is densely methylated, specifically repetitive sequences. CpG-islands form a prominent class of sequences exempt from methylation. In cancer cells, up to several hundred CpG-islands become hypermethylated (black dots), whereas repetitive sequences, in particular, are hypomethylated.

Hypermethylation of CpG-islands is almost invariably associated with stable silencing of the affected genes. Therefore, hypermethylation is a very efficient means of gene inactivation in cancers. It is now regarded as a mechanism of tumor suppressor gene inactivation comparable to mutation and deletion. For instance, the *CDKN2A* locus (→5.2) is inactivated in a wide variety of human cancers. In almost every cancer type, mutation, deletion, and promoter hypermethylation are all observed as mechanisms of inactivation, although their relative contributions vary. One important difference towards mutation and deletion, however, is that hypermethylation is in principle reversible by inhibitors of DNA methylation. Of course, in cancers with hundreds of genes inactivated by hypermethylation, not each one of them is a tumor suppressor like *CDKN2A*. Rather, gene silencing via DNA methylation may reflect a 'slimming' of gene expression.

By comparison, global hypomethylation would be expected to increase gene expression across the genome at large. There is, indeed, some evidence for hypomethylation to increase the level of 'transcriptional noise' and to cause inappropriate expression of certain sequences, e.g. of retrotransposon sequences and of 'cancer testis antigens', i.e. genes that are normally restricted to developing male germ cells. More importantly, perhaps, global hypomethylation is associated with enhanced chromosomal instability. The underlying mechanism are under investigation. Possibly, decreased methylation in pericentromeric repeats and interspersed repetitive sequences facilitates illegitimate recombination and chromosomal loss during mitosis.

**Table 8.2.** *A selection of hypermethylated genes in human cancers*

| Gene | Function | Cancers with hypermethylation | Chapter |
|---|---|---|---|
| *RB1* | cell cycle regulaior | retinoblastoma | 5.2 |
| *CDKN2A* | cell cycle regulator | many cancers | 5.3 |
| *CDKN2B* | cell cycle reguator | selected leukemias | 5.3 |
| *SOCS1* | signal transduction | selected cancers | 6.9 |
| *APAF1* | apoptosis | selected cancers | 7.3 |
| *CDKN1C* | cell cycle regulation | selected carcinomas | 8.2 |
| *RARB2* | retinoic acid signaling | many carcinomas | 8.5 |
| *CDH1* | cell adhesion, modulation of WNT signaling | many carcinomas | 9.2 |
| *SFRP1* | modulation of WNT signaling | colon cancers and others | 13.2 |
| *MLH1* | DNA mismatch repair | many carcinomas | 13.4 |
| *VHL* | regulation of response to hypoxia | renal clear cell carcinoma | 15.4 |
| *RASSF1A* | RAS antagonist | many carcinomas | 15.5 |
| *ESR1* | estrogen response | breast cancer and others | 18.4 |
| *ESR2* | regulation of estrogen response | breast cancer and others | 18.4 |
| *GSTP1* | detoxification, cell protection | prostate carcinoma and selected other carcinomas | 19.3 |
| *MGMT* | Reversion of base alkylation in DNA | various | 22.2 |

DNA methylation patterns are set up by specific enzymes, of which DNA methyltransferases are better characterized, whereas DNA demethylases remain stubbornly elusive. In somatic cells, DNA methyltransferase 1 (DNMT1) provides the major activity. The enzyme prefers hemimethylated over unmethylated DNA by a large margin. Hemimethylated DNA originates from symmetrically methylated DNA during replication (Figure 8.4), because DNA polymerases only insert unmodified cytidines. Its preference of hemimethylated DNA makes DNMT1 an ideal enzyme to stably propagate DNA methylation states during cell proliferation. Unmethylated sites remain unmethylated, whereas methylated sites become remethylated in both daughter strands. Accordingly, DNMT1 is tightly associated with the replisome, where it binds to PCNA, and is preferentially expressed in S-phase and in proliferating cells. Establishment of methylation at previously unmethylated sites, 'de-novo-methylation' (Figure 8.4), likely requires other enzymes. The methyltransferases DNMT3A and DNMT3B may be responsible for most de-novo-methylation occurring during development.

The mechanisms leading to demethylation are more obscure. One obvious mechanism is passive (Figure 8.4). If DNA methyltransferases are kept from remethylating newly replicated DNA, successive rounds of cell proliferation will yield unmethylated DNA. During embryonic development, this mechanism is likely supplemented by active enzymatic demethylation. Which protein is responsible is a moot point. One candidate is a methylcytosine-specific glycosylase acting similar to a repair enzyme (→3.1).

The known properties of DNMTs go some way to explain how changes in overall methylation could come about. They do, however, not account for patterns of methylation at specific genes and sites. Specifically, it is not clear how hyper- and hypomethylation in cancer cells are generated. It is likely that DNMTs are directed towards or prevented from certain sites and genes by interaction with other proteins. As discussed below, histone modifications and the proteins establishing and interpreting these may also direct DNMTs. Conversely, active transcription complexes appear to exclude DNMTs from promoter regions. This is apparently part of the mechanism by which CpG-islands are kept free of DNA methylation.

Cancer cells tend to over-express DNA methyltransferases. Compared to normal tissues, the expression of DNMT1 is almost always increased. However, since DNMT1 is regulated in parallel with DNA synthesis in normal cells, a large fraction of this increase may simply reflect increased proliferation. Although demonstratable in model experiments, it is questionable whether altered expression of DNMT1 as such is responsible for aberrant methylation in cancer cells. In contrast, increased expression of DNMT3A and DNMT3B observed in some cancers is certainly significant, since these enzymes are normally expressed at low levels in somatic cells. It is, however, still unclear, to which extent they are responsible for hypermethylation.

Maintenance of correct levels and patterns of DNA methylation is not only dependent on chromatin factors and DNA methyltransferases, but also on the availability of the methyl group donor S-adenosylmethionine (SAM, Figure 8.4). More precisely, the reactions catalyzed by DNA methyltransferases and many other methyltransferases in the cell are influenced by the ratio of the substrate and product, SAM and S-adenosylhomocysteine (SAH). SAM is synthesized from the essential amino acid methionine, which is recycled through several steps from SAH (→20.3). The efficiency of recycling is influenced by dietary factors, prominently the supply of folic acid and vitamin $B_{12}$ and varies between humans as a consequence of genetic polymorphisms in enzymes involved in folate metabolism and the 'methyl cycle' (→20.3). Deficiencies in the diet and genotype exert a synergistic effect on the DNA methylation levels in rapidly turning over cells. So, these prevalent polymorphisms may contribute to cancer predisposition (→2.3) and some alterations in DNA methylation observed in human cancers could be related to a synergistic effect of diet and genetic predisposition.

## 8.4 CHROMATIN STRUCTURE

While promoter methylation is a conspicuous mark of gene silencing, DNA methylation is certainly not the only mechanism involved. Many invertebrates achieve gene silencing without DNA methylation at all and even in mammals not all silenced genes become methylated. Rather, DNA methylation in general represents one of the last steps in a chain of events leading to stable gene repression (Figure 8.7) and often serves to fix the state of a gene silenced by other means. It has therefore been considered a 'lock-in' mechanism, but – to keep with the metaphor - in some cases, as during X-inactivation, the door seems already to have been locked by other mechanisms and DNA methylation acts as an additional bolt. Importantly, the chain of events presented in Figure 8.7 is not a gradual transition. Rather, feedback and feed-forward mechanisms tend to enforce and stabilize the active and inactive states of a gene.

**Heterochromatinization** →

Methylation at K4 ← Acetylation at K9 ← unmodified H3 → Methylation at K9 → DNA Methylation

← **Transcriptional activity**

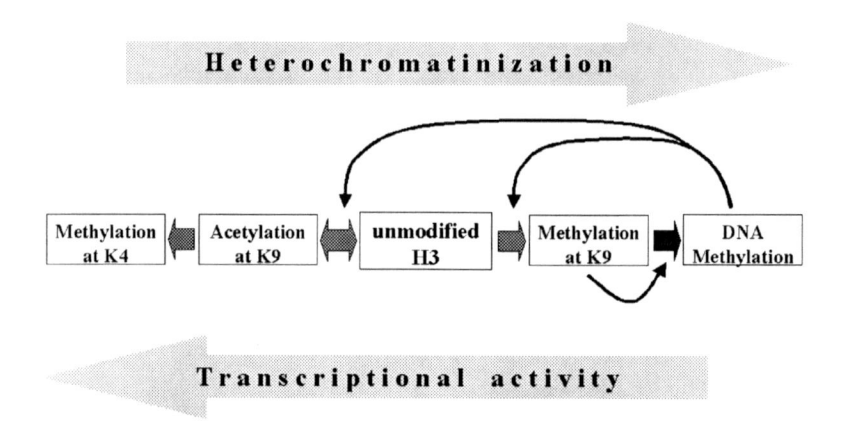

**Figure 8.7** *Interaction of histone H3 modifications and DNA methylation in the control of gene silencing and gene activation*
For simplicity, only some modifications of H3 and none at H4 are displayed. Also, none of the proteins establishing, removing, and recognizing the various modifications at H3, H4, and DNA are shown. These and further factors mediate the indicated feedback interactions (arrows), of which only those best established are illustrated.

The mechanisms leading to gene silencing involve histone modifications and changes in the binding of non-histone proteins to DNA (Figure 8.7). Transcriptionally active genes are characterized by acetylation of histones H3 and H4 as well as by methylation of H3 at Lys4 (K4). These changes tend to loosen the attachment of DNA to nucleosomes, allowing remodeling and the binding of transcriptional activators. Transcription activating factors recruit co-activators and exclude co-repressors. Typical co-activators exhibit themselves histone acetyl-transferase (HAT) activities or attract additional proteins which acetylate and methylate histones to yield an active nucleosome structure. In this way, the active state of a gene tends to be self-reinforcing. It is not entirely clear, how this active state is propagated through S phase and mitosis. Several mechanisms are discussed. Of these, histone phosphorylation at specific sites is best characterized.

The sequence towards gene silencing starts with loss of transcriptional activator proteins and/or their replacement by repressors. These changes elicit deacetylation by histone deacetylases (HDACs). Proteins acting as transcriptional repressors recruit deacetylases and histone methyltransferases (HMT) with SET domains which methylate e.g. H3 at the K9 position. Some co-repressors exhibit HDAC or HMT activities themselves. A specialized class of DNA-binding repressors is constituted by polycomb proteins. Two complexes of such proteins active in man, which exhibit HMT activity.

Acetylation and deacetylation are apparently quite rapid and are readily reversible, even in a non-proliferating cell. In contrast, histone methylation is certainly not as rapidly reversible and may require replacement of a nucleosome to be removed. This would normally occur in the course of DNA replication and cell division. So, histone H3 methylations at K9 and K4 appear to constitute already a quite stable mark of gene inactivation or activation, respectively. Methylation of H3 K9 is recognized by further chromatin proteins including HP1 which is a characteristic component of heterochromatin. The same modification may also serve to direct DNMTs to the DNA on that nucleosome. In turn, DNMTs may attract HMTs and HDACs leading again to reinforcement of the chromatin state – in this case an inactive one. Moreover, the assembly of proteins at one modified nucleosome may target neighboring nucleosomes. For this reason, inactive chromatin states, as DNA methylation, tend to spread.

Methylcytosine in DNA interferes with transcription in two ways. First, some transcriptional activators cannot recognize DNA, if CpGs in their binding site are methylated. Secondly, specialized repressor proteins recognize methylated DNA. These include MeCP2, MBD2 and MBD3 which bind preferentially to methylated DNA, blocking access and also recruiting chromatin-remodeling complexes and HDACs that buttress the repressed state (Figure 8.7). A silenced chromatin state established in this fashion amounts to facultative heterochromatin. It is normally stably inherited and not easily reversed without extensive remodeling of chromatin and demethylation of DNA which normally requires several cell cycles. Such extensive remodeling is mostly found during development. A further common consequence of heterochromatization, which may help to ensure its propagation, is a shift of replication into the later part of S phase.

The intricate and complex mechanisms involved in histone modification and chromatin remodeling are incompletely understood, even more so in humans. It is therefore likely that disturbances in these mechanisms contribute to human cancer to a much larger extent than is presently known. Alterations in DNA methylation are relatively easy to detect and have been found in a large variety of human cancers. As they represent the end of a sequence (Figure 8.7), related epigenetic mechanisms leading to aberrant gene silencing may be even more ubiquitous. An indication of the importance of these mechanisms is provided by the exquisite sensitivity of a variety of cancers to HDAC inhibitors.

Indeed, several proteins involved in chromatin modification – beyond the DNMTs – are aberrantly expressed or mutated in human cancers. HATs, HDAC, and HMTs have all been found as parts of fusion proteins in leukemias and lymphomas. Misdirection of co-repressor proteins is the crucial event, e.g., in promyelocytic leukemia (→10.5). Such misdirection may, of course, also be responsible for DNA hypermethylation in carcinomas.

Overexpression of the polycomb repressor protein EZH2 has been found, e.g., in prostate and breast cancers (→19.3). The polycomb protein BMI1 is also overexpressed in some cancers, prominently in acute leukemias arising from hematopoetic stem cells. Maintenance of the stem cell population in these cancers requires the BMI1 repressor. In particular, BMI1 prohibits the activation of $p16^{INK4A}$ transcription in response to the continuous proliferation of the cancer cells (→7.4).

Conversely, loss of repressors can also contribute to cancer development. A prominent repressor protein is RB1 (→5.2) which directs deacetylation to promoters by binding to E2F factors. RB1 also elicits remodeling at E2F-dependent promoters, which could in principle make inactivation of these promoters irreversible. Compared to other pocket proteins in the RB family, such as p130 and p107, RB1 may therefore cause more persistent and in some cases irreversible inactivation of such promoters, effectively establishing replicative senescence or enforcing terminal differentiation. These relationships could explain, why loss of RB1 function represents such a large step towards cancer, and why RB1 of the three closely related pocket proteins is by far the most frequently one mutated in cancer.

The precise functions and mechanisms of chromatin remodeling are only beginning to be understood. Certainly, chromatin remodeling, i.e., the reorganization of nucleosomes along a DNA sequence, is necessary when the activity state of a gene is fundamentally switched, in either direction. This mechanism could therefore be essential both in epigenetic overexpression and silencing in human cancers. The extent to which it is involved, is not known. One specific component of the SWI/SNF remodeling complex, hSNF5/INI1, which is encoded by a gene at chromosome 22q11, behaves as a tumor suppressor in rare childhood cancers of the nervous system. Apparently, in these cancers, correct differentiation fails because chromatin remodeling cannot be properly performed.

## 8.5 EPIGENETICS OF CELL DIFFERENTIATION

Genetic changes like the loss of RB1 function or that of a chromatin remodeling protein like hSNF5/INI1 cause cancer by obliteration of epigenetic mechanism in which these proteins are involved. It is perhaps not incidental that the cancers caused in these instances impress primarily as failures of differentiation. One could go one step further and ponder whether some cancers might be caused by purely epigenetic mechanisms and represent a specific, if aberrant form of cell differentiation. This idea is, in fact, the core of an older theory that is now obsolete as a consequence of the discovery of the multitude of genetic changes present in the great majority of human cancers. So, if any human cancers are caused purely by epigenetic mechanisms, they are rare. Most likely, they would be childhood cancers characterized by failed differentiation. Some cases of Wilms tumors may come close (→11.3).

However, the theory contained an important core of truth which is still relevant and may in fact now be understudied. Cell differentiation and cancer development resemble each other in that for a large number of genes 'cell-type-specific' patterns of expression, including strong activation as well as strict silencing must be stably established. It is plausible that the same mechanisms might be at work in both processes. These mechanisms are now partly understood for gene silencing in cancer cells (→8.4). By comparison, it is hardly known how overexpression of genes in cancer cells is established by epigenetic mechanisms.

For instance, overexpression of the EGFR is an important step in the progression of many carcinomas. In some cases, this overexpression is due to gain of chromosome 7p or regional amplification at 7p12, but these alterations are not generally found in cancers with overexpression of the protein. So, overexpression is likely caused by deregulation of gene expression secondary to alterations in other genes. For instance, degradation of the EGFR requires the CBL protein which could be lacking or the protein could be stabilized as a consequence of altered phosphorylation by PKC enzymes (→6.5). Still another possibility is that the *ERBB1* gene encoding the receptor is locked in an activated state by epigenetic mechanisms, comparable to the silenced state of a tumor suppressor gene established by promoter hypermethylation (→8.3).

For the establishment and maintenance of an active gene state across cell division, the continuous presence of transcriptional activators is necessary. Current understanding is that transcriptional activators binding to promoter and enhancer sequences recruit co-activators to form a large protein complex, a transcriptosome, which modifies the local chromatin and guides the actual transcriptional apparatus including RNA polymerase. This local chromatin state is transmitted through mitosis in a largely unknown fashion.

In any case, the differentiation state of a cell strictly depends on the pattern of transcriptional activators expressed. Cell differentiation often involves cascades of transcription factors which successively activate each other. These cascades often involve autoregulatory loops that make the process essentially irreversible.One well-studied example is myoblast differentiation (Figure 8.8). It is initiated and carried through by muscle-specific transcription factors (MSTFs or MRFs) that belong to the basic helix-loop-helix family (bHLH) like the MYC proteins (→4.3). Like these, they bind to specific DNA sequences called E-boxes. The best-known of these factors is MYOD. E-boxes are present in genes encoding the typical proteins of muscle cells, but also in the enhancers of the genes encoding the MRFs. So, activation of MRFs beyond a threshhold leads to an autocatalytic cascade, in which several MRFs become expressed at increasing levels until full differentiation is achieved. The threshhold may be determined by the expression of MYC factors, but also by specialized inhibitor proteins, called ID. The four small ID proteins belong to the same general class of proteins as MRFs and MYC proteins, but lack a DNA-binding domain. Rather, they heterodimerize with and block the action of cell-type specific bHLH transcription factors. Overexpression of ID proteins, sometimes as a consequence of gene amplification, is a common finding in human cancers, particularly in carcinomas.

Importantly, transcription factor cascades not only lead to expression of cell-type specific products during normal cell differentiation, but also turn off cell

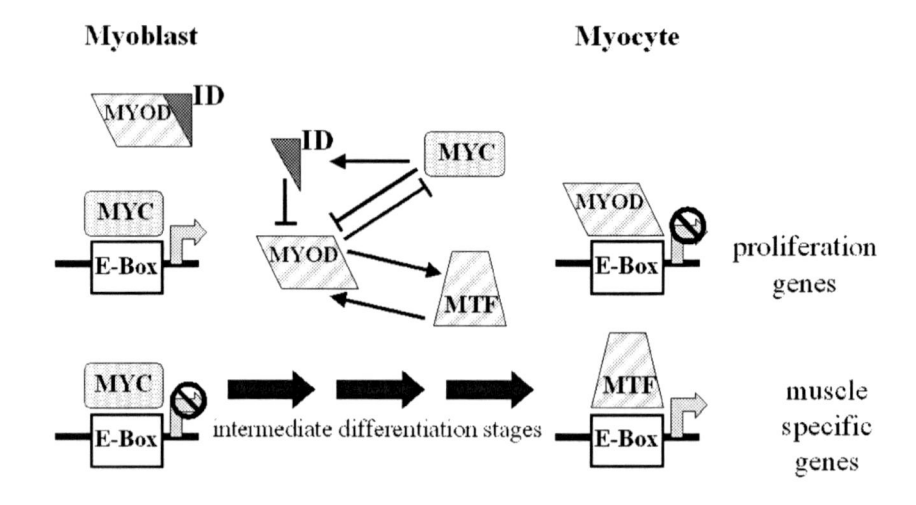

**Figure 8.8** *An autocatalytic transcription factor cascade during cell differentiation*
An interaction of pro-proliferative (MYC, IDs) and cell-type specific (MYOD, MTF) basic helix-loop-helix proteins regulates the terminal differentiation of myoblasts. The (simplified) autocatalytic loop between MYOD and other myoblast-specific transcription factors (MTF) at the center of the figure is responsible for the irreversibility of the process.

proliferation by interacting with cell cycle regulators. In muscle cells, cell proliferation and differentiation are mutually exclusive. MYOD not only competes with MYC, but also represses the transcription of the AP1 factors FOS and JUN in differentiated cells, while in proliferating myoblasts the reverse occurs. So this system tends to be either in one (proliferation) or the other (differentiation) state. In addition, RB1 is activated during muscle differentiation and inactivates E2F-dependent promoters required for cell proliferation, while supporting the action of MYOD.

Similar transcription factor networks are thought to act in the differentiation of other cell types. For instance, in the differentiation of hepatocytes, insulin-producing cells of the pancreas, and proximal tubule cells of the kidney, the transcriptional activators HNF4 and HNF1 may be organized in a mutually activatory autocatalytical loop stabilizing the differentiated phenotype.

In many cell types, retinoids contribute to differentiation and growth arrest. Retinoids activate one of several retinoic acid receptors, named RAR$\alpha$, $\beta$, or $\gamma$, which are organized in an autocatalytic cascade. The *RARB2* gene promoter contains a sequence, named a RARE (retinoic acid responsive element), to which retinoic acid receptors can bind and increase transcription of the gene. During differentiation induced by retinoic acid, one of the other receptors ($\alpha$ or $\gamma$, depending on the cell type) initiates transcription of RAR$\beta$, which amplifies its own induction. Loss of the initiating RAR or inactivation of the *RARB* gene blocks the cascade. Accordingly, a translocation linking RAR$\alpha$ to a repressor domain blocks the differentiation of promyeloid cells in an acute leukemia (→10.5). In different carcinomas, the retinoid activatory cascade is interrupted by hypermethylation of the *RARB2* promoter.

In summary, then, disruption of epigenetic mechanism leading to cell differentiation is an important component in cancer development. Conversely, cancer cells may set up their own epigenetic cascades that maintain a status of gene expression compatible with continuous tumor growth.

## 8.6 EPIGENETICS OF TISSUE HOMEOSTASIS

Stably inherited phenotypes can not only be achieved by mechanisms acting within one cell, but also by interactions between similar or distinct cell types. In many tissues, such interactions occur between the mesenchymal and the epithelial component. They exchange paracrine factors in the steady-state of the tissue. Upon wounding or infection, the steady-state is disturbed and the exchange intensifies leading to wound healing and immune responses.

In the skin, paracrine factors are exchanged between keratinocytes in the epidermis and fibroblasts in the dermis (Figure 8.9). Epithelial keratinocytes in the epidermis produce, a.o., the cytokine interleukin-1 (IL1), of which normally only a fraction reaches the underlying dermal tissue. Mesenchymal cells produce and secrete low amounts of the fibroblast growth factor 7 (FGF7), which is also called keratinocyte growth factor (KGF), because it stimulates the proliferation of

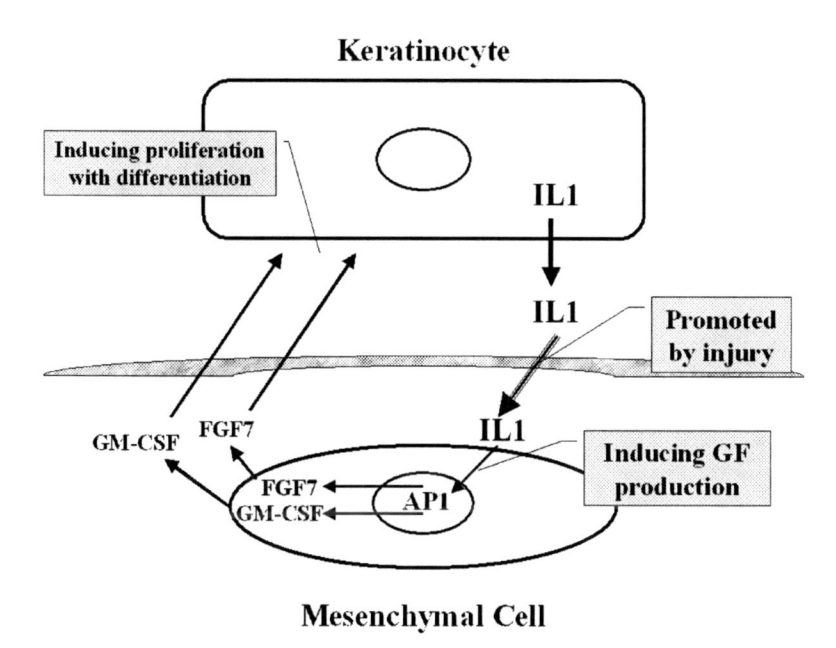

**Figure 8.9** *Paracrine regulation of wound healing in the skin*
GF: growth factor. See text for futher explanation.

keratinocytes and other epithelial cells. They likewise produce low amounts of GM-CSF, a factor stimulating the proliferation and maturation of myeloid cells. It acts on keratinocytes, too, promoting their proliferation, but more strongly their differentiation. Upon damage to the epidermis, IL1 is released and binds to its receptor on fibroblasts. This activates the JNK and p38 MAPK pathways (→6.2) and leads to an increased activity of JUN transcriptional activators at the promoters of the *FGF7* and *GMCSF* genes. Increased secretion of FGF7 and GM-CSF stimulates proliferation and differentiation of the keratinocyte compartment, until it is healed and the IL1 concentration returns to normal levels. This is, of course, a simplified description, as many more factors are involved. Moreover, during wound healing a substantial reorganization of the extracellular matrix takes place which is performed, e.g., by proteases secreted from fibroblasts and invading immune cells in response to IL1 and GM-CSF (→9.3). Immune cells are attracted by cytokines, chemokines and other factors from the activated fibroblasts. Another level of regulation is required to limit the ensuing inflammation.

The crucial argument in this example is that normal tissues use mutual paracrine interactions to achieve tissue homeostasis and to react appropriately to its disturbances. This could certainly be considered an epigenetic mechanism. Importantly, this type of interaction also takes place in cancers between tumor cells and the tumor stroma. However, in cancers, these interactions are grossly disturbed and do not lead back to a steady-state, but to a continued expansion of the tumor mass (→9.6). Disturbances of paracrine interactions are also crucially involved in the co-carcinogenic effect of HIV (Box 8.1).

A case in point is angiogenesis (→9.4), an essential process in many cancers,

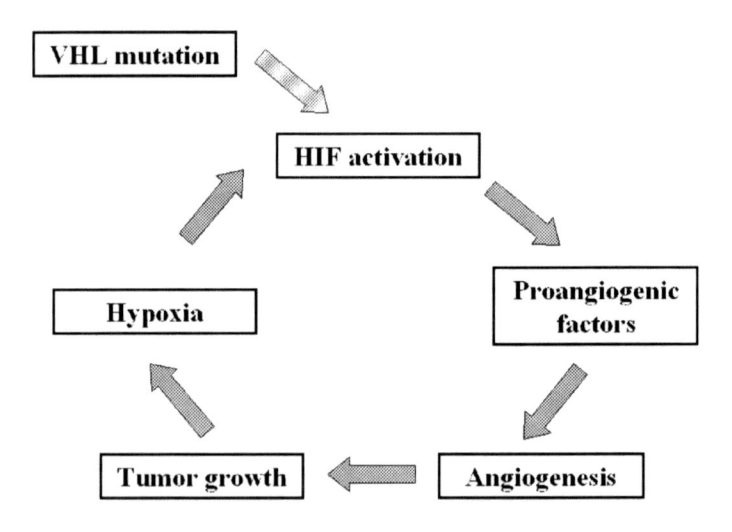

**Figure 8.10** *The vicious cycle of angiogenesis in cancer*
See text for further explanation.

that can be activated by genetic or epigenetic mechanisms (Figure 8.10). Some cancers carry mutations which lead to the constitutive production of pro-angiogenic growth factors such as VEGF or bFGF (FGF2) that stimulate branching of capillaries and proliferation of endothelial cells. In benign tumors and malignant renal cell carcinomas arising in the Von-Hippel-Lindau syndrome (→15.4), this constitutive production is due to mutations in a regulator of the cellular response to hypoxia. As a consequence, HIF (hypoxia-induced factor) transcription factors are overactive and enhance the production of pro-angiogenic growth factors. So, in this case, a specific genetic change is responsible for increased angiogenesis.

Other cancers do not carry according genetic defects. Instead, when the tumor mass has exceeded the size that allows sufficient supply of oxygen by diffusion, tumor cells become hypoxic. This elicits a normal physiological response, viz. induction of HIF transcription factors leading to the production of angiogenic growth factors. These stimulate angiogenesis. Accordingly, the supply of oxygen (and other nutrients) improves allowing further expansion of the tumor to the point where oxygen becomes limiting again leading to further induction of HIF, angiogenic growth factors and continued angiogenesis. This epigenetic vicious circle is often exacerbated by genetic defects in the cancer cell, e.g. TP53 mutations that diminish the production of anti-angiogenic factors.

It is important to realize that by such paracrine interactions the tumor cells change the character of the normal cells with which they interact. While these cells need not become genetically altered (although this has been occasionally reported), they are persistently activated, which can alter their properties considerably. Experimentally, it can be demonstrated that stromal cells from malignant tumors acquire an 'epigenetic memory', i.e., their activated state tends to persist even if the actual cancer cells are removed. This is plausible, if one considers the role of epigenetics in cell differentiation (→8.5).

Interactions between tumor cells and neighboring normal stromal cells are particularly important during metastasis. Setting up stable interactions is crucial for the survival and eventual expansion of metastatic cells. This presupposes a selection for those cancer cells which fit into the target tissue and their successful adaptation to the local environment. For instance, metastatic prostate cancer cells adapt so well to the microenvironment in the bone by interacting with local osteoblasts and osteoclasts that they have been termed 'osteomimetic' (→19.4). As in normal tissues, these mutual interactions are to a great deal mediated by exchange of paracrine growth factors, and to some extent by direct cell-to-cell interactions.

A final example of epigenetic mechanisms relevant to both normal tissues and cancer concerns stem cells. Stem cells are defined as cells with unlimited proliferation potential, the ability to generate differentiated derivatives, and the ability to do this by assymetric division generating another stem cell and a more differentiated daughter cell (Figure 8.11). Stem cells which can give yield to any cell-type (in principle) are called pluripotent. In a healthy adult human, two types of stem cells are present: those of the germ-line and tissue stem cells. Tissue stem cells are probably not pluripotent. Rather, they can give rise to a limited number of diverse cell types. They are therefore also labeled as 'tissue precursor' cells and as

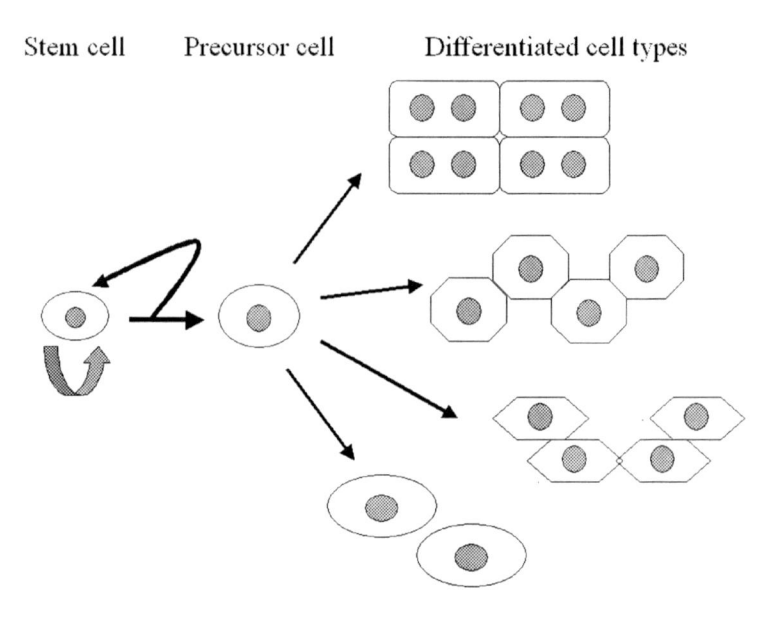

**Figure 8.11** *Properties of stem cells*
Stem cells are characterized by the abilities to self-renew, divide assymetrically, and give rise to diverse differentiated cell types.

'multi- or oligopotent'.[10] As the DNA of stem cells does not differ from that in (most) somatic cells, they must be defined by epigenetic mechanisms. In fact, both intercellular and intracellular mechanisms are involved.

Intracellular mechanisms include the expression of hTERT and of active telomerase, which allows the escape from replicative senescence (→7.4). The other mechanism inducing replicative senescence upon continued cell proliferation, viz. induction of CDK inhibitors (→7.4) is likewise repressed. Specifically, accumulation of p16$^{INK4A}$ appears to be prevented by a polycomb repressor complex with BMI1 (→8.4) as its crucial component. It is not precisely clear, how pluripotency is maintained. In germ cells, expression of specific transcription factors appears to be involved. For instance, the transcriptional activator OCT3/4, now systematically called POUF5, is expressed in the germ line and in the early embryo. Its expression is lost, when pluripotent cells in the epiblast become committed to specific tissues. It is also strongly expressed in germ cell cancers and is necessary for their continuous proliferation. Germ cells and perhaps tissue precursor cells also have patterns of DNA methylation different from those typical for somatic cells. These are accordingly reflected in germ cell cancers. In germ cells as in germ cell cancers, moreover, expression levels of both RB1 and TP53 may be relatively low.

---

[10] A note of caution: definitions in this field are often very loosely used. So, it is not unusual to find all basal cells in an epithelium summarily denoted as tissue precursor cells or tissue stem cells being called 'pluripotent' on the argument that they can give rise to distinct cell types.

This may prohibit a commitment to specific pathways of differentiation as well as replicative senescence.

In the testes and ovaries, respectively, primordial germ cells reside in an environment that allows their maintenance and organized differentiation towards mature oocytes and spermatozoa. Oogenesis is almost completed after the fetal period, whereas spermatogenesis continues throughout life and the stem cells remain present in the epithelia of the seminiferous tubules of the testes. There, they divide assymetrically to give rise to spermatozoa after several differentiation steps including meiosis. The location within the testicular epithelia is on one hand crucial for the survival of these cells, as they tend to undergo apoptosis outside this environment. On the other hand, primordial stem cells can principally develop into tumors when placed into the wrong environment. This is exemplified by experimental teratocarcinomas in specific mouse strains, which appear to arise by purely epigenetic mechanisms. However, germ cell tumors in humans do show chromosomal aberrations. In contrast, normal primordial germ cells contain the same amount and the same sequence of DNA as somatic cells. Thus, the mechanisms that make them immortal and pluripotent are purely epigenetic. The tubules of the testes are an example of a stem cell 'niche'. This niche is actively maintained not so much by the germ cells themselves, but by the surrounding testicular tissue. Cells in this tissue provide, e.g., SCF (stem cell factor), a ligand for the receptor tyrosine kinase KIT.

Tissue stem (precursor) cells are less obvious and in humans they are only beginning to become characterized. Like primordial germ cells, however, they do not differ in DNA amount or sequence from their differentiated progeny. Instead, their state appears to be determined by their location in particular niches within a tissue, e.g. near the basis of crypts in the intestine or near the root of hair bulbs in the skin. Maintenance of their state appears to be achieved partly by growth factors produced by the surrounding mesenchyme. WNT factors in the intestine and WNT factors as well as SHH in the skin are thought to be essential (→6.10). Less is known about the intracellular mechanisms which maintain their stem cell character. Very likely, telomerase expression is involved and may be stimulated via MYC through WNT or SHH-dependent pathways.

In contrast to primordial germ cells or cells of the epiblast, tissue-stem cells may not be pluripotent. Rather, they may be committed to a limited spectrum of differentiation fates, which may, however, involve quite different types of cells. For instance, stem cells of the large intestine are precursors of enterocytes, enteroendocrine, Paneth and goblet cells, which exhibit quite different functions. Moreover, if transplanted into a different environment, tissue stem cells may show a great deal of plasticity. Thus, bone marrow stem cells can not only give rise to many different types of blood cells and of the immune system, but even to some kinds of epithelial cells such as hepatocytes or to endothelial cells. Again, these different differentiation potentials must be imposed by epigenetic mechanisms that are at present insufficiently understood.

Clearly, most cancers likewise develop some kind of stem cell phenotype, by different mechanisms (Figure 8.12). In a number of cancers pathways involved in

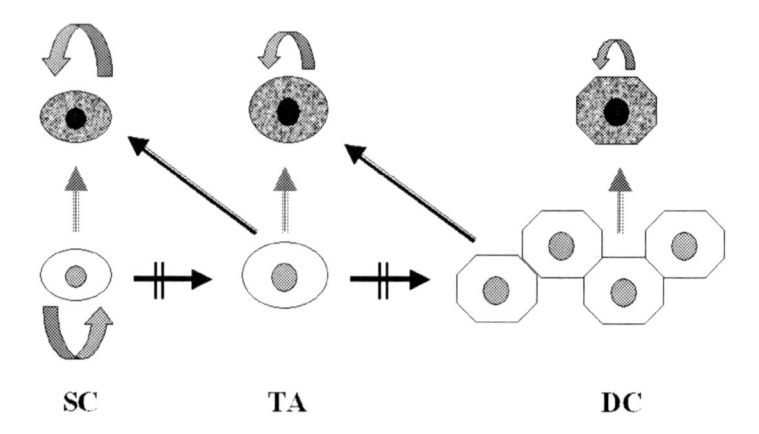

**Figure 8.12** *Relationship of cancer cells to stem cells*
Cancer cells can be derived directly from stem cells (SC) retaining their characteristics, but with at least partially blocked differentiation (left), from transient amplifying (TA) cells which do not terminally differentiate and/or secondarily acquire a stem cell phenotype (center), or from differentiated cell types that fail to turn off proliferation and/or revert to a less differentiated precursor stage (right).

the maintenance of tissue stem cells are activated by mutations in components of these pathways or by autocrine mechanisms. This mechanism is likely responsible for the precursor cell phenotype of colorectal cancers (→13.2) and basal cell carcinoma of the skin (→12.3). Chronic myelocytic leukemia (CML) is also clearly a stem cell disease caused by overactivity of 'cancer pathways' (→10.4).

Other cancers also resemble stem cells in possessing apparently unlimited proliferation potential, and most express telomerase. In some cancers, there is even evidence for a subpopulation which gives rise to a larger fraction of more differentiated tumor cells. However, it appears that in many cancers the stem cell properties are acquired secondarily. So, cancers do not necessarily develop directly from stem cells, although some certainly do. Others may develop from cells at a more differentiated stage that resume the phenotype of their precursor cell. Still others may acquire only selected aspects of a stem cell phenotype, such as telomerase expression. While the stem cell properties of cancers are often caused by genetic aberrations, a stem cell phenotype can be established by purely epigenetic mechanisms and this could well important in some cancers.

## Further reading

Wolffe A (1999) Chromatin: structure and function. 3$^{rd}$ ed. Academic Press
Szyf M (ed.) DNA methylation and cancer therapy. Landes Biosciences, online

Ross SA (2000) Retinoids in embryonic development. Physiol Rev. 80, 1021-1054
Fereira R (2001) The Rb/chromatin connection and epigenetic control: opinion. Oncogene 20, 3128-3133
Ehrlich M (2002) DNA methylation in cancer: too much, but also too little. Oncogene 21, 5400-5413
Geiman TM, Robertson KD (2002) Chromatin remodeling, histone modifications, and DNA methylation: how does it all fit together. J. Cell. Biochem. 87, 117-125
Jacobs JJL, van Lohuizen M (2002) Polycomb repression: from cellular memory to cellular proliferation and cancer. BBA 1602, 151-161
Johnstone RW (2002) Histone-deacetylase inhibitors: novel drugs for the treatment of cancer. Nat. Rev. Drug Discov. 1, 287-299
Jones PA, Baylin SB (2002) The fundamental role of epigenetic events in cancer. Nat. Rev. Genet. 3, 415-428
Li E (2002) Chromatin modification and epigenetic reprogramming in mammalian development. Nat. Rev. Genet. 3, 662-673
Mueller MM, Fusenig NE (2002) Tumor-stroma interactions directing phenotype and progression of epithelial skin tumor cells. Differentiation 70, 486-497
Reik A, Gregory PD, Urnov FD (2002) Biotechnologies and therapeutics: chromatin as a target. Curr. Opin. Genet. Devel. 12, 233-242
Costa RH et al (2003) Transcription factors in liver development, differentiation, and regeneration. Hepatology 38, 1331-1347
Esteller M (2003) Cancer epigenetics: DNA methylation and chromatin alterations in human cancer. Adv. Exp. Med. Biol. 532, 39-49
Herman JG, Baylin SB (2003) Gene silencing in cancer in association with promoter hypermethylation. NEJM 349, 2042-2054
Jaenisch R, Bird A (2003) Epigenetic regulation of gene expression: how the genome integrates intrinsic and environmental signals. Nat. Genet. 33 Suppl., 245-254
Jaffe LF (2003) Epigenetic theories of cancer initiation. Adv. Cancer Res. 90, 209-230
Lessard J, Sauvageau G (2003) Bmi-1 determines the proliferative capacity of normal and leukaemic stem cells. Nature 423, 255-260
McLaughlin F, Finn P, La Thangue NB (2003) The cell cycle, chromatin and cancer: mechanism-based therapeutics come of age. Drug Discov Today 8, 793-802
Ohlsson R et al (2003) Epigenetic variability and the evolution of human cancer. Adv. Cancer Res. 88, 145-168
Otte AP, Kwaks THJ (2003) Gene repression by Polycomb group protein complexes: a distinct complex for every occasion? Curr. Opin. Genet. Devel. 13, 448-454
Passaguè E et al (2003) Normal and leukemic hematopoesis: Are leukemias a stem cell disorder or a reacquisition of stem cell characteristics. PNAS USA 100, Suppl.1, 11842-11849
Sims III RJ, Nishioka K, Reinberg D (2003) Histone lysine methylation: a signature for chromatin function. Trends Genet. 19, 629-639
Walter J, Paulsen M (2003) Imprinting and disease. Semin. Cell Dev. Biol. 14, 101-110
Egger G et al (2004) Epigenetics in human disease and prospects for epigenetic therapy. Nature 429, 457-463
Nakayama M et al (2004) GSTP1 CpG island hypermethylation as a molecular biomarker for prostate cancer. J. Cell. Biochem. 91, 540-552
Roberts CWM, Orkin SH (2004) The SWI/SNF complex – chromatin and cancer. Nat. Rev. Cancer 4, 133-142

## Box 8.1 Carcinogenesis by HIV

Like HTLV-I, HIV1 contributes to human cancer development, and like HTLV-I it expresses a number of accessory proteins in addition to the usual set encoded by the *gag*, *pol* and *env* genes (cf. Fig. 4.1).

HIV1 does not cause tumors itself, but instead facilitates the development of a specific set of cancers through immunosuppression. In many of these, other viruses may be active. They include certain lymphomas (EBV-associated?), squamous carcinomas (HPV-associated?), and specifically, Kaposi sarcoma (KS). In addition to immunosuppression, HIV acts through one or several paracrine mechanisms.

The actual causative virus in KS is a 165 kb herpes virus, HHV8 or KHSV. Kaposi sarcoma consists of a mixture of mesenchymal cell types, which may partially be derived from an undifferentiated precursor cell. It is not quite clear which proteins of the virus are oncogenic. Some suppress apoptosis. Others act as cytokines and still others induce cytokine receptors. In KS, HIV may contribute more directly, beyond immunosuppression. HIV-infected cells release the viral transactivator protein tat, which may promote replication of KHSV and expression of viral proteins in cells harboring that virus. Moreover, HIV tat may induce secretion of the cytokine IL6 in uninfected cells, promoting the proliferation of KHSV-infected and other cells in the tumor.

HIV1 action in carcinogenesis might be schematically illustrated as follows:

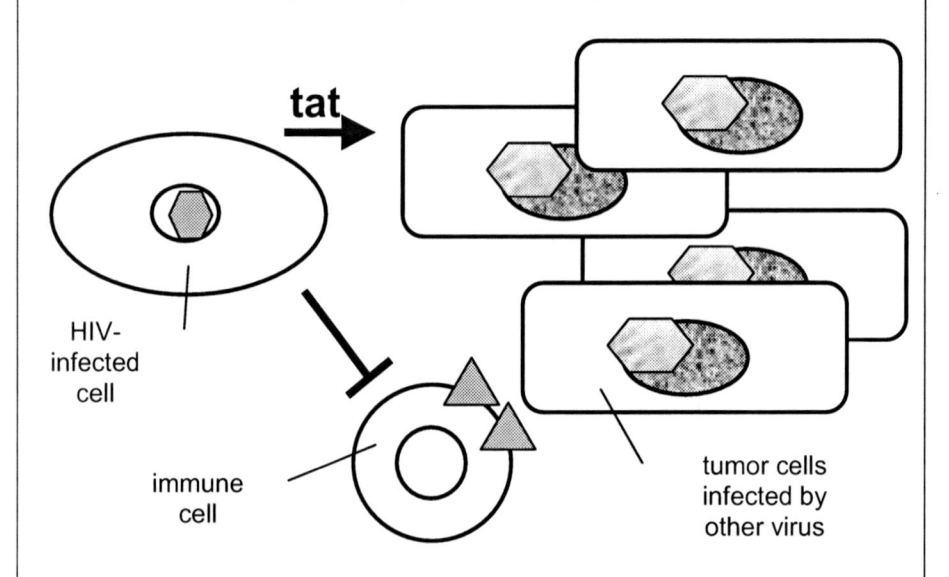

Bellan C, De Falco G, Lazzi S, Leoncini L (2003) Pathological aspects of AIDS malignancies. Oncogene 22, 6639-6645
Scadden DT (2003) AIDS-related malignancies. Ann. Rev. Med. 54, 285-303.

# CHAPTER 9

# INVASION AND METASTASIS

- The spread of solid cancers beyond the confinements of their tissue compartment into other parts of the same tissue and successively into neighboring tissues (invasion) and distant organs (metastasis) is the defining property of malignancy. Invasion and metastasis are decisive for the clinical course of most cancers.
- Invasion and metastasis are complex processes, particularly in carcinomas, since normals epithelia are strongly adherent and are confined by a basement membrane. Before or during invasion, carcinomas activate their surrounding connective tissue, eliciting inflammation and angiogenesis. Actual invasion by carcinoma cells involves decreased cell adhesion and increased motility as well as destruction of the basement membrane and remodeling of the extracellular matrix. Metastasis, in addition, requires cancer cells to enter blood or lymph vessels, to survive there, to extravasate, reattach, and proliferate in a different tissue. Furthermore, during invasion and metastasis, cancer cells need to evade cytotoxic cells of the immune system such as cytotoxic T-cells (CTL) and natural killer (NK) cells.
- Invasion and metastasis require an extensive reorganization of carcinoma cells, particularly of their cytoskeleton and their surface. Likely, this reorganization involves coordinated changes of gene expression and cell structure that impress as 'programs' for invasion or metastasis. These 'programs' are predominantly secondary consequences of mutations in oncogenes or tumor suppressor genes, but are aided by specific mutations, e.g. in cell adhesion molecules.
- Changes in cell surface molecules accompany altered cell-cell and cell-matrix interactions during invasion and metastasis. Molecules mediating homotypic interactions such as E-cadherin and connexins are down-regulated or mutated. Expression patterns of proteins mediating interactions with the extracellular matrix, such as integrins, are changed. Various other proteins on the cell surface including adhesion molecule, antigenic glycoproteins, and recognition proteins for immune cells are expressed at altered levels, alternatively spliced or processed, or mutated.
- Successful invasion depends crucially on interactions with non-tumor 'stromal' cells, i.e. tissue mesenchymal, endothelial and inflammatory cells, which also display changes in gene expression and behavior. The emergence of 'activated stroma' may distinguish highly malignant carcinomas.
- The destruction of the basement membrane and other extracellular matrix components in connective tissues surrounding a carcinoma is predominantly accomplished by proteases secreted from tumor and stromal cells, notably metalloproteinases and plasmin. Various members of the matrix metalloproteinase (MMP) family are over-expressed in tumor tissues and their inhibitors (TIMP) are often down-regulated. Proteases also activate latent growth

factors from their storage sites in the extracellular matrix that act on carcinoma and stromal cells.
- Angiogenesis is a prerequisite for the progression of many solid tumors. Migration and proliferation of endothelial cells which form new blood capillaries and lymph vessels is stimulated by several factors secreted by tumor cells and reactive stroma, such as VEGFs, PDGF, and certain FGFs.
- Recent evidence indicates that the changes in gene expression during tumor invasion are contingent on an overall activation of protein synthesis in tumor as well as stromal cells evident as an increase in expression of translation initiation factors and phosphorylation of translational proteins.
- Paracrine interactions between tumor and stromal cells are copious in carcinomas. At least during early progression stages, carcinoma cells may still depend to a large extent on growth factors from the stroma for their survival and proliferation. In turn, carcinoma cells stimulate production of growth factors from the stroma acting on carcinoma cells, stromal, immune and endothelial cells. Moreover, tumor cells secrete factors like TGFß which stimulate proliferation of connective tissue cells, but inhibit lymphocytes.
- Interactions with stromal cells are probably even more crucial in the establishment of metastases. While tissues with microcapillary systems, such as liver, lung, and bone, are obviously preferred targets for metastases for mechanical reasons, the actual pattern of metastases does not only depend on mechanical and anatomical factors. Rather, to survive and expand, metastatic tumor cells need to set up mutual interactions with stromal cells in the target tissue. This is the molecular basis of the 'seed-and-soil' hypothesis.
- Primary tumors and metastasizing tumor cells are potential targets of the immune system. Multiple mechanisms limit its ability to eliminate tumor cells. Theycomprise production of inhibitory cytokines, down-regulation of recognition molecules and of death receptors used by cytotoxic immune cells and even active counterattack by expression of death receptor ligands.

## 9.1 INVASION AND METASTASIS AS MULTISTEP PROCESSES

Even more than tumor expansion as such, the extent of local invasion and distant metastasis determine the clinical outcome of cancers, especially in solid cancers like carcinomas. Invasion and metastasis can be regarded as multistep processes, where invasion towards a certain stage constitutes a prerequisite for metastasis. Each step requires and selects for certain properties of the tumor cells. Therefore, overall, metastasis may be a very inefficient process. While details vary, invasion and metastasis of a carcinoma in general can be roughly described by the following sequence of steps (Figure 9.1):

(1) While the carcinoma proliferates and extends laterally and vertically within the epithelium, the tumor cells become less adherent to each other and to adjacent normal epithelial cells.

(2) The underlying stroma becomes activated and inflammation may occur. The basement membrane which separates the epithelium from the underlying mesenchymal connective tissue is partly or completely destroyed.

### Intraepithelial Growth

Angiogenesis — INVASION — Stroma Activation

Intravasation

Spread by Lymph or Blood

Continued Local Growth

Extravasation

Micrometastasis Formation

Angiogenesis — Stroma Activation

### Metastasis Growth

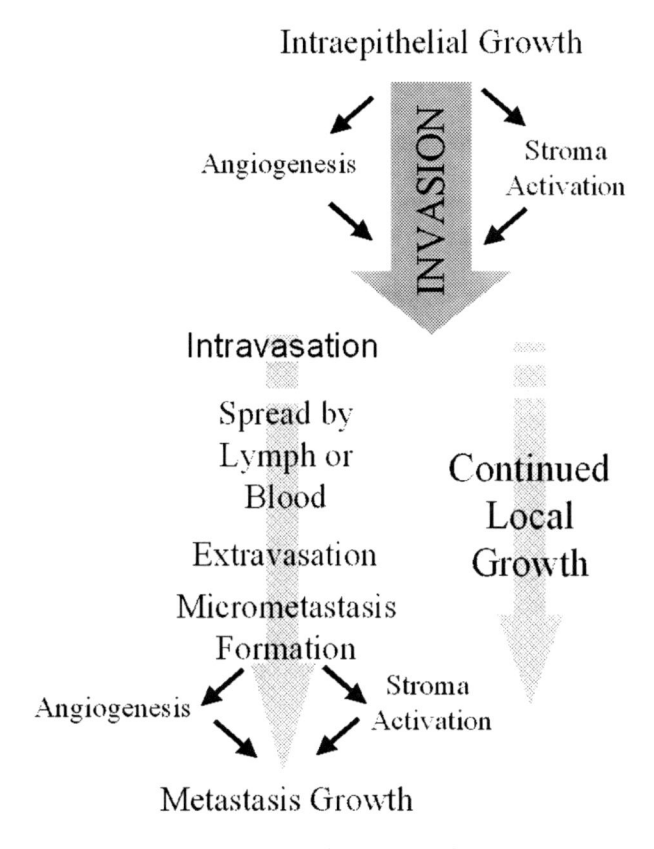

**Figure 9.1** *Steps of invasion and metastasis*

(3) The tumor continues its growth into the connective tissue. This is one of the more variable steps. Some carcinomas continue to grow as solid, coherent masses compressing the neighboring connective tissue or develop processes that spread into it, breaking up the extracellular matrix (ECM). From other carcinomas, small groups of cells or single cells split off and migrate into the underlying tissue, sometimes in an 'indian file' pattern and sometimes as adherent clusters. These migrations, even by single tumors cells, involve remodeling of the ECM.

(4) In fact, the invasion of stroma by tumor cells is not as one-sided as it may appear. It is accompanied by altered gene activity in stromal cells, with some changes promoting and others inhibiting invasion. The type of stromal reaction may be one of the most important factors determining the ability of a tumor to metastasize. Often, cells of the immune system are attracted, by signals emanating

from the stroma and from tumor cells. This can result in pronounced inflammation. Like the stromal reaction, the effect of inflammation is ambiguous; it may impede or promote invasion.

(5) A critical step in invasion is reached when the growing or migrating tumor cells encounter blood or lymph vessels and invade them. Like the previous steps, this can occur by a tumor mass growing through the vessel wall into the lumen or by single tumor cells squeezing through the vessel lining. By this 'intravasation' step tumor cells gain access to the circulation and can reach distal organs by 'lymphogenic' or 'hematogenic' routes. Gaining access to blood and lymph vessels is also not necessarily a one-directional process. Since many tumors induce angiogenesis, capillaries sprout from blood and lymph vessels into the direction of the tumor mass. Since these are often leakier than normal vessels, they may offer easier access to the circulation.

(6) Independent of whether metastasis occurs, invasion may continue into further layers of the organ from which the carcinoma arises, through a tissue capsule, into surrounding adipose tissue and into neighboring organs. An important spreading route for some cancers, e.g. of the kidney, liver, ovary and pancreas, is through the lumen of the retroperitoneum or the peritoneum ('transcoelomic metastasis').

(7) When tumor cells have entered into lymph vessels, they are transported to the filtering system of the local lymph nodes, where some may survive and start lymph node metastases. Cells from these metastases may eventually penetrate towards the main lymph vessels and eventually enter the blood by this route. Tumor cells or debris and signal molecules from tumor and stromal cells transported to the lymph nodes influence the immune reaction towards the primary tumor.

(8) Tumor cells having entered into blood vessels can theoretically spread to any part of the body. However, they are larger than normal blood cells and are not well adapted for survival in a moving liquid[11]. Survival in the blood to reach distant tissues may be limiting for metastasis.

(9) To form metastases, carcinoma cells must exit from the circulation by 'extravasation'. Most often, this appears to take place in organs with microcapillary systems, such as the liver, the lung, the kidney, and bone. Because of their size, carcinoma cells (and certainly cell clusters) get stuck in capillaries. This is not sufficient, however, to establish micrometastases.

(10) In the new tissue environment, carcinoma cells have to reattach to the matrix, survive, sometimes for extended periods, and eventually start to expand into micrometastases, which again can lie dormant for many years. After all the complicated previous steps, it may be surprising to learn that this step is by many considered the most critical, i.e. least efficient step in metastasis formation.

(12) The final step in metastasis is the expansion of micrometastases to actively growing tumors. This requires establishment of a sufficient nutrient supply and interaction with a different type of stroma, often including once more induction of angiogenesis and further local invasion.

Although invasion and metastasis are such important processes in the course of cancer progression, they are incompletely understood. This is largely owed to their

---

[11] In fact, carcinoma cells are often observed in blood as small cells clusters.

complexity, since at almost every step complex interactions between different cell types and extracellular tissue components are involved. In addition, the early steps of metastasis in humans can rarely be observed. Moreover, metastases specimens are not regularly available for investigation, particularly from carcinomas, since they are rarely treated by surgery. Therefore, much of our knowledge on this matter is inferred from experimental animal models, which are by themselves complex enough. At this stage, therefore, many individual factors involved in invasion and metastasis have been identified and selected interactions have been pinpointed (see below). However, there are considerable deficits in understanding the relative importance of this factors and how they interact with each other.

One particular important issue is what drives the overall process of invasion and metastasis. Two alternative ideas are entertained. One hypothesis maintains that tumor cells acquire one property after another as they proceed through the steps outlined above. At each step (or at least at many), the best adapted tumor cells are selected from the numerous variants created by inherent genomic instability. The alternative hypothesis holds that the 'invasion and metastasis' program is an inherent property of certain cancers that is expressed very early on in their development. This second hypothesis would predict that primary cancers are more similar to their metastases than to each other in their genetic alterations and gene expression patterns, and that primary cancers which metastasize show more similar alterations to one another than to those that do not. This has indeed been observed in some investigations using expression profiling methods, e.g. in breast cancers (→18.5). Current opinion therefore leans towards the second hypothesis. Interestingly, and perhaps not unexpectedly, the most striking differences between metastatic and non-metastatic cancers were found in the gene expression pattern of the stromal rather than the tumor cells. The distinction between these two hypotheses is important for cancer diagnosis (→21.4). The prognosis of a cancer can only be determined from a sample from the primary site, if the ability to metastasize successfully is reflected in molecular parameters of the primary tumor. It would be difficult, if the primary cancer was a mixture of cells with different abilities to metastasize and/or if this ability developed gradually.

## 9.2 GENES AND PROTEINS INVOLVED IN CELL-TO-CELL AND CELL-MATRIX ADHESION

Epithelial cells adhere to each other and interact with each other through several types of contacts (Figure 9.2). Morphologically distinct and functionally important contacts include adherens junctions, gap junctions, and tight junctions (occluding junctions).

Occluding junctions seal epithelia and define the apical and lateral membrane compartments of an epithelial cell. Therefore, their loss in a carcinoma cell is associated with the loss of cell polarity.

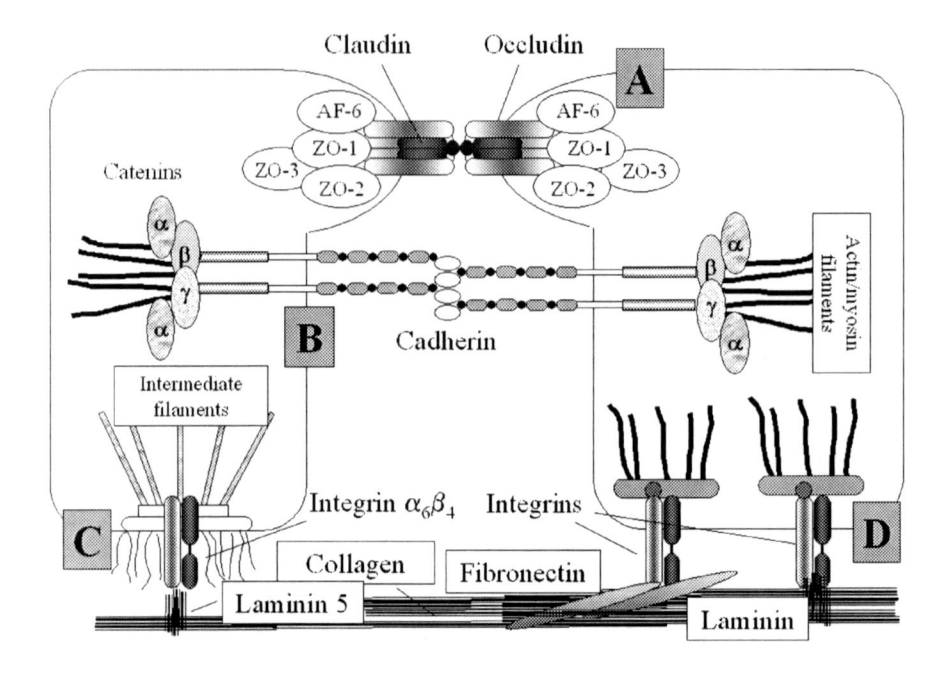

**Figure 9.2** *Cell-cell- and cell-matrix-contacts of epithelial cells*
A: Tight junctions; B: Adherens junctions; C: Hemidesmosomes; D: Integrin / extracellular matrix contacts. Different integrins prefer different matrix proteins.

Adherens junctions are arranged in a belt-like configuration (hence 'belt desmosomes') between adjacent epithelial and are intracellularly connected to actin filaments. The proteins actually mediating homotypic interactions between adjacent epithelial cells are cadherins, of which E-Cadherin is a typical representative (Figure 9.3). Interaction between E-Cadherin molecules is $Ca^{2+}$-dependent. On the cytoplasmic surface of the cell membrane, E-Cadherin is linked to actin filaments by α-Catenin and β-Catenin. Down-regulation or mutation of E-Cadherin is frequent in human cancers and often occurs during tumor progression. In some invasive tumors, E-Cadherin is replaced by other members of the family with adhesion properties more suitable for a migrating cell, e.g. N-Cadherin. This is labelled a 'cadherin switch'.

Spot desmosomes are connected to cytokeratin filaments in the epithelial cell by desmoplakin and plakoglobins. The actual contact between cells is again made by specialized cadherins.

Gap junctions also contribute to adhesion between cells, but primarily are communication channels that connect cells of the same type in epithelia or in other tissues, e.g. the nervous system. They allow the passage of small molecular weight molecules ($\approx$ 1000 Da), in particular of cations such as $Ca^{2+}$ and $K^+$, but also of other small signaling molecules such as cyclic AMP. Gap junctions are formed from one connexon in each partner cell, which consists of six connexin molecules. Different connexins are expressed in a tissue-specific fashion accounting for the specificity of the interaction. Gap junctions are temporarily closed, when cells separate from their neighbors for division. Closure is regulated by phosphorylation of connexins. Gap junctions are also sensitive to high $Ca^{2+}$ levels and accordingly shut when cells become apoptotic or necrotic. In cancers, gap junctions become as a rule inactive during tumor progression and tissue-specific connexin expression is down-regulated.

In addition to adhering to each other, at least the basal cells in an epithelium are attached to the basement membrane. This contact is made through integrins ($\rightarrow$ Figure 6.8) which bind fibronectin and laminin in the extracellular matrix. On the cytoplasmatic side they are attached to actin filaments via $\alpha$-actinin, vinculin and talin. Integrins are heterodimers consisting of each one $\alpha$ and $\beta$ subunit. There are >17 $\alpha$- and >8 $\beta$ subunits in humans. A typical integrin of an epithelial basal cell is $\alpha_2\beta_1$. A specific integrin, $\alpha_6\beta_4$, is a constituent of hemidesmosomes. These also attach epithelial cells to the basement membrane matrix, but are attached to the cytokeratin cytoskeleton on the cytoplasmatic face. The composition of integrins is cell-type specific, and changes in migratory cells. For instance, cells of the hematopoetic lineage down-regulate their expression of $\alpha_4\beta_1$ integrin, when they have become differentiated and leave the bone marrow. Tumor cells often express different integrins than their normal counterparts. Down-regulation of integrins accounts, e.g., for the failure of chronic myelogenous leukemia cells to remain in the bone marrow sufficiently long for complete differentiation ($\rightarrow$10.4). In carcinomas, too, changes in integrin composition may be essential for invasion and metastasis, with some integrins favorable and others inhibitory for invasion.

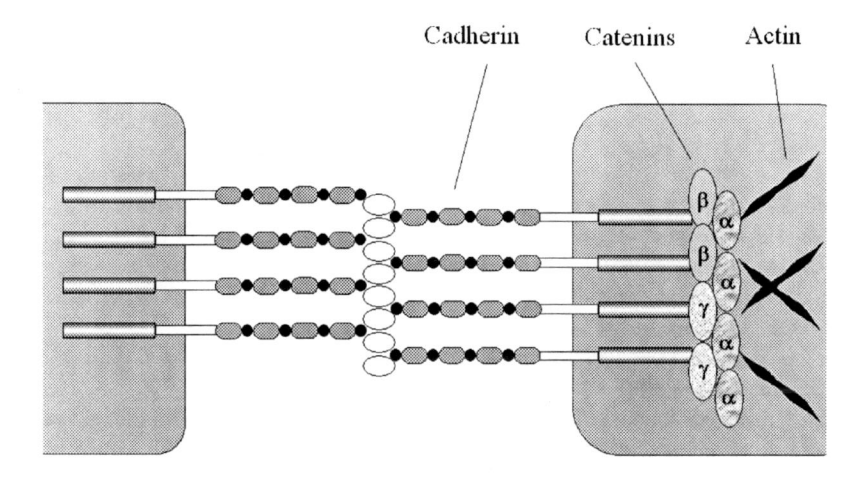

**Figure 9.3** *Structure and interactions of E-Cadherin*
See text for further explanation

Cell-to-cell adhesion and cell-matrix adhesion elicit signals within the cell, e.g. integrins through protein kinases like FAK, ILK, and SRC family tyrosine kinases that influence several pathways regulating cell growth and survival (→6.5). In epithelial cells, lack of attachment tends to cause anoikis, a specific kind of apoptosis. Alterations in signaling as a consequence of altered adhesion in carcinoma cells must therefore be compensated. This may be the reason, why altered SRC expression is a common finding in cancers.

Importantly, the ability to migrate is not simply acquired by loss of adhesive properties. Instead, cell migration requires a dynamic pattern of adhesion contacts that may overall be more complex than in a cell residing in a tissue. A single cell migrating through a tissue needs to adhere to the extracellular matrix to just the right extent, i.e. sufficiently tightly to pull itself through the tissue, but not so tightly as to become unable to extricate itself. Moreover, if a net movement is to be achieved, attachments need to be established at a leading edge and be broken at a trailing edge by well-localized proteolysis, and cytoskeletal contractions must be coordinated. Localized activation of proteases at certain points on the cell surface is therefore a prerequisite for migration in a tissue (→9.3). Some normal and tumor cells also move by a more amoeboid mode with less attachment and less activity of the actin cytoskeleton.

In fact, carcinoma cells more often migrate as cell clusters or cell files, or even expand like cell sheets during fetal development. These modes of migration nevertheless require reorganization of cell-to-cell adhesion and the cytoskeleton compared to cells in a resting epithelium. For instance, they may be accompanied by changes in the integrin isotypes and in the type of cadherin present. Some carcinomas also undergo a morphological change to a more mesenchymal phenotype, called an 'epithelial-mesenchymal transition' (sometimes abbreviated as EMT). This allows them to invade and migrate as single cells, which is otherwise more typical of hematological and soft tissue cancers. However, cell-to-cell and cell-matrix adhesion remain dynamic in tumor tissues, as a rule, and the EMT can be reversible.

The course of E-Cadherin expression in the course of prostate cancer development illustrates the dynamic nature of changes in cell adhesion in cancers. E-Cadherin becomes down-regulated during invasion of prostate cancer cells into the stromal tissue compartment, but is re-expressed in many metastases. A modulation of gene expression in this fashion is obviously more efficiently achieved by epigenetic than by genetic alterations. So, the promotor of the *CDH1* gene encoding E-Cadherin is hypermethylated in some carcinomas in a highly variable fashion that is associated with a variable expression of the protein.

In contrast, cancers with a diffuse-type growth pattern down-regulate specific cell adhesion molecules irreversibly. Again, E-Cadherin can serve as an example. Diffuse-type gastric cancer is characterized by highly invasive small groups of loosely adherent undifferentiated tumor cells. In this cancer type, E-Cadherin is regularly inactivated by mutation and deletion of the gene. In fact, germ-line mutations in *CDH1* are responsible for rare familial cases of this cancer (→17.2).

Mutations in E-Cadherin are among the few genetic changes that specifically alter cell adhesion molecules in carcinoma cells. A much wider range of changes on the surface of tumor cells (Table 9.1) are achieved by epigenetic mechanisms. These include aberrant methylation of genes like *CDH1* and others associated with more or less strong down-regulation of expression. In other cases, up-regulation, altered splicing, or altered processing of cytoskeletal and membrane proteins are observed. Often, these changes in invasive tumor cells give the impression of a coordinated switch in gene and protein expression towards an 'invasion program' rather than that of an accumulation of alterations in individual genes selected to yield an invasive phenotype.

Nevertheless, some among the myriad changes at the cell surface of cancer cells may be more important than others and may be decisive for invasion and metastasis. In particular, down-regulation of specific genes appears to be required for metastasis in certain cancers, although these genes do not influence the growth of primary tumors. Such genes have been designated 'metastasis suppressor genes'. Most of

**Table 9.1.** *Some cell surface proteins altered in human cancers*

| Protein(s) | Function | Alteration(s) in cancer |
|---|---|---|
| E-Cadherin | homotypic cell adhesion | mutation, down-regulation |
| H-Cadherin | homotypic cell adhesion | down-regulation |
| N-Cadherin | homotypic cell adhesion | up- or down-regulation |
| CAM(s) | homotypic cell adhesion | down-regulation |
| Connexins | formation of gap junctions | down-regulation |
| Integrins | adhesion to basement membrane and extracellular matrix | altered pattern |
| KAI1 | cell adhesion | down-regulation |
| CD9 | scavenging and presentation of growth factors | down-regulation |
| CD44 | attachment to hyaluronic acid, homing | altered splicing, down-regulation |
| MHC | interaction with immune cells, presentation of antigens | down-regulation |
| MICA, MICB | co-receptors for natural killer cells | down-regulation |
| FAS | receptor for cytotoxic FAS ligand | down-regulation of trans-membrane form, up-regulation of soluble form |
| uPAR | binding and activation of protease | up-regulation |
| MMP14 | protease, activation of pro-proteases | up-regulation |

them encode cell surface proteins, but some of them are protein kinases relaying signals from the cell surface, e.g. MEKK4.

For instance, KAI1 is a cell adhesion molecule encoded by a gene located on chromosome 11p12. Although this chromosomal region is subject to LOH in different carcinomas, the gene is never mutated or affected by promoter hypermethylation. Nevertheless, expression of KAI1 is down-regulated in many aggressive carcinomas.

Another gene in this category is CD9, which encodes a cell surface protein that sequesters EGF-like growth factors. Decreased expression of CD9 or KAI1 in primary tumors indicates a higher likelihood for the presence of metastases.

A more complex case is CD44. Like all proteins with the CD designation, it was first discovered as a lymphocyte surface antigen, but is expressed on many different cell types. It recognizes and binds to hyaluronic acid. In this fashion, it may function as an adhesion receptor directing lymphocytes to specific tissues. The *CD44* gene is spliced in several alternative ways. As a consequence, altered expression in carcinomas may take the form of outright down-regulation with promoter hypermethylation or that of expression of CD44 variants that could contribute to the 'targeting' of metastatic cells.

## 9.3 GENES AND PROTEINS INVOLVED IN EXTRACELLULAR MATRIX REMODELING DURING TUMOR INVASION

The structure of the extracellular matrix is generated by a backbone of fibrillar proteins and proteoglycans. This matrix is in a dynamic state. Structural proteins are synthesized by fibroblasts and other cells embedded in connective tissue. These same cells, together with cells of the immune system and from blood vessels, are responsible for the turnover of the extracellular material. Turnover and remodeling of the ECM are enhanced during inflammation and wound repair, as they are during tumor invasion. In fact, essentially the same mechanisms and enzymes are involved in each situation.

The key enzymes among the many involved in ECM remodeling are proteases, and among these the plasminogen activator (PA) protease cascade and the matrix metalloproteinases (MMPs). These proteases degrade components of the basement membrane as well as proteins and proteoglycans of connective tissue extracellular matrix. They process other proteases and enzymes as well as growth factors and they liberate 'latent' growth factors from their storage sites in the extracellular matrix. These growth factors then act on several different cell types, incuding or inhibiting proliferation or eliciting apoptosis in some cases. Among the factors that are activated in this fashion are FGFs (fibroblast growth factors), HGF (hepatocyte growth factor), and TGFβs (→6.7). Obviously, proteins on cell surfaces can also be substrates.

The end result of the plasminogen activator cascade (Figure 9.4) is localized activation of plasmin near the surface of specific cells. The inactive precursor of plasmin, plasminogen, is synthesized and secreted mainly in the liver and is present throughout the body. It is specifically cleaved to an active protease by plasminogen activator, abbreviated uPA (for 'urokinase-type plasminogen activator'). The activity of uPA is localized to cell surfaces by its binding to a specific membrane receptor, uPAR ('urokinase receptor'). The uPAR in turn is localized on the cell surface by dynamic interaction with specific integrins. Thus, cell-type dependent expression of the uPAR is one mechanism of directing plasmin activity to a specific location. Plasmin activity is further localized by a dynamic association of the uPAR with integrins on the cell surface.

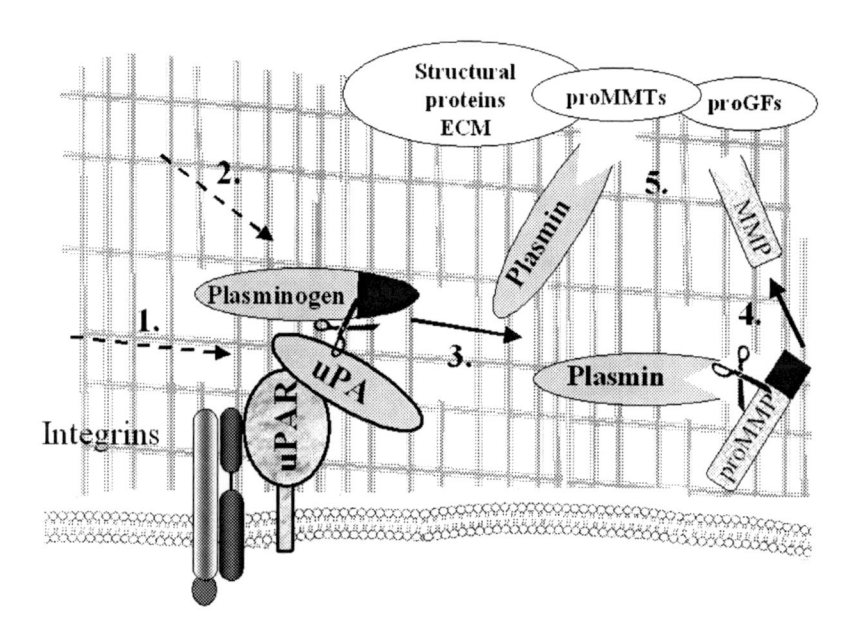

**Figure 9.4** *The plasminogen cascade*
The successive steps of the cascade (1. – 5.) lead to localized activation of plasmin and matrix metalloproteinase. See text for further explanations and cf. also Figure 9.5.

Conversely, the expression of plasminogen activator inhibitor 1 (PAI-1) prevents the cleavage of plasminogen and thus excludes plasmin from the environment of cells that express it. Furthermore, the protein protease inhibitor $\alpha_2$-antiplasmin limits the activity of plasmin in the extracellular fluid in general. Plasmin is best known for its ability to digest fibrin, but it is also capable of digesting a range of ECM proteins as well as of cleaving and activating latent growth factors, pro-MMPs, and pro-uPA.

An even wider range of substrates can be digested by matrix metalloproteinases (MMPs). There are at least 23 members of the MMP family in man (Table 9.2). Between them, they are able to degrade every protein in the extracellular matrix, although each MMP displays preferences for certain substrates. All MMPs have a similar basic structure with a $Zn^{2+}$ ion essential for catalytic activity. Accordingly, compounds that complex $Zn^{2+}$, such as hydroxycarbamates or tetracyclines, are inhibitors, although often not specific for one member of the family. More specific inhibitors have been designed and have been tested in clinical trials.

As usual for proteases, MMPs are synthesized as proproteases ('zymogens'), and since most of them are secreted, they contain initially a signal peptide. The membrane-type MMPs differ from all others by containing a transmembrane domain at their C-terminal end. This locates them to the cell surface. Since pro-MMPs, mostly MMP-2, are among their substrates, they contribute to precisely localized protease activity, as would be required at the edges of a migrating cell or the growth cone of a tumor mass.

The MMPs are not only regulated at the level of activation from pro-enzymes, but also substantially at the transcriptional level. Their basal expression is cell type dependent and various factors induce their expression. These (depending again on the cell-type) include interleukins and growth factors of the EGF, FGF and TGFβ families, as well as stress and adhesion signals. Intracellularly, these factors act through one of the ERK, JNK, or p38 MAPK cascades which increase transcription by activating AP-1 factors at a conserved binding site in the gene promoters.

As for plasmin, specific as well as general protease inhibitors limit the activities of MMPs. MMPs are confined to local activity by ubiquitous protease inhibitors in blood and extracellular fluids, mainly $\alpha_2$-macroglobulin and $\alpha_1$-antiprotease. Moreover, they are restricted as well as directed by specific inhibitors, the tissue

Table 9.2. *Some matrix metalloproteinases involved in human cancers*

| Family | Members | Structure* | Characteristic | Substrates |
|---|---|---|---|---|
| Collagenase | 1, 8, 13, 18 | S/P-C/D-HR-HS | | collagens |
| Gelatinase | 2, 9 | S/P-C/D-HR-HS | strongly variant catalytic domain | gelatins, laminin |
| Stromelysin | 3, 10, 11, 12 | S/P-C/D-HR-HS | differs from collagenases in pro-domain | various structural proteins, proteoglycan, pro-MMPs, protease inhibitors |
| Membrane-type | 14, 15, 16, 17 | S/P-C/D-HR-HS-TM | membrane-associated | pro-MMPs, collagen |
| Matrilysin | 7 | S/P-C/D-HR | lacks HS domain | gelatin, fibronectin |

* The basic structural domains are signal peptide/prodomain (S/P), zinc-binding catalytic domain (CD), hinge region (HR), hemopexin similarity (HS), transmembrane domain (TM)

inhibitors of metalloproteinases ('TIMPs'). There are four members of this family, TIMP-1 through TIMP-4. They bind to the substrate-binding site of the MMPs, but cannot be cleaved by their catalytic centres. TIMPs also block the activation of pro-MMPs. The TIMP inhibitors are produced by a broader range of cell types than the proteases, and at least TIMP-2 is often constitutively expressed. However, the expression of most TIMPs is additionally regulated at the transcriptional level. In general, the same factors that induce MMPs are involved, but additional cytokines such as TNFα as well as glucocorticoids and retinoids are active, at least in the regulation of TIMP-1.

It would be too simplistic to state that the activity of MMPs in a tissue depends on the relative levels of MMPs and TIMPs, for two reasons. (1) TIMPs exert a number of (not too well defined) effects on cells beyond inhibiting MMPs. TIMP-3, in particular, appears to act as a pro-apoptotic factor. It is accordingly down-regulated in many tumor cells, usually by promoter hypermethylation. (2) While antagonism between TIMPs and MMPs is the rule, there are clearly exceptions. One of the better defined ones is that TIMP-2 is a cofactor in the activation of MMP2 by membrane-type MMPs. This example illustrates the likely role of TIMPs in physiological tissue remodeling. MMPs are produced by specific cell types, but TIMPs by most others. This means that MMP activity is limited to specific sites within a tissue 'around' cells that perform the remodeling, while other cells remain protected (Figure 9.5). In fact, even the MMP activity 'around' an actively remodeling cell may not be homogeneously distributed, as particularly evident during cell migration.

The changes in protease and protease inhibitor expression taking place during tumor invasion must be considered on this background. In many human cancers, protease expression and activity are increased during tumor progression and invasion. Typically, mRNA and protein levels of several proteases are enhanced, e.g. of MMP-2, MMP-7, MMP-9, and MMP-14 as well as of uPA. Conversely, TIMPs tend to be down-regulated, most consistently TIMP-3. Enhanced expression of MMPs and of uPA/plasmin are almost generally associated with a worse outcome, i.e. primary tumors with increased expression are more likely to have metastasized than those with low expression.

The relationship between TIMP expression and tumor behavior is not as consistent. Whereas TIMP-3 down-regulation is often found in more aggressive cancers, decreased expression of the other inhibitors is not always predictive of a particular clinical course, and in some cancers the relationship is inverse. For instance, increased expression of TIMP-1 may indicate a worse prognosis in breast cancer. The explanation for this apparent paradox may lie in the question where precisely the proteases and the inhibitors are expressed. In many carcinomas, the increased expression of MMPs is found in the stromal part of the tumor. While tumor cells largely stain negative for many MMP mRNAs and proteins, often expressing only MMP-7, the stromal cells, fibroblasts and particularly invading macrophages and monocytes, express a full range of active proteases. Conceivably, in this situation, the expression of TIMPs may actually protect the carcinoma cells. This type of distribution of protease expression between carcinoma and stromal cells appears to be most pronounced in adenocarcinomas and other cancers arising from glandular structures, whereas other carcinomas, e.g. from squamous epithelia, express a wider range of MMPs themselves.

Thus, tissue remodeling during tumor invasion is far from a one-sided affair and in most cases requires an active contribution by the tumor stroma. This may be part of the explanation why an 'active stroma' gene expression 'signature' in expression profiling studies correlates with the presence of metastases.

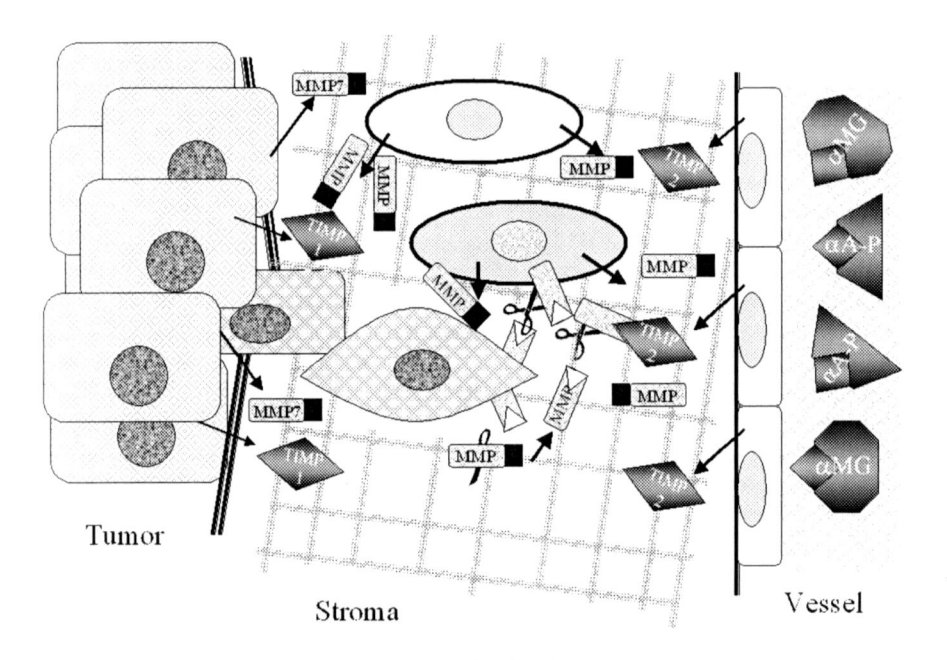

**Figure 9.5** *Mechanisms that localize MMP activity*
Some cells of a cancer (on the left) invading connective tissue (center) undergo an epithelial-mesenchymal transition (spindle-shaped cell in the center). The cancer cells in this case secrete MMP7 but also TIMP1. Endothelial cells (right) secrete TIMP2 which is (still) sufficient to neutralize MMPs secreted by stromal cells. These are only activated around the leading edge of the migrating cancer cell by membrane-type MMP. In the vessel (outside right) α-macroglobulin and α-antiprotease inactivate any MMP reaching the bloodstream.

## 9.4 ANGIOGENESIS

As tumors exceed a minimum size, nutrients such as glucose and amino acids as well as oxygen can no longer reach their center by diffusion. Conversely, end products of metabolism such as lactate, ammmonia and bicarbonate cannot diffuse out easily. In fact, this minimum size may be no more than a few cell layers in diameter, in the order of 70 μm (cf. Box 9.1). Beyond this size, primary tumors as well as metastases must lie dormant, disperse, or acquire a vasculature. During the progression of many tumors, the development of a vascular supply is a critical step designated as the 'angiogenic switch'. In primary tumors, this switch often takes place during or as a prelude to invasion. In metastases, it may often mark the crucial transition from micrometastases to actively growing tumors.

Most malignant tumors develop a vasculature by angiogenesis, i.e. the sprouting and proliferation of local capillaries. In some cases angiogenesis is supplemented by vasculogenesis, the de-novo-formation of blood vessels from hemangioblasts which are recruited from hematopoetic stem cell precursors. In other cases, tumors grow along existing blood vessels. In a few tumor types characterized by high cell plasticity, such as some melanomas and prostate cancers, cancer cells transdifferentiate into endothelial cells, forming 'tunnel-like' structures themselves or contribute endothelial cells to capillaries formed by angiogenesis.

Like many other individual events during tumor invasion, angiogenesis is not restricted to malignant tumors, but also takes place during inflammation and wound healing. Angiogenesis is also a normal process in the cyclically growing tissues of females, most of all in the endometrium, and during tissue hypertrophy, e.g. in muscle. However, like tissue remodeling during tumor invasion, tumor angiogenesis is a less orderly and often deregulated process. It often results in relatively incomplete, leaky capillaries which allow tumor growth at the periphery with continued undernutrition and hypoxia towards the tumor center, to the extent of necrosis.

Angiogenesis starts by extension of normal capillaries which thereby become leaky and allow plasma proteins like fibrinogen to escape into the perivascular space. This contributes to the laying down of a matrix which is used by perivascular cells ('pericytes') to migrate into the tissue. Concomitantly, the basement membrane surrounding the vessel is degraded, a.o. by MMPs from activated endothelial cells and stroma (→9.3). Endothelial cells then proliferate and the capillaries ramify extending into the surrounding tissue towards or into the tumor mass. While the proliferation of endothelial cells is transient during normal angiogenesis, it often remains permanent during tumor angiogenesis. Moreover, during normal angiogenesis, the newly emerging capillaries eventually become covered by pericytes and a vascular basement membrane once more, but this part of angiogenesis is not usually completed during tumor angiogenesis. Tumor capillaries also show irregular shapes and irregular orientations.

The irregular growth, shape and function of tumor capillaries compared to normal capillaries are likely consequences of the differences in which normal and tumor angiogenesis are regulated. Regulation of angiogenesis is achieved by a network of factors, some of which are pro-angiogenic and some anti-angiogenic (Table 9.3). Their interplay is often depicted as a balance, which the predominance of one or the other sort of factors tilts towards or against angiogenesis. However, it needs to be considered that angiogenesis (like cell migration) is a spatially ordered process. In order to achieve proper angiogenesis, these factors must act in a spatially ordered fashion. So both the overall balance and the local balance are relevant. Some of the pro-angiogenic and anti-angiogenic factors are produced in a localized fashion. For instance, among the pro-angiogenic factors are FGFs, e.g. bFGF (FGF2), which are deposited in the ECM and liberated by proteases such as MMP-2 and plasmin. Conversely, several break-down products of matrix collagens generated by MMPs and other proteases, such as endostatin and arrestin, are among the most potent inhibitors of angiogenesis and limit its extent in normal circumstances. The disorganization of tumor tissue and tumor stroma may therefore constitute one factor causing the disturbances in tumor capillaries.

A second difference between normal and tumor angiogenesis concerns the growth factors involved. It is thought that most tumors do not provide a similar 'balanced' assortment of growth factors for stimulation of angiogenesis as found in normal tissues. Tumors secrete to various degrees FGFs, EGF-like factors, and PDGF, as well as the angiopoetins ANG-1 and ANG-2 and cytokines like IL8 that stimulate capillary sprouting and endothelial cell proliferation, or they stimulate the production (synthesis and release) of such factors from stromal cells. Inflammatory cells, in particular, secrete prostaglandins which promote angiogenesis. This is associated with an increase in the prostaglandin biosynthetic enzyme cyclooxygenase 2 (COX2).

However, the main factor eliciting angiogenesis in tumors is VEGF (vascular endothelial growth factor), which is over-expressed in many human cancers (Figure 9.6). So, in cancers the angiogenic growth factor 'cocktail' may be biased towards VEGF, and assorted other factors like FGF1 and FGF2. In fact, there are several related VEGF genes, VEGFA through VEGFD. The VEGFC and VEGFD proteins are thought to be mainly responsible for induction of lymphangiogenesis by tumors. The VEGFA gene is translated into several different proteins, from alternatively spliced mRNAs. VEGFA proteins increase capillary permeability and endothelial cell proliferation, mainly through the VEGFR2 receptor tyrosine kinase expressed by endothelial cells. Some tumor cells also express VEGFR1 or VEGFR2, so VEGFs can additionally act as autocrine factors.

Enhanced expression of VEGFA in tumors can be a consequence of several genetic and epigenetic events. Some activated oncogenes, e.g. of the RAS family, induce VEGF expression. Conversely, TP53 down-regulates VEGFA and several other pro-angiogenic factors, at least under some circumstances, and also increases the expression of anti-angiogenic factors. Among the genes regularly induced by

Table 9.3. *Some important pro-angiogenic and anti-angiogenic factors*

| *Pro-angiogenic* | *Anti-angiogenic* |
|---|---|
| VEGFA, VEGFB-D | Thrombospondin (TSP1) |
| PDGF | Angiostatin (plasminogen fragment) |
| FGFs, particularly FGF2 and FGF1 | Arrestin (collagen IV fragment) |
| Angiopoetins ANG1, ANG2 | Canstatin (collagen IV fragment) |
| IL8 | Endostatin (collagen XVIII fragment) |
|  | Tumstatin (collagen IV fragment) |

activated TP53 is *TSP1* which encodes thrombospondin 1, one of the most potent inhibitors of angiogenesis. So, several of the most consistent changes in human cancers, such as RAS mutations and TP53 loss of function would be expected to augment VEGFA expression and angiogenesis.

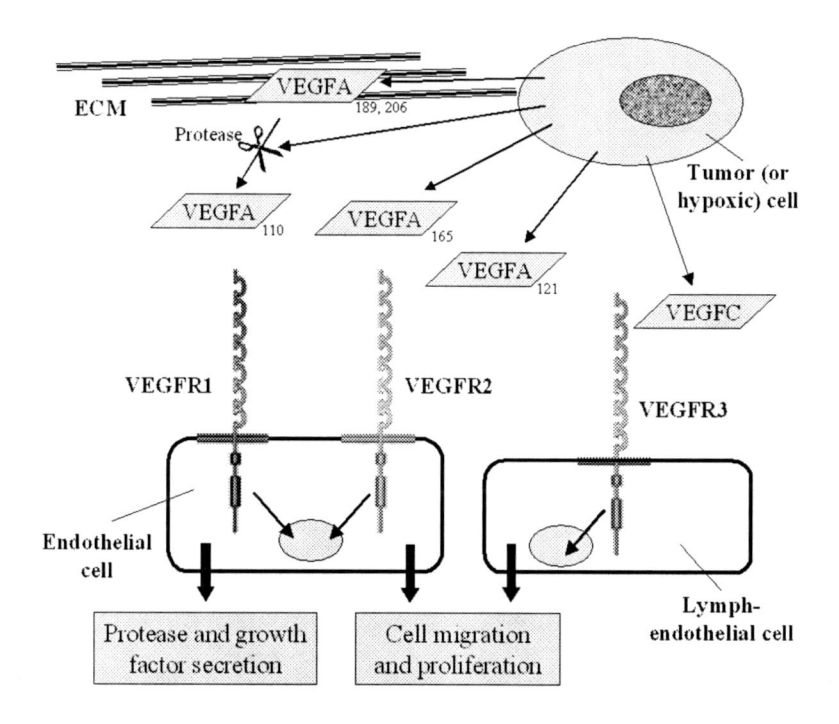

**Figure 9.6** *VEGF and its receptors*
Various forms of VEGFA (named according to the number of amino acids) translated from differently spliced mRNAs are secreted by hypoxic normal cells or by tumor cells. They bind to heparin and other components of the extracellular matrix (ECM) to different extents and can be processed and released by proteases, e.g. plasmin. All VEGFA forms as well as VEGFB bind to VEGFR1 and VEGFR2 receptors, predominantly expressed on endothelial cells. Note that activation of VEGFR1 or VEGFR2 has different consequences. VEGFR3 is mainly expressed on lymphendothelial cells and activated by VEGFC or VEGFD. All VEGFRs are similar in structure, with 7 extracellular immunoglobulin-like repeat domains and an intracellular 'split' tyrosine kinase domain.

In normal cells, VEGFA expression is controlled by the oxygen partial pressure. Hypoxia (Box 9.1) leads to increased stability of specific transcriptional activators, designated hypoxia inducible factors (HIF). The accumulating HIF1α or HIF2α factors (which depends on the cell type) combine with a cofactor, ARNT (HIFβ), and activate the transcription of a set of genes that contain a specific binding element in their basal promoters. One of the most responsive genes is VEGFA.

Constitutive activation of HIF factors by an inherited defect in their regulation is the cause of Von-Hippel-Lindau disease which predisposes to renal cell carcinoma and other, less malignant tumors (→15.4). However, in many other cancers HIF activity is also elevated, typically as a consequence of hypoxia, and VEGFA is induced as a consequence.

As angiogenesis is such a crucial step in the expansion of many cancers and of their metastases, it may constitute an excellent target for therapy. This idea is particularly attractive, as the proliferating cells in angiogenesis are largely normal endothelial cells, which lack genomic instability. It is therefore less likely that they develop resistance to the treatment than actual cancer cells. Several inhibitors of angiogenesis have been used or are currently investigated in clinical trials, including physiological protein inhibitors of angiogenesis like endostatin and angiostatin, antibodies to VEGFA or VEGF receptors, and chemical inhibitors of the VEGFR, PDGFR, and FGFR tyrosine kinases.

## 9.5 INTERACTIONS OF INVASIVE TUMORS WITH THE IMMUNE SYSTEM

As carcinomas spread beyond their compartment barrier provided by the basement membrane, they encounter cells of the immune system. These are attracted by chemokines and other signals emananating from the tumor stroma and from the tumor cells themselves. As a consequence, different types of immune cells, including macrophages, granulocytes, and various types of lymphocytes are found in the activated stroma of human carcinomas and even in between compact carcinoma masses. While the immune response to the tumor may indeed limit its growth, it is evidently overcome in advanced cancers, and in many cases its effect may even be ambiguous.

This is so because the presence of immune cells in tumor stroma and the tumor mass as such does not necessarily indicate an active response against the cancer cells, even though ample evidence indicates that immune responses against cancers are regularly initiated. In cancer patients, B-lymphocytes as well as T-lymphocytes are usually found that are directed against antigens present on the tumor cells. For instance, many patients with advanced stage cancers carry antibodies against mutated TP53 protein. Likewise, cytotoxic T-cells (CTL) and natural killer (NK) cells can be isolated from patients which specifically kill their tumor cells in vitro.

It is generally assumed that a cellular immune response, as opposed to a humoral response, is required for the successful rejection of a tumor. Tumor cell killing can be performed by cytotoxic T-cells directed against antigens presented by the MHC (major histocompatibility complex) and recognizing co-stimulatory molecules on the tumor cell surface. Cells that lack MHC expression or are marked by antibodies are, conversely, targets for NK cells. However, in vivo, these cytotoxic cells are often not successful in eliminating the cancer cells. While the reasons for this are not fully understood, several factors are known to be involved (Table 9.4).

Accessibility: Compact tumor masses may not be well accessible to immune cells. In this regard, the insufficiency of its vascular system (→9.4) may work in favor of the cancer. In stroma-rich cancers, the stroma may provide a buffer against the immune response.

Presentation by professional cells: Tumor antigens are present on many cancers (→12.5), but to be recognized by the immune system they need to be presented, best by 'professional' antigen-presenting cells, e.g. dendritic cells. These do not seem to function optimally in many cancers.

Presentation by the tumor cells: Cancer cells may down-regulate proteins required for the processing and presentation of antigenic peptides, e.g. proteasomal proteins, the TAP transporter, or frequently the MHC itself.

Activation of T-cells: This may one of the most ubiquitous mechanisms, as many tumor cells, including cancer and stromal cells, secrete cytokines which limit the activity of cytotoxic T-cells. Conversely, cytokines activating T-cells such as IL2 may be lacking or be sequestered in the tumor milieu. Many advanced carcinomas, e.g., are themselves unresponsive to TGFβ (→6.7)), but secrete high amounts of the factor, which are liberated from the extracellular matrix and processed towards its active form by proteases in the activated stroma. TGFβ stimulates proliferation of stromal cells and the production of matrix proteins, but also inhibits the activation and proliferation of T-cells. This contributes to the characteristic 'anergy' of T-cells isolated from actual tumor tissues.

Recognition of cancer cells: Down-regulation of MHC expression at the tumor cell surface diminishes recognition by cytotoxic T-cells. Cells with down-regulated MHC should be preferred targets for NK cells, but tumor cells might also down-regulate receptors for NK cells like MICA and MICB, while retaining inhibitory KIR receptors.

Efficiency of cell killing: Cancer cells become resistant to the mechanisms by which T-cells kill, which is usually some form of apoptosis (→7.3). For instance, many tumor cells down-regulate the FAS (CD95) death receptor which is activated by the FAS ligand expressed on the surface of cytotoxic T-cells, or express a soluble form that diverts the immune cells. Post-receptor defects like FLIP overexpression or Caspase 8 mutations may contribute, as might overactivity of the anti-apoptotic PI3K and NFκB pathways.

**Table 9.4.** *Mechanisms limiting the immune response to cancers*

| Mechanism |
|---|
| Accessibility of the tumor mass to immune effector cells |
| Antigen presentation and cytokine secretion by professional antigen-presenting cells |
| Antigen presentation by tumor cells |
| Anergy of cytotoxic T-cells |
| Recognition of tumor cells |
| Efficiency of cell killing |
| Counterattack of tumor against immune cells |

Counterattack: Tumor cells may not only contribute to anergy of T-cells, but even actively destroy them. For instance, some cancer cells secrete the FAS/CD95ligand themselves and mount a counterattack towards the immune cells (→7.3).

While the immune response towards the actual tumor cells is often blunted by these and further mechanisms, the presence of immune cells in the tumor and its stroma may contribute considerably to the destruction of the tissue. For instance, proteases secreted by macrophages and granulocytes destroy extracellular matrix, cytokines secreted by immune cells stimulate the proliferation of stromal and tumor cells as well as angiogenesis, and reactive oxygen species produced by macrophages may contribute to mutagenesis. The COX2 enzyme is thought to play a dual role in this regard, by producing prostaglandins and reactive oxygen at the same time.

The immune response is directed towards a tumor by several signals. Among the more important ones are chemokines (Figure 9.7). Chemokines are 8 – 20 kDa peptides, of which there are >50 in man. They are produced, e.g., by activated macrophages and granulocytes to attract further immune cells towards a site of infection or other tissue damage. It is therefore not surprising that chemokines are produced and secreted by stromal cells in many cancers. More surprisingly upon first thought is that cancer cells themselves also often produce chemokines, e.g., CXCL12. This chemokine activates a specific receptor, CXCR4, on the surface of immune cells, but also of hematopoetic and endothelial precursors. The receptor is also expressed on some cancer cells. So, production of CXCL12 may have several consequences, which in effect can favor cancer growth. (1) Attraction of immune cells may lead to tissue destruction favoring invasion and metastasis. (2) The factor may promote growth and survival of cancer cells expressing the CXCR4 receptor. (3) CXCL12 may promote the recruitment of precursor cells for vasculogenesis. (4) Activation of CXCR4 leads to strongly increased production of TNFα which itself may exhibit another array of effects.

Thus, like other normal physiological processes that are 'coopted' during tumor invasion, the immune response towards tumor cells is distorted and often seems to be diverted towards promoting cancer spread instead of achieving its destruction.

## 9.6 THE IMPORTANCE OF TUMOR-STROMA INTERACTIONS

In the course of cancer invasion, therefore, several normal cell types, fibroblasts, endothelial cells, and immune cells are influenced by the actual cancer cells and in effect contribute to growth and spread of the tumor. It is now thought that the activation of stroma is in many cases rate-limiting for tumor invasion.

Similar interactions may be repeated during the establishment of metastases. However, there is one crucial difference in that during invasion cancer cells interact with essentially the same cells as their normal counterparts. Thus, interactions between tumor and local stroma may resemble those during normal tissue homeostasis and particular those during wound repair, although they are deregulated and distorted. However, during metastasis, cancer cells enter a novel environment. To expand into secondary tumors they need to activate this new environment, e.g., to provide nutrients and growth factors. The ability to achieve this interaction depends on both the cancer cells and their new environment. Such interactions may determine the sites of metastases more than anatomical relationships.

For instance, mammary cancers and prostate cancers have a propensity for metastasis to bone which appears to be determined by their ability to set up

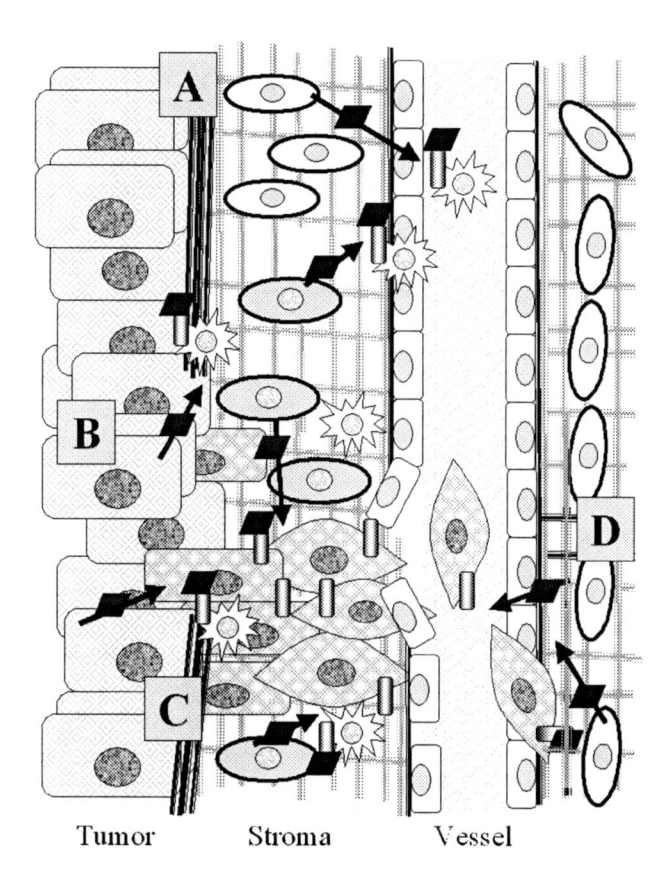

**Figure 9.7** *Functions of chemokines and chemokine receptors in human cancers*
A: As a consequence of inflammation or alerted by tumor cells (rectangular on the left), stromal cells (oval) secrete chemokines (black diamonds) which attract immune cells (exemplified by macrophage-like cells) that express appropriate chemokine receptors (cylinders). B: Tumor cells secrete chemokines which stimulate immune cells to contribute to the destruction of the basement membrane and extracellular matrix remodeling. C: Tumor cells themselves begin to express receptors for chemokines enabling them to react to clues from stromal cells and stimulating their own migration. D: At a distant site, tumor cells expressing chemokine receptors are attracted by chemokines and extravasate to form metastases.

paracrine interactions with osteoclasts or osteoblasts in this tissue (→19.4). Like many other aspects of metastasis, the precise relationships are not understood and are the subject of current investigations. Most studies are based on the hypothesis that metastasis is determined by an adequate interaction between tumor cells and target tissue. This is the current version of the 'seed-and-soil' hypothesis which was proposed by Paget already in the 19th century.

One unexpected example of interactions between metastatic cancer cells and their target tissue concerns chemokines. Unlike their normal counterparts, carcinoma cells express receptors for chemokines (Figure 9.6), which typically become induced during a hypoxic phase of cancer development. Breast cancers tend to express the CXCR4 receptor for CXCL12 (→9.5). Its ligand is normally mainly found in lymph nodes, lung, liver, and bone, which are the preferred sites of metastasis for this cancer. Melanomas, in comparison, often express the CCR10 receptor. Its ligand CCL27 is predominantly expressed in skin. So, the unusual propensity of melanomas to metastasize to skin may not only be due to a 'natural' affinity of melanocytes, but also to this relationship. Clearly, though, the presence of chemokines is only one factor of several preparing the 'soil' for metastases.

Another subject of current investigation is the behavior of stroma during local tumor growth. As described above, tumor stroma is no longer regarded as a a passive medium invaded by the tumor, but as an active contributor to the success of tumor invasion. If so, genetic and epigenetic alterations in stromal cells might contribute to cancer growth, in addition to alterations in the actual carcinoma cells. Indeed, cultured senescent fibroblasts, as they may occur in tissue of older persons (→7.4), secrete growth factors to which carcinoma cells, but not normal epithelial cells, respond. In similar experiments, fibroblasts from activated tumor stroma have been found to stimulate the proliferation of early carcinoma cells. In breast carcinomas, mutations of *PTEN* and *TP53* have been reported to occur in either stroma or carcinoma cells, in a mutually exclusive fashion. Such findings point towards a kind of vicious circle, in which mutations in the carcinoma drive alterations in the stroma that again promote the progression of the carcinoma, etc.

Yet another complication in this context is tumor 'mimicry'. As carcinomas progress, they often lose markers of differentiation, including the cytokeratins typical of epithelial cells as well as cadherins (→9.2), and cells change morphology. Some carcinoma cells, in vivo as in culture, assume a spindle-like shape which is quite similar to that of a typical mesenchymal fibroblast in the tumor stroma (Figure 9.8). This change, sometimes called EMT (→9.2), is particularly frequent in cells that have lost expression of E-Cadherin and other epithelial adhesion proteins. By their morphology alone and even by staining for cytokeratins and other epithelial markers, these cells are difficult to identify as being derived from the cancer. Moreover, some tumors present with a stem-cell phenotype and can differentiate into several different cell types. This property is typical of certain 'embryonic' tumors like teratocarcinomas and Wilms tumors (→11). Nevertheless, 'mimicry' of other cell types has been documented for a much wider range of cancers. So, tumor cells may appear as fibroblastoid components of stroma, as endothelial cells in tumor capillaries, or as osteoblast-like cells in bone metastases. This means that some cancers may not fully depend on recruitment of active stroma for invasion or metastasis, but generate it themselves by transdifferentiation 'mimicry'.

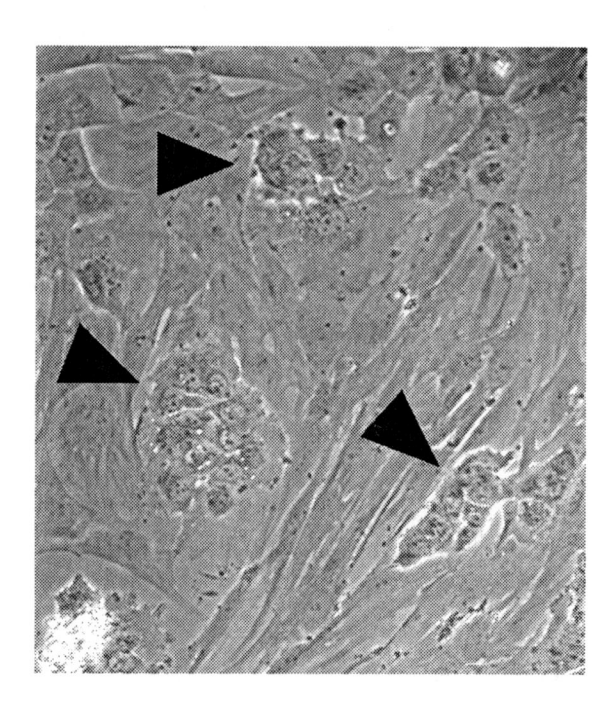

**Figure 9.8** *A bladder cancer cell culture undergoing EMT in vitro*
The cancer cells variously take on epithelial (arrowhead) or fibroblastoid morphology.
Courtesy: Andrea Prior

## *Further reading*

Weiss L (2000) Metastasis of cancer: a conceptual history from antiquity to the 1990s. Cancer Metastasis Rev. 19, 193-400

Hidalgo M, Eckhardt SG (2001) Development of matrix metalloproteinase inhibitors in cancer therapy. J. Natl. Cancer Inst. 93, 178-193

Liotta LA, Kohn EC (2001) The microenvironment of the tumour-host interface. Nature 411, 375-379

Murphy PM (2001) Chemokines and molecular basis of cancer. NEJM 354, 833-835

Chambers AF, Groom AC, MacDonald IC (2002) Dissemination and growth of cancer cells in metastatic sites. Nat. Rev. Cancer 2, 563-572

Coussens LM, Fingleton B, Matrisian LM (2002) Matrix metalloproteinase inhibitors and cancer: trials and tribulations. Science 295, 2387-2392

Hajra KM, Fearon ER (2002) Cadherin and Catenin alterations in human cancer. Genes Chromosomes Cancer 34, 255-268

Harris AL (2002) Hypoxia – a key regulatory factor in tumour growth. Nat. Rev. Cancer 2, 38-47

Almholt K, Johnsen M (2003) Stromal cell involvement in cancer. Rec. Res. Cancer Res. 162, 31-42

Bergers G, Benjamin LE (2003) Tumorigenesis and the angiogenic switch. Nat. Rev. Cancer 3, 401-410

Birchmeier C et al (2003) MET, metastasis, motility and more. Nat. Rev. Mol. Cell Biol. 4, 915-925

Brakebusch C, Fässler R (2003) The integrin-actin connection, an eternal love affair. EMBO J. 22, 2324-2333

Dvorak HF (2003) How tumors make bad blood vessels and stroma. Am. J. Pathol. 162, 1747-1757

Engbring JA, Kleinman HK (2003) The basement membrane matrix in malignancy. J. Pathol. 200, 465-470

Ferrara N, Gerber HP, LeCouter J (2003) The biology of VEGF and its receptors. Nat. Med. 9, 669-676

Friedl P, Wolf K (2003) Tumour-cell invasion and migration: diversity and escape mechanisms. Nat. Rev. Cancer 3, 362-374

Kauffman EC et al (2003) Metastasis suppression: the evolving role of metastasis suppressor genes for regulating cancer cell growth at the secondary site. J. Urol. 169, 1122-1133

Ramaswamy S, Ross KN, Lander ES, Golub TR (2003) A molecular signature of metastasis in primary tumors. Nat. Genet. 33, 49-54

Subarsky P, Hill RP (2003) The hypoxic tumour microenvironment and metastatic progression. Clin. Exp. Metastasis 20, 237-250

Balkwill F (2004) Cancer and the chemokine network. Nat. Rev. Cancer 4, 540-550

Cavallaro U, Christofori G (2004) Cell adhesion and signalling by cadherins and Ig-CAMs in cancer. Nat. Rev. Cancer 4, 118-132

Verheul HMW, Voest EE, Schlingemann RO (2004) Are tumours angiogenesis-dependent? J. Pathol. 202, 5-13

### Box 9.1 Tumor hypoxia and its consequences

The oxygen partial pressure (pO$_2$) in normal tissues is in the 10 – 40 mm Hg range. In tumors, it can be much lower, creating a transient or chronic state of hypoxia. Local and systemic factors are responsible, encompassing abnormal structure of macro- and microvessels (→9.4), limited perfusion, limited diffusion through several cell layers (>70 µm), and systemic anemia.

Hypoxia is operationally often defined as pO$_2$ < 2.5 mm Hg (1), but some of its consequences arise earlier.

➢ As pO$_2$ approaches 10 mm Hg, therapies that require the presence of oxygen like photodynamic therapy or radiotherapy fail or become less efficient.
➢ Below 10 mm Hg, metabolic processes become affected and cells begin to adapt. Closer to 0.1 mm Hg (2), prolyl hydroxylases that control HIF1/2α activity and degradation become inactive and gene expression changes more considerably in order to compensate (cf. 15.4). When pO$_2$ persists in this range for extended periods, in addition to adjusted blood flow, angiogenesis becomes induced.
➢ Additionally, at a consistent pO$_2$ < 1 mm Hg, cells may become apoptotic (3).
➢ If pO$_2$ decreases much further (4), even apoptosis is impeded, perhaps because of a failure by mitochondria to provide sufficient ATP. Cells may die by necrosis.
➢ A very problematic consequence of persistent pO$_2$ < 1 mm Hg [(3) and (4)] is increased genomic instability. Gene amplification and other structural chromosomal alterations arise and may be tolerated because of a decreased inability of hypoxic cells to undergo apoptosis.

The first hypoxic episode of may constitute a critical stage in the development of a cancer. It develops when the tumor has reached a certain size and is suboptimally supplied with oxygen as well as other nutrients, including precursors and coenzymes for DNA synthesis. Both deficiencies may favor genetic alterations and persistent genomic instability may become established.

A similar situation may arise again at an advanced stage, when a cancer is again subject to hypoxia as a consequence of anemia caused by its own effects on the host and/or by chemotherapy that suppresses erythropoesis and damages vessels. Again, genomic instability may be promoted and contribute to the emergence of resistant and even more aggressive tumor clones (→22.2). Conceivably, this effect might even complicate therapy by novel anti-angiogenic drugs which are aimed at the tumor vascular rather than the cancer cells themselves.

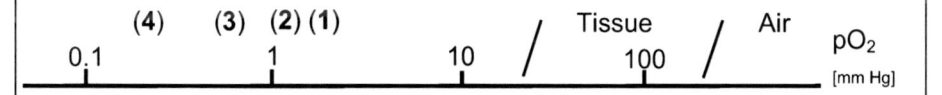

```
         (4)  (3) (2)(1)                    Tissue         Air
  0.1     |    |   | |       10       /     100      /          pO₂
   |_____|____|___|_|_____|_____/_____|_____/_____ [mm Hg]
```

Vaupel P, Thews O, Hoeckel M (2001) Treatment resistance of solid tumors: role of hypoxia and anemia. Med. Oncol. 18, 243-259
Harris AL (2002) Hypoxia – a key regulatory factor in tumour growth. Nat. Rev. Cancer 2, 38 - 47

# PART II

# HUMAN CANCERS

# CHAPTER 10

# LEUKEMIAS AND LYMPHOMAS

- Leukemias and lymphomas are cancers arising from cells of the hematopoetic lineage (hence: hematological cancers). Leukemias originate from hematopoetic stem cells or cells at different stages of myeloid or erythroid differentiation which spread throughout the body. Lymphomas also develop from more differentiated lymphoid cells in lymphoid organs and may present as localized cell masses. Hematological cancers can be classified by their derivation from erythroid, myeloid, or lymphoid cells at specific stages of development as determined by their morphology and by protein markers.
- Hematological cancers are also often characterized by recurrent chromosomal aberrations, which in their initial stages may represent the only evident genetic change. These aberrations also comprise chromosome gains and losses, but specific translocations are most distinctive.
- Translocations cause hematological cancers by either of two mechanisms. They activate a proto-oncogene by destroying its negative regulatory elements and/or by placing it under the influence of activating enhancers or they fuse two genes from the translocation sites that yield fusion proteins with novel properties. Either way, the gene product from the translocation site drives tumor development. During tumor progression, such as during the blast crisis of chronic myelogenous leukemias, further genetic and epigenetic changes accumulate, as commonly seen in carcinomas.
- Chronic myelogenous leukemia (CML) is a disease of hematopoetic stem cells characterized by hyperproliferation of often immature cells of the myeloid, megakaryocytic and erythroid lineages. Its diagnostic chromosomal change is the 'Philadelphia chromosome' resulting from a translocation between chromosomes 9 and 22 that creates the BCR-ABL fusion gene. The gene product is a fusion protein that retains the protein kinase activities of both original proteins, but is deregulated and mislocalized to the cytosol. There, the BCR-ABL protein activates several signaling pathways that promote cell proliferation, block apoptosis and decrease cell adhesion, thereby blocking maturation and causing release of immature cells into the blood. After several years, CML turns into blast crisis, a rapidly lethal disease resembling acute leukemias. This progression is promoted by interference of the BCR-ABL protein with the control of genomic stability. The acute phase of the disease is characterized by more pronounced chromosomal instability, with loss of tumor suppressors such as TP53 and $p16^{INK4A}$.
- CML is treated by interferon therapy, chemotherapy and allogeneic stem cell transplantation. An important component of current therapy is a specific inhibitor of the ABL kinase, imatinib, which represents one of the great successes of

- molecular cancer research. Cytogenetic and molecular methods which detect the BCR-ABL translocation are helpful in diagnosis and in monitoring of treatment.
- The efficacy of interferon therapy and allogeneic stem cell transplantation in CML have implications extending beyond this specific disease. They suggest that tumor cells can be recognized and kept under control through surveillance by the immune system. Highly sensitive molecular methods corroborate this assumption by proving the presence of Philadelphia chromosomes in healthy humans.
- Burkitt lymphoma (BL) is an aggressive cancer derived from B-lymphocytes endemic in certain tropical areas of the world, and sporadically occuring in immuno-compromised individuals. These lymphomas have an extremely high proliferation rate which is not compensated by a high rate of apoptosis. Three alternative, diagnostic chromosomal aberrations invariably bring the *MYC* proto-oncogene on chromosome 8q24.1 under the influence of immunoglobulin gene enhancers from chromosome 14 (major translocation), 22, or 2 (minor translocations). Often, these translocations also destroy negative regulatory elements constraining this potent oncogene. MYC activity can be further augmented by point mutations.
- Epstein Barr Virus (EBV), present in a few cells in most humans, but in the majority of BL clones, acts as a cofactor in this disease, perhaps by decreasing apoptosis. Additional genetic alteration in BL that diminish apoptosis severely aggravate the disease by counteracting the tendency of MYC to induce apoptosis in addition to cell growth. Beyond chemotherapy and stem cell transplantation, no specific treatment is available for BL.
- In contrast, promyelocytic leukemia (PML), an acute leukemia (hence also: APL), is an example for successful targeted molecular therapy. It arises most often from a translocation involving chromosomes 15 and 17, t(15;17)(q22;q21), that creates a fusion protein from the nuclear coordinator PML and the retinoic acid receptor $\alpha$ (RAR$\alpha$). The fusion protein blocks differentiation and apoptosis of partially differentiated myeloid cells from which the tumor is derived. Unlike RAR$\alpha$, it does not respond to physiological levels of retinoic acid. Fortunately, pharmacological doses of retinoic acid activate the fusion protein, eliciting tumor differentiation and apoptosis. This 'differentiation therapy' is not efficacious in cases of PML caused by fusion of RAR$\alpha$ with the chromatin repressor PLZF. However, since PLZF acts by recruiting histone deacetylases, combined treatment with retinoids and histone deacetylase inhibitors may be possible. PML illustrates the importance of chromatin alterations in the development of human cancers and the therapeutic potential of understanding tumor pathophysiology at the molecular level.

## 10.1 COMMON PROPERTIES OF HEMATOLOGICAL CANCERS

Hematological cancers comprise leukemias and lymphomas which arise from the hematopoetic lineage. All cells in this lineage originate from stem cells (→8.6) which in humans, after birth, are located in the bone marrow, after residing in the liver during the most of the fetal period. Here, stromal cells and an appropriate extracellular matrix support stem cell maintenance and help to regulate differentiation. Hematopoetic stem cells in the bone marrow and circulating in the blood can be identified by their expression of the surface antigen CD34.

The first step of differentiation of the rare pluripotent hematopoetic stem cells distinguishes the lymphoid lineage and the myeloid/erythroid lineage (Figure 10.1) and leads to a larger population of 'committed' progenitor cells. The lineages split further and lead to the various types of differentiated cells such as the B-cell and T-cell subtypes in the lymphoid branch, as well as the erythrocytes and platelets (erythroid and megakaryocytic lineage) and mast cells, granulocytes, monocytes and macrophages (myeloid lineage).

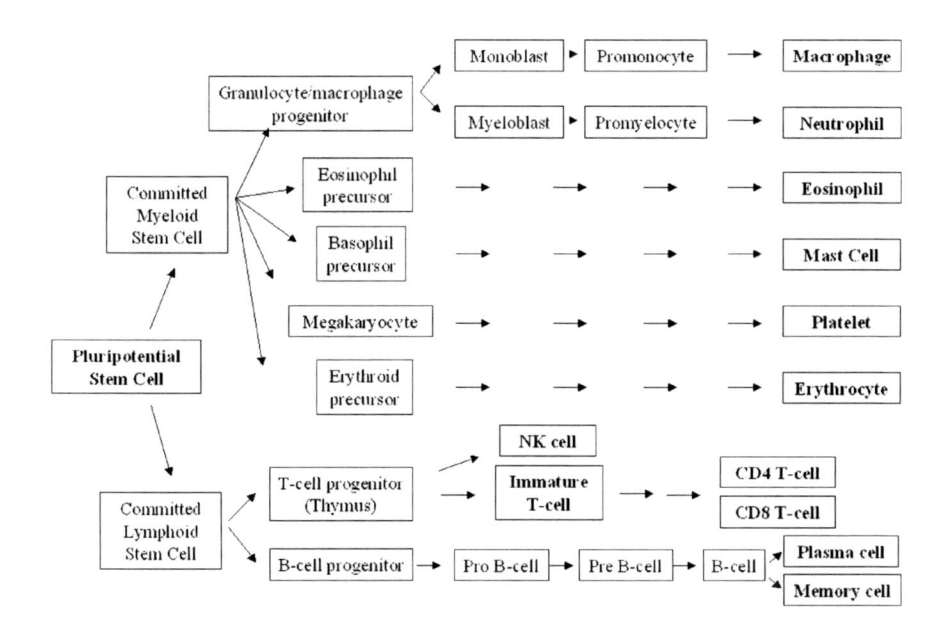

**Figure 10.1** *A sketch of cell lineages in the hematopoetic system*
See text for explanation.

Proliferation and differentiation of hematopoetic cells are directed and controlled by a panoply of growth factors present in the bone marrow environment (Table 10.1), and intracellularly by networks of lineage- and cell-type transcription factors (cf. 8.5). In each branch of the lineage, the options for differentiation are progressively restricted and cells become more and more committed towards a specific fate. In parallel, as a rule, their proliferative potential becomes more and more restricted, most obvious in case of erythrocytes which even lose their nuclei. Especially the later stages of differentiation in the myeloid and erythroid lineages are characterized by a tight coupling between differentiation and loss of proliferative potential, so that the number of cells produced is limited. For cancers to arise in this

Table 10.1. *Some important growth factors in the hematopoetic system*

| Growth factor | Molecular weight [Da]* | Origin | Target cells** |
|---|---|---|---|
| SCF | ≈30 | bone marrow stroma, other mesenchymal cells | stem and progenitor cells of multiple lineages |
| GM-CSF | 22 | macrophages, mast cells, T-cells, endothelium, fibroblasts | granulocyte and macrophage stem cells, many differentiated cell types |
| G-CSF | ≈22 | monocytes, macrophages | cells in neutrophil lineage |
| IL3 (M-CSF) | 28 | T-cells, mast cells | progenitors and precursors in many lineages |
| IL6 | ≈25 | activated immune and mesenchymal cells | macrophages, monocytes, T-cells, B-cells, epithelial cells |
| Erythropoetin | ≈25 | kidney | erythroid lineage precursors |
| Thrombopoetin | 76 | liver, kidney | megakaryocyte precursors |
| IL1 | 17 | many immune and mesenchymal cells | many immune, mesenchymal, and epithelial cells |
| IL2 | 15 | activated T-cells | T-cells, other immune cells and their precursors |
| IL8 | 7 (dimer subunits) | macrophages, monocytes, activated stroma | granulocytes, T-cells, endothelial cells and precursors |

*Molecular weights can vary because of glycosylation and other post-translational modifications; ** Effects vary, including stimulation of proliferation, differentiation, migration and induction of specific functions lineage, this dependency must be broken or the stimuli must be mimicked. Moreover, since differentiation diminishes proliferative potential, a carcinogenic change must block or at least substantially diminish the rate of cell differentiation.

The relationship between proliferation and differentiation is somewhat different in the lymphoid lineage since mature lymphocytes still remain competent for proliferation in response to infections. Their multiplication, however, is normally dependent on stimulation by antigens and by paracrine factors, mostly cytokines. Moreover, when an infection abates, lymphocyte numbers are restrained by apoptosis and only memory cells survive for longer periods. Controlled apoptosis is also involved in the selection procedure against self-reactive lymphocytes. Thus, cancers in the lymphoid lineage can arise not only in immature cells by failure along the path to differentiation, but also in differentiated cells by acquired independence of external proliferation stimuli or by evasion of apoptosis, or in the worst case both.

The phenotype of individual leukemias and lymphomas is therefore strongly dependent on which stage in which lineage is primarily affected. A hematological cancer may be derived from a stem cell such as in chronic myelogenous leukemia (→10.4), a partially differentiated cell as in acute promyelocytic leukemia (→10.5), or a cell at an advanced stage of differentiation as in multiple myeloma (a B-cell lymphoma). The many different subtypes of B-cells and T-cells, in particular, with different functions and tissue distributions, can give rise to a large variety of lymphomas.

The name of a hematological cancer often indicates the dominant cell type. Obviously, the cell of origin cannot always be straightforwardly identified, because cancer cells deviate from the original cell. In many types of leukemias and lymphomas, only the introduction of molecular markers in the last decades has allowed to determine their cell of origin and, in more than a few cases, to recognize morphologically similar diseases as different entities. Molecular markers used for this purpose comprise surface antigens characteristic for lineages and differentiation stages as well as typical chromosomal aberrations. An identification as precise as possible is, of course, necessary to predict prognosis and to choose appropriate treatments (→21.2).

## 10.2 GENETIC ABERRATIONS IN LEUKEMIAS AND LYMPHOMAS

On average, leukemias and lymphomas contain fewer genetic aberrations than carcinomas. Very often, their karyotype is near-diploid with few chromosomes altered. Those aberrations that are present, however, are characteristic for the particular type of leukemia or lymphoma. Although chromosomal losses and gains are also found, many hematological cancers are characterized by typical translocations (Table 10.2).

In some hematological cancers, these translocations are so characteristic that they allow to classify the disease accordingly. So, >90% of chronic myeloid leukemias contain a characteristic translocation between the long arms of chromosomes 9 and 22, t(9;22) (q34;q11). On the other hand, apparently similar diseases may result from disparate translocations. So, the same t(9;22) translocation is found in ≈25% of acute lymphoblastic leukemias as well. However, in this group of diseases, a variety of other translocations are found (Figure 10.2).

Unlike deletions such as those of 6q or 13q found in some leukemias, translocations do not simply destroy genes and their functions, but alter them. Two mechanisms can be distinguished (Figure 10.3, cf. 4.3).
(1) The coding region of the gene at the translocation site remains intact, but its regulation is altered, since the translocation removes some of its own regulatory regions and/or brings the gene under the influence of novel regulatory elements. In most cases, this leads to ectopic or deregulated expression of a gene product acting as an oncogenic protein. Burkitt lymphoma with its translocations activating the MYC gene is a case in point (→10.3). Another frequent translocation t(14;18) (q32;q21) leads to deregulation and overexpression of the anti-apoptotic regulator protein BCL2 (→7.2) in follicular lymphoma. Here, the *BCL2* locus is brought under the control of a heavy-chain immunoglobulin enhancer (cf. 10.3).

Table 10.2. *Some characteristic translocations in leukemias and lymphomas*

| Translocation | Genes involved | Consequence | Cancer type* |
|---|---|---|---|
| t(8;14) (q24;q32) t(2;8) (p12;q24) t(8;22) (q24;q11) | *MYC*, Ig loci | deregulation of MYC | Burkitt lymphoma |
| t(8;14) (q24;q11) | *MYC, TCRA* | deregulation of MYC | T-ALL |
| t(14;18) (q32;q21) | *IGH, BCL2* | deregulation of BCL2 | follicular lymphoma |
| t(11;14) (q13;q32) | *IGH, CCND1* | deregulation of Cyclin D1 | B-CLL |
| t(9;22) (q34;q11) | *ABL, BCR* | expression of BCR-ABL fusion protein | CML |
| t(15;17) (q22;q21) | *PML, RARA* | expression of PML-RARA fusion protein | APL |
| t(11;17) (q23;q21) | *PLZF, RARA* | expression of PLZF-RARA fusion protein | APL |
| t(7;9) (q35;34.3) | *TCRB, NOTCH1* | overexpression of truncated NOTCH1 protein | T-ALL |
| t(3;14) (q27;q32) t(2;3) (p12;q27) t(3;22) (q27;q11) | *BCL6*, Ig loci | deregulation of BCL6 | B-cell lymphoma |
| t(8;16) (p11;p13) | *MOZ, CBPA* | expression of MOZ-C/EBPα repressor fusion protein | AML |

ALL: acute lymphoblastic leukemia, AML: acute myelocytic leukemia APL: acute promyelocytic leukemia, CLL: chronic lymphoblastic leukemia, CML: chronic myeloid leukemia

(2) The translocation creates a fusion protein expressed from a novel gene formed from the segments of two genes flanking the translocation sites. The fusion protein may have a different expression pattern, intracellular localization, regulation, and activity (or a combination of these properties) compared to the original components. Not infrequently, the destruction of one or both original genes also matters. Two examples out of many for this mechanism are the BCR-ABL fusion protein in chronic myeloid leukemia (→10.4) and the retinoic acid receptor fusion proteins in acute promyelocytic leukemia (→10.5).

The characteristic translocations are causally involved in the development of each specific cancer subtype. An impressive illustration of this relationship is found in acute myeloid leukemia (AML). Translocations in AML often interrupt the genes encoding transcription factors involved in the differentiation of myeloid cells, such as that encoding C/EBPα, and create fusion protein, which instead of activating target genes repress them. Some cases lack these characteristic translocations. In these, point mutations in the *CBPA* gene have been found which lead to a truncated protein that acts as a dominant-negative inhibitor of the protein expressed from the remaining intact allele.

AMLs belong to those hematological cancers that behave quite heterogeneously in the clinic. The course of the disease and the response to therapy can to a certain degree be predicted from the type of translocation (or mutation) present. In general,

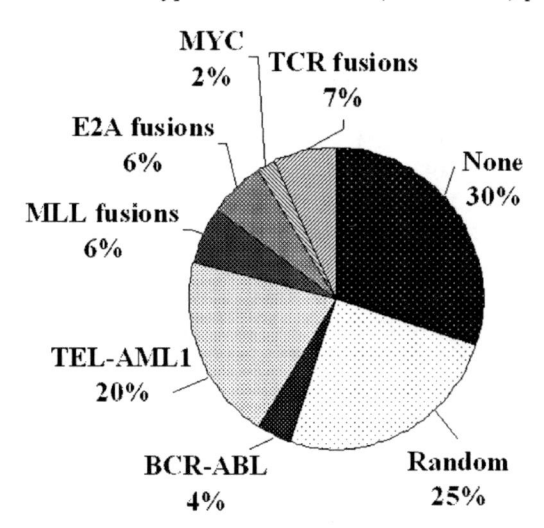

**Figure 10.2** *Characteristic translocations in acute lymphocytic leukemias*
The areas in the circle reflect the proportion of ALLs associated with the various alterations. TCR is the T-cell receptor; E2A is a transcriptional activator, MLL is a transcriptional repressor, TEL and AML1 are likewise chromatin proteins; MYC and BCR-ABL are discussed in the text. ALL with MYC translocations are Burkitt lymphomas. Note that ALLs with BCR-ABL harbor the same translocation as CMLs, but are a different disease.

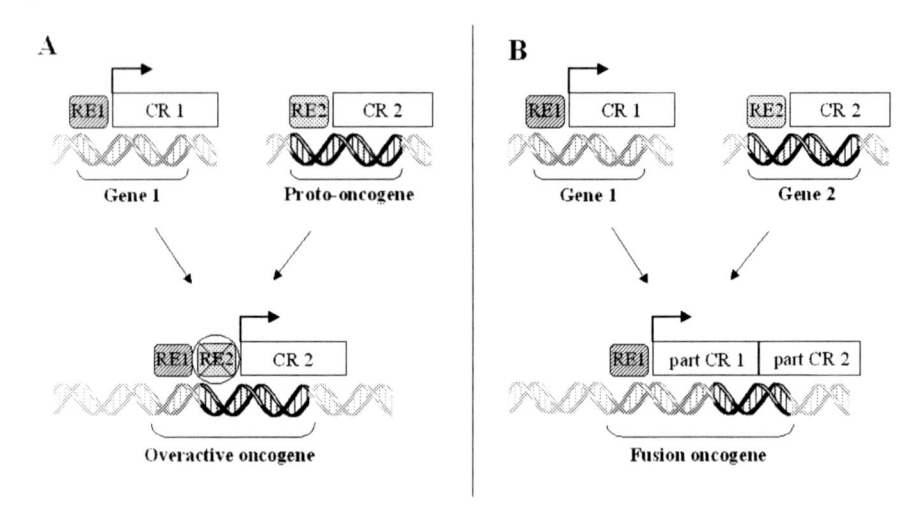

**Figure 10.3** *Principle mechanisms in translocations*
A: Gene deregulation; B: Formation of a fusion gene; RE: regulatory elements; CR: coding region. See text for further explanation.

determining the characteristic translocations in a leukemia or lymphoma is very helpful in the diagnostis and treatment of the disease (cf. 21.2).

It is not entirely clear, which further genetic changes in addition to the typical translocations are present in the initial stages of leukemias and lymphomas and to what extent they are required for their development. Certainly, however, many leukemias and lymphomas progress to a state of more generalized genomic instability with aneuploid karyotypes, additional point mutations and changes in DNA methylation. Genes affected at this stage also comprise some of the oncogenes and tumor suppressors important in carcinomas, such as *RAS* genes, *TP53*, *RB1*, *CDKN2A*, and also *CDKN2B*. This genomic instability not only confers a more aggressive character, but also presents a significant obstacle to chemotherapy by allowing the development of resistant cell clones, e.g. through point mutations or gene amplification (→22.2).

## 10.3 MOLECULAR BIOLOGY OF BURKITT LYMPHOMA

Burkitt lymphoma (BL) is an aggressive malignancy consisting of highly proliferative B-cells which infiltrate lymph nodes and other organs. The tumor cells are distinguished as B-cells by their expression of IgM and κ or λ light chains and by specific surface markers such as CD19 and CD20. The high proliferative rate is obvious from numerous mitotic figures. Staining for markers such as Ki67 or PCNA, which are characteristic of cycling cells, shows that almost all tumor cells are participating in proliferation. The rapid proliferation in this cancer is only partly compensated by a high apoptotic rate, evident most straightforwardly from interspersed macrophages phagocytosing the apoptosed tumor cells. (Figure 10.4)

BL was first described as a tumor in young people in equatorial Africa where it is endemic in some regions. It is much rarer in Europe and North America, although more frequent in the context of AIDS and in other immunocompromised patients. In tropical areas, most BLs harbor Epstein-Barr virus (EBV), which is not as consistently associated with the disease in countries of the temperate zones. Treatment by chemotherapy and stem cell transplantation[12] can be successful in some cases.

The three characteristic translocations in BL all involve the *MYC* locus at 8q24.1. The most frequent, 'major' translocation joins the gene to the *IGH* locus at 14q32 which encodes the immunoglobulin heavy chain. The 'minor' translocations involve the *IGL* (Igλ) locus at 22q11 or the *IGK* (Igκ) locus at 2p12 (Figure 10.5). These loci encode the two immunoglobulin light chains, either of which can be used in B cells. One of these three translocations is found in each case of BL. The diagnosis of BL therefore is made contingent on their presence. The converse is not true, i.e. these translocations are also found in other lymphomas, all of which are also aggressive.

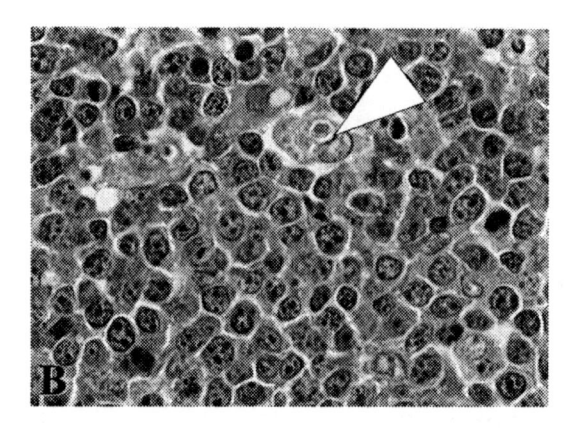

**Figure 10.4** *Histology of Burkitt lymphoma*
Note the macrophage phagocytising an apoptotic cell.

In each translocation, the MYC gene is brought under the influence of an immunoglobulin gene enhancer which is strongly active at this stage of B-cell differentiation. BL is thus an important example of the gene activation mechanism by translocations (Figure 10.3A). The translocations in BL result in deregulation and overexpression of a potent oncogene, which normally ought to become down-regulated in mature B-cells. While the overall result is the same, the mechanisms leading to deregulation differ in detail between the various translocations (Figure 10.6).

---

[12] Stem cells were formerly transferred by bone marrow transplantation; today they can be harvested from peripheral blood following administration of G-CSF. They can be identified by their CD34 marker.

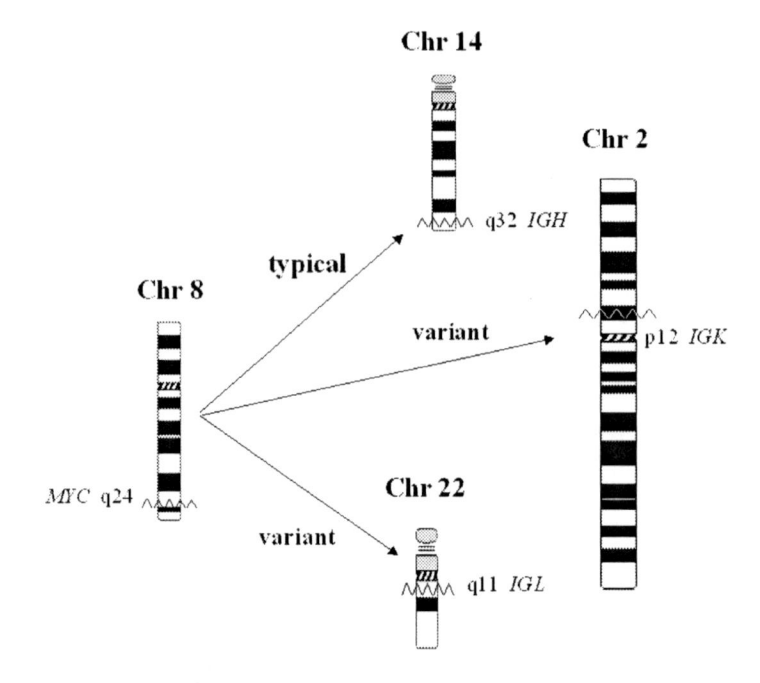

**Figure 10.5** *Translocations in Burkitt lymphoma*
See text for further details.

In most major translocations, the translocation breakpoint lies at some distance upstream of the *MYC* gene, which remains intact. The translocated *IGH* locus is oriented in inverse orientation, with the enhancer positioned towards the *MYC* gene. This configuration appears to result in a 'classical' enhancer activation mechanism, in which the strong enhancer activates the unchanged MYC promoters.

In some of the major translocations, the breakpoints are located in the first intron of the *MYC* gene, so that the first exon with the P1 and P2 promoters is removed. The gene is transcribed from the otherwise 'cryptic' (i.e. unused) P3 promoter located in the first intron. In this configuration, all elements in the *MYC* upstream region controlling transcription are deleted. Specifically, the translocation inactivates an attenuation mechanism in the first intron which makes transcription of the gene contingent on continuous stimulation. In these translocations, the placement of the *IGH* enhancer next to the gene does not seem to be absolutely essential for *MYC* activation, but the presence of an active *IGH* gene is required, perhaps to guarantee an open chromatin structure.

In the minor translocations, the breakpoints are located downstream of the *MYC* gene, sometimes at a considerable distance (>100 kb). The regulatory regions of the *MYC* gene remain largely intact, with the possible exception of a negative regulatory element presumed to lie downstream of the gene. The light chain loci are located in inverse orientation with their downstream enhancers oriented towards MYC. These seem to be mainly responsible for driving MYC over-expression. In the *IGK* locus, further elements may contribute, including an intronic enhancer.

In all translocations, the translocated *MYC* gene becomes susceptible to mutations that augment the effects of the enhancers. In translocations retaining the *MYC* regulatory regions, point mutations are found which inactivate negative regulatory elements in exon 1 and the first intron, thereby exacerbating overexpression. In all translocations, mutations are also found in the N-terminal coding region of MYC. In general, these stabilize the protein. For instance, a common mutation in Thr58 prevents a regulatory phosphorylation that targets the gene for proteosomal degradation. Interestingly, an according mutation is also common in the v-myc oncogenes of retroviruses (→Figure 4.1).

The genetic alterations at the *MYC* locus found in Burkitt lymphoma can be understood as accidents during the maturation of B-cells. During the differentiation of B-cells, V(D)J recombination of the *IGH* and either *IGL* or *IGK* immunoglobulin genes is employed to generate the antibody repertoire necessary for the immune response to many different antigens. Later, a further recombination event accompanies the switch from expression of IgM to that of IgG. Furthermore, in the germinal centers of the lymphoid organs, somatic hypermutation is directed to the rearranged immunoglobulin genes to generate additional antibody variants with

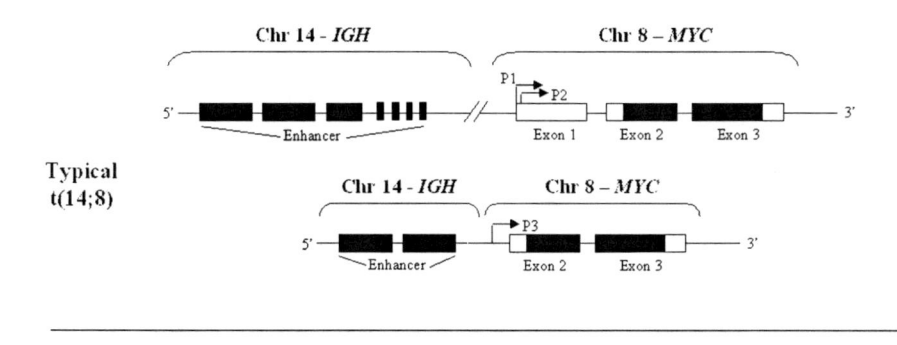

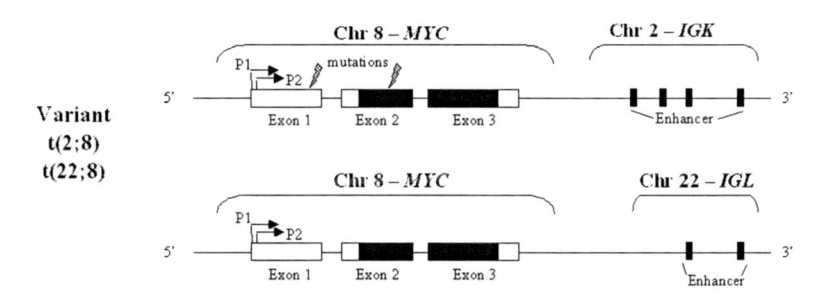

**Figure 10.6** *Mechanisms of MYC activation in Burkitt lymphoma*
A more detailed illustration of the mechanisms leading to *MYC* deregulation. Note that the figure is not to scale.

higher affinities and improved specifity towards the encountered antigens. These processes are recognizable in BL, although deviant.

The RAG recombinases mediating V(D)J rearrangements recognize particular sequence motifs in the immunoglobulin genes which are found in the vicinity of the breakpoints in the translocated immunoglobulin genes, although not usually in the *MYC* gene. It is therefore possible that the translocations are initiated by the physiological introduction of double-strand breaks into the immunoglobulin genes by B-cell specific recombinases. By accident, these breaks are then wrongly connected to the *MYC* gene. Likewise, placement of the *MYC* gene into an immunoglobulin gene cluster may set it up as a target for somatic hypermutation. The accidents that initiate BL are likely to be very rare, even if favored by the unknown agents that cause endemic BL. However, activation of *MYC* provides a strong proliferation stimulus that may provide a large pool of cells from which the lymphoma arises, likely by further alterations that are less conspicuous than the translocations. Since BL is more frequent in immunocompromised individuals, an immune defense may exist which normally eliminates B cells with aberrantly rearranged immunoglobulin genes.

Activation of *MYC* to an oncogene accounts for many of the properties of BL cells. MYC stimulates many aspects of cell proliferation and cell growth (Table 10.3). In particular, in many cell types overexpression of MYC is sufficient to maintain them active in the cell cycle, independent of growth factors. It may therefore account for the high proliferative fraction and rate in BL. MYC, through its action on telomerase, may also contribute to immortalization. Moreover, some proteins regulated by MYC are involved in cell adhesion, e.g. LFA-1 in lymphocytes. This may lead to altered interactions of BL cells with other immune cells.

**Table 10.3.** *Effects of MYC on various cellular functions*

| Function | Some proteins regulated by MYC* |
|---|---|
| Cell growth | ↑ RNA polymerase, nucleolar proteins, ribosomal proteins, splice factors, eIF proteins, CAD, polyamine biosynthesis (ODC, spermidine synthase), chaperones, |
| Cell proliferation | ↑ Cyclin D2, Cyclin B1, CDK4, CDC25A, E2F1, CUL1, hTERT |
|  | ↓ $p21^{CIP1}$, $p27^{KIP1}$, MYC |
| Cell differentiation | ↓ cell-type specific bHLH transcriptional activators, ID proteins |
| Metabolism | ↑ LDH, PFK, enolase |
| Adhesion | ↓ LFA1, PAI1 |
| Apoptosis | ↑ E2F1 ($p14^{ARF}$), BAX |

* ↑ up-regulation or activation ↓ down-regulation or direct interference

The oncogenic potential of MYC is normally restrained by two factors. (1) The gene and the protein are tightly regulated by a variety of mechanisms which are subverted by the translocations and mutations in BL (see above). (2) MYC is a strong inducer of apoptosis. This induction is, at least partly, mediated by induction of p14$^{ARF}$ by E2F1 which is a transcriptional target of MYC. p14$^{ARF}$ then blocks HDM2/MDM2 activating TP53 to elicit apoptosis (→5.3). The high apoptotic rate found in BL suggests that a mechanism of this sort may indeed be active, but not sufficiently so as to compensate for the increase in proliferation.

It is therefore thought that the development of BL requires at least one further genetic alteration which limits apoptosis. Different recurrent genetic alterations, each observed in a fraction of BLs, may account for this requirement. About one third of all cases have mutations in *TP53* and LOH at 17p, others have lost TP73, a related protein and a potential mediator of TP53-induced apoptosis. The *BCL6* gene at 3q27 is affected in BL, as well as in other B-cell lymphomas, by translocations and mutations, which may also stem from somatic hypermutation. The function of the transcriptional repressor BCL6 is probably specific to B-cells at a certain stage of differentiation in germinal centers. It is down-regulated during terminal differentiation. The various translocations, which are typically promoter substitutions, appear to keep the gene active beyond its time. Importantly, BCL6 does not stimulate cell proliferation, but influences the expression of specific B cell proteins and decreases apoptosis.

Since EBV is found in so many BL, it is tempting to speculate that this herpes virus may provide a complementary function to that of MYC in the development of this tumor. The relationship has, however, emerged as more roundabout. EBV occurs in >90% of all adults, but only in a fraction of B-cells. It is capable of immortalizing cells of this type. This ability is therefore used in the laboratory to generate permanent lymphoblastoid lines. The virus exists as an episome, whose maintenance and replication require minimally the expression of the EBNA-1 protein (Figure 10.7). In lymphoblastoid lines and in long-term infected cells *in vivo*, further genes are expressed which down-regulate apoptosis and help to avoid immune detection. These genes are obvious candidates for synergizing with MYC in BL, but most of them are more weakly expressed in the actual disease than normally. The EBER RNAs of the virus are also good candidates for cooperating with MYC. Conversely, there is evidence that strong MYC expression may support the maintenance of the virus.

Several alternative plausible explanations are therefore considered for the evident association of the virus with the disease. For instance, EBV-infected cells may be more susceptible to accidents during attempted V(D)J recombination or the BL cells may be initially protected from apoptosis and elimination by the immune system by EBV proteins, while later on the same functions would be provided by other alterations such as translocations or mutations of *BCL6*.

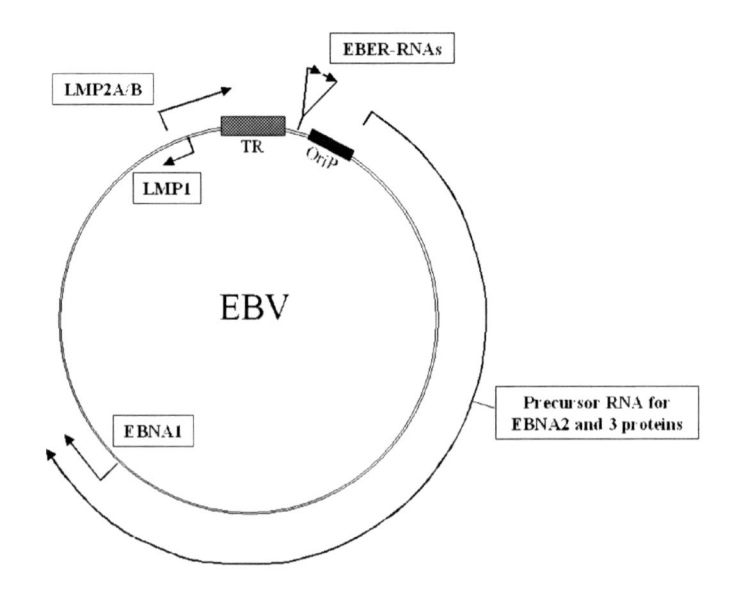

**Figure 10.7** *Structure of the EBV genome*
The 172 kb circular DNA genome of the virus with selected important features. Specifically, only the transcription units most important in stably infected cells are shown. TR: terminal repeat; LMP: latent membrane protein; OriP: viral origin of replication.

## 10.4 MOLECULAR BIOLOGY OF CML

Chronic myeloid leukemia (also: chronic myelogeneous leukemia, commonly abbreviated CML) is a moderately frequent leukemia occurring in adults, most often in their 40's or 50's. While its causes are in general unknown, it has been observed following exposure to high doses of ionizing radiation.

CML originates from hematopoetic stem cells. As a consequence, the production of many different cell types is increased overall and mature cells as well as progenitor cells of many or all hematopoetic lineages are present in the blood, prominently the myeloid cells for which the disease is named. Almost all cases are caused by one specific genetic alteration, a BCR-ABL fusion gene, usually present in an aberrant, smaller appearing chromosome 22 termed the Philadelphia chromosome.

If left untreated, CML continues over several years with rather unspecific symptoms, but then progresses through an accelerated phase into blast crisis in which highly aberrant leukemic cells kill the patient within months.

Until recently, CML was treated by chemotherapy and/or by interferon $\alpha$ (IFN$\alpha$) which lead to hematological remission, i.e. a normalization of the cell distribution in blood, in about 70% of the cases. In 10% of the cases, 'cytogenetic' remissions are achieved, i.e. no metaphases with Philadelphia chromosomes are detected in bone marrow aspirates. However, these treatments do not usually lead to a cure and the cancer recurs eventually.

A better chance of a cure is provided by allogeneic stem cell transplantation. Here, hematopoetic stem cells from an HLA-matched donor are transplanted. These provide normal cells necessary for function of the blood and immune system, but also generate an immune response towards the tumor. However, in some cases the immune cells from the bone marrow graft attack not only the tumor cells, but also normal tissues, leading to chronic or acute graft-versus-host-disease (GVHD), which can be lethal. Therefore, although better selection of donors supported by genotyping helps, stem cell transplantation remains risky.

Stimulation of an immune response against the tumor cells also appears to be responsible for the efficacy of IFNα in the treatment of CML. Like the increased incidence of Burkitt lymphoma in immuno-compromised individuals, it points to a role of the immune system in prevention of cancer by some sort of immune surveillance. This effect can apparently be recruited for therapy in some cases (cf. 22.5). The evidence from CML for a tumor-specific immune surveillance is actually a bit stronger than that from BL, because a virus might be involved in the pathogenesis of the latter disease, so the immune response may actually be directed against an exogenous antigen. There is no evidence for an infectious agens in CML.

Many properties of CML can be explained as consequences of a single genetic change, i.e. the creation of the BCR-ABL gene by a translocation between chromosomes 9q34 and 22q11. This is an example of the mechanism in Figure 10.3B. The translocation creates a novel gene which expresses a fusion protein with properties different from those of the original genes.

In most cases of CML, the translocation is reciprocal and the resulting chromosomes 9q+ and 22q- (or Ph') are the only aberrations in the karyotype of many CML cells in the chronic phase. Of course, being balanced, the translocation creates two fusion genes, but for all we know only the *BCR-ABL* fusion is relevant. Here, the promoter of the *BCR* gene from chromosome 22 drives the expression of a fusion protein consisting of a large N-terminal region from BCR and most of the ABL protein originally encoded on chromosome 9 (Figure 10.8). The breakpoints in *BCR* and *ABL* can be located in different exons leading to several different fusion genes and proteins of different molecular weights. They all, however, concur in retaining the serine/threonine kinase activity of BCR and the tyrosine kinase activity of ABL, while lacking domains that control the activities of these kinases in the original proteins. Moreover, the fusion proteins also localize primarily to the cytoplasm instead of to the nucleus. The functions of the normal BCR and ABL proteins are not overly well characterized. The ABL kinase is normally involved as a signaling molecule in DNA repair and apoptosis, interacting a.o with DNA-PK. Although the precise function is unclear, it is certainly not a growth-stimulatory protein. Of note, an abl gene has also been recruited by the Abelson leukemia virus as an oncogene (→4.1). In this murine virus, the N-terminal domain of the cellular gene is replaced by a part of the viral gag protein.

The BCR-ABL fusion protein acts in a pleiotropic fashion changing the activity of several signaling pathways and protein complexes in the cell (Figure 10.9). It

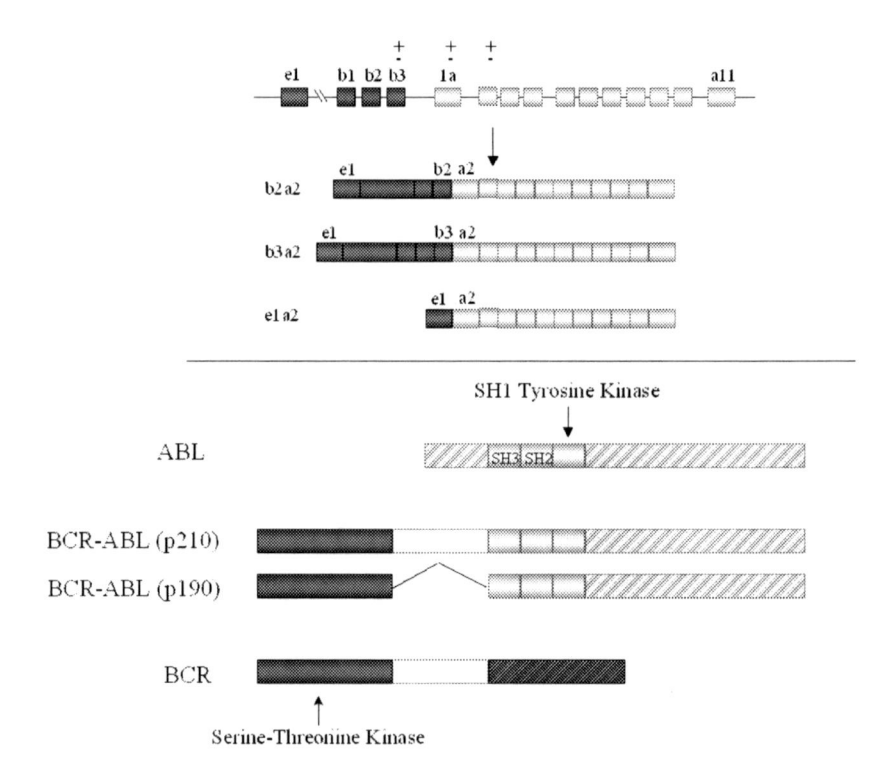

**Figure 10.8** *Fusion genes and proteins in CML*
Top: Exons of *BCR* and *ABL* involved in various translocations. Bottom: Proteins arising from the gene fusions.

interferes with the control of proliferation, apoptosis, adhesion, and DNA repair. RAS is activated and the MAPK pathway (→4.4) stimulated by GRB/SOS binding to a phosphorylated tyrosine (Y177) in the BCR domain, which is generated by the activity of the ABL tyrosine kinase. This kinase also phosphorylates and activates JAK2 and thereby the STAT pathway (→6.8) which stimulates proliferation in the hematopoetic lineage. Phosphorylation of STAT5 may also contribute to inhibition of apoptosis. This effect may be enhanced by activation of PI3K (→6.3) which is otherwise inhibited by normal ABL.

An important factor in the pathogenesis of CML is decreased adhesion, which explains the appearance of hematopoetic precursors in blood. By being lost from their proper environment, the immature cells are removed from exposure to growth factors and interactions with the ECM of the bone marrow which guide their differentiation. Altered adhesion is caused by the misdirected activity of the ABL tyrosine kinase which in the cytosol is bound to actin and phosphorylates proteins that organize the cytoskeleton like paxillin and focal adhesion kinase (FAK).

Integrin function is also impeded and CBL is activated leading to altered turnover of membrane proteins.

Moreover, there is evidence that BCR-ABL also interferes with DNA repair, acting largely in a fashion opposite to the normal ABL protein, i.e. down-regulating TP53 and ATM signaling. This last activity of the oncoprotein may tile the path towards progression of the disease to its accelerated phase and blast crisis.

If perhaps not the only genetic alteration in chronic phase CML, the formation of a *BCR-ABL* fusion gene is certainly a consistent event and likely obligatory. This is helpful in the diagnosis of the disease and even more for monitoring of its treatment (→21.2). It also provides a clear target for therapeutic intervention (→22.4).

Detection of BCR-ABL can be made by several techniques, with different sensitivities. Cytogenetically, the Ph' chromosome can be detected in metaphases or more sensitively by FISH (fluorescence in situ hybridization) showing a closed apposition of the normally separate genes in interphase nuclei. Molecular techniques are even more sensitive, especially detection of a fusion transcript by RT-PCR.

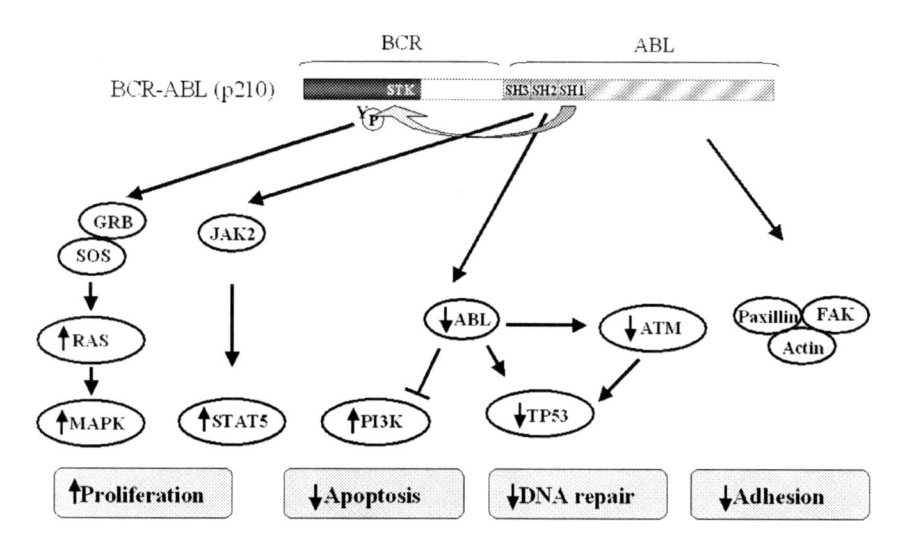

**Figure 10.9** *Effects of the BCR-ABL fusion protein*
See text for details.

Prior to therapy, the billions of leukemia cells in the blood are obvious under the microscope. Following therapy, their number decreases and hematopoetic precursors disappear. This is called 'hematological remission'. Unfortunately, it does not amount to a cure, because the cancer often recurs after some time, in CML after several years. Such patients harbor 'minimal residual disease'. Indeed, cytogenetic methods often detect tumor cells in the bone marrow of patients with an apparent normalization of the cell numbers and composition in the blood. If this is not the case, one speaks of 'complete cytogenetic remission'. However, even this does not signify a certain cure, because millions of descendents of the tumor stem cells in the bone marrow may remain hidden among the normal cells in the blood. Descendents of these cells are detectable by RT-PCR in peripheral blood samples. So, the notion of 'molecular remission' has entered modern hematology denoting the situation in which even this sensitive method detects no more leukemia cells. CML patients with 'molecular remission' are usually cured.

So, very sensitive methods have been developed which are capable of detecting very few cells with Ph' chromosomes. Surprisingly, they reveal their presence in persons without evidence of disease and no evidence of their numbers increasing. The interpretation of this finding is controversial. It may mean that a second genetic event is needed for CML development after all, or provide further evidence for successful immune surveillance.

The elucidation of the role of BCR-ABL in CML has also led to the development of a specific drug, an inhibitor of the ABL kinase, imatinib (STI-571, marketed under the name of 'Gleevec' or 'Glivec'). This drug has recently proven very efficacious against CML, inducing remissions even in blast crisis CML and used as first-line therapy in chronic phase patients. However, while remissions are long-lasting in a large proportion of chronic phase patients (the cure rate will only be ascertained after longer observation periods), resistant tumors develop in many acute phase patients. Importantly, the mechanisms underlying this resistance show that the target was well chosen. In some cases, the resistance is caused by point mutations in the ATP binding domain of the ABL kinase in the BCR-ABL fusion protein which diminishes the impact of the inhibitor. In other cases of resistance, the fusion gene has become amplified.

The different rates of resistance against imatinib in blast crisis patients vs. chronic phase patients illustrates the development of genomic instability during the progression of CML towards blast crisis. Blast crisis cells are aneuploid with multiple chromosomal aberrations including chromosomal losses, gains, further translocations and amplifications.

Among the genes affected by these changes are *RB1* and *TP53* as well as *CDKN2A* which are inactivated by gene loss, mutation or, in the case of *CDKN2A*, also by promoter hypermethylation. Loss of TP53 by mutation and deletion certainly contributes to resistance against chemotherapy in this disease, as in cancers in general. Another remarkable instance of hypermethylation is that of the remaining intact *ABL* gene, underlining the tumor suppressor function of the normal gene. In contrast, the *MYC* gene is over-expressed as a consequence of trisomy 8 or outright gene amplification in blast crisis, but not in the chronic phase.

Acute myeloid leukemias other than those developing from CML also display characteristic translocations that lead to fusion proteins. One of them is a t(3;21) event which creates the AML/EVI1 fusion gene and protein. This translocation is also prevalent in blast crisis leukemia developing from CML. Like MYC overexpression, it is thought to further stimulate proliferation, but also to exacerbate the block to differentiation, which is relatively weak in chronic CML. Finally, a typical indicator of the acceleration phase that precedes blast crisis is an increased copy number of the *BCR-ABL* gene itself, typically becoming evident as a duplication of the Ph' chromosome. Once again, this highlights the crucial importance of this genetic alteration in CML, even at its final stage.

## 10.5 MOLECULAR BIOLOGY OF PROMYELOCYTIC LEUKEMIA

Promyelocytic leukemia is, like the final phase of CML, an acute leukemia and therefore alternatively abbreviated as APL or PML (or most systematically as AML M3). It is, however, not a disease of stem cells, but of cells from an intermediate stage of differentiation towards granulocytes, denoted promyelocytes (Figure 10.1). These cells are obviously arrested at this stage and continue to proliferate. Compared to CML, PML is a rare disease comprising only a fraction of all AML. Like CML, however, PML has become paradigmatic for the understanding of leukemias and beyond, primarily as an example of successful 'differentiation therapy' and, more recently, of chromatin alterations in cancer.

A distinctive t(15;17) (q22;q21) translocation found in >90% of all PMLs fuses the *PML* gene at chromosome 15q22 to the *RARA* gene at 17q21. Variant translocations always include *RARA* as well, but fuse it to the *PLZF* (promyelocytic leukemia zinc finger) gene at 11q23, to the *NPM* (nucleophosmin) gene at 5q35, or *NUMA* (nuclear matrix associated) gene at 11q13 (Figure 10.10).

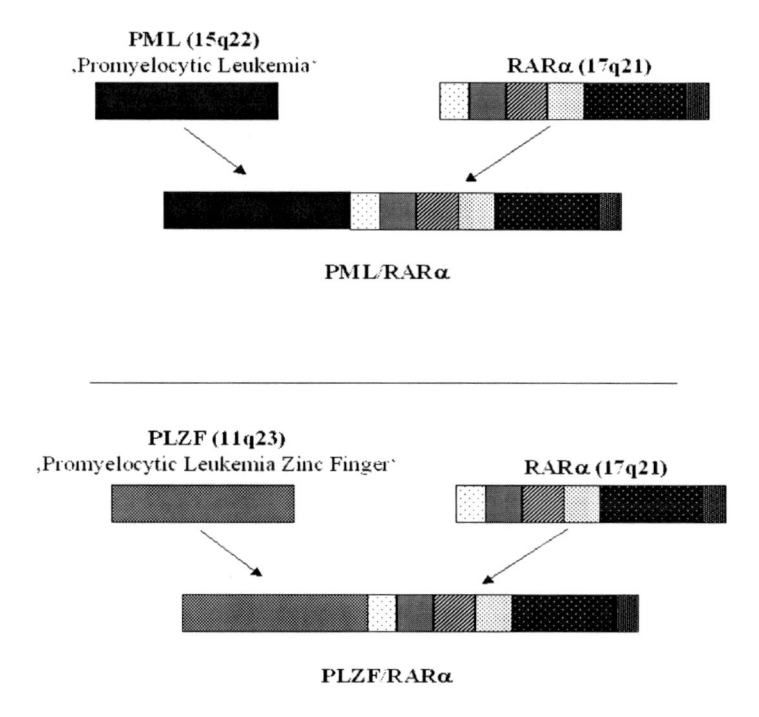

**Figure 10.10** *The two most important translocations in APL*

This general involvement of the retinoic acid receptor RARα in PML translocations is striking, as already in the 1980's its ligand all-trans retinoic acid (ATRA) was found to induce often dramatic hematological remissions of PML. Responsiveness to ATRA is specific for PML, and not observed in other AMLs. PML also responds better to certain drugs which are more generally used in chemotherapy, such as anthracyclins. So, precise diagnosis is a prerequisite for optimal treatment of this disease. Moreover, as in CML, cytogenetic and molecular techniques are employed to monitor treatment, detect minimal residual disease and select patients for stem cell transplantation.

Both certain cytostatic drugs and ATRA are efficacious in PML, and are used in combination. Cytostatic drugs act through apoptosis and cell cycle arrest in PML, as in other leukemias (→22.2). In contrast, ATRA relieves the block to differentiation in the disease, allowing terminal differentiation towards granulocytes, with a concomitant decrease in proliferation. Some degree of apoptosis also occurs. This type of treatment has therefore been called 'differentiation therapy'.

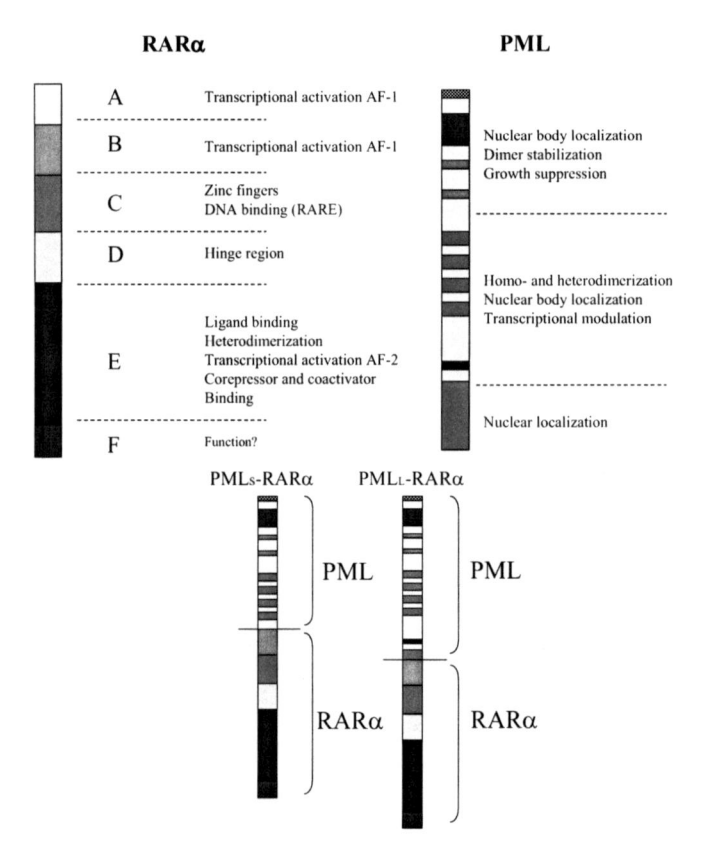

**Figure 10.11** *Structures of RARα, PML, (top) and fusion proteins(bottom) in APL*
See text for further explanation.

The PML-RARα fusion protein (Figure 10.11) in PML acts as an inhibitor of both proteins from which it originates. It may also have gained additional, novel functions. RARα, retinoic acid receptor α, belongs to a group of closely related receptors for retinoic acids which are part of the steroid hormone receptor superfamily. Retinoic acid is produced from vitamin A as a locally acting hormone regulating growth and differentiation of many tissues, prominently the skin, reproductive organs and some glands. Accordingly, the closely related RARβ and RARγ proteins are frequently inactivated in other cancers, e.g. RARβ in breast and prostate cancer (→19.3).

The basic structure of RARα resembles that of steroid hormone receptors (Figure 10.11, cf. 18.1). An N-terminal transcriptional activation domain (AF-1) is followed by the DNA-binding domain, an hinge region, and the ligand binding domain (LBD) containing a second transcriptional activation function (AF-2), with a short C-terminal region. The LBD/AF-2 domain is the most crucial one in the protein, since the ligand bound in this region elicits heterodimerization and recruitment of transcriptional co-activators such as CBP in exchange for the co-repressor N-CoR/SMRT. The DNA-binding domain interacts with a specific recognition site in DNA called RARE (retinoic-acid responsive element) which has the typical motif for steroid hormone receptor-like transcriptional activators, AGGTCA. RARE sites actually used consist of a direct repeat of this sequence separated by five arbitrary nucleotides. The second half-site is occupied by one of the more distantly related RXR receptors which form heterodimers with RARs, as they do with several other receptors of the superfamily. Binding of RARα to a gene promoter usually leads to opening of the chromatin structure by remodeling and histone acetylation (→8.4) as well as to transcriptional activation.

In hematopoesis, gene regulation by retinoic acid appears to be particularly important during granulopoesis at the stage of promyelocyte differentiation. At this differentiation step, growth-promoting genes such as *MYC* need to be down-regulated and inhibitors of the cell cycle such as $p21^{CIP1}$ must be up-regulated. The genes of transcription factors more specifically involved in myeloid differentiation such as C/EBPε and HOXA are also RARα targets. Most interestingly, the RARα gene is activated by its own product. This auto-regulatory loop may provide a crucial amplifying step for the signal that makes differentiation irreversible (→8.5).

The PML protein, named after the disease, is normally localized in 10-30 'PML nuclear bodies', a subcompartment of the nucleus consisting of morphologically recognizable protein assemblies sized 0.2 - 1 μm. The assembly of ≈ 1,000,000 proteins in these nuclear bodies is dependent on PML which may be their principle organisator and regulator. The function of the nuclear bodies is not entirely clear. Most likely, they act as a reservoir and interaction structure for proteins involved in response to cellular stress, DNA repair, apoptosis and chromatin structure. Proteins found to associate with PML and nuclear bodies include the proapoptotic DAXX protein, the Bloom syndrome helicase BLM (→3.3), and TP53 (→5.3), but also the mRNA-cap-binding protein EIF-4E. Interestingly, formation and function of nuclear bodies may involve SUMOlation, i.e. covalent attachment of a SUMO protein to PML and other constituents. Another regulator of nuclear bodies is the JNK1 kinase

(→6.2). PML itself may then mediate the SUMOlation of other proteins, including TP53 (→5.3). The function of PML is reflected in its structure which contains several motifs involved in protein-protein interactions, such as a RING finger and a coiled domain.

The PML-RARα fusion protein retains most parts from both partners. PML mostly lacks its C-terminal Ser-Pro rich domain, while RARα lacks part of the AF-1 interaction domain (Figure 10.11). However, these changes suffice to turn the fusion protein into a transcriptional repressor at physiological concentrations of retinoic acids, and to destroy nuclear bodies and interfere with their function. Transcriptional repression may be aided by dimerization of the fusion protein which then binds to a dimeric RARE promoter site and recruits co-repressors to repress rather than activate genes crucial for myeloid differentiation.

Fortunately, the variant protein can still be turned into a transcriptional activator by micromolar concentrations of retinoic acid, i.e. at pharmacological concentrations, 100-1000-fold above the physiological level. In response to ATRA treatment, the co-repressors dissociate allowing transcription of the genes required for granulocyte differentiaton. The fusion protein itself is rapidly degraded following ATRA treatment, so transcription may also or even predominantly be promoted by normal RARα, which is itself induced during differentiation. Like the function of RARα, that of PML seems also to become restored as evidenced by reappearance of nuclear bodies.

The overall effect of ATRA treatment at the cellular level is differentiation towards a more mature granulocytic cell, with arrest of proliferation, but also significant apoptosis. Of note, the ensuing differentiated cells are no more cancerous, but are not normally functioning and can pose problems to the host. Moreover, retinoic acid at supraphysiological concentrations has some side-effects. So, while differentiation therapy is efficacious and may seem more elegant than chemotherapy, it is strenuous and dangerous to the patient.

Like the predominant type of PML caused by t(15;17) translocations, those with the variant t(5;17) and t(11q13;17) translocations respond to ATRA treatment. However, the rare t(11q23;17) translocation, which creates the PLZF- RARα fusion gene, does not. The difference lies in the fusion partner. PLZF is itself a transcriptional repressor capable of recruiting co-repressors and histone deacetylases. Therefore, in this fusion protein, not only is RARα locked in a repressor conformation, but also its fusion partner contributes to repression to at least the same extent (Figure 10.12). At least experimentally, however, it has been possible to still reactivate RARα by a combined treatment with ATRA and inhibitors of histone deacetylases. These may obliterate the repressor function of PZLF. As more and better such inhibitors become available, there seems to be a chance of successful treatment for patients with this translocation as well.

Finally, the obvious question is whether the RARα translocations are sufficient to induce PML. This cannot be definitely answered at present. Since vitamin A deficiency and other maladies that cause inefficient function of RARα only moderately affect granulopoesis, with at most a mild hyperplasia of promyelocytes, not only the loss of retinoic acid responsiveness, but the formation of the fusion protein is certainly crucial. Although the details are controversial, a function of PML in the regulation of apoptosis and replicative senescence is generally accepted. Moreover, the fusion protein obviously has a dominant-negative effect on at least some aspects of PML function. So, inhibition of differentiation and of apoptosis could well be oncogenic effects provided by the fusion protein. This leaves room for a genetic change that causes hyperproliferation. The best candidate for this is mutation of *FLT3*. The FLT3 protein is a receptor tyrosine kinase activated by the hematopoetic growth factor M-CSF which is particularly active in the myeloid lineage (Table 10.1). Mutations in the *FLT3* gene are observed in several different subtypes of AML and in at least a third of PML patients.

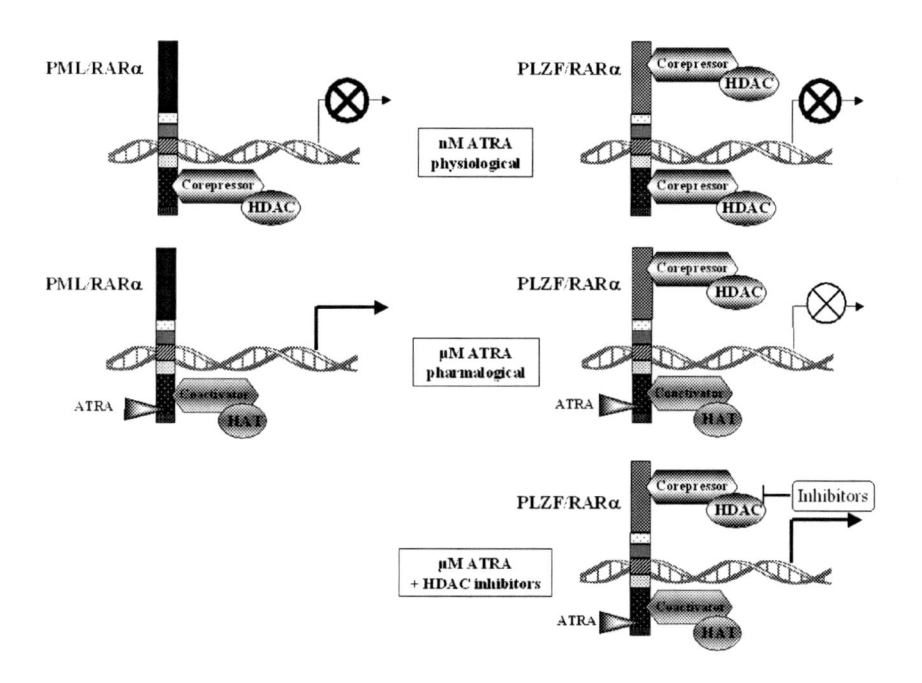

**Figure 10.12** *Differences in the reaction to all-trans retinoic acid (ATRA) between standard (PML-RARα-, left) and variant (PZLF-RARα, right) APL*
See text for further explanation.

## *Further reading*

Melnick A, Licht JD (1999) Deconstructing a disease: RARα, its fusion partners, and their roles in the pathogenesis of APL. Blood 93, 3167-3215

Faderl S et al (1999) The biology of CML. NEJM 341, 164-172

Deininger MW, Goldman JM, Melo JV (2000) The molecular biology of chronic myeloid leukemia. Blood 96, 3343-3356

Hecht JL, Aster JC (2000) Molecular biology of Burkitt's lymphoma. J. Clin Oncol. 18, 3707-3721

Melnick A (2002) Spotlight on APL: controversies and challenges. Leukemia 16, 1893ff (editorial, see also following articles)

Salomoni P, Pandolfi PP (2002) The role of PML in tumor suppression. Cell 108, 165-170

Shet AS et al (2002) CML: Mechanisms underlying disease progression. Leukemia 16, 1402-1411

Mistry AR et al (2003) The molecular pathogenesis of acute promyelocytic leukemia: implications for the clinical management of the disease. Blood Rev. 17, 71-97

Passaguè E et al (2003) Normal and leukemic hematopoesis: Are leukemias a stem cell disorder or a reacquisition of stem cell characteristics. PNAS USA 100, Suppl.1, 11842-11849

Pitha-Rowe I et al (2003) Retinoid target genes in acute promyelocytic leukemia. Leukemia 17, 1723-1730

Tenen TG (2003) Disruption of differentiation in human cancer: AML shows the way. Nat. Rev. Cancer 3, 89-101

Hennessy BT, Hanrahan EO, Daly PA (2004) Non-Hodgkin lymphoma: an update. Lancet Oncol. 5, 341-353

# CHAPTER 11

# WILMS TUMOR (NEPHROBLASTOMA)

- Wilms tumors are nephroblastomas arising in young children from nephrogenic rests, parts of the developing kidney that have failed to complete differentiation. Accordingly, the tumor mass consists of several components resembling tissue structures in the fetal kidney, such as blastema, mesenchymal stroma, and tubular structures. Most cases can be cured by a combination of chemotherapy and surgery.
- Like retinoblastoma, Wilms tumors can be unilateral or bilateral, and the latter situation occurs more often with germ-line mutations. However, the disease is genetically heterogeneous and the 'two-hit' model does not apply strictly.
- Wilms tumors are normally rare, occuring in 1:10,000 children, but are more frequent in the context of syndromes which disturb the development of the genitourinary tract at large, such as the WAGR, Denys-Drash, and Beckwith-Wiedemann syndromes.
- In many cases, a limited number of genetic changes are observed and the tumor cells remain near-diploid or diploid. Chromosomes 1, 11, 16, and 22 are most often affected. Tumors with multiple chromosomal aberrations are more prevalent in older children. In younger children, germ-line mutations are more often found. Mutations in TP53 characterize a small group of anaplastic (poorly differentiated) tumors
- Mutations at several loci predispose to Wilms tumors. The best characterized locus is *WT1* on chromosome 11p13 which possesses many properties of a classical tumor suppressor gene, although with incomplete penetrance. Germ-line mutations of *WT1* predispose to early and bilateral development of Wilms tumors, in which the second allele of the gene is also inactivated by loss or recombination. Some sporadic cases also show inactivation of both *WT1* alleles. However, due to low penetrance, *WT1* is responsible for only a fraction of familial cases. These are instead ascribed to loci on 17q and 19q which are not yet identified. Another apparent 'Wilms tumor locus' called *WT2* is located on 11p15.5.
- *WT1* encodes a transcription factor which is involved in the development and differentiation of the genitourinary tract. In particular, WT1 is expressed and its function is necessary at the time when the metanephric mesenchyme forms renal tubuli under the influence of the branching ureter bud. The understanding of the function of WT1 is complicated by the presence of several different isoforms resulting from differential splicing and translational initiation. Accordingly, consequences of mutations in the *WT1* gene vary depending on their location. While some mutations predispose to Wilms tumor only, others cause malformations in the genitourinary tract.

➤ While *WT1* may already seem a complex locus, *WT2* may not be a single locus at all. *WT2* is related to a locus that causes Beckwith-Wiedemann syndrome, a condition of fetal overgrowth, metabolic disregulation, and predisposition to childhood tumors, although the precise relationship is unclear. Some cases of this syndrome are caused by translocations in the 11p15.5 region, others arise as a consequence of uniparental disomy or apparently independently of chromosomal changes. In either case, the expression of imprinted genes located in the region is disturbed. Actually, two different gene clusters are involved, one consisting of the *H19/IGF2* tandem locus and one comprising, among others, the *CDKN1C* and the *BWR1C* genes. Disturbances in the imprinting of IGF2 may be the relevant alterations causing Wilms tumor, while alterations in *CDKN1C* are likely more pertinent to other overgrowth symptoms in Beckwith-Wiedemann syndrome. However, the relationship is complex. Intriguingly, it cannot be excluded that some cases of Wilms tumor are initiated by epigenetic changes only.

## 11.1 HISTOLOGY, ETIOLOGY AND CLINICAL BEHAVIOR OF WILMS TUMORS

Development of the kidney is among the more unusual processes in mammalian ontogeny because it involves transition of a mesenchymal to a predominantly epithelial structure. The ureter bud invading the condensed metanephric blastema induces the formation of tubuli with progressively differentiating epithelial cells. Concomitantly, the ureter branches to form the segmented renal pelvis and the collecting ducts. This process involves a sequence of concerted gene expression changes in the blastema and the ureter cells regulated by paracrine factors and cell-cell-contacts which couple their further proliferation to step-wise differentiation. It is neither surprising that this marvellous reorganization is a favorite issue in developmental biology, nor that it occasionally goes wrong, in various ways. As a consequence, congenital malformations of the kidney and ureter are relatively frequent in man. Some appear to be accidental, while others are expedited by inherited gene defects. This statement applies to Wilms tumor in particular.

Wilms tumor is a 'nephroblastoma' found in one of ten thousand young children, presenting very rarely after the age of 6 years. The tumor mass typically consists of an undifferentiated proliferating blastema (therefore 'nephroblastoma') mixed to various extents with differentiated components, which are typically partly tubular epithelia and partly mesenchymal stroma (Figure 11.1). In some tumors the stroma contains ectopic structures recognizably resembling other mesenchymal tissues, e.g., muscle, cartilage, bone, and adipose tissue. Often, the tumors are associated with nephrogenic rests, i.e. residual metanephric blastema that has not differentiated. Thus, Wilms tumors are apparently caused by renal precursor cells which do not differentiate and instead continue to proliferate. Wilms tumors are more likely to arise in the context of several syndromes characterized by malformations of the genitourinary tract and inappropriate growth (Table 11.1), underlining their origin as a consequence of defective development.

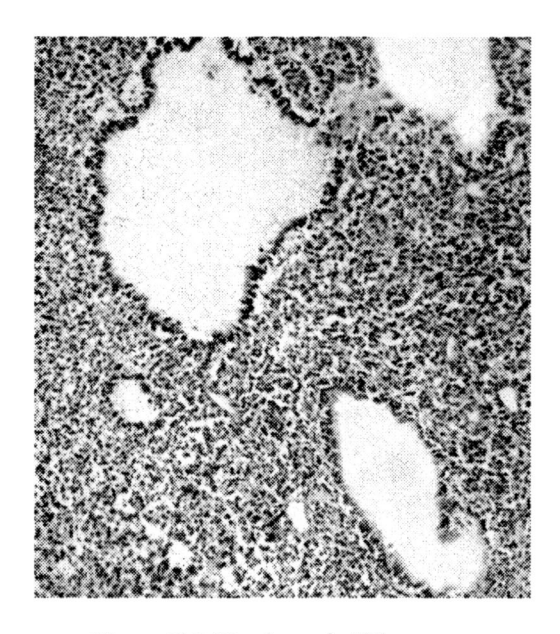

**Figure 11.1** *Histology of a Wilms tumor*
This is a tumor of the subtype in which immature blastema predominates, in this case surrounding cystic structures.

Histologically, one can distinguish different subtypes of Wilms tumor, depending on which tissue components they contain. The most prevalent subtype is triphasic, i.e. consists of blastema, stroma and tubular epithelia. This subtype arises more often in association with intralobular nephrogenic rests, as do stroma-rich tumors with ectopic mesenchymal tissues. Wilms tumors from perilobular rests probably originate at a later stage of development and consist mostly of blastema. Anaplastic tumors, the most aggressive subtype, show no indication of differentiation. Today, Wilms tumor is regarded as a separate entity from other pediatric renal cancers which develop from connective tissue components, such as rhabdomyosarcoma or clear-cell sarcoma. Like Wilms tumors these are characterized by failure to differentiate. For instance, rhabdomyosarcomas develop from precursors of muscle cells.

As a result of several large international studies, the treatment of Wilms tumors is based today on a combination of chemotherapy and surgery. More than 85% of all children can be cured, although the cancer remains lethal in some cases because of local complications and of metastases.

Like retinoblastoma (→5.2), Wilms tumor can be unilateral or bilateral, i.e. develop in one or both kidneys. Also like retinoblastoma, some Wilms tumors occur in families, although familial cases constitute only a few percent overall. Conspicuously, the age vs. incidence curve appears biphasic with a first peak around 2 years and a second one around 4 years of age. Familial cases often fall into the first group. Despite such parallels to retinoblastoma, the molecular genetics of

Wilms tumors is more complicated and the Knudson model (→5.1) applies only partly.

**Table 11.1.** *Some syndromes with increased incidence of Wilms tumors*

| Syndrome | Other symptoms | Gene Mechanism | Wilms tumor risk |
|---|---|---|---|
| WAGR* | aniridia, genitourinary defects, mental retardation | WT1/PAX6 deletion | 50% |
| Frasier | kidney defects, gonadal dysgenesis and tumors | WT1 splice mutation | low |
| Denys-Drash | kidney defects, incomplete formation of genitals | WT1 missense mutation in zinc fingers | 90% |
| ICNS/DMS** | kidney defects | WT1 missense mutation | high |
| Beckwith-Wiedemann | overgrowth, deregulation of glucose metabolism, selected morphological defects | 'WT2' gene cluster at 11p15.5 including *CDKN1C* translocation, epigenetic | up to 5% |

\* Wilms tumor, aniridia, genitourinary defects, mental retardation
\*\* isolated congenital nephrotic syndrome, also designated diffuse mesangial sclerosis

## 11.2 GENETICS OF WILMS TUMORS AND THE WT1 GENE

One would presume that tumor development in a young child cannot result from an accumulation of multiple genetic alterations, unlike cancers in older people (→2.5). Indeed, Wilms tumors as a rule contain a limited number of evident genetic aberrations. Many have diploid or near-diploid karyotypes. In the other cases, chromosomes 1, 11, 16, and 22 are most often affected. On chromosome 1, the short arm tends to be lost, while the long arm is often gained. This can be explained by the formation of an isochromosome 1q with loss of 1p (Figure 11.2). 16q and 22q are subject to losses. The mutations in *TP53* and loss of heterozygosity (LOH) at its locus at 17p are found in a few cases, usually with anaplastic histology. They bode a poor response to chemotherapy. Chromosome 11 is also affected by losses, while chromosomes 6, 7, 8, 12, or 13 are gained in some cases.

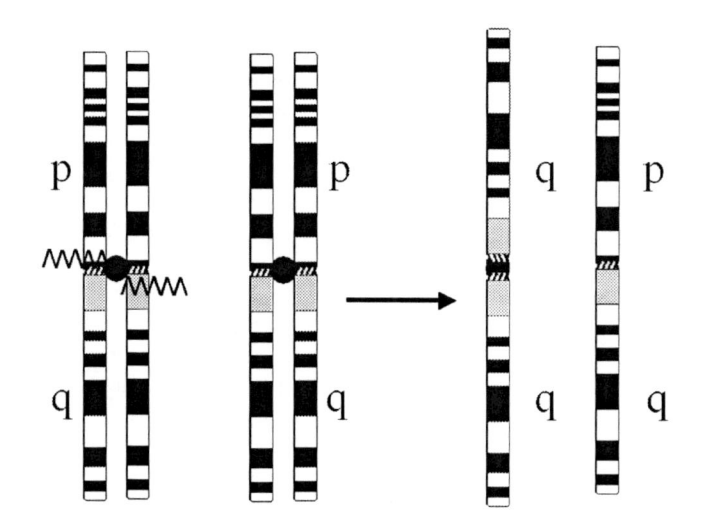

**Figure 11.2** *Formation of an isochromosome 1q*
Note the large juxtacentromeric heterochromatic region 1q11 marked in grey.

The causes of these chromosomal alterations are not known. Loss or gain of whole chromosomes points to defects in chromosome segregation during mitosis. The frequently afflicted chromosomes 1 and 16 are distinguished by harboring a particularly large proportion of the GC-rich satellite sequences (*SAT1* – *SAT3*) in the human genome. They are arranged as tandem repeats in juxtacentromeric heterochromatin at 1q and 16q. The CpG-rich satellite DNA is normally highly methylated in somatic cells (→8.3), but is hypomethylated in Wilms tumors, which may make it more susceptible to breakage and recombination.

Several individual loci are involved specifically in Wilms tumor, of which WT1 is best characterized. The WT1 gene is located at 11p13 and was first identified in patients suffering from WAGR syndrome (Table 11.1). The acronym WAGR stands for Wilms tumor, aniridia (lack of an iris), genitourinary malformations, and mental retardation. The syndrome is caused by deletions of several consecutive genes on chromosome 11p13 including *WT1*, which explains the genitourinary malformations and Wilms tumor, and the paired-box transcription factor gene *PAX6*, which explains aniridia. Specifically, aniridia patients in whom the deletion does not encompass the *WT1* gene do not develop Wilms tumors. In Wilms tumors from WAGR patients, the second allele of *WT1* is also inactivated, often by nonsense mutations.

The WAGR syndrome is rare, but in ≈15% of all Wilms tumors as well, both alleles of the *WT1* gene are inactivated. Most often, this occurs by mutation of one allele and loss of the second by recombination, which is detectable by LOH at 11p13. In at least one third of these patients, the first mutation is present in the germ-line and the patients tend to develop tumors in both kidneys. In the others, the first mutation is also somatic and their tumors are unilateral. Thus, in this regard *WT1* behaves like a classical tumor suppressor gene such as *RB1* (→5.2). However, until now, tumors caused by *WT1* mutations have not been observed in successive generations. The main reason for this seems that *WT1* mutations show much lower penetrance than mutations of *RB1*. Also, germ-line mutations in WT1 do not imply an increased risk for other tumors.

The reasons for these differences are not understood. They may relate to the different functions of the proteins and a difference in the 'window of opportunity' for the cancers to arise. The WT1 protein seems to be required at one specific stage in renal development, i.e. for the blastema to begin its mesenchymal-epithelial transition. Once the cells have moved beyond this stage, loss of WT1 function may not matter, and the gene is, in fact, down-regulated. So, in a carrier of a germ-line mutation, the second 'hit' must have been acquired by that time. In contrast, the requirement for functional RB1 may be most crucial at the very end of retinoblast differentiation, to permit the permanent exit from the cell cycle, i.e. terminal differentiation. Accordingly, retinoblasts might have more time to acquire the decisive 'second hit' (Figure 11.3). A second, more speculative explanation is that *RB1* hemizygosity - unlike that of WT1 - already increases tumor risk somewhat, perhaps because of its additional functions in maintaining genomic stability (→5.2).

Like RB1, however, WT1 is crucially involved in directing differentiation, certainly in the developing kidney, but likely also in the gonads, spleen and mesothelium, where WT1 expression is also found. In the adult kidney, its

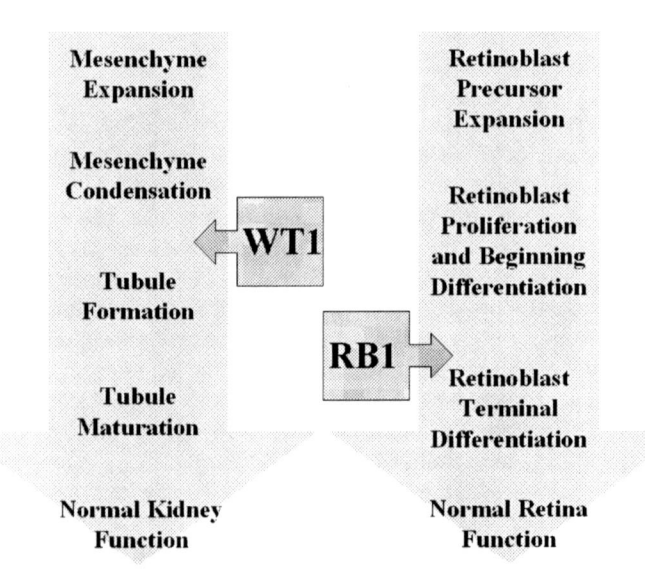

**Figure 11.3** *A hypothesis explaining the different penetrance of inherited* WT1 *and* RB1 *mutations*
See text for further exposition.

expression is restricted to podocytes, epithelial cells which are an essential component of the glomerulus. Importantly, during development of the kidney WT1 expression is most prominent in the condensing nephroblastema at the onset of the mesenchymal-epithelial transition. In suitable mesenchymal cells, transfection of WT1 can induce a transition of this type. Thus, the time and place where WT1 is needed in development fit well with the morphological appearance and location of Wilms tumors that have lost its function.

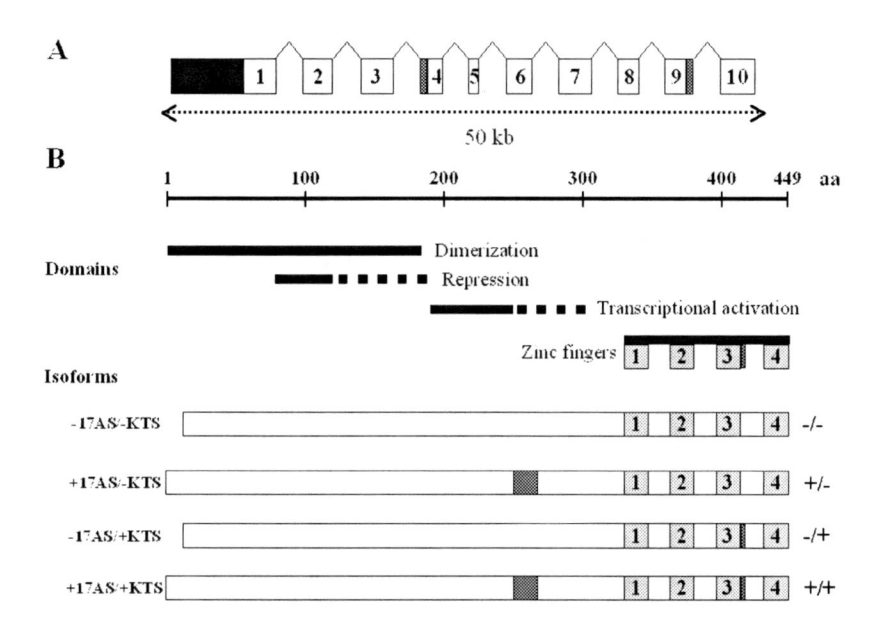

**Figure 11.4** *The WT1 gene and its encoded proteins*
A: Structure of the gene; only the exons are drawn to size (relative to each other); alternatively spliced parts are marked in dark grey. B: Domains and isoforms of the WT1 protein. The sequence resulting from differential splicing are marked as in A. See text for further explanations.

The 50 kb *WT1* gene encodes transcription factor proteins (Figure 11.4) with a four zinc-finger DNA binding domain that is similar to those in the widely distributed EGR transcriptional activators and recognizes the same DNA sequence motif. WT1, however, contains both an activator and a repressor domain. Moreover, the protein binds to RNA and may regulate splicing. There are, in fact, several isoforms of the WT1 protein which differ with respect to these abilities. Translation of WT1 can initiate at two different codons. In addition, two sites are used for alternative splicing. The presence or absence of exon 5 alters the size of the activation domain by 17 amino acids and likely its function, too. Use of two alternative splice donors 9 bp apart at the end of exon 9 leads to two isoforms which differ by three amino acids in zinc finger 3 and are denoted as +/-KTS. Moreover, the WT1 transcript is subject to editing. Thus, overall, the WT1 gene encodes up to

16 proteins with 52 – 56 kDa MW which differ in their ability to repress or activate transcription and with respect to DNA binding. They are also expressed in different patterns in different tissues.

The significance of many variants is unknown, but at least the KTS variant is important, since splice mutations causing loss of this sequence are found in children suffering from Frasier syndrome (Table 11.1). Denys-Drash-syndrome, by comparison is caused by de novo germ-line missense mutations in exons 8 or 9 of the *WT1* gene. These lead to changes in those amino acids in the WT1 zinc fingers that interact with DNA. Characteristics of the syndrome are defects in the glomeruli causing kidney failure, incomplete differentiation of the genitals, and an increased risk of Wilms tumors. The developmental defects are caused by the germ-line mutation alone, whereas formation of Wilms tumors requires inactivation of the second allele as well. Missense mutation in the WT1 zinc fingers are also found in sporadic Wilms tumors, but in these truncating mutations are more prevalent. Conversely, truncating mutations are not found as a cause of Denys-Drash-syndrome, suggesting that the characteristic zinc-finger missense mutations in this disease act as dominant-negatives or by a 'gain of function' mechanism.

So, in summary, while WT1 mutations may be passed on in the germ-line in some families, and sometimes cause Wilms tumors, they are not responsible for the majority of familial cases. Instead, pertinent loci have been mapped by linkage analyses to chromosomes 17q12-21 and 19q13 and have been named tentatively *FWT1* and *FWT2* (for: familial Wilms tumor). Previously suspected loci at 16q and 11p15 have now been exculpated as a cause of familial Wilms tumors, but they are certainly important in some sporadic cases.

## 11.3 EPIGENETICS OF WILMS TUMORS AND THE 'WT2' LOCUS

Beckwith-Wiedemann syndrome (BWS), which conveys an increased risk for Wilms tumors and other pediatric malignancies, is caused by genetic or epigenetic defects in the chromosomal region 11p15.5. This region comprises two clusters of imprinted genes (→8.2), including the twin loci *IGF2/H19* and a larger gene cluster apparently controlled by an imprinting center (IC) located in the *KCQN1* gene. Children suffering from BWS are oversized and overweight at birth, often presenting with overgrowth of only one side of the body ('hemihyperthrophy') or of visceral organs such as the liver. Furthermore, glucose homeostasis can be dangerously deregulated perinatally (around birth). The syndrome is caused by one of several defects: (1) balanced translocations or inversions in the 11p15.5 region on the chromosome inherited from the mother; (2) uniparental disomy, in which this region is derived from the father on both chromosomes 11; (3) point mutations in the *CDKN1C* gene encoding the CDK inhibitor p57$^{KIP2}$.

Most changes found in BWS have in common that *CDKN1C* function is compromised. Since the gene is significantly transcribed only from the maternal allele (→8.2), chromosomal aberrations and uniparental disomy strongly diminish its expression, whereas point mutations inactivate the gene product, the CDK inhibitor protein p57$^{KIP2}$ (→5.2).

However, the phenotype of BWS is very likely not only due to *CDKN1C* loss of function. Specifically, Wilms tumors do not occur in children in which BWS is caused by mutations in this gene. Indeed, most alterations affect other genes as well, often from both imprinted gene clusters at 11p15.5. For instance, some translocations separate the imprinting center IC2 in the *KCQN1* gene from the *CDKN1C* and other genes, disturbing their maternal imprinting pattern.

Overall, there are six or more imprinted genes in the *KCQN1/CDKN1C* cluster (Figure 11.5). Several of them may have functions which could contribute to the phenotype of BWS such as regulation of transcription (*ASCL2/MASH2*), cell-cell interaction (CD81), signal transduction (*TSSC3/IPL/BWR1C*), and regulation of apoptosis (*TSCC5/BWR1A*). Obviously, their interaction may be difficult to disentangle, even more so, since animal models are of limited use, because the imprinting patterns of some genes differ between mammalian species, in particular between mouse and man. Still, *CDKN1C* 'knockout' mice show several developmental defects resembling those in BWS including overgrowth. The *IGF2/H19* genes are also disturbed in many cases of BWS. For instance, uniparental disomy with both chromosomes derived from the father leads to an increase in IGF2 expression, as would be expected for two active gene copies compared to one.

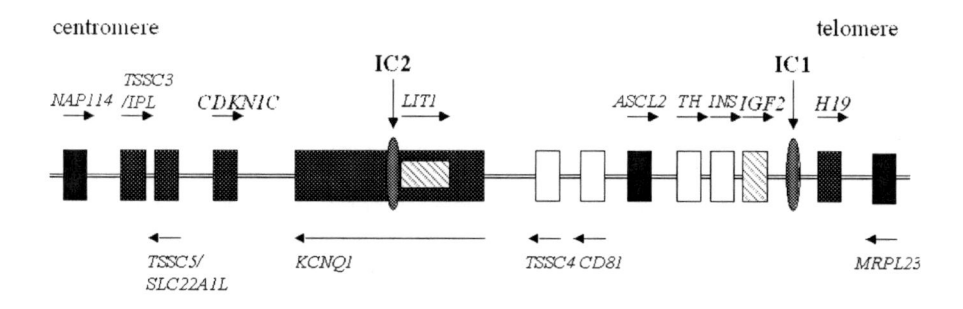

**Figure 11.5** *Imprinted genes at the 'WT2' locus at 11p15.5*
Stippled: maternal allele expressed; hatched: paternal allele expressed; black: not imprinted; white: data not definitive. cf. also Figures 8.1 and 8.2

In Wilms tumors, loss of imprinting in the *IGF2/H19* cluster is frequent, not only in those tumors arising in BWS patients. The expression pattern becomes paternal for both alleles, with an increase in IGF2 peptides and a decrease in H19 RNA. Increased expression of IGF2, a potent growth and survival factor (→4.4), could well account for the overproliferation of nephroblastoma cells. Moreover, *IGF2* may be a target gene of WT1 and vice versa, suggesting a kind of pathway involved in the genesis of Wilms tumor. However, the data in this regard are controversial and it is possible that the down-regulation of H19 may after all also be significant. For instance, overexpression of H19 RNA in a nephroblastoma cell line was found to arrest its growth.

There are further epigenetic events affecting the development and progression of Wilms tumor. (1) As mentioned above, a frequent genetic change in Wilms tumors is loss of chromosome 1p. While often the entire arm is involved, mapping of the common region of deletion, in Wilms tumor and other pediatric cancers, points to a tumor suppressor gene in the 1p36 region, where some genes are also imprinted. So, loss of a single copy of this region may be sufficient to inactivate a tumor suppressor gene. (2) While the *WT1* gene itself is not imprinted, it overlaps partly with an RNA gene named *WIT1* transcribed in the opposite direction, but only from the paternal allele. LOH involving chromosome 11p, at large, is biased towards loss of the maternal copy, which diminishes the function of the maternally imprinted genes at the tip of the chromosome. However, this also implies an increased dose of *WIT1*. Its functional significance is, however, unknown.

Since so many cases of Wilms tumors remain diploid, it is an interesting question whether epigenetic changes may in some cases be sufficient to cause formation of this tumor type. This is not known for sure, but if this mechanism occurs in any human cancer at all, it could be Wilms tumor.

## 11.4 TOWARDS AN IMPROVED CLASSIFICATION OF WILMS TUMORS

The histology and clinical behavior of Wilms tumors vary, and its genetic and epigenetic causes are heterogeneous. While one would presume that the latter may explain the former, the relationship does not seem to be straightforward. Nevertheless, a consensus appears to emerge which may help to cure even more children and select an optimal treatment for each patient.

For a while already, it has been known, that an anaplastic histology is a bad prognostic sign and TP53 mutations presage a poor response to chemotherapy. The same seems to be true for cancers with monosomy of chromosome 22, for reasons that are not understood. It has also been claimed that tumors with WT1 mutations may be more aggressive than others, too, even though they tend to have fewer chromosomal alterations. It is, however, not clear, whether this is a property of tumors with WT1 mutations or not – more likely – a property of the subtype to which they belong. Mutations in WT1 are almost exclusively found in tumors with a high stromal content that are associated with intralobular nephrogenic rests. These tumors appear to fail epithelial differentiation completely and then, in a more or less stochastical fashion, differentiate into some mesenchymal lineage. They arise earlier than the blastema-rich tumors associated with perilobular nephrogenic rests. These are very often characterized by loss of imprinting and overexpression of IGF2 and appear to respond better to chemotherapy. In contrast to the former group, WT1 is normally expressed indicating that the block in differentiation is caused by a different defect.

Our understanding of Wilms tumors is certainly at present incomplete. However, researchers and clinicians in the field agree that the pieces of the Wilms tumor puzzle begin to fall in place. This implies the prospect that this once often lethal cancer may eventually be cured with still fewer side-effects in an even larger proportion of the afflicted children.

## *Further reading*

Lee SB, Haber DA (2001) Wilms tumor and the WT1 gene. Exp. Cell Res. 264, 74-99

Malik K et al (2001) Wilms' tumor: a paradigm for the new genetics. Oncol. Res. 12, 441-449

Feinberg AP, Williams BRG (2003) Wilms' tumor as a model for cancer biology. Methods Mol. Biol. 222, 239-248

Menke AL, Schedl A (2003) WT1 and glomerular function. Semin. Cell. Dev. Biol. 14, 233-240

Schumacher V et al (2003) Two molecular subgroups of Wilms' tumors with or without *WT1* mutations. Clin. Cancer Res. 9, 2005-2014

Shah MM et al (2004) Branching morphogenesis and kidney disease. Development 131, 1449-1462

# CHAPTER 12

# CANCERS OF THE SKIN

> The skin is the largest organ and the most frequent site of cancers in humans. The life-time risk of skin cancer for fair-skinned individuals may now add up to 40%. Fortunately, the two most frequent subtypes, basal cell carcinoma (BCC) and squamous cell carcinoma (SCC), are rarely life-threatening. The rarer melanoma is and its incidence appears to increase alarmingly.

> The most important carcinogen in the skin is UV radiation. It is incriminated by a clear correlation of incidence with exposure and by typical mutations found in skin cancers, i.e. point mutations at dipyrimidine sequences. This is a paradigmatic example for molecular epidemiology, which aims at determing the causes of cancer from the mutation patterns observed. In skin cancer, UV radiation acts not only as a mutagen, but also by altering gene expression in epidermal keratinocytes and mesenchymal dermal cells. It also modulates immune responses in skin.

> The incidence of skin cancer is strongly influenced by inherited genetic variation. At the population level, skin pigmentation is the most obvious factor modulating risk. Individuals with inherited deficiencies in the repair of UV-damaged DNA are at even greater risk, prominently xeroderma pigmentosum patients. In a different fashion, *PTCH1* in BCC, certainly, *ESS1* in SCC, perhaps, and *CDKN2A* in melanoma, in rare cases, behave as classical 'gatekeeper' tumor suppressors predisposing to skin cancer.

> Basal cell carcinoma consists of undifferentiated keratinocytes resembling basal cells of the epidermis. They often carry *TP53* mutations which show the diagnostic signature of being induced by UV radiation. Other consistent mutations lead to constitutive activation of the hedgehog pathway. This pathway normally helps to maintain basal cells in a precursor state. Its activation explains the morphological appearance of BCC. BCC cells exhibit mutations in either an inhibitory or an activatory component of the hedgehog pathway. Most inactivating mutations affect PTCH1, which is a classical tumor suppressor. Germ-line mutations in *PTCH1* occur in the Gorlin syndrome which predisposes to BCC and selected other cancers. This syndrome encompasses developmental defects which occur without mutations in the second copy of *PTCH1*, i.e. by haploinsufficiency. Alternatively, hedgehog pathway overactivity in BCC is caused by mutations that activate the SMO membrane protein. So, *SMO1* behaves as an oncogene.

> In normal skin, the main reservoir of proliferating keratinocytes consists of basal cells from which the upper protective layers of the skin are formed by terminal differentiation. This 'transient amplifying' fraction of cells can be replenished from stem cells. Normally, hedgehog activity is restricted to precursor cells. Its

constitutive activation in BCC appears to confer a precursor cell phenotype to the tumor cells.
- In SCC, keratinocyte differentiation is more advanced but proliferation does not cease. Regularly, UV-induced *TP53* mutations are found, with LOH on 17p. Additional LOH is found on chromosomes 9 and/or 13q, and may signify loss of cell cycle control by p16$^{INK4A}$ and RB1. Together, these changes could account for blocked cell cycle exit and immortalization in SCC.
- Melanoma develops from melanocytes which synthesize protective pigments. They have a different developmental origin than keratinocytes, viz. neural crest cells which migrate to the epidermis during fetal development. Melanomas are often highly invasive and form metastases much more readily than BCC and SCC. Melanoma cells maintain the expression of differentiation antigens specific for melanocytes such as gp100 and tyrosinase. In addition, they express 'cancer testis antigens' which occur otherwise only in male germ cells. These antigens may be targets of host anti-tumor immune responses and are exploited in (experimental) immunotherapy, for which melanoma seems a promising target.
- Although epidemiology does not link the causation of melanoma as clearly to UV exposure as that of SCC and BCC, typical point mutations are found in many cases. They activate either the *NRAS* or the *BRAF* gene, which likely stimulate cell proliferation and cell migration. These mutations may be complemented by mutations in *TP53*, *CDKN2A*, *CDK4*, and *PTEN* that may inactivate the mechanisms by which inappropriate proliferation induces apoptosis or replicative senescence. Altered responses to melanocyte-specific growth factors may also contribute to melanoma development.

## 12.1 CARCINOGENESIS IN THE SKIN

The skin is the largest organ in humans. As it covers and protects our external surface, it is subject to mechanical damage and is exposed to a variety of potential carcinogens, chemicals, infectious agents, and radiation alike. Thus, a strong capacity for repair and regeneration is mandatory. This requirement is met by continuous turnover of the epidermis in a structural arrangement that minimizes the impact of carcinogenic agents and protects the body as a whole and the skin itself from cancer development (Figure 12.1).

The outmost layer of the epidermis ('stratum corneum') is composed of crosslinked dead keratinocytes filled with filamental proteins. This layer forms a barrier that rejects many infectious agents, reacts with chemicals and absorbs radiation. The underlying layers ('stratum granulosum' and 'stratum spinosum') are formed by living cells which are irreversibly committed to terminal differentiation (→7.1). Therefore, genetic changes afflicted upon these cells become rarely permanent, because the cells are destined to becoming incapable of proliferation, losing their nuclei, and being eventually eliminated by shedding. Even the basal layer of epithelial cells in the skin, to which proliferation activity is largely confined, consists mostly of cells with a limited replicative potential. They form the transient amplification compartment. The actual stem cells of the skin are rare and are thought to proliferate normally very slowly, except during wound repair. Nevertheless, even then, the brunt of expansion is borne by the transient amplifying fraction.

Another factor in the protection against infection and carcinogenesis is the immune system of the skin. Langerhans cells are dendritic cells which present antigens for recognition by T-cells to elicit immune response against infectious agents and cancer cells.

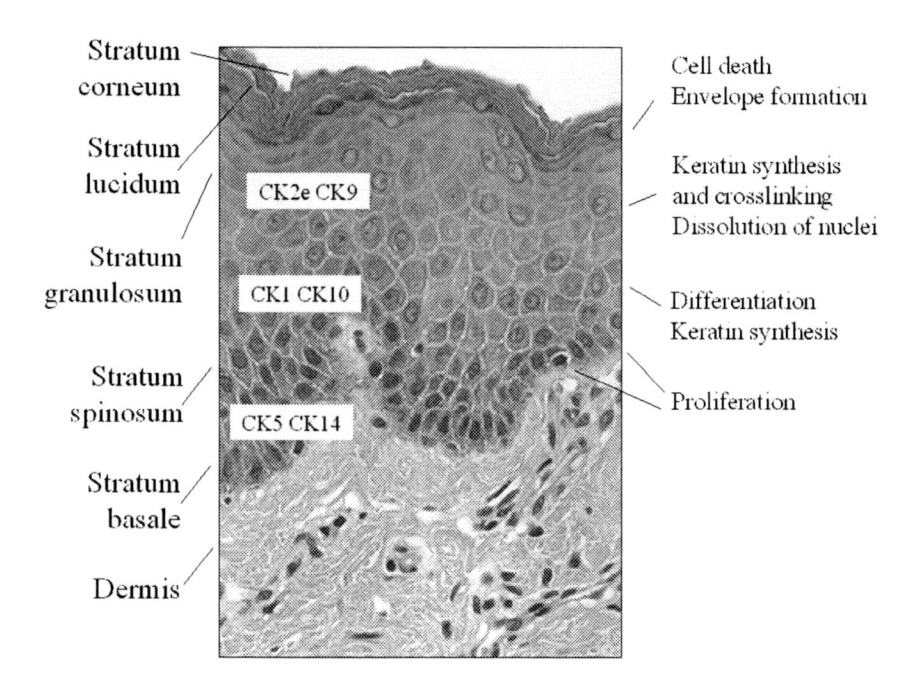

**Figure 12.1** *The protective organization of the skin*
See text for further details.

Melanocytes help to protect specifically against light by producing melanin pigments which are deposited in the keratinocytes. As they differentiate, they carry the pigments to the upper layers of the skin. The extent of pigmentation is the most obvious factor modulating skin cancer risk.

In spite of its intricate protective system, the skin is the most frequent site of cancers in humans. The combined life-time risk for all skin cancers is estimated as 30-40% for lightly pigmented (Northwestern) Europeans, albeit lower for others according to pigmentation. Three different types of skin cancer are prevalent, in decreasing order basal cell carcinoma (BCC), squamous cell carcinoma (SCC), and melanoma (Figure 12.2). Conversely, melanoma is by far the most lethal of these cancers, in ≈20% of all cases. SCC metastasizes only rarely, and BCC almost never.

The incidences of all three types of cancer have increased over the last decades, often steeply. This is particularly worrying in the case of the life-threatening

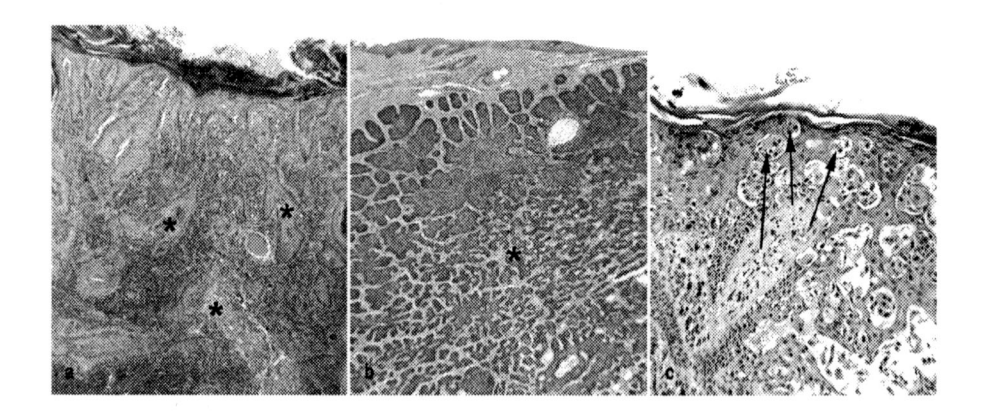

**Figure 12.2** *Histologies of skin cancers*
Histological aspects of A: squamous cell carcinoma, B: basal cell carcinoma, C: melanoma. In C, tumor cells within the epidermis are indicated by arrows.

melanoma. The presumed cause of the rise is an increased exposure to UV-rich sunlight. While the risk of skin cancers is modulated by a variety of genetic factors, short wavelength light is the most important exogenous carcinogen in the skin. Accordingly, most skin cancers, SCC and BCC, and to a lesser degree melanoma, develop in light-exposed areas of the skin.

UV light is categorized by wavelength into UVC (200-280 nm), UVB (280-320 nm), and UVA (320-400 nm). UVC cannot penetrate the upper layer of the skin, but ≈0.4% of UVB and a few percent of UVA reach the basal layer of the epidermis (Figure 12.3). Some UVA even penetrates into the dermis, as does most visible light. How much UVA and visible light reaches the deeper layers of the skin depends on the intensity of pigmentation.

Exposure to sunlight has a range of effects on the skin. At moderate doses, it is beneficial, while higher doses of UV-rich light are problematic. In DNA, UVC can induce single-strand and double-strand breaks and even ionization of bases by direct action. Like ionizing radiation, it also generates reactive oxygen species like hydroxyl radicals that damage DNA by reaction with bases and with the deoxyribose-phosphate backbone. The typical consequence of UVC exposure encountered by living cells is therefore death.

The effects of UVB are more subtle and therefore more dangerous. UVB can be absorbed by DNA, albeit weakly, and lead to photoproducts like thymine dimers and thymine-cytosine 6-4 adducts (Fig. 3.5). These are normally removed by nucleotide excision repair (→3.2). Hereditary defects in nucleotide excision repair cause xeroderma pigmentosum which is characterized by greatly enhanced photosensitivity and risk of skin cancers (→3.4).

Extensive DNA damage by UV radiation elicits activation of cellular checkpoints, and activates TP53 which can initiate apoptosis (→5.3). This happens at a large scale during 'sunburns'. In many skin cancers, inactivation of *TP53* therefore appears to be a necessary initiating step, which has to take place before further mutations can be acquired. Indeed, *TP53* mutations can be detected in morphologically altered, but non-cancerous regions of the skin.

UVA is only weakly mutagenic, but induces cellular reactions in different skin cell types such as modulation of immune responses, altered cytokine production, and activation of stress responses and proliferative signaling pathways. It is now thought that alterations of cellular interactions also contribute to the carcinogenic effect of UV radiation and are elicited by UVA as well as UVB. Diminuation of immune surveillance by inhibition of dendritic Langerhans cells and cytotoxic T-cells may be particularly important. In fact, even manifest carcinomas of the skin often still respond to treatment with stimulators of T-cell responses such as 'imiquimod'.

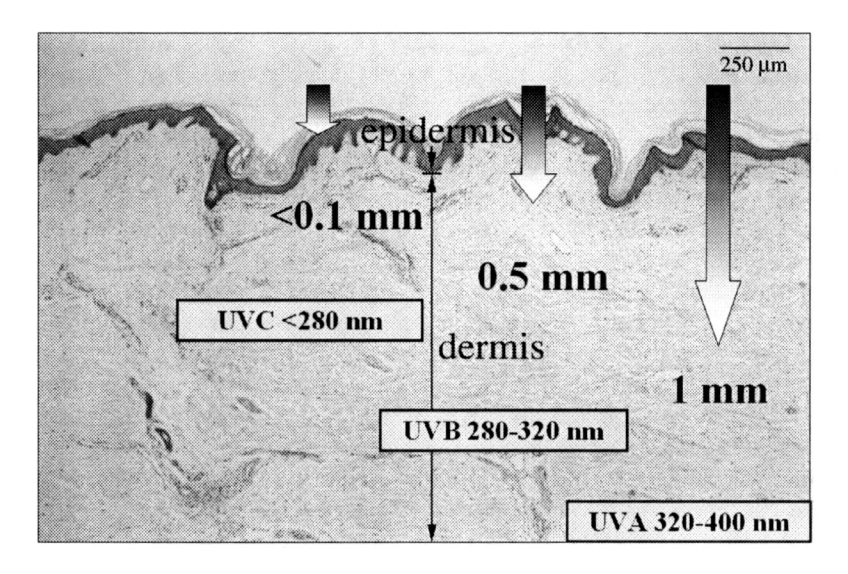

**Figure 12.3** *Action of different wavelength UV radiation in the skin*
See text for detailed explanation. Courtesy: Prof. V. Kolb-Bachofen

The complex mode of UV action illustrates that carcinogenesis requires more than mutations in DNA. Typically, it involves an altered tissue environment, in which tumor cells can more easily proliferate, escape from interactions with normal neighboring cells, and evade immune responses. Strong carcinogens, therefore, elicit such tissue reactions as well as a high rate of mutations. For instance, tobacco smoke inhaled in the lung also acts by several means including, of course, mutagenesis by polyaromates like benzopyrene, but also cell death with tissue repair, chronic inflammation induced by tar and particles, and a direct inhibitory effect of (non-mutagenic) nicotine on normal cells in the tissue.

## 12.2 SQUAMOUS CELL CARCINOMA

Squamous cell carcinomas (SCC) are composed of differentiated epithelial cells, which show many typical markers of differentiated keratinocytes like keratins, involucrin, and other structural proteins. However, unlike normal keratinocytes SCC cells do neither cease to proliferate nor undergo cell death, and the tissue becomes severely disorganized. Squamous cell carcinomas develop from precursor lesions such as actinic keratoses. They sometimes progress to invasive and even metastatic tumors.

Almost invariably, SCC show mutations in the *TP53* gene and often LOH at 17p, where the gene is located. Such mutations are also highly prevalent in actinic keratoses. Activation of TP53 is responsible for the apoptotic death of epidermal cells during sunburns. The cells that survive may do so, because they have acquired mutation in the gene. This may set the stage for the eventual development of skin cancers.

Typically, *TP53* mutations in skin and skin carcinomas such as SCC are located at pyrimidine-pyrimidine dinucleotides identifying them as caused by UV radiation. Absorption of UVB by DNA causes Py-Py dimers, either cyclobutane or 6-4 photoproducts. These are usually removed by nucleotide excision repair ($\rightarrow$3.2), but in rare cases they are bypassed by error-prone DNA synthesis, e.g. using DNA polymerase η. In this type of repair, if the damaged dipyrimidine cannot be read, the sequence AA is inserted by default. This makes sense in so far, as thymine-thymine dimers are most frequent. However, TC and CC dinucleotides are misrepaired in this strategy. Therefore UVB induces TC$\rightarrow$TT or CC$\rightarrow$TT mutations at pyrimidine dimer sites. These are exactly the mutations most frequently found in *TP53* in SCC and other skin carcinomas, implicating UVB as the responsible carcinogen.

This relationship is an example of molecular epidemiology. If a particular molecular alteration can be related to a specific carcinogen, it is possible to determine to which extent cancers are caused by that carcinogen. For instance, carcinogenic aflatoxins add to a specific guanine in the *TP53* gene, causing 249G$\rightarrow$T mutations. By following this particular change, it can be proven that aflatoxins contribute to a large proportion of hepatomas in Africa and Asia, but not in Europe or North America ($\rightarrow$16.1). This approach has its limits, however, because many carcinogens do not leave specific fingerprints or several carcinogens can leave the same one. For instance, it was thought for a while that C$\rightarrow$T mutations at CpG sites were mostly due to an increased rate of mutation of methylated cytosines following spontaneous deamination ($\rightarrow$8.3), which would exculpate exogenous carcinogens. More recently, however, methylated cytosines were shown to preferentially react with several activated carcinogens, re-opening this issue to debate.

Inactivation of TP53 in SCC is thought to lead to diminished apoptosis and to favor genomic instability. A second consistent change in this cancer is loss of chromosome 9, in particular of the short arm harboring the *CDKN2A* gene. Loss of p16$^{INK4A}$ is clearly involved in the progression of actinic keratosis to carcinoma. It may contribute to deregulation of the cell cycle leading to continuing proliferation and failure of terminal differentiation of the tumor cells. In some SCC, losses of 13q are found which would be suspected to lead to decreased or abolished RB1 function with basically the same biological outcome as p16$^{INK4A}$ loss (→5.2). Loss of either p16$^{INK4A}$ or RB1 is found in many human carcinomas, particularly those with a more differentiated phenotype (cf. 14.2).

While *CDKN2A* is generally accepted to represent the critical target of 9p loss in SCC, the relevant gene affected by the likewise frequent loss of 9q is not known. In basal cell carcinoma of the skin and a limited range of other cancers, the loss contributes to a decreased function of *PTCH1* (see below). This change is clearly not involved in the development of SCC. Instead, a locus tentatively named *ESS1* in the 9q22-31 region has been implicated through linkage analysis in a small number of families and seems to behave as a classical tumor suppressor.

Several further prevalent alterations in SCC (Table 12.1) are characteristic of a wider range of epithelial cancers.

(1) Loss of 3p is not only frequent in SCC of the skin, but also in SCC of the head and neck, which has a different etiology, and in lung cancers. In these cancer types, it may be the frequent chromosomal change. Loss of 3p is even diagnostic for clear-cell carcinoma of the kidney where it contributes to lack of the *VHL* gene function that constitutes the gatekeeper event in this tumor type (→15.4). *VHL* remains functional in other cancers, including skin SCC. Therefore, other potential tumor suppressor genes on 3p are thought to be more important in SCC. Good candidates are *RASSF1A* (→6.2) and *RARB2* (→8.5). RARB2 is a receptor for retinoic acid which is known to regulate proliferation and differentiation particularly of the epidermis.

(2) Either *KRAS* or *HRAS* are mutated in SCC, at moderate frequencies. Perhaps, inactivation of RASSF1A which relays a signal limiting RAS action (→6.2), occurs as a complementary change to these mutations. RAS activation by mutation is certainly pleiotropic (→Figure 6.2) and may yield a proliferative signal mainly through the MAPK pathway and/or an anti-apoptotic signal through the PI3K pathway. In either case, it may contribute to independence of external growth factors.

**Table 12.1.** *Prevalent genetic changes in different cancers of the skin*

| Squamous cell carcinoma | Basal cell carcinoma | Melanoma |
|---|---|---|
| *TP53* mutation and LOH | *TP53* mutation and LOH | *TP53* mutation and LOH |
| *CDKN2A* inactivation | *PTCH1* inactivation or oncogenic *SMO* activation | *CDKN2A* inactivation |
| Chromosome 9q loss | *HRAS* mutation | *HRAS* or *BRAF* mutation |
| Chromosome 13q loss |  | Chromosome 6q loss |
| Chromosome 3p loss |  | *PTEN* inactivation |
| *HRAS* or *KRAS* mutation |  | Loss of TGFβ response |
| EGFR overexpression |  |  |

(3) Overactivity of the EGFR as a consequence of gene amplification or autocrine stimulation by EGF-like factors would be expected to have similar consequences (→4.4). Again, this is a widespread change in human carcinomas, occuring in some at an early stage of development, while being associated with progression and metastasis in others. SCC of the skin appears to belong to the latter group. As in general, it is not clear why this is so. One speculation goes as follows. Proliferation in normal epidermis is regulated by paracrine interactions with mesenchymal cells in the underlying dermis (→8.5). Increased EGFR activity may diminish the dependence of the cancer cells on paracrine stimulation and enable them to survive in a wider range of environments, thus creating one prerequisite for invasion and metastasis.

In summary, thus, while skin SCC appears less malignant than most other carcinomas, it is prototypic, showing an assembly of alterations also found in these cancers. These alterations comprise inactivation of the TP53 and RB1 regulatory systems, activation of RAS-dependent pathways, EGFR overactivity, and 3p loss.

## 12.3 BASAL CELL CARCINOMA

Among those carcinomas that are very different from SCC of the skin is basal cell carcinoma of the skin. Whereas the tumor cells in SCC resemble differentiated keratinocytes, those in basal cell carcinoma look similar to those of the basal layer and indeed carry their typical markers such as basal cell cytokeratins (CK5 and CK14) and integrins ($\alpha_2\beta_1$). However, while they proliferate like basal cells, they do not differentiate. Furthermore, proliferation in the tumor is no longer restricted to cells in the basal layers.

The crucial genetic alterations in BCC lead to constitutive activation of the hedgehog (SHH) pathway (→6.10). Normally, this pathway is activated in the skin by binding of SHH (sonic hedgehog) to the PTCH1 (patched) protein (Figure 12.4). This relieves the inhibition of SMO (smoothened) by PTCH, turning the SHH intracellular signal cascade on. The steps following SMO activation are not precisely known, but may involve an activating G-protein and be subject to some form of regulation by protein kinase A. The next components definitely known are Fused and SUFU (suppressor of fused) which are part of a protein complex located at microtubules. This complex binds GLI transcription factors maintaining them in an inactive state.

Three GLI proteins are known in humans. Of these, GLI1 and GLI2 are activated by SHH in basal cells of the skin. The factors migrate to the nucleus to activate target genes of the pathway. Among the proteins induced – directly or indirectly – are activators of cell growth and proliferation like Cyclin D1 and MYC, and the transcription factor gene FOXM1, but also PTCH1 and another membrane protein HIP1, which serves as a feedback inhibitor of the pathway together with PTCH1. GLI1 and GLI2 are also regulated by SHH signaling.

The activity of the SHH pathway is normally restricted to a subset of the basal cells in the epidermis and is thought to define these cells as keratinocyte precursor cells with stem cell character (→8.6). It is not clear how the restriction to this subset is achieved. Perhaps, the NOTCH pathway, which is capable of repressing hedgehog signals (→6.10), may be involved. If so, it would be plain why the NOTCH pathway tends to be inactive in BCC.

Constitutive activation of the SHH pathway in BCC can occur by one of several mechanisms. Most frequently, the function of *PTCH1* is lost by inactivation of both alleles. The *PTCH1* gene is located at 9q22. Many BCC show LOH in this region and inactivating mutations in the remaining allele. Not surpringly, as UVB is also implicated as a carcinogen in BCC, some mutations in *PTCH1* show the characteristic signature of this carcinogen.

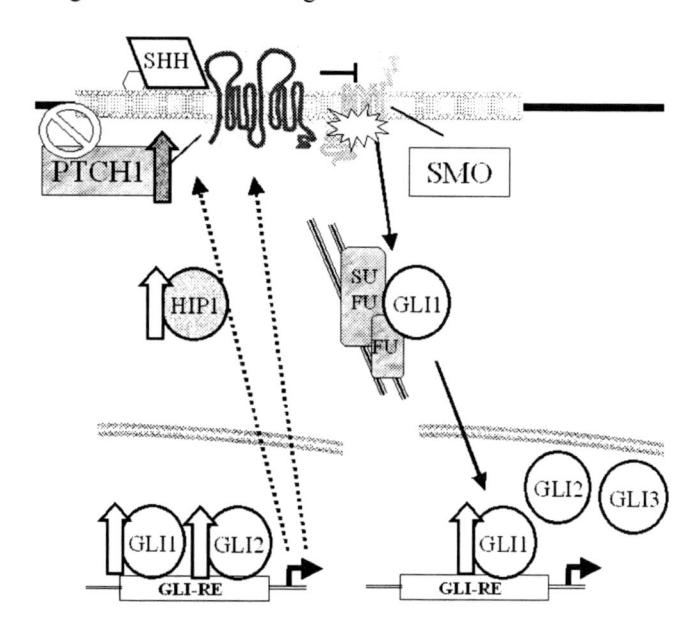

**Figure 12.4** *Alterations in the SHH pathway in basal cell carcinomas of the skin*
Almost all skin BCC contain either mutations constitutively activating SMO (exploding star) or inactivating both alleles of *PTCH1* (prohibition sign). As a result, pathway activity is increased and target genes are induced, including *GLI1*, *GLI2*, *HIP1*, and *PTCH1* (arrows). Compare also Figure 6.13.

Since PTCH1 inactivation leads to increased activity of the SHH pathway, of which it itself is a transcriptional target, BCC overexpress the mRNA of mutated *PTCH1*. In fact, a hallmark of SHH pathway activation is the overexpression of several target genes including those encoding GLI factors and HIP1.

*PTCH1* is also the gene mutated in the germ-line of patients with nevoid basal cell carcinoma syndrome (NBCCS or Gorlin syndrome), a rare autosomal dominantly inherited disease. Patients develop multiple nevi and BCC at a relatively young age, even during childhood or more frequently around puberty. Like sporadic BCC, those in Gorlin syndrome also preferentially develop in light-exposed areas.

Moreover, LOH at 9q22 is found in these tumors. Thus, *PTCH1* is a classical tumor suppressor gene in the Knudson definition and belongs to the 'gatekeeper' class (→5.4). Gorlin syndrome patients are also at risk for a selected range of other cancers, including medulloblastoma. Indeed, many sporadic cases of this type of brain tumor, but not of others such as glioma, also contain mutations in SHH pathway genes.

One pecularity sets Gorlin syndrome apart from other 'gatekeeper' cancer syndromes. Patients often present with a characteristic range of developmental abnormalities, mostly of the skeleton. These do not seem to require inactivation of the second *PTCH1* allele or a dominant-negative function of the mutated gene product. Hedgehog signaling is crucially important during development. So, in some organs, expression of one intact *PTCH1* allele may not suffice to ensure proper regulation of the pathway. Thus, *PTCH1* may display haploinsufficiency with regard to its function in human development. Clearly, this begs the question whether haploinsufficieny of the gene may also occur in the context of tumor development.

The second most frequent alteration leading to constitutive SHH signaling in BCC – and actually in medulloblastomas as well – are point mutations in *SMO1*. These mutations occur in specific sites and lead to constitutive activity of the protein. In particular, they abolish regulation of SMO by PTCH1. For a while, it was thought that the mutations affected the interaction between the two proteins. Now, it is assumed that PTCH1 regulates SMO indirectly and in a non-stochiometric fashion by directing a low-molecular inhibitor towards it. So, the mutations may prevent binding of this – unknown – inhibitor. As *PTCH1* has to be regarded as a tumor suppressor, *SMO* is obviously a proto-oncogene which is activated by specific point mutations.

Inactivation of *PTCH1* and oncogenic activation of *SMO* are very likely not the only alterations which can lead to SHH overactivity in BCC. Mutations in *PTCH2* have already been observed, and others are expected, since the pathway is not yet fully elucidated. Moreover, distinct modes of activation may take place in other cancers. For instance, amplification of a region at 12q where *GLI1* is located close to *HDM2* may not only lead to overexpression of HDM2, but also of GLI1.

An entirely different mechanism appears to be at work in small-cell lung cancer (SCLC). SCLC could be regarded also as sort of a stem-cell cancer, in so far as it is composed of neuroendocrine cells who are thought to be produced by activation of tissue stem cells. In this case, however, an autocrine SHH loop may be responsible for over-activity of the pathway. This mechanism has also been observed in some pancreatic carcinomas.

Of course, activation of the SHH pathway is not the only alteration in BCC (Table 12.1). Other genetic changes do occur and are probably required for development as well as progression of this cancer. As in SCC, RAS mutations are observed and TP53 is very frequently inactivated. It may be significant, however, that alterations in the RB1 pathway do not seem to be as essential in BCC as they are in SCC.

In addition to BCC and SCC, there are further cancers derived from the wider keratinocyte lineage. Keratinocytes share precursor cells with the cells of the hair bulges and the sebaceous glands from which rarer tumors can arise. These are again characterized by different patterns of genetic alterations, such as activation of the WNT pathway (→6.10) in hair follicle tumors.

This suggests two alternative possibilities how the diverse tumors may arise (Figure 12.5). One possibility is that each type of cancer is formed by different mutations in the common tissue precursor cell: Mutations impeding the RB1 regulatory system might lead to SCC because keratinocytes can differentiate, but not exit from the cell cycle, whereas mutations activating the SHH pathway would arrest the development of keratinocytes at the basal cell stage, etc. The second possibility

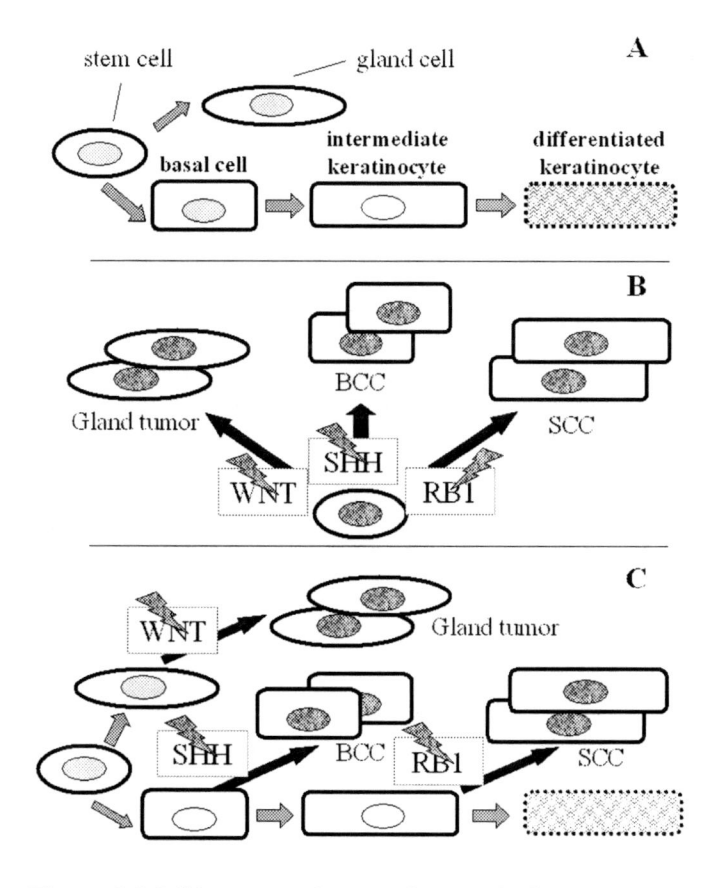

**Figure 12.5** *Cell lineages, pathways and origin of different skin cancers*
A: Cell lineages in the normal skin (simplified). One hypothesis illustrated in B suggests that alterations in different regulatory pathways lead to different cancer types, e.g. WNT pathway alterations cause gland tumors, SHH pathway alterations lead to basal cell carcinoma (BCC), whereas disturbances of RB1 function elicit differentiated squamous cell carcinoma. An alternative hypothesis shown in C suggests that alterations in these pathways occur in different branches resp. at different stages of the lineage.

is that cells at a more advanced stage of commitment and differentiation might require different types of genetic alterations to turn them into tumor cells. For instance, activation of the SHH pathway would not be possible or would not be able to induce tumor formation once a certain stage of keratinocyte differentiation has been reached. Whichever of these alternatives holds, the elucidation of the molecular basis of skin cancers has made very clear that specific molecular therapies need to be directed at different targets in BCC, SCC, and other skin cancers.

## 12.4 MELANOMA

The third most frequent cancer of the skin, melanoma, is derived from melanocytes. These cells belong to a distinct cell lineage as keratinocytes. During development, melanocyte precursors migrate from the neural crest to the basal layer of the skin. Melanocytes specialize in synthesizing the pigment melanin from tyrosine. The enzymes of melanin biosynthesis, such as tyrosinase, are only expressed in this cell type. The insoluble pigment is transported by dendritic processes to surrounding epidermal keratinocytes and deposited in them. Differentiating keratinocytes transport melanin to the upper layers of the skin where it absorbs visible and UV light, protecting the living cells of the skin.

Skin pigmentation in man is highly variable. It ranges from a complete lack due to mutations in tyrosinase ('albino') to very intense, e.g. in populations from equatorial Africa. In most humans, pigmentation is inducible in response to sun exposure. It is regulated by interaction of melanocytes with the neighboring keratinocytes and by the hormone αMSH (melanocyte stimulating hormone).

The hormone acts specifically on melanocytes because they express the MC1R receptor (Figure 12.6). This is a classical 'serpentine' receptor coupled to trimeric $G_s$ proteins that activate adenylate cyclase. As in all cell types, increased cAMP activates protein kinase A which induces transcription through the CREB transcription factor. Specifically in melanocytes, CREB induces another transcriptional activator, MTF, which is actually responsible for the induction of melanocyte-specific genes and increased production of melanin. In normal melanocytes proliferation is also dependent on αMSH.

The differences in pigmentation and its inducibility are categorized as 'skin types' by dermatologists. For instance, persons with skin type I have low pigmentation and very little inducibility, whereas persons with skin type II have also relatively pale skins, but develop some pigmentation in response to sun-light. Skin type I is prevalent in Northern and Northwestern European populations, presumably as an evolutionary adaptation to less intense sun exposure. Therefore, travel and migration of Northern European to areas with intense sun exposure is a major factor in the alarming rise of skin cancer incidence in these countries and in states like Australia with a high proportion of immigrants from Northern Europe. Skin types are mostly caused by polymorphisms in the αMSH receptor MC1R. *MC1R* is thus a cancer predisposition gene (→Table 2.4).

Melanoma is a much more lethal cancer than BCC and SCC, because of its stronger invasive and metastatic potential. However, many genetic alterations in this cancer are not too different from those in SCC, in particular (Table 12.1). Frequent chromosomal losses affect chromosomes 9p and 17p and contribute to inactivation of *CDKN2A* and *TP53*. The second alleles of these genes are subject to point mutations. These carry occasionally the signature of UVB carcinogenesis, but not as regularly as in BCC and SCC. In many cases, *CDKN2A* is inactivated by homozygous deletions which obliterate expression of $p14^{ARF1}$ in addition to that of $p16^{INK4A}$ and in many cases $p15^{INK4B}$ as well. The overall effect of these changes would be loss of function of the TP53 and RB1 regulatory systems contributing to immortalization, loss of cell cycle control, and genomic instability.

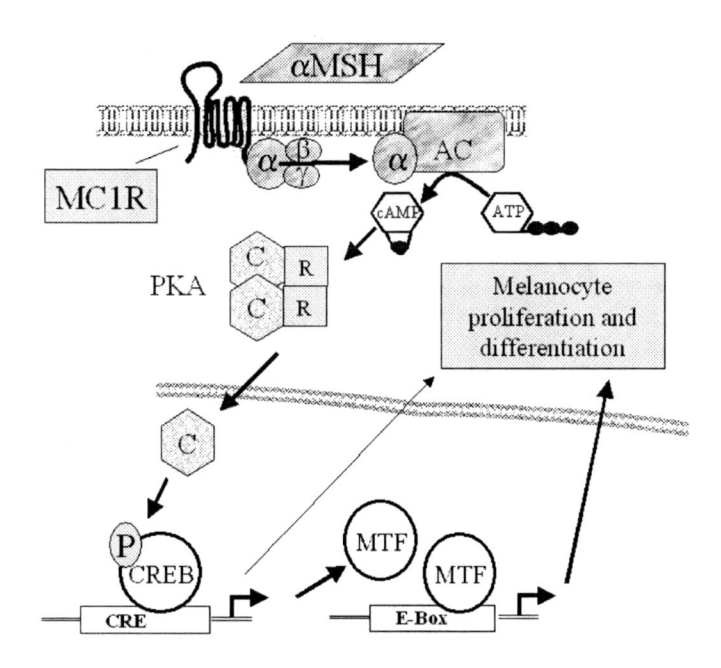

**Figure 12.6** *Mechanisms of αMSH action*
C and R: catalytic and regulatory subunits of protein kinase A (PKA). See text for further explanation.

In melanoma, loss of $p16^{INK4A}$ function may be particularly important. In some of the families prone to melanoma development, germ-line mutations in *CDKN2A* were found which inactivate $p16^{INK4A}$ or at least diminish its function as an inhibitor of CDK4 in biochemical assays. In fact, in a few families with predisposition to melanoma, mutations were detected in *CDK4*, always affecting the part of the kinase to which the $p16^{INK4A}$ inhibitor binds. Of note, in melanoma-prone families, an association between UV exposure and cancer development is maintained. So, this may be regarded as a border-line case between high-risk gene mutations and mutations modulating the sensitivity to exogenous agents (cf. 2.3).

Germ-line mutations in *CDKN2A* have also been observed in patients with pancreatic cancers, and melanomas, too, have occurred in these families. However, germ-line mutations in *CDKN2A* do not lead to a severe generalized cancer syndrome such as the Li-Fraumeni-syndrome caused by TP53 mutations. This is certainly unexpected in view of the frequent inactivation of the locus in a wide range of human cancers (→6.4).

Further chromosomal losses in melanoma affect chromosomes 3p, 6q, and 10q. At 10q, the *PTEN* gene is likely involved, and as in many other human cancers, this may an important step during progression (→6.3).

A further set of alterations in melanoma activate the canonical MAPK pathway. Activating mutations in the *NRAS* gene occur in a subset of melanomas. In a complementary fashion, the *BRAF* gene harbors mutations that lead to an increased activity of the protein kinase. There are multiple pathways emerging from RAS and three different RAF genes and proteins in humans (→6.2). The complementarity of these mutations indicates that the MAPK pathway is particularly important in tumor formation by activated RAS and that BRAF, among the three protein kinases, is the one transducing the most relevant signals. However, since this constellation has, so far, been found in a restricted number of cancers, it could reflect a cell-type specific 'wiring' of the MAPK signaling network.

There is also some evidence for altered responses to melanocyte-specific growth factors like αMSH in melanoma.

Finally, the most pressing question is of course, what makes melanoma so much more invasive than SCC and BCC?[13] Several points are discussed in this regard. (1) Melanocytes are ontogenetically derived from a highly mobile and migratory cell type. So, they may easily fall back into a 'fetal' pattern of behavior. (2) Melanocytes are very peculiarly located. They sit as single cells at the basis of the epithelium and while they maintain cell-cell-contacts with keratinocytes, these are flexible and certainly not comparable to the multiple junctions between proper epithelial cells (→9.1). So, it may be less complicated for them to dissociate from the epithelium and grow or migrate into the dermis. (3) There are more molecular changes in melanoma typically associated with invasion and metastasis. Melanoma appear to express higher levels of certain metalloproteinases (→9.3), they exhibit more consistently down-regulation of responsiveness to TGFβ (→6.7), and they express chemokine and cytokine receptors that may allow 'homing' to certain metastatic sites (→9.6).

---

[13] One might, of course, argue that cancers arising in the interior of the body could be no more aggressive than SCC and BCC, but simply become detected much later. This may indeed be so in some cases. However, very "big" BCC in particular do not metastasize, whereas relatively small melanomas can give rise to metastases, although the risk is clearly related to their size and particular to their depth of growth.

## 12.5 TUMOR ANTIGENS

Diminished immune responses to tumor cells are thought to constitute one component in the initial carcinogenic action of UV radiation (→12.1). Immune responses remain important, however, during tumor progression. In the case of melanoma, the tumor cells may be particularly immunogenic, although this does obviously not present an insurmountable obstacle to tumor progression. However, immune cells directed against tumor cell antigens can be detected in melanoma patients. Several different types of tumor cell antigens can be recognized (Table 12.2).

Cell-type specific antigens: Melanocytes express specific proteins which are not found in any other cell-type. These comprise, of course, the enzymes and proteins involved in melanin production such as tyrosinase, but also surface proteins like gp100.

Viral antigens: Some tumor cells harbor viral genomes. SCC of the skin, e.g., often contain HPV genomes and consequently present viral antigens on their surface. While the role of HPV in SCC etiology is debated, the expression of viral proteins may aid in containment of these cancers.

Oncofetal gene expression: Many cancers express ectopic proteins. Some of these are normally expressed only in the respective fetal tissues and are therefore called oncofetal antigens. Carcinoembryonic antigen (CEA) in gastrointestinal tumors and $\alpha$-fetoprotein (AFP) in liver cancers belong to this category.

Cancer-testis antigens: Melanoma cells strongly express a somewhat different type of ectopic markers, called 'cancer testis antigens'. These proteins, comprising several MAGEs, GAGEs, etc. each, are otherwise only expressed in testicular tissue in adult humans. It is not clear, why these genes are also activated in melanoma, or in some other cancers. Reactivation is often associated with promoter hypomethylation, but this is probably not sufficient.

In the majority of cases, obviously, immune cells directed against such antigens cannot eliminate the cancer completely. It is thought, however, that they limit its progression and it is hoped that the immune response towards cancer antigens can be harnessed for the purpose of 'immune therapy' (→22.5).

**Table 12.2.** *Classes of tumor antigens*

| Type of tumor antigen | Example |
| --- | --- |
| Cell-type specific | tyrosinase, melanocyte surface protein gp100 |
| Viral | papilloma virus proteins |
| Oncofetal | carcino-embryonic antigen, $\alpha$-fetoprotein |
| Cancer testis | melanoma antigens (MAGE) |

## Further reading

Weedon D et al (eds.) Pathology and genetics of tumours of the skin. IARC Press, 2004

Taipale J, Beachy PA (2001) The Hedgehog and Wnt signalling pathways in cancer. Nature 411, 349-354
Fuchs E, Raghavan S (2002) Getting under the skin of epidermal morphogenesis. Nat. Rev. Genet. 3, 199-209
Ruiz i Altaba, Sánchez P, Dahmane N (2002) Gli and Hedgehog in cancers: tumours, embryos and stem cells. Nat. Rev. Cancer 2, 342-350
Zabarovsky ER, Lerman MI, Minna JD (2002) Tumor suppressor genes on chromosome 3p involved in the pathogenesis of lung and other cancers. Oncogene 21, 6915-6935
Alonso L, Fuchs E (2003) Stem cells of the skin epithelium. PNAS USA 100 Suppl 1., 11830-11835
Bastian BC (2003) Understanding the progression of melanocytic neoplasia using genomic analysis: from fields to cancer. Oncogene 22, 3081-3086
Bataille V (2003) Genetic epidemiology of melanoma. Eur. J. Cancer. 39, 1341-1347
Chin L (2003) The genetics of malignant melanoma: lessons from mouse and man. Nat. Rev. Cancer 3, 559-570
Giglia-Mari G, Sarasin A (2003) TP53 mutations in human skin cancers. Hum. Mutat. 21, 217-228
Norgauer J et al (2003) Xeroderma pigmentosum. Eur. J. Dermatol. 13, 4-9
Perez-Losada J, Balmain A (2003) Stem-cell hierarchy in skin cancer. Nat. Rev. Cancer 3, 434-443
Reifenberger J, Schön MP (2003) Epitheliale Hauttumoren: Molekularbiologie und Pathogenese-basierende Therapie. Hautarzt 54, 1164-1170
Olivier M et al (2004) TP53 mutation spectra and load: a tool for generating hypotheses on the etiology of cancer. IARC Sci Publ. 157, 247-270

# CHAPTER 13

# COLON CANCER

- Carcinomas of the colon and rectum (here summarized as colon cancer) are among the most prevalent cancers in Western industrialized countries. Together with lung cancer, breast cancer in females and prostate cancer in males, they constitute the majority of lethal cancers in these countries and a major public health problem. Like breast and prostate cancer, colon cancer is much less prevalent in the developing world, but also in the richer Asian countries and Southern Europe. These differences point to important life style factors in the etiology of the disease. Good evidence exists for the involvement of dietary factors.
- Several distinct precursor stages can be distinguished in histology. The stages of cancer development thus range from hyperplasia through various benign adenomas and locally invasive carcinoma to systemic metastatic disease.
- Several hereditary cancer syndromes increase the risk for colon carcinoma more or less specifically, among them familial adenomatous polyposis coli (FAP), hereditary non-polyposis colon carcinoma (HNPCC), Peutz-Jeghers syndrome (PJS), and Cowden's disease. The study of these moderately prevalent to very rare syndromes has greatly contributed to the understanding of sporadic colon cancer.
- While it is a general tenet that human carcinomas arise by alterations in several distinct genes and regulatory pathways, in colon cancer, many of these alterations have actually been identified. Moreover, they can be assigned to specific stages of tumor development and related to specific histological changes. This molecular model of colon carcinogenesis has promoted understanding of the underlying biological processes. Attempts to transfer this model have often stimulated research on other cancers. In this sense, colon cancer has become paradigmatic for molecular oncology.
- The fundamental 'gatekeeper' step in colon cancer development leading primarily to adenomas is a constitutive activation of the WNT signaling pathway. It is usually caused by loss of function of the classical tumor suppressor APC, less frequently by mutations in *CTNNB1* oncogenically activating β-Catenin, and rarely by other changes in the pathway. The activation appears to confer a kind of 'stem-cell' phenotype to the carcinoma cells.
- Further steps in the progression of the cancer are associated with mutations activating KRAS, loss of function of TP53, and inactivation of the cellular response to TGFβ.
- Colon cancer exemplifies that tumor development can be driven by several different mechanisms. Molecular research has revealed that the accumulation of molecular alterations in individual colon cancers can alternatively be driven predominantly by an increased rate of fixed point mutations, by chromosomal

instability, or an increased rate of epigenetic alterations. These molecularly defined subclasses differ only moderately in their morphology and clinical behavior, but the classification should become very helpful for diagnostics and for definition of therapeutic targets.

➢ Specifically, a subgroup of colon cancers is characterized by an increased rate of point mutations, which is most evident as frequent alterations in the length of microsatellite repeats. This 'microsatellite instability' is caused by defects in the DNA mismatch repair system. It can arise spontaneously by genetic or epigenetic inactivation of genes encoding components of this system or as a consequence of inherited mutations in these genes in the dominantly inherited HNPCC syndrome.

➢ Colon cancer also occurs in the context of chronic inflammatory bowel diseases such as ulcerative colitis. Non-steroidal antiinflammatory drugs diminish colon cancer risk in such patients, but seem to be efficacious also in others at risk for colon cancer. Thus, colon cancer also is also a pioneer example for chemoprevention.

## 13.1 NATURAL HISTORY OF COLORECTAL CANCER

Colorectal cancer can develop in all segments of the large intestine. Although its location somewhat influences the prognosis and obviously the approach for surgical treatment, from a biological point of view, it can be considered under one heading, i.e. colon cancer. Interestingly, cancers of the small intestine are very rare and certainly represent a distinct disease.

There is good morphological and molecular genetical evidence that colon carcinoma develops through several precursor stages (Figure 13.1). The earliest recognizable preneoplastic changes result in hyperplastic or dysplastic crypts. There is some debate whether one or the other of these, or both can give rise to adenomas and carcinomas. In contrast, it is generally agreed that adenomatous polyps are a precursor stage for many carcinomas. They are usually found as single benign tumors protruding into the lumen of the bowel and consist of a thickened, more or less disorganized epithelium. Multiple polyps are found in certain circumstances, e.g., in the familial cancer syndrome 'familial adenomatous polyposis coli' (FAP) discussed below. These polypous tumors are considered a type of adenoma and can be categorized into several substages according to the degree of growth and dysplasia.

Invasion of tumor cells through the basement membrane into the underlying mesenchyme is the diagnostic mark of carcinoma. Several tumor stages are distinguished in routine pathology according to the extent of invasion and the spread of the tumor mass. Localized colon carcinomas without metastases can often be cured by surgery. In many cases, 'adjuvant' chemotherapy is applied after surgery to kill remaining tumor cells. The prognosis of colon carcinoma becomes worse and the treatment much more difficult, if the tumor has spread to local lymph nodes or metastasized to the liver, lung and other organs. Surgery and chemotherapy can still be curative, but often only prolong survival.

The incidence of colon cancer varies considerable across the world, with the highest incidences in Western industrialized countries. Colon cancer is usually a disease of older people. However, the incidence remains different between countries with on average younger or older populations even after adjustment for age. The causes for these differences in incidence are not really understood. The best evidence points to dietary factors being responsible (→20.3).

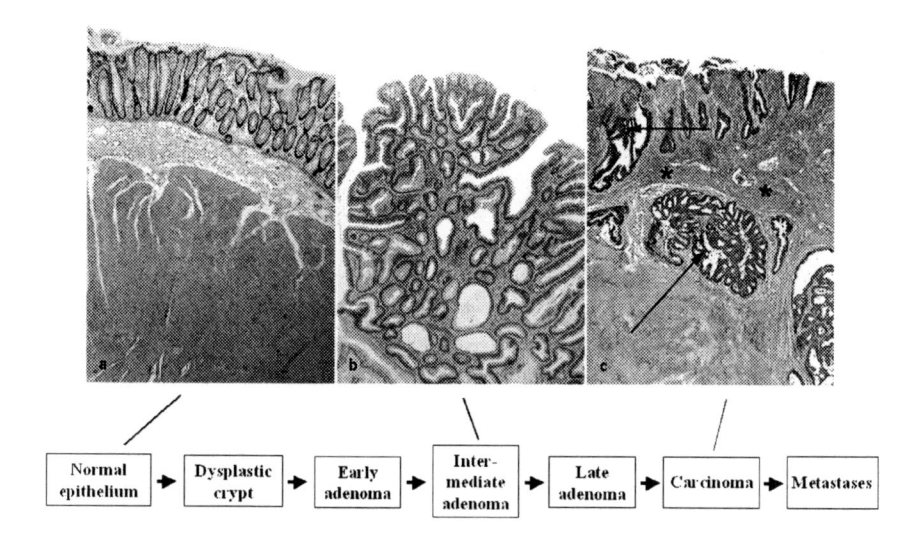

**Figure 13.1** *Stages in the development of colorectal cancers*
The histologies show normal mucosa (left), an adenomatous polyp with moderate dysplasia (center), and invasive adenocarcinoma with tumor glands (arrows) in the submucosa (right).

## 13.2 FAMILIAL ADENOMATOUS POLYPOSIS COLI AND THE WNT PATHWAY

Familial Adenomatous Polyposis Coli (FAP) is the most prevalent of several related syndromes which carry a strongly enhanced risk for colon cancer (Table 13.1). Related afflictions include Gardners syndrome, Turcots syndrome, and an attenuated form of FAP. These syndromes are distinguished from each other by the kind of tumors and developmental defects that appear in other organs besides the large intestine.

Patients with standard FAP characteristically develop hundreds of adenomatous polyps in their colon, rectum, and duodenum as well as gland polyps in the stomach, early in life. Congenitally, they display hypertrophy of the retinal pigment epithelium. Although the polyps are adenomatous, i.e. benign tumors, eventually one or the other progresses to malignancy and carcinomas develop, typically in the third or fourth decade of life. The life-time risk of carcinoma development approaches 100%. Therefore, removal of the colon is used as a preventive measure (see also 13.6). FAP is inherited in an autosomal-dominant fashion. With its constellation of multiple tumors in the same organ, carcinoma development much before the usual age, and autosomal-dominant mode of inheritance, FAP is a prime example of an inherited tumor disease corresponding to the 'Knudson' model (→5.1).

The gene mutated in FAP was identified in 1991 and was named *APC*, for 'adenomatous polyposis coli'. Positional cloning (→Box 13.1) in families with FAP and the related Gardner's syndrome was successful, strongly aided by a key patient with a cytogenetically recognizable deletion in chromosome 5q21 where the *APC* gene resides. *APC* turned out to be a large gene (Figure 13.2) comprising

Table 13.1. *Some hereditary syndromes predisposing to colon cancer*

| Syndrome | Colon symptoms | Other tumors | Gene affected |
|---|---|---|---|
| Familial adenomatosis polyposis coli (FAP) | multiple adenomatous polyps, carcinoma | stomach and duodenal polyps and cancers, retinal hyperplasia | *APC* |
| Gardner | multiple adenomatous polyps, carcinoma | stomach and duodenal polyps and cancers, retinal hyperplasia, desmoid tumors, osteoma | *APC* |
| Attenuated FAP | fewer polyps, increased risk of carcinoma | | *APC* |
| Hereditary non-polyposis colorectal cancer (HNPCC) | carcinoma, often multifocal, with few polyps | cancers of the endometrium, ovary, stomach, liver, upper urinary tract, brain | *MSH2*, *MLH1*, *PMS2*, others |
| Cowden | frequent hamartomas, increased risk of carcinoma | hamartomas in many organs, breast, thyroid and other cancers | *PTEN* |
| Juvenile polyposis coli* | hamartomas or polyps, increased risk of carcinoma | hamartomas, polyps and increased cancer risk in the stomach | *SMAD4* |
| Peutz-Jeghers* | hamartomatous polyps, carcinoma | hamartomatous polyps and carcinoma in all parts of the gastrointestinal tract | *LKB1* (*STK11*) |

* see OMIM database for details on these syndromes

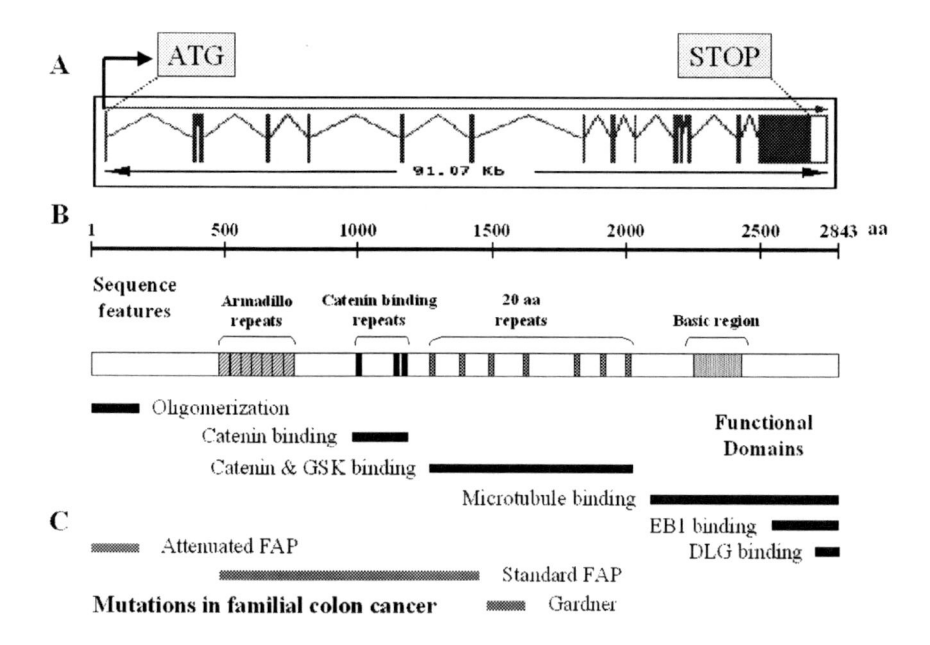

**Figure 13.2** *The* APC *gene and its encoded protein*
A: Organization of the gene as shown in the ensembl data base. There is a second, alternative promoter with a non-coding additional first exon ≈15 kb upstream of exon 1 shown here. B: Sequence features and functional domains of the APC protein. C: Localization of germ-line (nonsense) mutations in familial colorectal cancers. Point mutations in sporadic cases are most prevalent in the region affect in Gardner syndrome.

binding several different proteins and of oligomerization. The oligomerization domain is located near the N-terminus. Several repeats are recognizable along the length of the protein, of which the catenin-repeats and the 20 aa-repeats are involved in assembling a protein complex containing β-Catenin, glycogen synthase 3β (GSK3β), and the scaffold protein Axin or its relative Conductin (or Axin2). This protein complex is involved in regulating WNT signaling (see below). A basic region in APC located further towards the C-terminus interacts with microtubules. The actual C-terminal region binds EB1 and DLG, two proteins interacting with the mitotic spindle. There is some evidence that one function of APC lies in chromosome segregation. Loss of this function may favor the development of aneuploidy in colon cancer.

Patients with FAP inherit a mutation in one *APC* allele. Most are nonsense mutations leading to a truncated or instable protein and almost all are located in the first half of the gene. As predicted by the 'Knudson' model (→5.1), adenomas and carcinomas either contain somatic mutations in the second allele as well or have lost the intact allele by 5q deletion or by recombination, as recognizable by LOH in the 5q21 region.

Not unexpectedly, the gene mutated in the Gardner and Turcot syndromes as well as in the attenuated variety of FAP is also *APC*. The various syndromes are in general distinguished by the location of the mutation, which appears to determine the spectrum of organs in which hyperplasia and tumors arise. For instance, truncating mutations between codons 463 and 1387 (cf. Figure 13.2) are associated with the congenital hypertrophy of the retinal pigment epithelium characteristic of standard FAP, while mutations between codons 1403 and 1578 are often found in families with Gardner syndrome. Interestingly, truncating mutations within the first 150 aa cause attenuated FAP, i.e. a milder phenotype. Nonsense mutations closer to the translational start would be expected to usually cause a complete loss of a protein, while mutations in later exons would be thought to more frequently yield truncated, but stable proteins. It is therefore conceivable that the truncated protein produced by mutations further downstream in the *APC* gene may somehow aggravate the disease, perhaps by interfering with the function of the normal APC protein encoded by the intact allele. Although these mechanistical relationships are still incompletely understood, cataloguing of such genotype-phenotype relationships is important to optimize the treatment and counseling of the affected patients and families.

Since deletions and LOH of chromosome 5q are among the most frequent alterations in colorectal cancer overall, following the identification of *APC* as the gene mutated in FAP, sporadic carcinomas were screened extensively for mutations in the gene. Today, it is assumed that both alleles of the gene are inactivated by point mutation and deletion/recombination, or occasionally promoter hypermethylation, in 70-80% of colon and rectal cancers, irrespectively of whether they are familial or sporadic. Moreover, the frequency of *APC* mutations is almost the same in early and late stage tumors. Thus, APC is a prototypic tumor suppressor. Since its inactivation appears to be almost mandatory for the development of colorectal tumors and most probably takes place at an early stage, the designation of 'gatekeeper' is appropriate (→5.4).

The crucial role of APC is underlined by the analysis of the 20% or so colon cancers, which retain a fully functional APC protein at normal expression levels. Almost all of these contain mutations in other components of the APC/β-Catenin/Axin/GSK3β complex. Most often, mutations in the *CTNNB1* gene encoding β-Catenin are found. More precisely, then, it is a disturbance of WNT signaling (cf. 6.10) that is so crucial for development of colorectal cancers.

There are >23 different WNT proteins in man. They are usually produced and secreted by mesenchymal tissues and act on neighboring epithelial cells in a paracrine manner. Most are agonists, but some may be antagonists. In the canonical pathway (Figure 13.3), WNT factors bind to one of several cell surface receptors named Frizzled (FZD). These receptors belong to the large 'serpentine' class of receptors characterized by seven transmembrane helices and interaction with trimeric G proteins. Activation of a FZD receptor activates the DSH (dishevelled) protein which in turn inhibits GSK3β. In a normal cell, this protein kinase is assembled together with Axin and APC in a cytosolic protein complex. This complex binds β-Catenin. In the absence of a WNT signal β-Catenin is phosphorylated by GSK3β at several sites near its N-terminus. This phosphorylation allows β-Catenin to be recognized by a ubiquitin ligase complex in which βTRCP performing the actual recognition.

Inhibition of GSK3β as a consequence of a WNT signal leads to accumulation of β-Catenin. The protein migrates into the nucleus, where it binds and activates TCF transcription factors, specifically TCF4 in colon cells. There are four known members of the TCF family, TCF1, LEF1, TCF3, and TCF4, which in the inactive

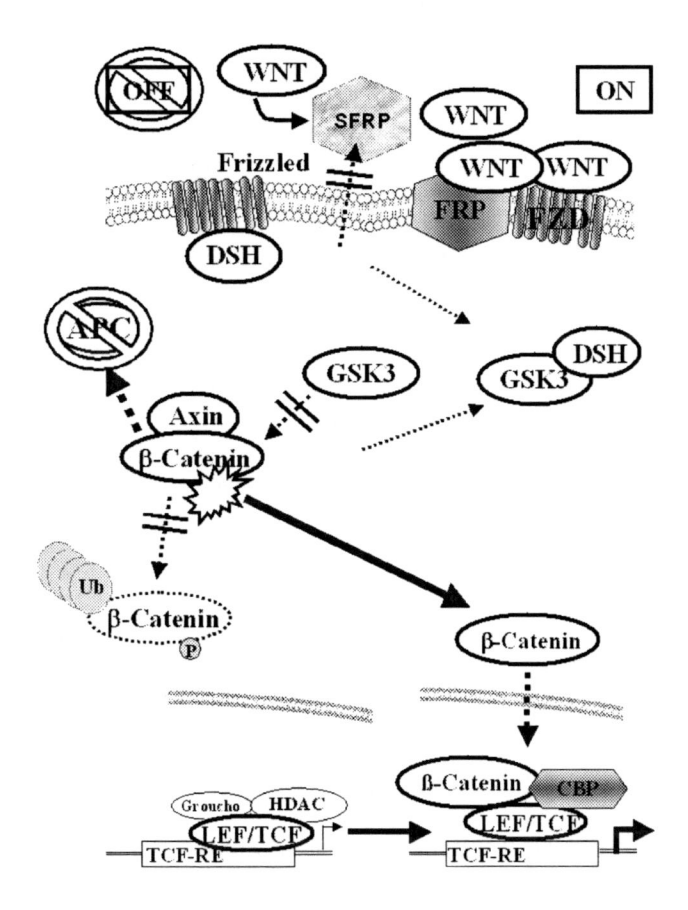

**Figure 13.3** *Alterations of the WNT pathway in colon cancers*
The WNT pathway is most often activated in colon cancers by mutations in APC (prohibition sign) that leads to disaggregation of the complex phosphorylating β-Catenin. In other cases, β-Catenin mutations (exploding star) prevent its inactivation. Constitutive activation of the pathway may often be exacerbated by loss of SFRPs that normally compete with FRPs for co-binding of WNT factors. See also Figure 6.12.

state are complexed with transcriptional repressor proteins, usually from the Groucho family. This repression is alleviated by β-Catenin and transcription of TCF target genes resumes. Among these target genes are several that act directly on cell proliferation and survival, such as *CCDN1* and *MYC*.

Before becoming implicated in WNT signaling, β-Catenin had been known for a long time as a cytoskeletal protein binding to E-Cadherin which mediates homotypic cell interactions (→9.2). Therefore, the concentration of active β-Catenin is also modulated by E-Cadherin. This may allow to integrate information from WNT signals and cell adhesion (→9.2). Several further factors modulate WNT activity, e.g. SFRPs (secreted frizzled-related proteins), which interfere with WNT binding to receptors, and LRPs which support WNT binding to FZD receptors.

This intricate network is fundamentally disturbed in colon cancer (Figure 13.3). Obliteration of APC function, the most frequent alteration, causes a permanent WNT signal in the nucleus, since the regulatory protein complex cannot be assembled and β-Catenin does not become phosphorylated. As a result β-catenin accumulates and causes a constitutive activation of TCF target genes. Mutations in *CTNNB1* alter the amino acids in the recognition sequence of GSK3β prohibiting phosphorylation of β-Catenin, with essentially the same consequence. The rarer deletions of Axin likewise disturb efficient phosphorylation of β-Catenin by removing the platform on which the proteins interact. Inactivating mutations have also been reported for βTRCP impeding the ubiquitin ligase that initiates the proteolytic breakdown of β-Catenin. These central alterations in the WNT pathway may be compounded by alterations in modulating factors. During tumor progression many carcinomas lose E-Cadherin expression which may exacerbate the accumulation of β-Catenin. Likewise, WNT signaling may be further enhanced by down-regulation of SFRPs.

Many other cancer types besides colorectal cancer have meanwhile been investigated for mutations in *APC* and its interacting genes. Two important conclusions can be drawn.

(1) Mutations in components of the WNT signaling pathway are found in many other carcinomas, with varying frequencies, although in none, it appears, with the same regularity. High to moderate prevalences are observed in hepatoma (→16.2), medulloblastoma, and breast cancer, lower frequencies in prostate carcinoma, renal carcinoma and glioblastoma. Of note, some of these cancers are also more frequent in the context of the Gardners and Turcots syndromes. Clearly, however, some cancer types lack mutations in the pathway; although it is difficult to exclude the occurrence of mutations with current techniques. Moreover, one promoter of the *APC* gene tends to hypermethylated in many carcinomas; it is not clear, however, whether this an indication of differential promotor use or overall diminished transcription.

(2) While *APC* is the major target in colorectal cancer, in other cancers mutations in *CTNNB1* predominate. It is not at all clear, why this is so. One possibility is that only colorectal carcinogenesis requires obliteration of other APC functions as well, such as the one presumed in mitosis. Another possibility is that the difference is related to the mechanisms involved in carcinogenesis in different

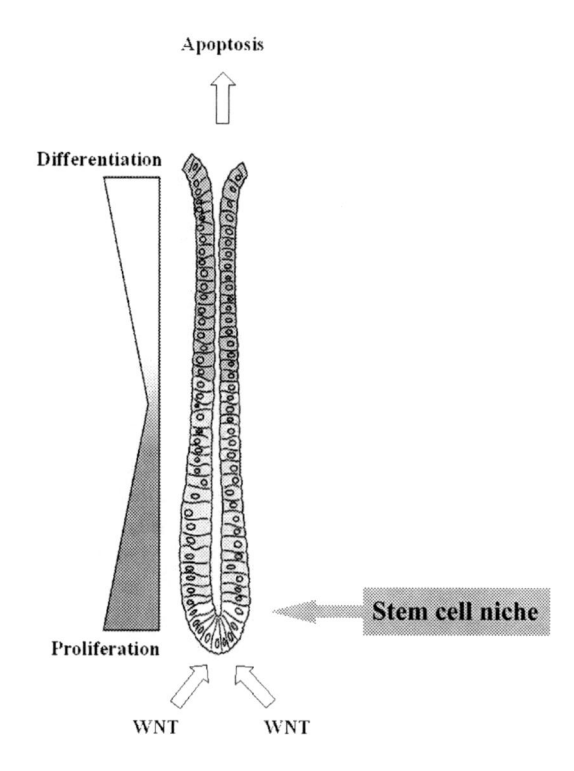

**Figure 13.4** *Organization of proliferation and differentiation in colon crypts*
See text for detailed explanation.

organs. Distinct carcinogens involved may preferentially target certain positions in certain genes.

The requirement for constitutive activation of the WNT pathway in colon cancer may be closely related to the organization of this tissue (Figure 13.4). The colon epithelium is a constantly renewing tissue. The cells of the colon epithelium are derived from a small number – maybe five or so - of tissue stem cells located near the bottom of the crypt in a stem cell niche (→8.6). When these cells divide asymetrically, one daughter cell retains the stem cell character while the other is committed to differentiation. It moves up the crypt, gradually differentiating and concomitantly losing its ability to proliferate. Cells that have arrived at the surface are sloughed off or die by apoptosis.

It appears that the WNT pathway is central to the regulation of this renewal strategy. WNT factors are produced by mesenchymal cells near the stem cell niche allowing reproduction and maintaining immortality of the stem cell population. As cells committed to differentiation move away from the growth factor source, differentiation and apoptosis programs are turned on. Cell-associated signaling proteins called Ephrins and their receptors are involved in establishing the differentiated state and are down-regulated by WNT signaling. Thus, colon epithelial cells not exposed to WNT signals may enter a default state of differentiation or apoptosis. Constitutive WNT signaling caused by APC loss of function or β-Catenin over-activity would prohibit differentiation and apoptosis and establish a stem-cell like state independent of position in the tissue.

A similar effect is thought to be exerted by activation of the Hedgehog pathway in basal cell carcinoma of the skin (→12.3). It is tempting to speculate that the range of cancers in which mutations of the WNT pathway are observed may be related to the proliferation strategy of the respective tissues. There is still too little data on this question, though, for a judgement on this hypothesis.

## 13.3 PROGRESSION OF COLON CANCER AND THE MULTI-STEP MODEL OF TUMORIGENESIS

As the FAP syndrome demonstrates, inactivation of both APC alleles is very likely sufficient for the formation of benign adenomatous polyps in the colon. For the transition to malignancy, i.e., invasion and metastasis, further genetic changes are required. Indeed, advanced colon cancers harbor a multitude of genetic and epigenetic alterations in addition to losses of chromosome 5q and mutations of *APC* or *CTNNB1*. Three of these alterations are consistent and well-characterized. In fact, they often seem to be associated with particular steps in the progression of colon cancer (Figure 13.5). They are (1) activation of KRAS by point mutations, (2) inactivation of TP53 by point mutation and allele deletion, (3) loss of responsiveness to TGFβ signaling, by loss of function of SMAD transcription factors or by mutations affecting the TGFβ receptor subunits.

(1) RAS proteins act as signal transducers in several pathways. They relay signals from mitogenic growth factors via the MAPK cascade to the nucleus, regulate the structure of the cytoskeleton, and act on the PI3K signaling pathways augmenting cell growth and survival (→6.2). In colon cancer, the *KRAS* proto-oncogene is activated by typical point mutations in codons 12, 13, or 61 that impede the activity of the RAS GTPase and prolong the active state during which the protein

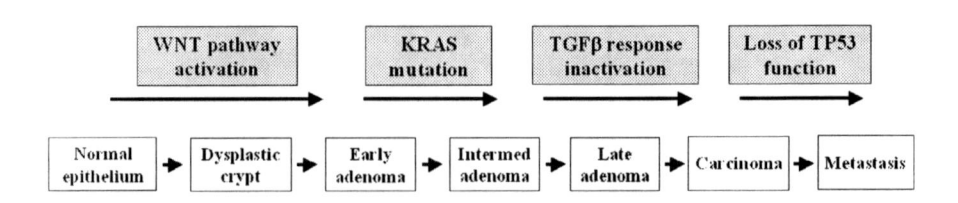

**Figure 13.5** *Histological progression of colon cancer and accumulation of genetic alterations*
See text for further explanation.

relays signals (→4.4). These mutations are found in 30–70% of all colon cancers and are more frequent in carcinomas than in adenomas. It is likely that these KRAS mutations further stimulate the proliferation of colon tumor cells, favoring the progression from adenoma towards carcinoma. Moreover, the effects of RAS signaling on the cytoskeleton and protein synthesis may contribute to the invasive potential of the transformed cells.

(2) Inappropriate activity of genes like *RAS* and *MYC* is counterbalanced in normal cells by induction of cellular senescence or apoptosis (→7). One important mechanism involves induction of p14$^{ARF1}$ by these and other oncogenes (→5.3) which leads to stabilization and activation of TP53. It is therefore not surprising that many advanced colon carcinomas have lost functional TP53. Usually, one copy of the *TP53* gene is inactivated by point mutations, predominantly in the central DNA binding domain of the protein (→5.3), and the second copy is deleted or exchanged by allelic loss, detectable as LOH at 17p. Loss of TP53 function also compromises the cellular response to other types of cellular stress, including the responses to DNA damage, aneuploidy, and nucleotide imbalances (→5.3). In colon cancer, loss of TP53 function appears to be associated with tumor progression. Likely, the loss of TP53 relieves a check on KRAS mutations and facilitates the development of increased genomic instability in advanced colon cancers.

(3) In epithelial cells, the pathway activated by TGFβ (→6.7) decreases cell proliferation, e.g., by inducing CDK inhibitors, and promotes cell-matrix interactions. Accordingly, it is very frequently inactivated in highly invasive human cancers. In addition, secretion of TGFβ by carcinoma cells modulates their interaction with stroma cells and immune cells at the primary tumor site and during metastasis (→9.6). In colon cancer, several genetic alterations may alternatively decrease TGFβ signaling. Truncating mutations of the TGFβRII receptor are frequent in the HNPCC subtype of colon cancer (→13.4). In many advanced stage colon cancers, allelic loss is found at chromosome 18q in the region where the SMAD2 and SMAD4 genes reside. In fact, loss of 18q is one of the most consistent chromosomal changes in advanced colon carcinomas, together with loss of 5q and of 17p. In accord with expectations, point mutations and even deletions of the remaining SMAD2 and SMAD4 alleles have been found. However, their frequency seems rather low and questions remain about their importance in colon cancer.

So, during the progression of colon carcinoma to an invasive, metastatic cancer, a number of genetic alterations accumulate. The minimal requirements may comprise constitutive activation of the WNT pathway, activation of RAS signaling, loss of TP53 function, and inactivation of TGFβ signaling. Each of these changes may require more than one genetic or epigenetic alteration. For instance, WNT pathway activation in a sporadic colon cancer may require a point mutation in *APC* and loss of 5q, with additional changes in SFRPs or other modulators. Thus, a minimum of 5 - 7 genetic changes may be needed, but the actual number is likely higher.

## 13.4 HEREDITARY NONPOLYPOSIS COLON CARCINOMA

In FAP and related hereditary diseases, multiple polyps develop in the colon, from which individual carcinomas emerge. In other families with early onset colon cancer inherited in an autosomal-dominant fashion, no polyposis is observed. Moreover, the spectrum of associated tumors in other organs is distinct from that in the various syndromes resulting from inherited *APC* mutations (Table 13.1). Instead, it includes endometrial and ovarian tumors in women as well as cancers of the stomach, liver, gall bladder and the upper urinary tract in both genders (in this approximate decreasing order). This syndrome was designated hereditary nonpolyposis colon carcinoma (HNPCC) and is now known to be considerably more frequent than FAP and its variant syndromes. Like FAP, its investigation has yielded important insights into the mechanisms of tumorigenesis not only in the colon, but also in other human cancers.

Colon cancers in the context of HNPCC are more often located in upper segments of the colon than in FAP and in sporadic cases and may - on average - have a more favorable prognosis. The major difference, however, lies in the type of mutations found in HNPCC carcinomas. Almost all genetic alterations in HNPCC carcinomas result from point mutations, base exchanges, small deletions and insertions, while chromosomal alterations are comparatively infrequent.

Mutations in this syndrome are most easily detected in microsatellite sequences which consist of repeats of one, two or three nucleotides. Since microsatellites are highly polymorphic, individuals are normally heterozygous for them at a given locus. Microsatellites can therefore be used as allelic markers in linkage studies, but

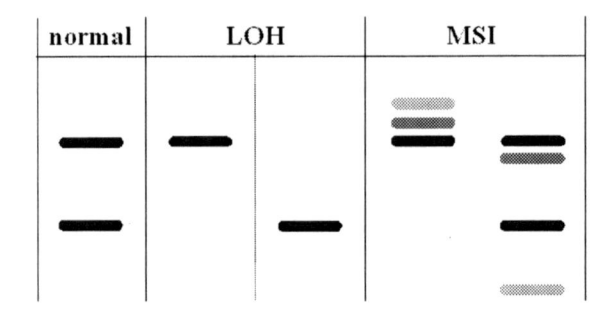

**Figure 13.6** *Microsatellite alterations in cases with LOH or with microsatellite instability* The normal pattern is shown in the leftmost lane. If LOH occurs, one or the other allele is lost, leading to loss of the according band (lanes 2 and 3). If microsatellite instability occurs, larger or smaller bands are found, which do not occur in the normal tissue. Normal bands may be retained (lane 5) or disappear (lane 4).

also to follow chromosomal loss or recombination in tumor cells (Figure 13.6). Typically, allelic loss resulting from deletion or recombination is seen as loss or strongly diminished intensity of one band resulting from PCR amplification of a microsatellite. In HNPCC tumors, additional bands appear in addition to the two present in normal tissue. These result from the expansion or contraction of a microsatellite repeat, corresponding to an insertion or deletion mutation. This phenomenon has been termed 'microsatellite instability' (abbreviated MSI).

Microsatellite instability and increased incidence of point mutations in HNPCC result from inactivation of genes involved in mismatch repair. Mismatches between DNA strands can occur as a consequence of several mechanisms and are repaired by according mechanisms (→3.1). One repair system most active during DNA replication consisting of several different proteins recognizes base mismatches and differences in the number of bases between opposite DNA strands, such as occuring by base misincorporation or 'slipping' of the DNA polymerase. These mismatches are first recognized by protein heterodimers consisting of MSH2 and MSH6 or MSH3 and MSH6, respectively. Further components of the complex, including PMS1, PMS2, and MLH1, are then directed towards the mismatch and remove the mismatched nucleotides (cf. Figure 3.4). A repair DNA polymerase is recruited to synthesize the correct sequence and a DNA ligase seals the corrected strand.

Patients with HNPCC carry mutations in one gene encoding a component of this repair system. Mutation in the *MSH2* and *MLH1* genes are most frequent, more rarely the *PMS1*, *PMS2* or *MSH6* genes are mutated. In rarer cases, the *MYH* or *MBD4* genes (→3.1) are affected, but in ≈30% of all patients showing the characteristics of HNPCC, the underlying mutations are not known. Obviously, they could affect additional, still unknown components of the repair system.

Importantly, the mutation inherited in one allele is not sufficient to confer the MSI phenotype. Rather, mutation or epigenetic inactivation of the remaining allele has to take place before the function of the repair system is seriously compromised. So, the first step in tumor formation is probably the accidental loss of this second allele. In the affected cell, this creates a state of greatly enhanced mutability, specifically a strongly enhanced rate of point mutations. Eventually, these point

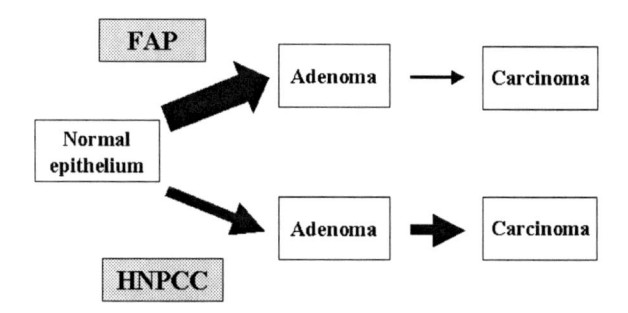

**Figure 13.7** *A comparison of cancer development in FAP and HNPCC*
The width of the arrows reflects the degree of acceleration of the process.

mutations will affect genes crucial for the development of colon cancer. This sequence of events would explain convincingly why only single or a few carcinomas arise in HNPCC and the process does not stop in most cases at the stage of adenomas as in FAP (Figure 13.7).

Interestingly, MSI is also observed in sporadic cancers arising in patients without a family history. These cancers are also defective in mismatch repair. In many such cases, the inactivation of the mismatch repair system is caused by inactivation of the *MLH1* gene through promoter hypermethylation.

Conceptually, one can consider the genes harboring inherited mutations in HNPCC as 'caretaker' tumor suppressors (→5.4), since mutations in these genes do not lead to tumor growth directly, but increase the risk of mutations in genes that control cellular proliferation, differentiation and survival. To some extent, therefore, HNPCC cancers contain mutations in the same genes as other colon carcinomas. For instance, the tumor suppressor genes *APC* and *TP53* are inactivated and the proto-oncogenes *KRAS* and *CTNNB1* are activated by point mutations.

However, some genes are preferential mutation targets in colon cancers arising in the context of HNPCC and of sporadic MSI compared to other colon cancers. A prominent example is the *TGFBRII* gene which contains eleven successive adenines in its coding regions. Typical slippage mutations in this minirepeat leading to frameshift mutations are highly prevalent in HNPCC cancers. Another gene frequently affected by frameshift mutations encodes the pro-apoptotic protein BAX (→7.2).

## 13.5 GENOMIC INSTABILITY IN COLON CARCINOMA

It is estimated that about 15% of all colon carcinomas overall, familial and sporadic together, display an MSI phenotype. Development and progression of these tumors are driven by an increased mutation rate. Importantly, only one particular type of mutations occurs at an increased frequency in these cancers, i.e. point mutations, which would be removed in normal cells by the mismatch repair system.

In contrast, chromosomal aberrations in MSI cancers are less frequent than in many other human carcinomas and some MSI cancers even remain diploid. One (hypothetical) explanation for this observation is that illegitimate recombinations are suppressed in these cancers. Such recombinations often involve homologous, but not identical sequences. Therefore, during recombination, small mismatches between these sequences appear. Unless these are corrected by mismatch repair, recombination may be inefficient.

In contrast, other colon cancers become aneuploid early during progression, losing or gaining whole chromosomes or chromosome parts. These changes result in increased or decreased gene copy numbers as well as chromosomal rearrangements that drive tumor progression, together with a smaller number of point mutations. This type of genetic instability is called chromosomal instability and sometimes abbreviated CIN. It is the predominant type of genetic instability in colon cancer, but also in many carcinomas of other organs. Nevertheless, the mechanisms causing CIN are not as clearly understood as those causing the MSI phenotype. In colon cancer, specifically, loss of APC function may contribute to the CIN phenotype, since the protein is associated with the mitotic spindle and the domains near the APC C-terminus involved in these interactions are often deleted by the prevailing truncating mutations. If correct, this explanation would account for the predominance of mutations in *APC* over those in *CTNNB1* in colon cancer. Intriguingly, a function in the control of mitotic segregation has also been postulated for RB1 (→5.2). Of course, loss of TP53 is at least 'permissive' for aneuploidy.

So, there are at least two types of genomic instability in colon carcinoma and other human cancers, one leading to an increased rate of chromosomal alterations and one leading to an increased rate of point mutations. An important hypothesis states that the multistage development of human tumors requires some sort of increased mutation rate, a 'mutator phenotype', which in these two types of cancer would be provided by an increased rate of point mutations or chromosomal changes, respectively. After all, at least 5 – 7 genetic alterations in one cell line are mandatory for colon cancer development, and this is likely a very conservative estimate (→13.3).

An interesting addition to this story has emerged more recently. Upon studying DNA methylation changes in colon cancer, it was noted that these were much more prevalent in individual colon cancers than in others. So, a subset of colon cancers showed a sort of 'hypermethylator' phenotype. The mechanism causing this phenotype is unknown. Moreover, the pattern of genes affected by mutations in this subset of colon cancers again seemed to be particular. In addition to a high frequency of *KRAS* changes, a high rate of inactivation of the *CDKN2A* gene by promoter hypermethylation was observed. This gene is otherwise not as frequently altered in colon cancers as in other human cancers (→5.3). This type of colon cancers has been designated CIMP+ for CpG-island methylator phenotype. Of note, this subset is not really distinct from MSI and CIN cancers. The reason for this is that hypermethylation can inactivate genes required for mismatch repair or chromosomal stability. Notably, the MLH1 gene is a frequent target of hypermethylation.

The classification of colon cancers according to the mechanisms of genetic – or epigenetic – instability is more than just intellectually appealing. It may identify different subclasses of cancers with different clinical prognoses. Thus, on average MSI cancers may be somewhat less aggressive than CIN cancers. However, the classification appears most meaningful in predicting the reaction of a particular cancer to therapeutic intervention. Thus, MSI and CIN cancers may differ in their response to chemotherapy and CIMP+ cancers would be predicted to be particularly sensitive to inhibitors of DNA methylation (→8.3).

## 13.6 INFLAMMATION AND COLON CANCER

An increased incidence of colorectal cancer is also observed in patients with chronic inflammatory bowel diseases. The increase in risk of colon cancer in patients with colitis ulcerosa has been variously estimated to be 4 – 20-fold increased.

The reasons for the increase in risk in these diseases may be complex. For instance, chronic damage to the mucosa may accelerate the turnover of the epithelial cells. Enhanced proliferation activity of tissue stem cells likely augments their risk of malignant transformation. Facilitated accessibility for mutagens from the intestine could also be relevant. However, inflammation as such is usually considered the main culprit for the increased risk.

Inflammatory cells like activated macrophages and lymphocytes release potentially mutagenic reactive oxygen species and a plethora of cytokines, proteases and other enzymes that modulate local tissue structure, cell proliferation and apoptosis. It was therefore not entirely surprising when epidemiological studies showed that common antiinflammatory drugs like acetyl-salicylic acid (aspirin) diminished the incidence of colorectal cancer. However, the magnitude of the effect suggested, up to 40%, came as a surprise, and it emerged that it was not restricted to patients with chronic bowel inflammation.

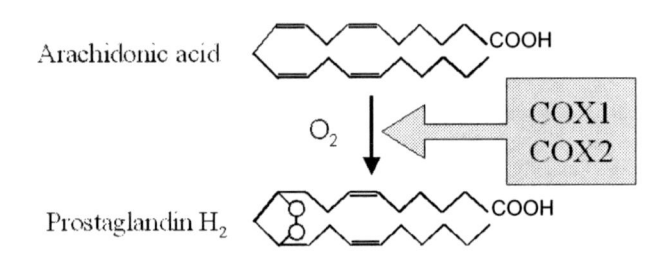

**Figure 13.8** *The main reaction catalyzed by cyclooygenases*

The most important targets of acetyl-salicilitic acid and related drugs, summarized as non-steroidal antiinflammatory drugs (NSAIDs) are cyclooxygenases, enzymes which convert arachidonic acid to prostaglandins (Figure 13.8). These reactions also yields reactive oxygen species as a side product. Prostaglandins act as paracrine signaling molecules on stromal as well as epithelial cells.

In fact, there are two kinds of cyclooxygenases. COX1 is a constitutively active enzyme, whereas COX2 is the enzyme induced by cytokines during inflammation. The promoter of the *COX2* gene is activated, a.o., by JNK and p38 MAPK (→6.2) and by NFκB (→6.9) pathways. COX2 can also be induced by mutated RAS. Drugs targeting specifically the inducible COX2 isozyme have turned out to be most effective. COX2 has been found to be induced in colon cancers, mostly in stromal cells like monocytes, fibroblasts and endothelial cells, but also in some carcinoma cells. Strong expression of COX2 appears to alter cell adhesion and to decrease apoptosis of emerging and established colon carcinoma cells. This effect may be mediated by paracrine effects.

Newer drugs such as sulindac and celecoxib have proven remarkably efficacious in diminishing the incidence of colorectal cancer and to slow down its progression in populations at risk including FAP patients. The action of sulindac is, however, not only due to its inhibition of COX-2, as it induces apoptosis in colon carcinoma cells that lack the enzyme. These relationships are under intense study, which is understandable given their potential for prevention (→20.4)

This line of research is clearly not at an end and may yield even better means of intervention in the near future, although the results from ongoing trials are already very promising. Another interesting notion from studies on anti-inflammatory drugs and colon cancer is that plants may contain compounds that act in a similar fashion to NSAIDs, e.g., the bright yellow spice component curcumin. So the protective effect of vegetables apparent from epidemiological studies (→20.3) might partly be due to their effect.

### Further reading

Kinzler KW, Vogelstein B (1996) Lessons from hereditary colorectal cancer. Cell 87, 159-170

Gupta RA, DuBois R (2001) Colorectal cancer prevention and treatment by inhibition of cyclooxygenase-2. Nat. Rev. Cancer 1, 11-21

Thiagalingam S et al (2001) Mechanisms underlying losses of heterozygosity in human colorectal cancers. PNAS USA 98, 2698-2702

Grady WM, Markowitz SD (2002) Genetic and epigenetic alterations in colon cancer. Annu. Rev. Genomics Hum. Genet. 3, 101-128

Van de Wetering M et al (2002) The β-Catenin/TCF-4 complex imposes a crypt progenitor phenotype on colorectal cancer cells. Cell 111, 241-250

Reddy BS, Rao CV (2002) Novel approaches for colon cancer prevention by cyclooxygenase-2 inhibitors. J. Environ. Pathol. Toxicol. Oncol. 21, 155-164

Giles RH, van Es JH, Clevers H (2003) Caught up in a Wnt storm: Wnt signaling in cancer. BBA 1653, 1-24

Loeb LA, Loeb KR, Anderson JP (2003) Multiple mutations and cancer. PNAS USA 100, 776-781

Lucci-Cordisco E et al (2003) Hereditary nonpolyposis colorectal cancer and related conditions. Am. J. Med. Genet. 122A, 325-334

Lustig B, Behrens J (2003) The Wnt signaling pathway and its role in tumor development. J. Cancer Res. Clin. Oncol. 129, 199-221

Grady WM (2004) Genomic instability and colon cancer. Cancer Metast. Rev. 23, 11-27

Koehne CH, Dubois RN (2004) COX-2 inhibition and colorectal cancer. Semin. Oncol. 31 Suppl. 7, 12-21

Nelson WJ, Nusse R (2004) Convergence of Wnt, beta-catenin, and cadherin pathways. Science 303, 1483-1487

## Box 13.1 Positional cloning of tumor suppressor genes in hereditary cancers

There is now a plethora of methods to identify genes involved in inherited human diseases (1). These are also employed for the identification of human 'cancer genes'. The elucidation of the human genome sequence and its ongoing annotation have further facilitated this endeavour.

However, many genes involved in hereditary human cancers were identified before this sequence was available. The identification of the APC gene as a 'gatekeeper' in colorectal carcinoma is a good example (2).

It was known that ≈40% of colon carcinomas (sporadic or familial) contain losses of chromosome 5q. These were also found at similarly high frequencies in adenomas suggesting they represented an early event. Studies in families showed a genetic linkage between 5q and FAP, i.e. polymorphic markers on 5q (restriction fragment length polymorphisms at the time) co-segregated at statistically significant rates with the appearance of the disease. Together, these findings suggested that the gene responsible for FAP was located at 5q and that it was a tumor suppressor gene, i.e. inactivating mutations in one copy were inherited in the germ-line of affected families. One patient with FAP presented with a cytogenetically detectable alteration in 5q in leukocytes, viz. a deletion in 5q21. This fitted with further detailed linkage studies. In colon cancers from various patients, a consensus region for losses within 5q21 was defined that contained four expressed regions, i.e. genes. One of these, which was then named *APC*, was found to be mutated in normal cells of FAP patients in one allele. Importantly, many of these mutations were nonsense mutations leading to a truncation of the predicted protein (→2.2). This made it very likely that the gene was a tumor suppressor and indeed, inactivation of its second allele was regularly found in tumors arising in FAP patients.

Many of the genes listed in Table 2.2 involved in inherited human cancers were identified by a similar strategy relying on linkage analysis in affected families, studies of chromosomal alterations in the respective tumors narrowing down the gene region, and mutational analysis of candidate genes, aided by clues from key patients (2).

This strategy has worked more or less well in cancers where familial clusterings are caused by mutations in one or a few genes. It encounters difficulties when too many genes are involved (prostate cancer →19.3), familial and sporadic cancers show different alterations (breast cancer →18.3), or familial cases do not occur (bladder cancer →14.2).

(1) Carlson CS, Eberle MA, Kruglyk L, Nickerson DA (2004) Mapping complex disease loci in whole-genome association studies. Nature 429, 446-452
(2) Vogelstein B, Kinzler KW (eds.) The genetic basis of human cancer. McGraw-Hill, 2nd ed., 2002

# CHAPTER 14

# BLADDER CANCER

- 'Bladder cancer' is a somewhat imprecise generic name for carcinomas of the urothelium, a specialized epithelium lining the urinary tract from the renal pelvis into the urethra. A more precise name is therefore urothelial cancer.
- Most bladder cancers retain markers of the characteristic transitional cells of the urothelium and are therefore called transitional cell carcinoma (TCC). One type of TCC forms papillary structures growing predominantly into the lumen of the urinary tract. Papillary TCC does not readily become invasive, but tends to recur. The more aggressive invasive TCC typically develop from the flat, dysplastic carcinoma in situ. A different histological subtype prevalent in countries with endemic schistosomiasis has lost urothelial differentiation and presents as squamous cell carcinoma. This is an example of metaplasia. Due to the diversity in the type and clinical course of bladder cancers, an important clinical problem is choosing the optimal treatment for each individual patient.
- Urothelial cancers are often caused by chemical carcinogens, notably aromatic amines, whereas squamous carcinomas of the bladder typically arise in the context of chronic inflammation. Genetic polymorphisms in genes involved in carcinogen metabolism modulate bladder cancer risk dependent on exposure, whereas no high-risk hereditary syndrome is known that specifically predisposes to this cancer.
- Urothelial cancers tend to develop at multiple sites, at the same time or successively. This is an example of 'field cancerization'. In fact, several factors are responsible, including true oligoclonality, migration of cancer cells within the epithelium, and spreading through the urine.
- Invasive bladder cancers show pronounced chromosomal instability and alterations in several 'cancer pathways'. Almost invariably, two crucial regulatory systems are incapacitated. (1) Cell cycle regulation is disrupted by loss of RB1, loss of $p16^{INK4A}$, or amplification of *CCND1*, augmented by down-regulation of CIP/KIP CDK inhibitor proteins. (2) The control of genomic integrity by TP53 is disturbed through mutations in its gene, loss of $p14^{ARF1}$, or overexpression of HDM2. These changes appear already in carcinomata in situ or high-grade papillary cancers.
- In contrast, well-differentiated papillary TCCs show a limited range of genetic abnormalities, e.g. loss of chromosome 9, RAS mutations, or mutations in the FGFR3 growth factor receptor that lead to its constitutive activation. Distinctive differences towards invasive cancers, in particular FGFR3 mutations, characterize tumors with a low risk of progression.
- The most consistent genetic change throughout all subtypes of bladder cancer is chromosome 9 loss, which is found in more than half of all cases. This loss targets the *CDKN2A* tumor suppressor on 9p encoding the two important

regulator proteins $p16^{INK4A}$ and $p14^{ARF1}$. However, LOH and deletions are also regularly found on chromosome 9q, strongly suggesting that additional tumor suppressor genes are located there. It has proven surprisingly arduous to identify them. This illustrates that the route from a recurrent chromosomal change to the identification of a relevant gene is not always straightforward.

➢ Papillary bladder tumors may grow mainly as a consequence of overactivity of those pathways which stimulate the proliferation of normal urothelial cells during compensatory growth or tissue regeneration. In contrast, invasive bladder cancers have largely uncoupled their growth regulation from extrinsic signals. In addition, the mechanisms maintaining chromosomal stability during cell proliferation are upset in invasive bladder cancers. The differences in clinical behavior of bladder cancers are thus based on differences in their molecular characteristics.

## 14.1 HISTOLOGY AND ETIOLOGY OF BLADDER CANCER

From the renal pelvis through the urinary bladder into the urethra, the urinary tract is lined by a specialized 'transitional' epithelium called 'urothelium' (Figure 14.1), whose structure is in several respects different from that of squamous epithelia in the skin and other organs. The urothelium forms a low permeability barrier that prevents the components of the urine, even water, from seeping back into the body. In a transitional epithelium, cells from several layers retain contact with the basement membrane. This allows them to shift across each other depending on the filling state. The top cellular layer forms the actual barrier and consists of terminally differentiated 'umbrella' cells linked by tight junctions. The low permeability in the urothelium is achieved by a dense protein array in their apical membrane composed of uroplakins. These proteins are specific markers of urothelial differentiation. The

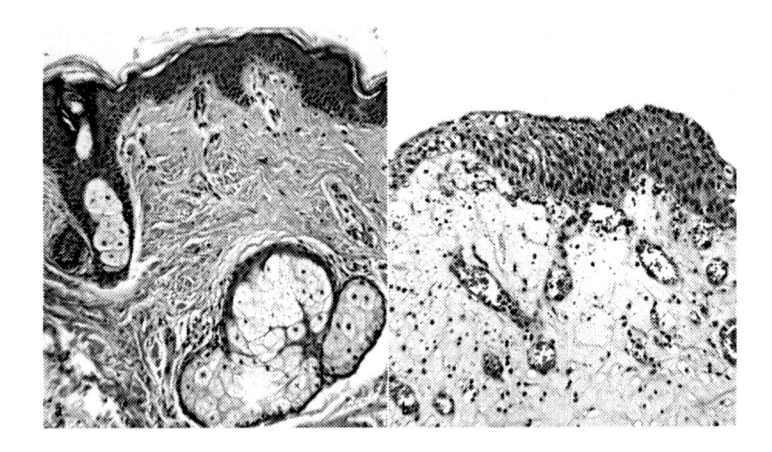

**Figure 14.1** *Histological comparison of a squamous and a transitional epithelium.*
Epidermis (left) and urothelium (right). Note also the sebacous glands in the left figure.

urothelium normally turns over very slowly, but can proliferate rapidly and extensively in response to injury or to bacterial infections to replace damaged areas.

Bladder cancer is a generic term for carcinomas developing from the urothelium. Indeed, most urothelial cancers grow in this part of the urinary tract, but as those in other segments have similar properties, the designation is often loosely used. Most carcinomas arising from urothelium retain morphological and biochemical markers of its original structure. In particular, they express urothelial differentiation markers such as uroplakins and specific cytokeratins (e.g. CK7). These cancers are accordingly categorized as transitional cell carcinomas (TCC).

Bladder cancer is the ≈fifth-most frequent cancer and more prevalent in males. Transitional cell carcinoma represents the predominant histological type in industrialized countries. In countries with endemic schistosomiasis, a second type of bladder cancer, designated squamous cell carcinoma, is more prevalent (Figure 14.2). Although originating as well from cells of the urothelium, this carcinoma consists of – sometimes well-organized - layers of cells that resemble a squamous epithelium. Accordingly, the cancer cells express markers of such epithelia, but not of urothelium, e.g. the cytokeratin CK14 or even involucrin. This is a clear instance of metaplasia, since no squamous epithelium exists in the normal urinary tract upwards of the distal part of the urethra.

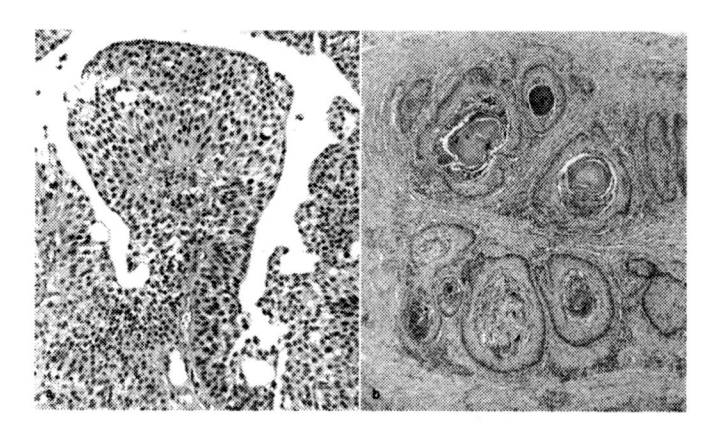

**Figure 14.2** *Histology of transitional cell carcinoma (left) and squamous cell carcinoma (right) arising from the urothelium*

Transitional cell carcinoma can be induced by chemical carcinogens. This was first realized by the surgeon Ludwig Rehn who at the end of the 19th century treated workers making azo-dyes in a chemical plant in a suburb of Frankfurt in Germany. These people were literally drenched in aniline, benzidine, and the dyes made from them, and developed bladder cancers early in life and at a high rate. Today, in spite of much better precautions, occupational risks for bladder cancer remain in the chemical industry and in other branches where workers are exposed to aromatic amines or their metabolizable derivatives.

Other chemicals such as nitrosamines, nitro-aromates, polyaromates, and the cytostatic drug cyclophosphamide, as well as arsenic are also established or very likely bladder carcinogens. A cocktail of carcinogens is inhaled with tobacco smoke and the risk of bladder cancer is consequentially increased approximately 4-fold in smokers. The rate of bladder cancer is also enhanced in Eastern Europeans who have incorporated radioactive cesium from the Chernobyl accident.

The fact that carcinogenicity by aromatic amines shows such a strong organ preference has facilitated its understanding. To become carcinogenic, aromatic amines must be activated by hydroxylation at the amino group (Figure 14.3). This

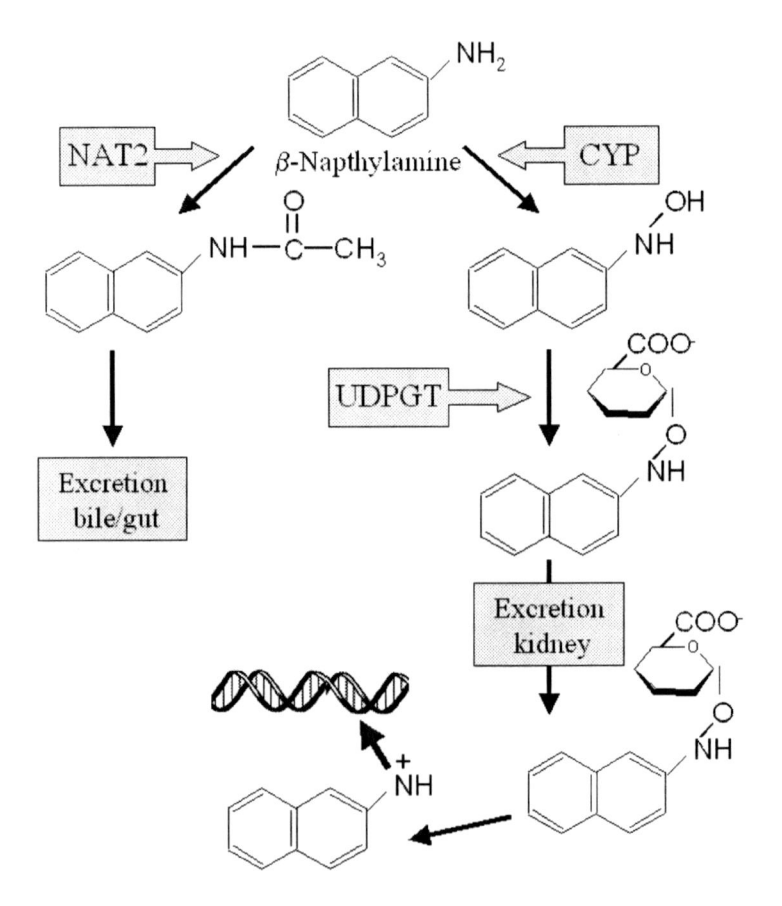

**Figure 14.3** *Metabolic activation of β-naphtylamine to a carcinogen*
This is a simplified scheme that ignores a.o. that acetylated arylamines can still be N-hydroxylated, even though less efficiently. Perhaps more importantly, the role of NAT1 is not illustrated, because it is debated. It is suggested that in kidney and urothelial tissue this enzyme promotes carcinogenesis by acetylating glucuronidated N-OH-arylamines.

reaction is performed by isoenzymes of the P450 monooxygenase family, whose genes are designated *CYP*. Protonation of this N-hydroxyl group leads to dissociation of a water molecule and formation of a highly reactive arenium ion. Hydroxylation at the amino group is prevented by its acetylation. This is catalyzed by N-acetyltransferases, mostly by NAT2 enzymes. Moreover, the efficiency of excretion of hydroxylated amines into the urine vs. the gut depends on the extent of glucuronylation and sulfatation performed by UDP-glucuronyl transferases and sulfotransferase, respectively.

As many of the enzymes involved in the metabolism of arylamines are polymorphic in humans (Table 14.1), the carcinogenicity of aromatic amines in individual humans depends not only on their level of exposure, but also on their genetic constitution. In chemical workers exposed to aromatic amines, *NAT2* appears to constitute the dominant genetic factor. Depending on dozens of combinations of different alleles, humans display two different phenotypes, 'slow' and 'rapid' acetylators, which are also important in the response to a range of medical drugs. Rapid acetylators are much less susceptible to bladder carcinogenesis by aromatic amines, although they may excrete more metabolites into the gut, leading to a somewhat increased risk for colorectal cancer.

Among bladder cancer patients in general, the *NAT2* genotype tends to be a less dominant factor because other carcinogens and endogenous processes contribute. However, the risk of smokers to develop bladder cancer is even greater in those who lack the GSTM1 glutathione transferase, which metabolizes chemical carcinogens related to benzopyrene. This lack is caused by homozygosity for the deletion null-allele of the GSTM1 gene ($\rightarrow$2.3, cf. Table 14.1).

While transitional cell carcinoma is often induced by chemical carcinogens, squamous cell carcinoma typically arises after chronic inflammation of the bladder. This is most obvious in schistostoma-induced bladder cancer. Schistosoma parasites enter the human body from contaminated water and establish themselves in the lung, liver, and urinary bladder. In the bladder, schistosoma mansoni causes a chronic inflammation (bilharziosis) with permanent tissue damage and regeneration. This prepares the ground for the development of squamous carcinoma. This relationship makes bladder cancer one of the most frequent cancers in warmer countries with endemic bilharziosis caused by this species of trematode parasites.

In industrialized countries of the North, $\approx$90% of all bladder cancers display transitional cell carcinoma histology, while most of the rest are squamous cell carcinoma. Transitional cell carcinoma comprises two subtypes with different properties. The most frequent type is a papillary tumor which grows predominantly into the lumen and remains well recognizable as being derived from urothelium. Although they are malignant, only $\approx$20% of these tumors actually progress to an invasive stage and further to metastasis. Most, but not all of the tumors that will eventually become invasive are less than well-differentiated initially. Papillary tumors can usually be removed by local resection, but tend to recur at different localizations in the urothelium, sometimes having progressed to a less differentiated or more invasive state.

The reasons for this behavior are not entirely clear. Bladder cancer is often regarded as an example of 'field cancerization' because multiple tumors arise in different places at the same time or successively, i.e. synchronously or metachronously. This could mean that the entire tissue has been transformed towards a kind of preneoplastic stage, perhaps as a consequence of exposure to carcinogens, chronic irritation or inflammation, or as a consequence of factors released by the actual cancer cells.

Table 14.1. *Polymorphic genes and enzymes in carcinogen metabolism*

| Gene | Polymorphism | Enzyme activity | Substrate |
| --- | --- | --- | --- |
| CYP1A1 | I462V, *Msp*I site in 3'-UTR | monooxygenase (hydroxylase) | polyaromatic hydrocarbons |
| CYP1A2 | regulatory element in intron 1 | monooxygenase (hydroxylase) | arylamines, heterocyclic amines |
| CYP2D6 | >16 alleles including one with gene duplication | monooxygenase (hydroxylase) | debrisoquine (test drug) and wide range of therapeutic and recreational drugs |
| CYP2E1 | flanking region, inducibility | monooxygenase (hydroxylase) | ethanol, others |
| GSTM1 | deletion | glutathione transferase | epoxides from polyaromatic hydrocarbons and others |
| GSTP1 | V104I | glutathione transferase | broad range of electrophilic compounds |
| GSTT1 | deletion | glutathione transferase | short alkenes, especially chlorinated |
| mEH | Y113H | epoxide hydrolase | epoxides from metabolism of aromatic and unsaturated compounds |
| NAT1 | variant polyadenylation site | N-acetyl transferase | arylamines |
| NAT2 | >20 variants in coding region | N-acetyl transferase | arylamines |
| NQO1 | base/aa change with protein instability | quinone reductase | quinones |
| DHEAST (SULT) | D186V | sulfotransferase | hydroxylated compounds including N-oxides |

Molecular comparisons between multiple synchronous and metachronous cancers using microsatellite markers (Figure 14.4) indicate that several factors combine to yield the impression of a field change. (1) Some bladder cancers are indeed oligoclonal, i.e. tumors at different sites or different times have mutually exclusive genetic changes (Figure 14.4) indicating an independent origin, in line with the original idea of a field change. (2) Some cancers have so closely related microsatellite patterns that they must represent descendents from the same clone. In this case, they may have spread by migration within the mucosa or through the lumen. (3) Specifically, some recurrent tumors can be shown to be identical to the original tumor, whereas others are clearly independent. In the first group, resection may have been incomplete or the tumor may have been spread during the surgical procedure.

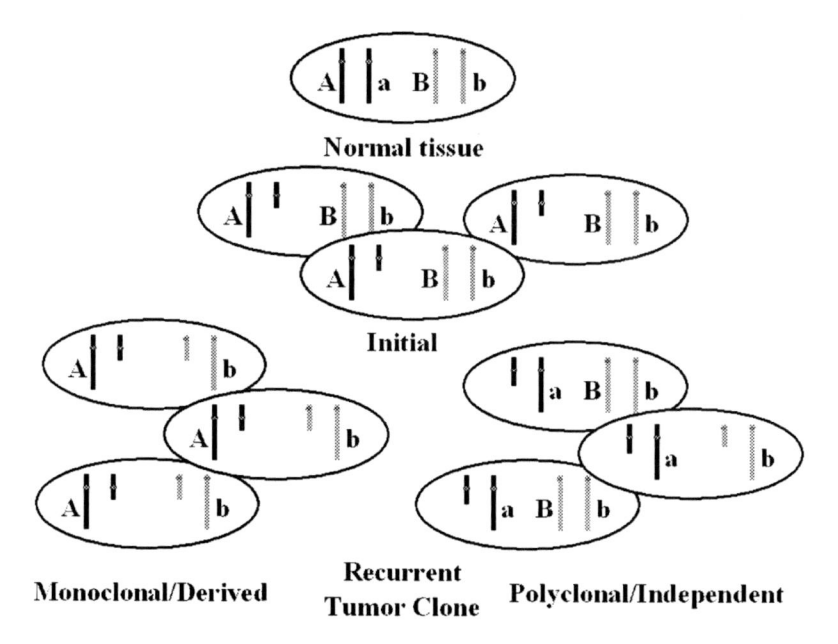

**Figure 14.4** *Use of microsatellite markers to determine the clonality of bladder cancers* cf. also Figure 8.3

Recurrences can be partly prevented by instillation of cytostatic drugs like mitomycin C or of BCG, a tuberculosis vaccine. This consists of inactivated mycobacteria which appear to induce an immune reaction that kills residual tumor cells along with parts of the urothelium. The normal urothelium is then replaced by regeneration. In spite of such 'adjuvant' preventive measures, patients need to be monitored for recurrences for many years after surgery.

A more aggressive form of bladder cancer is found in ≈20% of patients upon first presentation. In these, the tumor grows less extensively into the lumen and more into the deeper layers of the tissue and beyond them, with a propensity to metastasize (cf. Figure 1.6). These invasive cancers likely develop from carcinoma in situ, a flat, severely dysplastic lesion. Dysplasia in this case relates to both the cell morphology, to the tissue structure, and to the nuclei, which are highly polymorphic and aberrant, suggesting a substantial degree of aneuploidy. Invasive bladder cancers need to be treated more radically by removal of the urinary bladder ('cystectomy'), as soon they have progressed into the muscular layers of the tissue. Cystectomy represents a major surgical intervention associated with a low risk of mortality, but a significant rate of complications ('morbidity') and often severe consequences for the quality of life. Still, it is not invariably successful in curing the cancer. In some cases, chemotherapy is applied in addition to surgery (i.e. adjuvantly). In some cases where surgery is unadvisable, it is administered as the only treatment. Overall, cytostatic chemotherapy in bladder cancer is moderately efficacious, and certainly not curative in cases of metastatic disease.

So, compared to colorectal cancer, whose development by and large seems to follow a linear sequence, that of bladder cancer branches out into different varieties with quite different properties. In the clinic, this creates the question of how to treat each patient optimally (Figure 14.5). For cancer research, the challenge is to identify which biological mechanisms underlie the differences between the varieties. For 'translational' research, the tasks are to identify markers for the different varieties and to predict their clinical behavior and response to therapy, but also to find out which molecular targets they present for novel therapies.

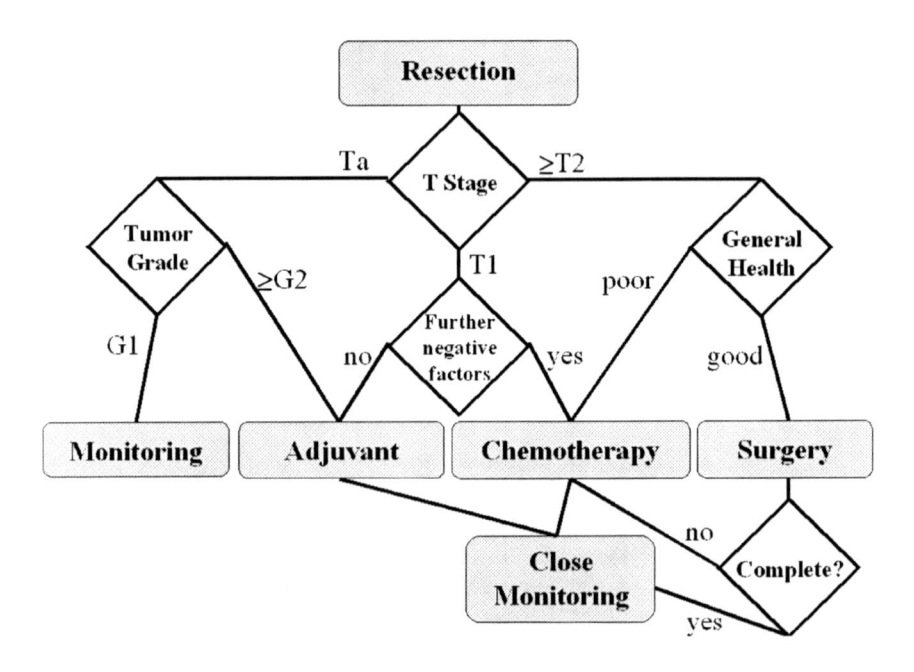

**Figure 14.5** *A tree of therapeutic options in the treatment of bladder cancers*

## 14.2 MOLECULAR ALTERATIONS IN INVASIVE BLADDER CANCERS

All invasive bladder cancers are apparently defective in two important regulatory systems, i.e., the regulation of the cell cycle by RB1 (→6.4) and the control of genomic integrity by TP53 (→6.6). Combined, these two defects appear to allow uncontrolled proliferation, increased genomic instability and growth beyond the limits posed by replicative senescence (→7.4). Indeed, telomerase is generally activated in invasive bladder cancers. This constellation of changes is typical of many advanced human cancers, i.e. the same statement could be made for lung cancer, pancreatic cancer, squamous carcinomas of the head and neck, or glioblastoma as well as to squamous cell carcinoma of the skin (→12.2). Of note, it would not as generally apply to colorectal cancers (→13.3). In the various cancer types, different mechanisms leading to inactivation of the two regulatory systems contribute to various extents (Figure 14.6).

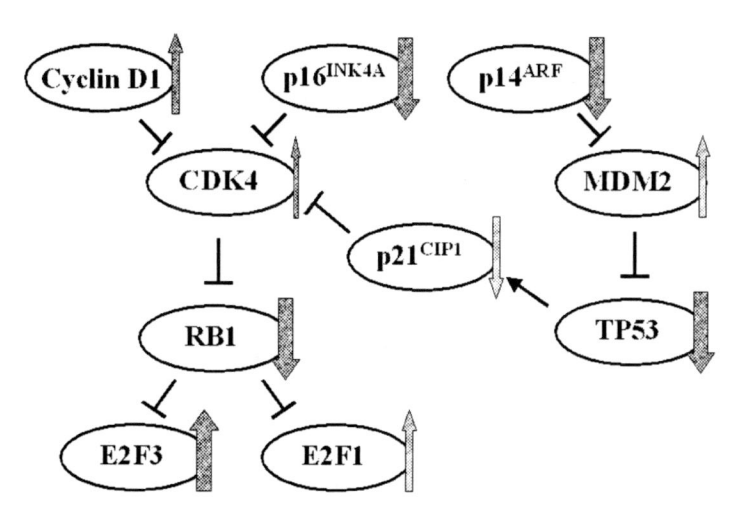

**Figure 14.6** *Changes in the RB1 and TP53 networks in bladder cancer*
The width of the arrows distinguishes frequent and infrequent changes. ↑: activated by overexpression or mutation; ↓: inactivated by deletion, mutation, and/or hypermethylation.

In advanced bladder cancers, inactivation of the RB1 pathway is achieved by loss of RB1 or of p16$^{INK4A}$, less frequently by constitutive overexpression of Cyclin D1 due to *CCND1* gene amplification, and in rare cases by overexpression of CDK4 due to amplification of its gene. Of these changes, loss of RB1 itself, usually by loss of one allele and deletions or truncating mutations in the other one, appears to implicate the most severe consequences. Bladder cancers with loss of RB1 protein, as detected by immunohistochemistry, take the most aggressive course. Moreover, the CDK inhibitors p21$^{CIP1}$, p27$^{Kip1}$, and p57$^{KIP2}$, which are all expressed in normal proliferating urothelial cells and in many superficial cancers, tend to disappear in advanced bladder cancers. Their down-regulation likely exacerbates the defects in cell cycle regulation. The mechanisms underlying these disappearances are poorly understood, but are probably to a large degree epigenetic.

Loss of TP53 function in bladder cancers is most frequently caused by point mutations in its gene and loss of the second functional allele by recombination or deletion affecting chromosome 17p. The TP53 point mutations are spread out through the central part of the gene and do not present an obvious clue to the carcinogens involved, although typical 'tobacco' mutations may be somewhat over-represented. Overexpression of MDM2/HDM2 is observed in some cases. More frequently, $p14^{ARF1}$ is lost. Like loss of $p16^{INK4A}$, this occurs most often by homozygous deletions encompassing the *CDKN2A* locus. It is technically difficult to ascertain the true frequency of homozygous deletions in tumor tissues, but a reasonable estimate is 40% deletions of both *CDKN2A* alleles in advanced bladder cancers, in addition to smaller fractions of cases with missense mutations or promoter hypermethylation.

Interestingly, loss of $p14^{ARF}$ and TP53 mutations are not mutually exclusive, but loss of $p16^{INK4A}$ and RB1 are. This has been observed in various human cancers, although the reasons are not understood. It is clear, however, that loss of RB1 causes upregulation of $p16^{INK4A}$ expression, because RB1 represses its promoter. So, increased expression of the cell cycle inhibitor $p16^{INK4A}$ in human cancers, as detectable by immunohistochemistry, is often a bad sign, since it reflects the (usually) more severe inactivation of RB1. Equally paradoxically, higher levels of TP53 protein in a cancer often indicate its inactivity, since only accumulated mutant protein can be detected by standard immunohistochemistry (→5.3). Accordingly, bladder cancers with immunohistochemically detectable TP53 appear to take a more aggressive course than the average. Clearly, the group with the worst prognosis is that with TP53 mutated as well as RB1 lost.

A notable feature of bladder cancer distinguishing it from colorectal cancer (→13.2) is the virtual absence of mutations activating the WNT/β-Catenin pathway. APC expression and function seem unaffected and mutations activating β-Catenin are very rare. Of course, target genes of the canonical WNT pathway such as *CNND1*, *MYC*, and *MMPs* are also induced in bladder cancers, but obviously by other means. There is, likewise, little evidence for an important role of the SHH pathway that is so crucial in basal cell carcinoma of the skin (→12.3), although the *PTCH1* gene is often hemizygous (cf. Box 14.1). Instead, the genetic alterations in bladder cancer resemble those in squamous cell carcinoma of the skin (→12.2).

There are two, not mutually exclusive hypotheses to account for these differences. (1) Colorectal cancer and basal cell carcinoma appear to arise from tissue precursor cells by the constitutive activation of the respective pathways that maintain their stem cell character. Bladder cancers may originate from more differentiated cells in which these pathways are not active anymore. This idea is, of course, compatible with the expression of markers of advanced urothelial differentiation such as uroplakins or the cytokeratins CK7 and CK20 in transitional cell carcinomas. Squamous cell carcinoma metaplasia remains enigmatic in this hypothesis. (2) The urothelium as a tissue is organized in a very different fashion from the gut and skin. While these tissues turn over continuously, urothelium proliferates significantly only in response to injury. The organization of proliferation and differentiation in urothelial tissue may therefore involve regulatory systems other than the WNT/β-Catenin and SHH pathways.

Nevertheless, loss of cell cycle regulation and of TP53 function are certainly not the only genetic alterations in invasive bladder cancers. Most invasive bladder cancers are highly aneuploid and almost every chromosome is subject to numerical or structural aberrations in one or the other case. In the nomenclature developed for colon cancers, they would be assigned to the CIN class (→13.5). Dysfunction of cell cycle checkpoints as a consequence of RB1 and TP53 inactivation is certainly one factor favoring this chaotic behavior. Additional, more specific defects in the maintenance of chromosomal stability may be involved, but as in many other carcinomas, they are not clearly defined.

In bladder cancer, however, it is clear that genomic instability arises very early in a subfraction of tumors, since many changes are already detectable in carcinoma in situ and in early invasive stages. In contrast, many papillary cancers present with a limited number of chromosomal alterations (→14.3). Compared to colon carcinoma, microsatellite instability is very rarely found in bladder cancers. Urothelial cancers do occur in HNPCC patients at increased rates, but somewhat mysteriously only in the upper urinary tract, i.e. in the renal pelvis and the ureter. There is no good explanation for this specificity, the more so, as genetic alterations in general are similar between cancers from different regions of the urothelium. Perhaps, the concentration of carcinogens or the length of exposure is higher in the renal pelvis and ureter.

As a consequence of widespread genomic instability in bladder cancer, it is difficult to define those alterations that are crucial for tumor development, as opposed to those that have occurred as a consequence of genomic instability and are propagated as 'passenger alterations' in a successfully expanding tumor cell clone. One approach to this problem is to focus on those chromosomal alterations that recur in many different cases or characterize specific subsets of cancers. For instance, loss of 13q and 17p is frequent in advanced cancers, and is likely related to the inactivation of RB1 and TP53.

Further recurrent chromosomal losses during progression concern 8p, 3p, 16q, and 11p, each in a subset of cases. One can presently only guess at which genes might be affected by these losses. *CDH1* encoding E-Cadherin on 16q and *CDKN1C* encoding p57$^{KIP2}$ on 11p are good candidates. Conversely, frequent gains of chromosomes 7p and 8q may relate to *ERBB1* encoding the EGFR and *MYC*, respectively, since in rarer cases more selective amplifications of 7p12 and 8q24.1 are observed. The consequences of the often concomitant gains or amplifications at 5p, 6p, and 20q are again enigmatic.

There is a practical and conceptual problem involved here in bridging the gap between chromosomal alterations and identifying the crucial genes activated or inactivated by them. This problem is alleviated, if an inherited high-risk syndrome points the way and positional cloning (→Box 13.1) can be used to limit the range of candidate genes. These can then be screened for inactivation mutations (in the case of a tumor suppressor) or activating mutations (in the case of an oncogene) in normal cells of patients from affected families.

In cancers where no such syndrome is available, the task is more difficult, and further exacerbated by genomic instability. In cancers with pronounced genomic instability like invasive bladder cancers, the first step involves sorting relevant from passenger alterations. This already involves a conceptual difficulty, because there may be a considerable 'grey zone' of alterations that affect properties of the cancer, although they might not be essential for its growth (cf. 4.4 and 5.4). However, the main problems arise from the fact that genetic (and epigenetic) alterations in cancers are not only functionally relevant and selected by their functional impact on the expansion of a tumor cell clone, but are also determined and restricted by the mechanisms causing them, especially by the type of genomic instability present in a cancer. Two examples from invasive bladder cancers can serve to illustrate this quite complex argument.

(1) The most frequent amplification in invasive bladder cancers concerns a region at chromosome 6p22. Gains (→2.2) of 6p are observed in a larger fraction of advanced tumors, and ≈25% display true amplifications of a smaller segment of 6p22. The segment that is amplified is still variable, typically comprising several genes (Figure 14.7). One of them encodes the transcriptional activator E2F3, which can activate genes during S-phase and is regulated by RB1 (→5.2). Therefore, its amplification could lead to overexpression and loss of cell cycle control by RB1. However, the amplified region in some cases also encompasses the gene *SOX4* which encodes a non-histone chromatin protein known to regulate cell fate and differentiation, or *DEK* encoding a protein kinase regulating chromatin structure and involved as a partner in oncogenic fusions in hematological cancers, as well as additional genes. So, which of these genes is responsible, or maybe several?

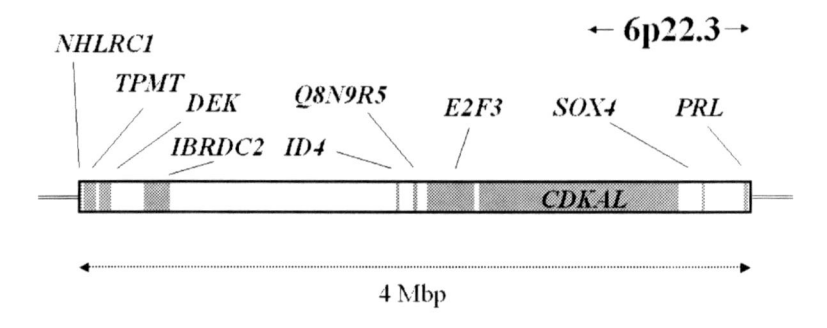

**Figure 14.7** *The 6p22 amplicon in bladder cancer*

Genes are indicated in grey, approximately to scale. Different segments of this region are amplified to various degrees in a substantial fraction of advanced stage bladder cancers. Moreover, chromosome 6p is often gained as a whole.

Similar questions have been encountered in other cases of amplifications in other cancers (→4.3), e.g. the 12q13 amplicon encompassing *HDM2/CDK4/GLI1* and the 11q13 amplicon comprising *CCND1/GSTP1/FGF1*. The reason for the difficulty stems from the mechanism underlying amplifications. They are likely generated as a consequence of chromosomal breaks that do not arise completely at random. Following an initial break, a segment of the DNA is replicated, which must have a minimal size. The maximum size may, in fact, also be influenced by structural features; it is possible that the size of an amplicon corresponds to some sort of DNA replication unit. Typically, amplicons are several Mbp in size. Since in the human genome, there is on average one gene per 100 kb, this may correspond to several dozen genes. So, in a human cancer, amplification of a single gene earmarking it as an oncogene is the exception rather than the rule.

(2) In bladder cancer, quite independently of its subtype, LOH at chromosome 9 occurs in >50% of all cases. Usually, it reflects chromosome deletions. Doubtless, these changes contribute to inactivation of *CDKN2A* located at 9p21. However, the long arm of chromosome 9 is no less frequently affected, suggesting strongly that at least one further important tumor suppressor gene resides there. However, it has not been identified to date. Some candidate genes are discussed in Box 14.1.

Identification of a 'classical' tumor suppressor presupposes consistent inactivation of both alleles of a gene. Since no familial syndrome is known for bladder cancer, changes in cancer tissues must be investigated. In many bladder cancers, LOH at 9q is caused by loss of one entire chromosome 9, i.e. monosomy, or of the whole arm. These cases are not helpful. Even limited LOH at 9q often stretches across a larger sequence, varying between different cancers (Figure 14.8). Ideally, a tumor suppressor gene would be located in the smallest region of overlap. However, since most regions of LOH are large, the actual definition of the region of overlap relies on the exceptional cases with small deletions, which could well be untypical. This type of strategy could lead to those sites where chromosome breaks occur most often rather than to those where the functionally most important gene is located, particularly in cancers with pronounced chromosomal instability.

The success of the strategy therefore depends crucially on finding alterations that activate the second allele of the presumed tumor suppressor. This means tedious searches for point mutations or promoter hypermethylation. Again, however, not all genes inactivated by promoter hypermethylation in a cancer may be functionally relevant (→8.3).

Chromosome 9q in bladder cancer is certainly an especially obstinate case, but similar problems have dogged the identification of tumor suppressors in many human cancers. It is therefore advisable to regard the identification of 'novel' tumor suppressors with due caution.

302                                   CHAPTER 14

Finally, it is quite conceivable, that the two-hit model for tumor suppressor function does not apply for each case (→5.4). Indeed, for chromosome 9q in bladder cancer, 'haploinsufficiency' is discussed, i.e. that deletion of one allele of a gene suffices to promote tumor development. Another possibility is that the predominance of larger regions of LOH may have a functional reason. Perhaps, these larger deletions or recombinations target two or more genes at once. This would then be a case of 'tumor suppressor cooperativity'.

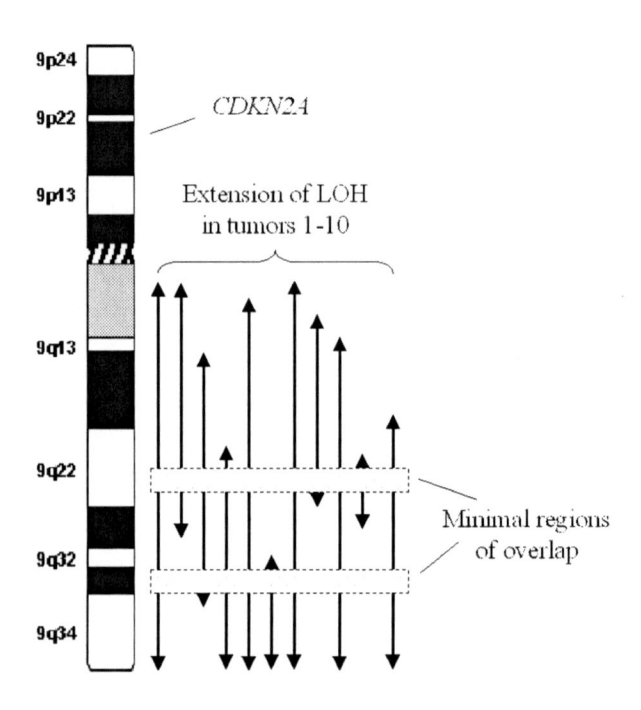

**Figure 14.8** *LOH at 9q in bladder cancer*
An illustration of typical results of LOH analysis at 9q in a series of bladder cancers. In most of these cases LOH would also be found at 9p and encompass the *CDKN2A* locus.

## 14.3 MOLECULAR ALTERATIONS IN PAPILLARY BLADDER CANCERS

In contrast to invasive bladder cancers, papillary transitional cell carcinomas contain a limited number of chromosomal aberrations. This is particularly true for well-differentiated tumors which are usually not invasive, although clearly hyperplastic and with diminished terminal differentiaton. On average, the number of genetic changes in this tumor type increases with tumor grade (→1.4).

Independent of tumor grade, however, loss of chromosome 9 is very frequent. As a result, $p16^{INK4A}$ and $p14^{ARF}$ may not be functional, and the enigmatic tumor suppressor gene at 9q (→14.2) could be affected. Expression of Cyclin D1 as well as MYC is usually increased. These are more likely indicators of increased cell cycle activity than its cause. In a small percentage of cases, mutations activating HRAS are detected, which could be responsible for the evident hyperproliferation. TP53 mutations are much rarer than in invasive bladder cancers and are found predominantly in high-grade tumors on the brink of becoming invasive.

Normally, urothelium is a quiescent tissue with a very low turnover. Following tissue damage, regeneration is stimulated by growth factors produced by the underlying mesenchyme, e.g. FGFs. Urothelial cells also produce autocrine factors, predominantly heparin-binding epidermal growth factor and related peptides of the EGF family (→18.4). These act in normal urothelium through the EGFR which is mainly expressed in basal cells. Members of the FGF family may also regulate the thickness of the epithelial layer during development of the tissue.

Many urothelial cancers overexpress the EGFR, e.g. as a consequence of chromosome 7p gain. The strength of EGFR overexpression rather closely parallels the rate of proliferation detected by immunohistochemical markers such as the DNA replisome subunit PCNA or the protein Ki67, which is thought to be needed for increased nucleolar activity in proliferating cells. In bladder cancers, the EGFR is rarely activated by mutation.

In contrast, ≈60% of papillary urothelial cancers contain missense mutations in the FGFR3 receptor leading to its constitutive activity. The mutations in FGFR3 occur at very specific sites (Figure 14.9) and appear to prolong the half-life of the receptor and of its activated state. The mutations in the FGFR3 extracellular ligand-

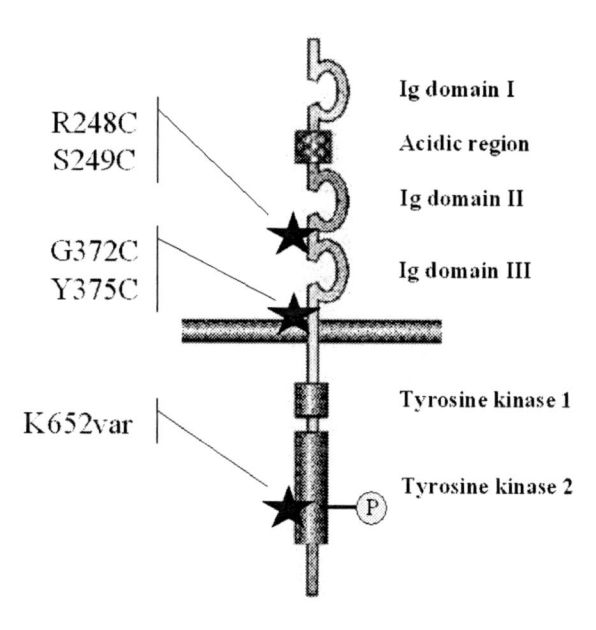

**Figure 14.9** *Mutations of FGFR3 frequent in papillary bladder cancers*

binding domain may also alter its specificity for the >20 members of the FGF family. Mutations in FGFR3 are otherwise only found in cervical cancers. However, these and further mutations have also been encountered in hereditary achondroplasia. In this syndrome, specific point mutations in the receptor lead to shortened bones, particularly in the thighs and upper arms, because over-activity of FGFR3 causes premature differentiation of cartilage tissue in the growth zones. The same mutations in bladder cancers increase proliferation rather than differentiation of urothelial cells.

Surprisingly, bladder cancers with FGFR3 mutations have a distinctly lower risk of progression than those without. Thus, these mutations can serve as molecular markers for cancers that can be treated conservatively, i.e. solely by resection and by monitoring (cf. Figure 14.5). In fact, the discrimination can be improved by using molecular markers associated with the opposite behavior. Mutations of TP53, which in bladder cancer can relatively reliably be detected by accumulation of mutant protein, or the proliferation index as determined by PCNA staining have been proposed.

## 14.4 A COMPARISON OF BLADDER CANCER SUBTYPES

So, what distinguishes superficial papillary from advanced invasive bladder cancer? For one, genomic instability. Once it has set in, tumors may continuously generate variant clones of which some acquire the potential for invasion and metastasis (→Fig. 1.5). Indeed, high-grade papillary tumors, particularly those with TP53 mutations, do accumulate multiple chromosomal aberrations and have a greater potential for invasion.

A second difference may lie in the mechanisms driving tumor cell proliferation. Many papillary tumors seem to grow by mechanisms very similar to those acting during development and regeneration of normal urothelial tissues. These appear to be predominantly stimulated by EGF-like peptides and members of the FGF family. Further factors may aid to induce differentiation. These same mechanisms appear to be over-active in papillary cancers. This mode of tumor growth, however, appears to be self-limiting or at least retain checks that limit growth to the epithelial layer. It is possible that limitations on proliferation imposed by TP53 and RB1 (→6.4, →6.6) are only partly compromised in these cancers, i.e. by loss of function of $p16^{INK4A}$ and $p14^{ARF}$, but not of TP53 and RB1 themselves.

These checks are obviously lost in invasive tumors in which growth seems to occur more independently of exogenous signals and even of endogenous signaling from the cell membrane. This is at least partly a consequence of abolition of cell cycle checkpoints by loss of RB1 and perhaps TP53. The open questions are, whether this abolition is sufficient to drive tumor progression and which mechanisms prohibit invasion in papillary tumors that proliferate as a consequence of growth factor receptor activation. Constitutive activation of the PI3K pathway (→6.3) and decreased responsiveness to the TGFβ (→6. 7) pathway are good candidates for these differences. Indeed, they appear to correlate with tumor progression in bladder cancer as they do in many other carcinomas.

Finally, it is not clear at all which changes direct urothelial cells to transdifferentiate during formation of squamous cell carcinoma. So far, no qualitative differences in genetic alterations have been found that might distinguish TCC and SCC.

In spite of these open questions, it appears that a molecular classification of bladder cancers allowing improved prognosis and treatment selection for many patients (Figure 14.5) may have come within reach.

### Further reading

Ebele JN et al (eds.) Pathology and Genetics of Tumours of the Urinary System and Male Genital Organs. IARC Press, 2004

Cordon-Cardo C, Cote RJ, Sauter G (2000) Genetic and molecular markers of urothelial premalignancy and malignancy. Scand. J. Urol. Nephrol. Suppl. 205, 82-93

Gonzalgo ML, Schoenberg MP, Rodriguez R (2000) Biological pathways to bladder carcinogenesis. Semin. Urol. Oncol. 18, 256-263

Brandau S, Bohle A (2001) Bladder cancer. I. Molecular and genetic basis of carcinogenesis. Eur. Urol. 39, 491-497

Knowles MA (2001) What we could do now: molecular pathology of bladder cancer. Mol Pathol. 54, 215-221

Hein DW (2002) Molecular genetics and function of NAT1 and NAT2: role in aromatic amine metabolism and carcinogenesis. Mutat. Res. 506, 65-77

Liu MC, Gelmann EP (2002) P53 gene mutations: case study of a clinical marker for solid tumors. Semin. Oncol. 29, 246-257

Bakkar AA et al (2003) FGFR3 and TP53 gene mutations define two distinct pathways in urothelial cell carcinoma of the bladder. Cancer Res. 63, 8108-8112

Grimm MO, Burchardt M, Schulz WA (2003) Perspektiven der molekularen Diagnostik am Beispiel des Harnblasenkarzinoms. Urologe 42, 650-659

Quek ML et al (2003) Molecular prognostication in bladder cancer: a current perspective. Eur. J. Cancer 39, 1501-1510

Thier R et al (2003) Markers of genetic susceptibility in human environmental hygiene and toxicology: the role of selected CYP, NAT and GST genes. Int. J. Hyg. Environ. Health 206, 149-171

van Rhijn BW et al (2004) FGFR3 and P53 characterize alternative genetic pathways in the pathogenesis of urothelial cell carcinoma. Cancer Res. 64, 1911-1914

## Box 14.1: Tumor suppressor candidates at 9q in bladder cancer

There are several genes on chromosome 9q that make good tumor suppressor candidates, because they are located in regions most consistently affected by LOH, have already been identified as tumor suppressors in other cancers, and/or are subject to genetic and epigenetic alterations of their second allele in some bladder cancers (TCC). They include the following:

➢ *TGFBRI* is located at 9q22 in the region most frequently affected by LOH in TCC. It encodes a subunit of the TGFβ receptor whose activation in epithelial cells causes inhibition of cell proliferation. Accordingly, loss of responsiveness to TGFβ is a typical change in carcinomas (→6.7). No mutations are, however, found in the remaining *TGFBRI* allele in TCC. Nevertheless, down-regulation of the protein has been observed, but not consistently.

➢ Inactivation of *PTCH1* at 9p22.3 is a crucial step in the development of basal cell carcinoma of the skin (→12.3) and a few other cancers. PTCH1 is an established tumor suppressor. It limits the activity of the SHH pathway (→6.10) which when overactive may confer a precursor phenotype to cancer cells. However, TCC do not exhibit a precursor cell phenotype, but express differentiated markers of the urothelium. Indeed, while one allele of *PTCH1* is often lost, the remaining one appears to be intact and expressed.

➢ *DBCCR1* ('deleted in bladder cancer candidate region 1') was isolated from an LOH consensus region at 9q33.2 (hence its name). Individual TCC even show homozygous deletions in this region. No mutations were found, but *DBCCR1* is often more strongly methylated in TCC, although apparently not completely silenced (→8.3). Its function is essentially unknown.

➢ The TSC proteins limit the effect of PI3K activation on cell growth (→6.3). Inherited mutations in *TSC1* or *TSC2* underlie the tumor syndrome tuberous sclerosis, in which TCC, however, are not particularly prevalent. Still, *TSC1* located at 9q34.1 is frequently affected by loss of one allele, and ≈10% of all TCC harbor mutations in the gene. However, LOH at 9q is found in >50%.

The list does not stop there. The crucial question in a case like this is whether one should focus on likely candidates or also analyze seemingly unlikely candidate genes for mutations or epigenetic inactivation. The dilemma is perhaps most dramatically illustrated by a cancer syndrome named 'hereditary leiomyoma with renal cell carcinoma' (→15.3 and Table 15.3). In this familial syndrome, the gene encoding fumarate hydratase, an enzyme in the citric acid cycle, was unequivocally identified as a 'classical' tumor suppressor. If one had investigated sporadic cancers, without the guidance of a linkage analysis (→Box 13.1), this gene would almost certainly have been at the bottom of the list of candidates.

Knowles MA (2001) What we could do now: molecular pathology of bladder cancer. Mol. Pathol. 54, 215-221

# CHAPTER 15

# RENAL CELL CARCINOMA

➢ Several cancer types originate from the adult kidney. In addition, the organ is a site for metastases from other tumors. Within the kidney, tumors can be derived from different cell types. Tumors of mesenchymal origin can be benign or malignant, e.g. angiomyolipoma vs. sarcoma,. In the renal pelvis, urothelial carcinomas resemble those in the bladder. Actual renal cell carcinomas (RCC) are derived from various segments of the nephron. They are distinguished by their histological appearance and accordingly designated chromophobic, chromophilic, clear-cell, papillary etc. The clear-cell variety is most common (hence also: 'conventional' RCC). There is also a benign epithelial tumor, oncocytoma.

➢ Molecular and cytogenetic studies have shown that different histological subtypes of RCC carry typical chromosomal alterations. This recognition has helped to sort out ambiguous cases. Thus, modern classification of renal cell carcinoma is based on a combination of histological and molecular parameters.

➢ Several hereditary syndromes confer an increased risk for renal carcinoma, prominently von-Hippel-Lindau (VHL) disease and hereditary papillary renal cell carcinoma (HPRCC). Genes affected in these syndromes are in general tumor suppressors, but the mutation underlying standard hereditary papillary carcinoma activates the *MET* proto-oncogene. This is one of a few exceptional instances, in which a hereditary tumor syndrome in man is caused by a dominant oncogene mutation.

➢ The *VHL* gene is not only involved in most inherited cases of clear-cell carcinoma, but also in sporadic RCC of this subtype. It is therefore a 'classical' tumor suppressor gene. While inactivation of *VHL* alleles can occur through various mechanisms, including deletion, point mutation and promoter hypermethylation, as a rule one allele is deleted by loss of chromosome 3p. This loss is characteristic for clear cell RCC. Its consistency may be facilitated by the presence of a major site of chromosomal instability (*FRA3B*) at 3p14.1, but likely also indicates the presence of a second tumor suppressor gene on this chromosome arm.

➢ The elucidation of *VHL* function has yielded important insights into tumor biology. The VHL protein is a pleiotropic regulator of cell physiology and growth. Most importantly, it is involved in the cellular response to hypoxia. Loss of VHL function causes a switch in cellular physiology and gene expression which normally occurs transiently during hypoxia to become constitutive. This explains many of the morphological and biological features of the clear cell carcinoma phenotype, including a metabolic switch to glycolysis and prominent angiogenesis. VHL is the target recognition module of a E3 ubiquitin-ligase protein complex, which becomes disfunctional in renal cell carcinoma. This

finding emphasizes how crucial disturbances of protein degradation are for the development of human tumors.
- ➤ RCC remains a challenge for therapy. While localized tumors can be cured by surgery, chemotherapy for metastasized cancers has proved woefully inefficacious. Cytokines like interferon α and interleukin 2 induce remissions in some cases and, strikingly, spontaneous remissions have been documented in individual cases. Since these are thought to be caused by an immune response against the tumor, RCC is a major target of experimental immunotherapy.

## 15.1 THE DIVERSITY OF RENAL CANCERS

Cancers of the kidney account for a few percent of all human malignant tumors. There is a large variety of tumors in this organ, which can serve well to illustrate the diversity of human cancers (Table 15.1).

Most cancers in the kidney are carcinomas, which have traditionally been distinguished by their histological appearance (Figure 15.1). The incidence of renal cell carcinomas (RCC) in industrialized countries is ≈10:100,000/year. Most are thought to originate from different parts of the nephron. Clear-cell carcinomas, the most frequent subtype, are thought to originate from proximal tubuli, which are located in the renal cortex. Because it is the most common subtype (70 – 80% of all cases), it is also called 'conventional renal cell carcinoma'. Papillary cancers also arise from the same region, perhaps from nephrogenic rests persisting from fetal development. In this regard they resemble Wilms tumors (→11), but papillary renal carcinomas are derived from more mature cells at a later stage of development.

Table 15.1. *Some benign and malignant tumors found in the kidney*

| Tumor type (remarks) |
|---|
| Angiomyolipoma (benign) |
| Chromophobic carcinoma (malignant) |
| Clear cell renal carcinoma (malignant) |
| Fibrosarcoma (malignant, adult) |
| Hyperproliferative cysts (benign) |
| Lymphoma (malignant) |
| Medullary carcinoma (malignant) |
| Metastases (malignant) |
| Metanephric adenoma (benign) |
| Oncoytoma (benign) |
| Papillary carcinoma type I (malignant) |
| Papillary carcinoma type II (malignant) |
| Papillary-tubular adenoma (benign) |
| Rhabdomyosarcoma (malignant, pediatric) |
| Urothelial carcinoma of the renal pelvis (malignant) |
| Wilms tumor (malignant, pediatric) |

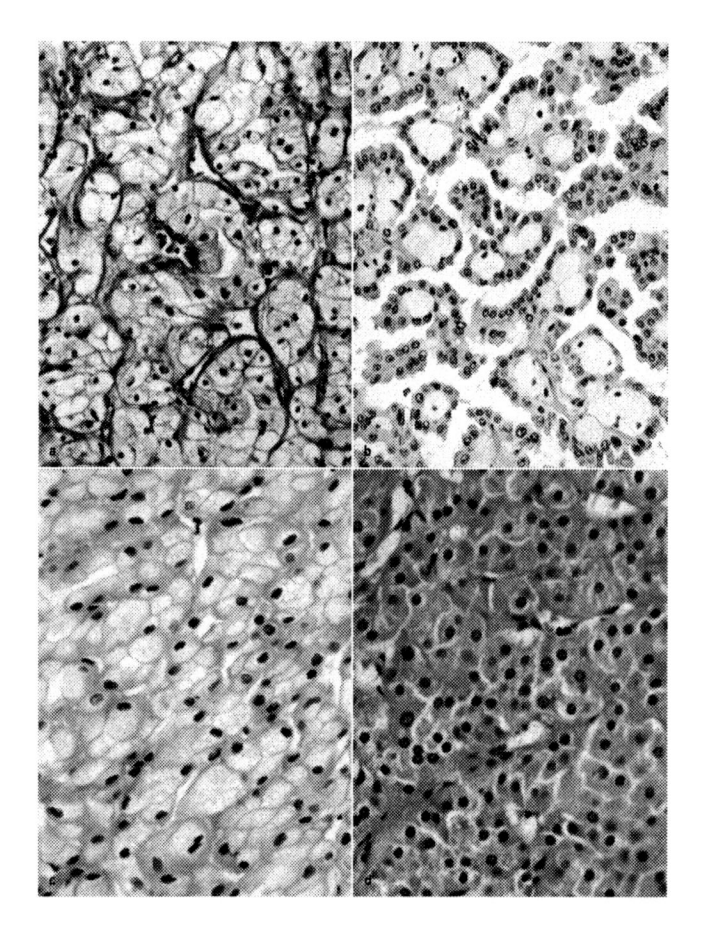

**Figure 15.1** *Histologies of renal cancers*
Histological aspects of clear cell (top left), papillary (top right), and chromophobic (bottom left) renal carcinomas and of non-malignant oncocytoma (bottom right).

Chromophobic carcinomas very likely originate from the distal tubuli, as do benign tumors named oncocytomas. The rare aggressive Ductus-Bellini carcinomas are derived from the collecting duct in the medulla of the kidney. Since the renal pelvis is lined by urothelium, like the bladder, urothelial cancers can also develop in the kidney (→14). There are also tumors from mesenchymal tissues, such as moderately frequent benign angiomyolipomas and rarer malignant sarcomas. Moreover, as an organ with an extended capillary system, the kidney is a site for metastases from carcinomas from distant organs and for lymphomas.

The variety of cancers in the kidney creates a certain predicament for therapy. Kidney tumors are nowadays usually detected by physical imaging techniques such as ultrasound or computer tomography. These give a good indication of the extension of the tumor, but no definitive information on the tumor type. This has to rely on histological and molecular investigations. Since biopsies on kidney cancers are regarded as problematic, non-invasive molecular assays for this purpose would certainly be very helpful (→21.4).

Traditionally, when a tumor was present, the entire kidney was removed. Today, partial nephrectomy is often performed, if histological and molecular markers indicate that a tumor recurrence is unlikely. This is evidently a good option for benign tumors like oncocytoma. Partial nephrectomy can also be performed in patients with small malignant tumors and an increased likelihood of failure of the other kidney, e.g. in the context of an inherited cancer syndrome (→15.3). For the choice of treatment, precise identification of the tumor variety is, of course, imperative.

For localized renal carcinomas, removal of the tumor by surgery is usually curative. In contrast, chances for a cure become minimal, once renal carcinomas have metastasized, since all varieties responding badly to currently available chemotherapy. In individual patients, therapies directed at stimulating the immune system can lead to excellent responses, but in a rather unpredictable fashion (→15.6).

## 15.2 CYTOGENETICS OF RENAL CELL CARCINOMAS

The various subtypes of renal cell carcinoma (RCC) were traditionally distinguished only by their histological appearance. Indeed, this is in many cases distinctive (Figure 15.1). Renal clear cell carcinoma is characterized by large, clear cells with small, densely stained nuclei, and an abundance of blood capillaries indicating pronounced angiogenesis. Papillary carcinoma is, as its name suggests, distinguished by its pattern of growth as finger-like connective tissue structures lined by strongly staining small epithelial tumor cells with prominent nuclei. Chromophobic carcinoma at first glance resembles clear cell carcinoma, but upon closer inspection the cells are more irregular, with more internal structure and less dense nuclei. Also, blood vessels are not as abundant.

Of course, these descriptions refer to textbook cases, and in everyday practice the distinction cannot be made for each case with such certainty. For instance, older RCC classifications distinguish a subtype characterized by spindle-shaped cells. This subtype is now thought to be generated from other types by an epithelial-mesenchymal transition (→9.2). Neither can other types of cancer (e.g. metastases) always be excluded without further molecular characterization.

In these circumstances it is helpful that cytogenetic investigations, corroborated by molecular assays, have revealed distinctions between the subtypes. Each type of renal cell carcinoma displays characteristic patterns of chromosomal losses and gains (Table 15.2). Such a close correspondence between karyotype and histological subtype is remarkable in carcinomas, because characteristic chromosomal alterations in carcinomas are often obscured by a cornucopia of other changes. Evidently, recurrent diagnostic chromosomal alterations point to a causal relationship.

Moreover, it follows as a corollary that pronounced chromosomal instability tends to be a relatively late event in renal carcinomas.

The characteristic change in clear cell RCC is loss of chromosome 3p, which is found in >90% of all cases. A typical site of breakage, albeit not the only one, is the major fragile site at 3p14.1 (*FRA3B*), which is prone to breakage under conditions suboptimal for DNA replication. Particularly during progression, 3p loss is accompanied by further chromosomal changes, both gains and losses.

In contrast, papillary carcinoma typically shows predominantly gains of specific chromosomes, prominently of 7, 17, and 3 (!), as well as 8, 12, 16, and 20, with loss of sex chromosomes.

Chromophobic carcinoma is characterized by a preponderance of whole chromosome losses affecting chromosomes 1, 2, 6, 10, 13, 17, and - in males - Y.

Oncocytoma, as a benign tumor that almost never metastasizes, would be expected to contain a limited number of genetic alterations. Indeed, the tumor is characterized by specific losses of chromosomes 1 and 14 and by translocations, like t(14;11). Some of these may activate the *CCDN1* gene.

**Table 15.2.** *Typical chromosomal alterations in different subtypes of renal cell carcinoma*

| Subtype* | Presumed origin | Chromosomal alterations |
|---|---|---|
| Clear cell renal carcinoma | proximal tubules | -3p, others |
| Papillary carcinoma | proximal tubules (nephrogenic rests?) | +7, +17, +3 |
| Chromophobic carcinoma | distal tubules | -1, -10 also -2, -6, -13, -17 |
| Oncocytomas (benign) | distal tubules | -1, t(11;14)? |
| Ductus-Bellini carcinoma | collecting ducts | highly aneuploid |

*The subtypes are ordered by decreasing incidence from top to bottom

## 15.3 MOLECULAR BIOLOGY OF INHERITED KIDNEY CANCERS

Several inherited syndromes in man predispose to kidney cancers, with varying degrees of penetrance (Table 15.3). Likewise, there are substantial differences in how specifically these syndromes predispose to cancers of the kidney.

<u>Autosomal dominant polycystic kidney disease (ADPKD)</u>: ADPKD may represent the most frequent autosomally dominant hereditary disease throughout the world with an incidence of ≈1:1000 and is therefore a major cause of kidney failure. In the course of their life, the patients develop multiple fluid-filled cysts lined by hyperproliferative, aberrantly differentiated renal epithelial cells, which eventually lead to kidney failure. Defects are also found in the liver and the vascular system. Renal carcinomas arise at an increased frequency in this syndrome, but are only as one among several complications. The disease may predispose to RCC in an indirect

fashion, i.e. it may increase the risk of renal carcinomas by its disturbance of the tissue architecture. Perhaps, the incompletely differentiated epithelial cells in the cysts provides a precursor cell population from which malignant tumors develop by further mutations.

Tuberous sclerosis (TSC): The kidney is often affected in tuberous sclerosis (→6.3), although renal carcinoma is not as frequent as benign tumors such as angiomyolipoma and renal cysts. This autosomal dominant syndrome is caused by inherited mutations in the *TSC1* or *TSC2* genes. Their respective products 'hamartin' and 'tuberin' interact with each other and inhibit the activation of protein synthesis by the PI3K pathway which is required for increased cell proliferation (→6.3).

Hereditary papillary renal carcinoma (HPRC): Unlike ADPKD and TSC, HPRC seems to predispose rather specifically to papillary tumors of the kidney. In this highly penetrant dominantly inherited syndrome, hundreds of papillary carcinomas may develop in both kidneys. They are, fortunately, not prone to early metastasis. It is therefore possible to watch them carefully and remove larger tumors by partial nephrectomy, thereby delaying the eventual removal of both kidneys and the ensuing requirement for regular dialysis.

Table 15.3. *Inherited syndromes predisposing to renal tumors*

| Syndrome | Gene(s) | Cancer type(s) in the kidney | Other symptoms |
|---|---|---|---|
| Autosomal dominant polycystic kidney disease (ADPKD) | PKD1, PKD2 | carcinomas usually clear cell | Multiple renal cysts, liver cysts, vessel defects |
| Tuberous sclerosis | TSC1, TSC2 | angiomyolipoma, renal carcinoma | benign tumors in various non-epithelial tissues |
| Hereditary papillary renal carcinoma (HPRC) | MET | papillary carcinoma, type I | rare carcinomas in the liver |
| Hereditary leiomyoma renal cell carcinoma | FH | papillary carcinoma, type II | frequent leiomyoma in the uterus |
| Burt-Hogg-Dubé (BHD) | BHD | oncocytoma, carcinoma | multiple skin papules (fibrofolliculomas) |
| Hereditary clear cell renal carcinoma (HRC) | t(3;8) (p14;q24) | clear cell carcinoma | other carcinomas (?) |
| Von-Hippel-Lindau | VHL | clear cell carcinoma, adenomas | adrenal cancers, angiomas, hemangioblastomas, adenomas and cysts of other organs, e.g., pancreas |

The gene affected in this disease is *MET* located at 7q31. It encodes the receptor for hepatocyte growth factor (HGF) which is a receptor tyrosine kinase (Figure 15.2). Notwithstanding its name, HGF (Figure 15.3) stimulates the proliferation not only of hepatocytes, but of many other cell types, in particular of epithelial origin. HGF is important during kidney organogenesis by stimulating the proliferation of epithelial precursor cells during the formation of renal tubuli. It remains essential in adult life for tissue maintenance and repair in the kidney. In fact, the therapeutic administration of HGF to prevent fibrosis after kidney injury is contemplated. However, while HGF can promote tubule formation and regeneration, under certain conditions it evokes the dissociation of epithelial cells from each other. This function has earned HGF the designation 'scatter factor' (hence also: HGF/SF). It may likewise contribute to its function in tissue morphogenesis.

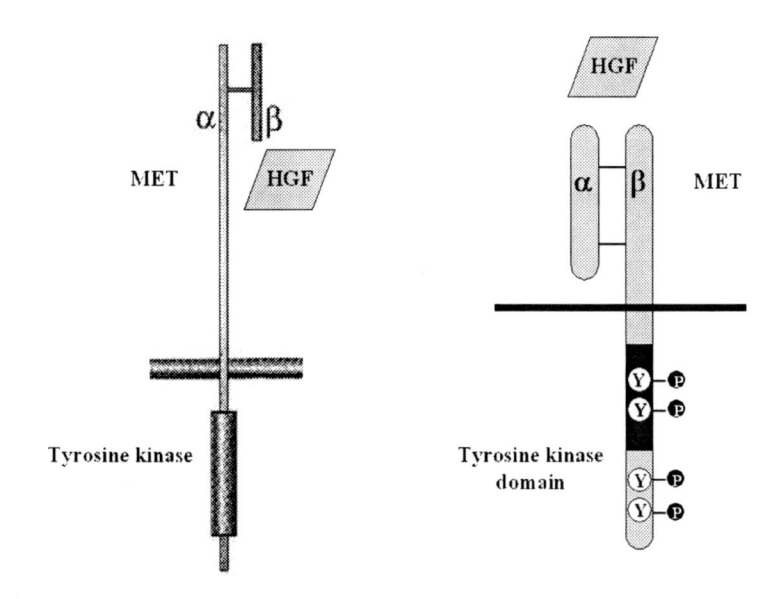

**Figure 15.2** *The MET receptor tyrosine kinase*
Two representations of the protein customary in the literature. The representation on the right also indicates auto-(cross-)phosphorylation sites. The subunit structure of the receptor is unusual compared to other receptor tyrosine kinases.

The diverse effects of HGF are all mediated through MET which activates different intracellular signaling pathways such as the MAPK pathway (→6.2) and the PI3K pathway (→6.3). In addition, it also regulates proteins involved in cellular adhesion and motility, evidently more strongly than other receptor tyrosine kinases. The MET mutations found in HPRC are located in the kinase regulatory loop of the intracellular domain and lead to constitutive activation of its tyrosine kinase (Figure

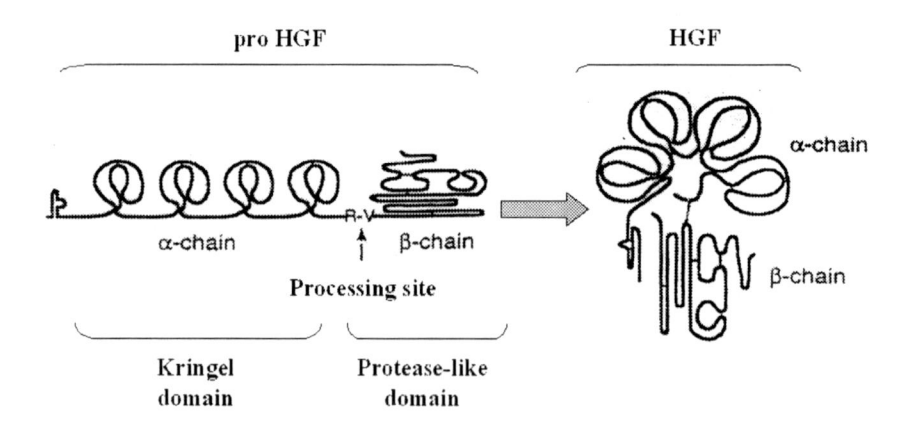

**Figure 15.3** *Structure and activation of the hepatocyte growth factor (HGF)*

15.2). This is a mechanism generally found in oncogenic mutated receptor tyrosine kinases (→4.4).

In HPRC, activation of MET appears to block the terminal differentiation of renal tubular cells and creates instead the aberrant structures of connective tissue papillas lined by proliferating epithelial cells that characterize the disease. Because no 'hit' in the second MET allele is required, this is one of a few autosomal-dominant cancer syndromes caused by mutations activating an oncogene (→5.1). However, since in HPRC patients the kidneys (and all other organs) show essentially no developmental defects, it is likely that further events are needed to initiate tumor formation, in addition to the inherited MET mutation. Likely, one event is gain of chromosome 7 leading to an increased dose of the MET oncogene as well as of other growth factors and their receptors such as EGFR. Apparently, the effect of a single mutated *MET* allele can be largely overruled by other homeostatic signals. Gain of chromosome 7 is also frequent in non-papillary kidney cancers (→15.2).

Hereditary leiomyoma renal cell carcinoma (HLRCC): Several features distinguish HPRC from another syndrome, HLRCC. HLRCC also includes a risk for papillary carcinomas, but with a slightly different morphology. These are designated type II and are more aggressive. Importantly, they develop metastases at a much earlier stage. In HLRCC syndrome, type II papillary renal carcinomas are accompanied by frequent leiomyomas in the uterus. HPRC, in contrast, is largely specific for the kidney, with a somewhat enhanced risk of hepatomas. The gene at 1q42.1 mutated in HLRCC behaves as a classical tumor suppressor (→5.1) and its second allele is inactivated in tumors developing in this syndrome. Surprisingly, the HLRCC syndrome gene encodes fumarate hydratase, a citric acid cycle enzyme. It is not at all clear, why loss of function of this enzyme should lead to renal carcinomas and leiomyomas.

Burt-Hogg-Dubé-syndrome (BHD): In BHD syndrome, the kidney is one of several organs in which tumors arise at an increased rate. In families afflicted by the disease, carcinomas as well as oncocytomas have been observed. It is caused by mutations in a gene in the 17p11 pericentromeric region.

Hereditary clear cell renal carcinoma (HCCRCC or simply HRC): Inherited susceptibility to clear cell renal cell carcinoma is typically conferred by mutations altering chromosome 3p, as one might expect (→15.2). In some families and individuals, clear-cell RCCs are caused by inherited or congenital balanced translocations involving this chromosome, typically a t(3;8)(p14;q24). The translocation sites on 3p cluster at the fragile site *FRA3B* (Figure 15.4).

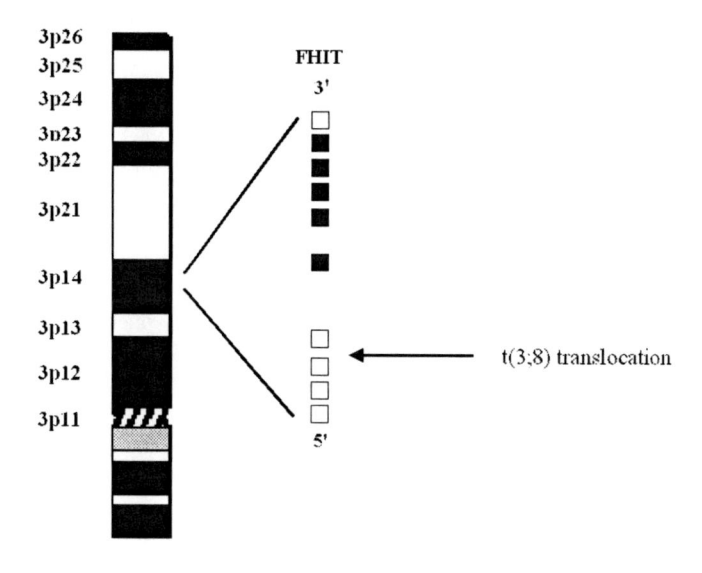

**Figure 15.4** *The* FRA3B *fragile site and the* FHIT *gene*

This site is spanned by the *FHIT* gene which is therefore disrupted by the translocations. FHIT stands for fragile site histidine triad, indicating its location and a characteristic motif in the protein. The FHIT protein resembles Ap$_4$A phosphorylases, which are enzymes responsible for metabolic switches in microorganisms. FHIT loss of function does seem to facilitate cell proliferation, but it is not well understood how. So, it is a moot point whether alterations of FHIT in renal cell carcinoma are causative events or 'bystanders'. This question extends beyond RCC, because deletions and translocations of 3p involving FRA3B also occur in a range of other cancers, prominently in lung carcinomas. The 8q24 gene affected by the t(3;8) translocations is *TRC8*, which has similarities to *PTCH1* (→6.10). Its altered function could well be relevant.

Alternatively to or in addition to changes in genes at the translocation site, tumor formation in HRC may involve an unusual mechanism: Since translocated chromosomes are more easily missegregated during mitosis, the frequency of 3p loss could be enhanced. Since loss of 3p appears to be a rate-limiting step in the formation of clear cell RCC (→15.5), the translocation may accelerate the development of clear cell carcinoma simply by facilitating this chromosomal loss.

Von-Hippel-Lindau syndrome (VHL): In the kidney, this syndrome predisposes exclusively to clear cell carcinoma. Its elucidation has not only improved our understanding of renal carcinomas, but of human cancers in general (→15.4).

## 15.4 VON-HIPPEL-LINDAU SYNDROME AND RENAL CARCINOMA

Von-Hippel-Lindau (VHL) syndrome is inherited in an autosomal-dominant fashion, with high penetrance, but variability in its phenotype. The most distinctive lesions in this multi-organ syndrome are angiomas and hemangioblastomas in the retina and cerebellum, respectively. Frequently, the patients also develop adenomas and cysts of the pancreas, epididymis, and the kidneys. All these tumors are benign, but ≈30% of VHL patients develop RCC, which can be bilateral and multifocal. It is always of the clear-cell type. A subtype of VHL disease now designated 'type II' is distinguished by an enhanced risk of pheochromocytoma, a tumor originating from the adrenal medulla and retaining the ability to produce catecholamines. Uncontrolled secretion of adrenalin and noradrenalin by pheochromocytomas causes increased blood pressure and metabolic disturbances which can be life-threatening.

Although a considerably variety of tumors can develop in von-Hippel-Lindau

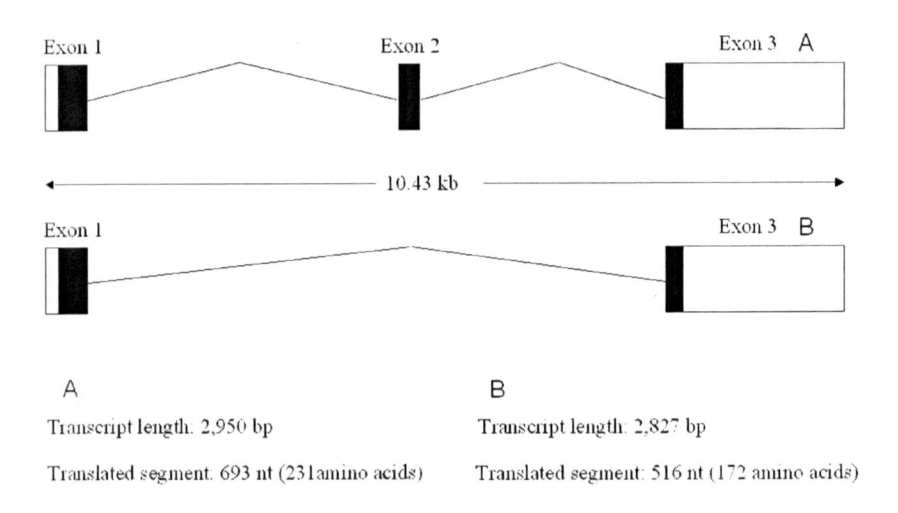

A
Transcript length: 2,950 bp
Translated segment: 693 nt (231 amino acids)

B
Transcript length: 2,827 bp
Translated segment: 516 nt (172 amino acids)

**Figure 15.5** *The VHL gene and its products*
The *VHL* gene gives rise to two (A, B) major mRNAs and proteins by differential splicing.

syndrome, they share characteristic common properties, i.e. strong vascularization and the emergence of a clear cell component characterized by an increased rate of glycolytic metabolism and enhanced storage of glycogen and lipids. These are evidently properties observed in clear-cell renal carcinoma in general, i.e. also in sporadic cases.

The von-Hippel-Lindau syndrome is caused by germ-line mutations in a gene located at 3p25, appropriately designated *VHL*. It behaves like a classical tumor suppressor gene (→5.1). Thus, each individual tumor arising in the VHL syndrome has lost the function of the second *VHL* allele as well, either by mutation, promoter hypermethylation, recombination or, typical of clear-cell RCC, 3p deletion.

The *VHL* gene encodes in three exons 6.0 and 6.5 kb transcripts containing a much smaller coding region which is translated predominantly into a 213 amino acid protein (Figure 15.5). Shorter protein variants are formed through use of an alternative start codon and alternative splicing of exon 2.

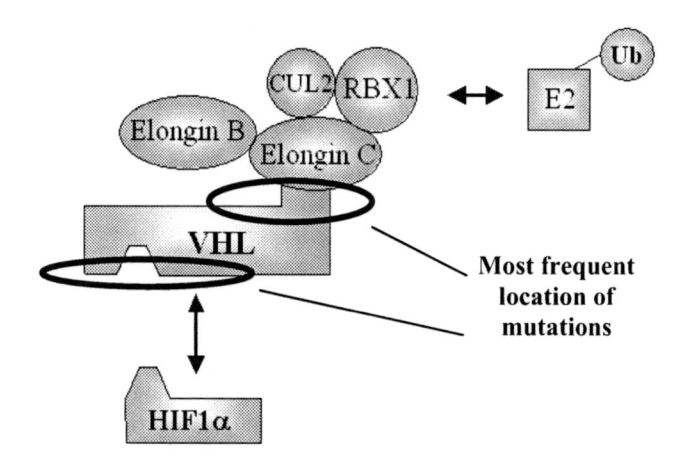

**Figure 15.6** *The VHL ubiquitin ligase complex*
See text for further explanations.

The VHL protein forms an essential part of the substrate recognition module of a specific E3 ubiquitin ligase protein complex (Figure 15.6). Further members of the complex are Elongin B and Elongin C, Cullin 2, and RBX1, which binds the E2-ubiquitin component. There are several different E3 complexes in a cell, each specific for a different range of substrates. While all have a similar basic composition, their individual components vary, most decisively the substrate recognition module, after which they are therefore (usually) named.

The VHL E3 protein complex is by far not the only ubiquitin ligase important in human cancer (Table 15.4). The SCF-$\beta^{TRCP}$ E3 complex directs the breakdown of $\beta$-Catenin. Its failure is crucial in colon carcinoma (→13.2) and in many liver cancers (16.2). The proto-oncogene product HDM2 constitutes the recognition protein of the

E3 ubiquitin ligase that degrades TP53 and its function is mimicked by the HPV E6 oncoproteins (→5.3). The SCF$^{SKP2}$ E3 ubiquitin ligase mediates the degradation of p27$^{KIP1}$ that relieves the inhibition of the CDK2/Cyclin E holoenzyme at the end of the G1 phase and allows progression of the cell cycle into S phase (→6.4). Overexpression of either Cyclin E, SKP2, or MYC appears to increase its activity inappropriately in several human tumors (including renal carcinomas).

As far as we know, the main substrates of the VHL E3 ubiquitin ligase are two closely related proteins, HIF1α and HIF2α, where HIF stands for 'hypoxia-inducible factor'. Missense mutations in the VHL protein cluster in two regions, i.e., in one face that makes contact with the HIFα proteins and in the opposite face that interacts with the Elongins. The type of mutations inherited in the VHL gene bears some relation to the disease subtype. For instance, pheochromocytoma (i.e. type II VHL syndrome) is only found in patients with certain point mutations, such as 505C>T or 712C>T. In contrast, insertion and deletion mutations (which would be expected to lead to frameshift mutations) increase the risk for RCC.

The HIFα proteins are transcription factors that induce a particular pattern of gene expression in response to low oxygen, i.e. during hypoxia (Figure 15.7). Under normal conditions, i.e. normoxia (→Box 9.1), HIF1α and HIF2α proteins undergo rapid turnover, with a half-life of a few minutes. The prevailing oxygen partial pressure is signaled through specific proline hydroxylases which hydroxylate one proline in the HIFα 'oxygen-dependent degradation domain' (ODD). These kind of enzymes belong to the EGLN family and are different from those involved in collagen biosynthesis, but likewise contain ferrous iron and require oxygen and 2-oxo-glutarate as cosubstrates. With its proline hydroxylated, the HIFα ODD domain is recognized by VHL and the HIFα proteins are ubiquitinated and targeted for degradation by the proteasome.

If the oxygen partial pressure decreases, the proline in the ODD domain remains non-hydroxylated and the HIFα proteins accumulate in the cytosol. They combine with HIF1β, also known as ARNT, which is the dimerization partner for several nuclear factors, including the dioxin receptor AhR. The HIFα/ARNT heterodimer enters the nucleus and binds to specific recognition sites in promoters to activate genes whose products help the cell adapt to low oxygen supply and to increase

**Table 15.4.** *Some ubiquitin ligases important in human cancers*

| Ubiquitin ligase | Function |
|---|---|
| CBL | degradation of receptor tyrosine kinases |
| CDH1 | mitotic regulation (anaphase promoting complex) |
| MDM2/HDM2 | degradation of TP53 |
| SKP2 | degradation of p27$^{KIP1}$ |
| βTRCP | degradation of β-Catenin |
| VHL | degradation of HIFα proteins |

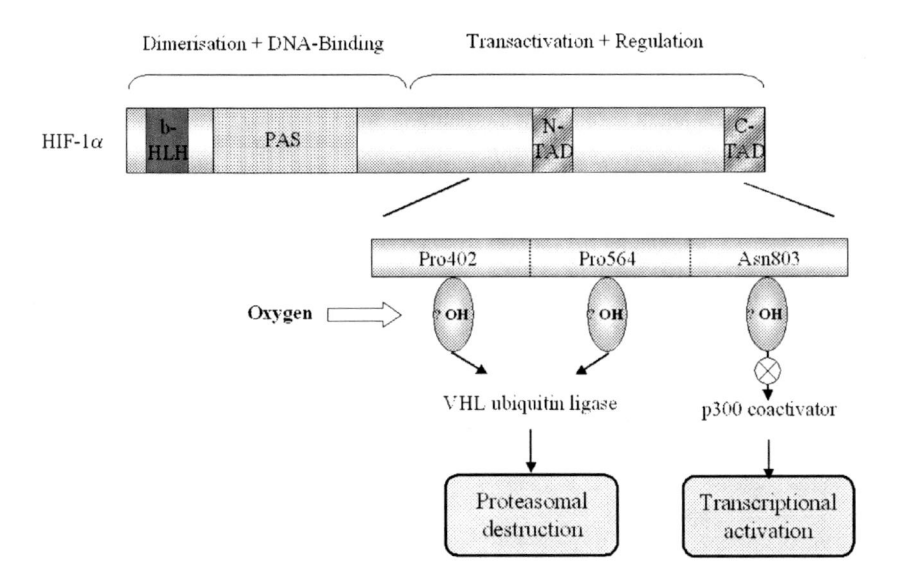

**Figure 15.7** *Structure and regulation of HIFα proteins by the oxygen concentration*
TAD: transactivation domain; bHLH: basic helix-loop-helix; PAS: Per-ARNT-Sim homology domain. See text for further explanations of the mechanisms involved.

oxygen availability. These HIF-binding sites are called hypoxia responsive elements or for short HRE[14]. The activity of HIFα factors is further regulated by oxygen-dependent hydroxylation of an asparagine in their transcriptional activation domain (Figure 15.7), which prevents binding of the p300/CBP co-activator protein.

When oxygen partial pressure normalizes, the HIFα proteins are degraded and the transcriptional response to hypoxia is terminated. In von-Hippel-Lindau disease, therefore, loss of VHL leads to accumulation of HIFα proteins and constitutive activation of a transcriptional program for hypoxia response, independent of oxygen supply.

The genes that are activated by the HIFs are intriguing, when considered in the context of clear-cell carcinoma and other tumors arising in the VHL syndrome (Table 15.5).

Oxygen supply: Improved oxygen supply is achieved by several mechanisms. For instance, endothelin-1 and inducible NO synthase (iNOS) lead to increased blood flow and VEGF and PAI-1 stimulate angiogenesis (→9.4). This provides a straightforward explanation for the enhanced vascularization in tumors arising in the VHL syndrome. Iron transport proteins like transferrin and its receptor are also induced. Moreover, hypoxia also stimulates erythropoetin production in suitable

---

[14] They are different from heat-shock response elements, which are also often abbreviated as HRE; there are only so many three-letter acronyms...

cells. This growth factor stimulates the proliferation and maturation of erythroid precursors in the bone marrow leading to 'polycythemia'. Erythropoetin production is regulated by the kidney in response to changes in oxygen partial pressure, e.g., at high altitudes. Constitutive activation of the hypoxia response in renal tumor cells can therefore lead to increased erythropoesis, which is one of several 'paraneoplastic' symptoms found in renal carcinoma patients. Such symptoms can be confounding in diagnosis, because they suggest completely different diseases.

Glucose and lipid metabolism: HIF transcription factors change the metabolism of glucose and of lipids. Target genes in this regard include the glucose transporter GLUT1 and many glycolytic enzymes, such as LDH, aldolase, and PGK. For many glycolytic enzymes, several isoenzymes are expressed in a tissue-specific fashion, depending whether the tissue performs glycolysis or gluconeogenesis. Hypoxia invariably induces the glycolytic over the gluconeogenic isoenzyme. The kidney, particularly in its proximal tubules, is capable of gluconeogenesis, like the liver and the intestine. Therefore, in VHL clear-cell carcinoma proximal tubule cells switch from gluconeogenesis to glycolysis as well as to glycogen and lipid storage. Much of this metabolic switch, like increased angiogenesis, can be accounted for by constitutive HIF activation.

Growth factors: In addition to VEGF, active HIFs induce further growth factors acting in a paracrine fashion, e.g. on endothelial cells, as well as in an autocrine fashion on the producing cell itself. These include PDGF, TGFβ factors, and – perhaps most significant in renal epithelial cells - TGFα, an important ligand of the EGFR. Within the cell, the PI3K pathway (→6.3) may be activated, independent of growth factor signals.

Apoptosis: While hypoxia elicites potential growth signals, it also promotes apoptosis. The fate of a hypoxic cell may therefore be decided by the relative balance of these signals. TP53 is activated and would be expected to not only promote apoptosis via BAX upregulation, but also to antagonize the pro-angiogenic response (→6.6). The proapoptotic NIX and NIP3 proteins (Table 7.1) may constitute specific mediators of hypoxia-induced apoptosis.

Regulation of pH: The carbonic anhydrases CA9 and CA12 are among the most strongly HIF1-inducible proteins. These proteins are mainly located at the cell membrane and regulate pH by accelerating the reaction between $CO_2$ and $H_2O$ (in both directions). In a hypoxic environment, they favor the establishment of a low pH through the dissociation of carbonic acid.

**Table 15.5.** *A selection of proteins induced by HIFs*

| Function | Examples |
| --- | --- |
| Oxygen supply and angiogenesis | ET1 (endothelin-1), ET2, iNOS, COX2, heme oxygenase, ferritin, transferrin, transferrin receptor, VEGF, PDGF, ANG-2, VEGFR1, TIE1 (angiopoetin receptor), IL8, PAI1, erythropoetin |
| Glucose metabolism | LDH, pyruvate kinase, PGK, GAPDH, aldolase, PFK (glycolytic isoenzymes), hexokinase, GLUT1, GLUT3 |
| Growth factors and growth factor signaling | TGFα, TGFβ, HGF, PDGF, FGF3, VEGFA, EGFR, PI3K activity, IL6, IL8, p27$^{KIP1}$, p21$^{CIP1}$ |
| DNA repair and cell protection | HSF (heat-shock factor), HIF1α, HIF2α, GADD153, HAP1 (AP-endonuclease), thioredoxin, KU70, KU80 |
| Apoptosis | NIX, NIP3, PI3K activity, IGFBP3, NF$_κ$B, spermidine acetyltransferase |
| pH regulation | Carbonic anhydrases IX and XII (CA9, CA12) |
| Cell adhesion and extracellular matrix remodeling | Specific collagens, metalloproteinases, specific CAMs (cell adhesion molecules), proline 4-hydroxylase, uPAR, vimentin, specific integrins |

## 15.5 MOLECULAR BIOLOGY OF CLEAR CELL RENAL CARCINOMA

The recurrent chromosomal alterations found in diverse subtypes of RCC indicate a causal relationship. Nevertheless, their significance is not really well understood. For instance, papillary RCC exhibit consistent gains of chromosomes which contain genes for growth factors and their receptors, such as HGF/MET, EGF-like growth factors/EGFR/ERBB2, located on chromosomes 7 and 17. However, in sporadic cases, these receptor tyrosine kinases are rarely mutated, unlike MET in HPRCC. So, one would have to postulate that increased gene dosage of both receptors and ligands could create autocrine growth factor loops that could drive the growth of papillary renal carcinoma. Thus, quantitative rather than qualitative changes in gene expression could be important for the development of this cancer. Quantitative changes as a cause of cancer are difficult to ascertain, but evidence from model systems also supports their importance. For instance, growth of several lymphoma types depends crucially on the dosage of *MYC* which itself may elicit quantitative rather than qualitative changes in gene expression.

In clear-cell renal carcinoma, the loss of 3p is evidently crucial, since it is found in >90% of all cases. One gene clearly targeted by this loss is *VHL*. Loss of 3p removes one copy of the gene. The second allele is inactivated in ≈80% of sporadic cases of renal cell carcinoma by point mutations, small deletions, or by promoter hypermethylation. So, *VHL* is a classical tumor suppressor, even to the point that bilateral renal clear cell carcinoma is predominantly found in cases of hereditary VHL disease.

Moreover, the function of the VHL protein explains several conspicuous phenotypical aspects of this tumor type, in particular its intense vascularization and its distinctive clear-cell appearance. However, the relationship of VHL inactivation to other properties of clear cell RCC is not as straightforward (cf. Table 15.5). VHL loss of function does appear to impede cell cycle regulation. It may also lead to over-production of TGFα and its receptor EGFR. Even MET has been suggested as a target of HIFα, or at least to be induced secondary to hypoxia-induced gene activation. However, it is far from clear whether these changes are sufficient to drive the proliferation of clear-cell carcinoma, especially its progression to advanced stages.

Likewise, HIFs affect the synthesis of the extracellular matrix and VHL appears to interact directly with fibronectin. Overall, however, loss of VHL may promote deposition of ECM. While ECM synthesis may aid vascularization, it would be thought to inhibit rather than to enhance invasion. By comparison, increased activity of membrane carbonic anhydrases like CA9 and CA12 may indeed aid invasion by altering the extracellular pH. One characteristic of clear cell RCC is a pronounced change in the pattern of cell adhesion molecules, which is difficult to fully ascribe to the changes in VHL/HIF only.

The perhaps most crucial issue concerns apoptosis. Since the cellular response to hypoxia involves the induction of pro-apoptotic proteins (Table 15.5), its constitutive activation by VHL loss would be expected to eventually lead to cell death. In VHL tumors and sporadic renal cell carcinoma, this branch of the hypoxia response is evidently not fully effective.

Therefore, in the development of clear cell renal carcinoma, the loss of VHL and the constitutive expression of the cellular response to hypoxia are likely compounded by additional genetic and epigenetic alterations. Some alterations are known and target tumor suppressors. Frequent LOH at chromosomes 17p13 and 9p21 points to the involvement of *TP53* (→6.6) and *CDKN2A* (→6.4), respectively, and additional mutations in these genes are indeed observed in advanced cases. Down-regulation and loss of PTEN, an antagonist in the PI3K pathway (→6.3) due to 10q loss may be another significant change in progressive cases. As in papillary RCC, certain chromosomal gains may lead to increased expression of growth factors and their receptors, especially of EGFR and MET. Conversely, responses to TGFβ are diminished, most likely through down-regulation of the TGFβRII. As a rule, in RCC anti-apoptotic proteins such as the IAP survivin are frequently over-expressed and death receptors like FAS/CD95 are down-regulated (→7.3). However, the mechanisms underlying these changes in the apoptotic balance are not understood.

In spite of these changes, each in a fraction of clear cell RCC, the by far most consistent change in clear-cell renal carcinoma is loss of 3p. It is now thought unlikely that loss of one *VHL* allele is its only relevant consequence. Most researchers agree that the loss of 3p is not only a consequence of frequent breaks at

*FRA3B*, but is also functionally selected for because it leads to the loss of a second tumor suppressor. There is less agreement on which (Figure 15.8).

(1) *FHIT* at the fragile site is an obvious candidate. (2) The *RASSF1A* gene located at 3p21 encodes a protein activated by RAS which may limit the cellular responses to RAS activation, perhaps by blocking mitosis (→6.2) or as a feedback inhibitor of RAS-induced gene expression. It is transcriptionally inactivated in many different human cancers, including RCC, by promoter hypermethylation. The RASSF1 locus is complex. In addition to the RASSF1A protein it encodes RASSF1B and RASSF1C which may counteract RASSF1A. (3) The gene for the retinoic acid receptor RARβ2, closely related to the RARα protein crucially involved in acute promyelocytic leukemia (→10.5), is located at 3p24. This member of the family appears to relay and amplify differentiation signals in epithelial cells by a typical autoregulatory loop (→8.5). The *RARB2* gene contains a RARE sequence to which retinoic acid receptors can bind. Binding by a ligand-activated RARγ, e.g., can activate *RARB2* transcription which increases the concentration of the RARβ2 receptor, further increasing gene activity until target genes that induce terminal differentiation can be turned on. Like *RASSF1A*, *RARB2* is found hypermethylated in several carcinoma types. As in the case of RASSF1A, the locus is complex and other isoforms may also matter. Of course, until further clarification, the list of candidates does not end here.

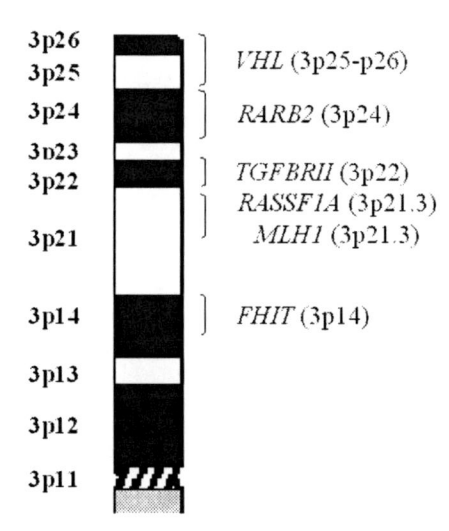

**Figure 15.8** *Some potential tumor suppressor genes at chromosome 3p*

## 15.6 CHEMOTHERAPY AND IMMUNOTHERAPY OF RENAL CARCINOMAS

Renal cell carcinomas are notoriously difficult to treat by chemotherapy or radiation therapy. Several factors may contribute to this 'primary' resistance.

(1) Renal cell carcinoma are not particularly fast growing, presenting poor targets for therapies targeting highly proliferating cells (→22.2).

(2) Excretion of toxic compounds is an important functions of the normal kidney. Tumor tissues from this organ retain some of the protective systems of the kidney and in particular the excretion system involving the 'PGP glycoprotein' MDR1 (Figure 15.9). MDR1 is an ATP-dependent transport protein ('ABC' transporter) which helps to exchange lipids between the inner and outer leaflet of the cell membrane. This reaction also allows the excretion of a broad range of lipophilic cytostatic drugs from the cell. In this fashion, the protein contributes to multi-drug resistance in many human cancers. In other cancers, its expression is acquired or induced only after exposure to chemotherapy, leading to 'secondary' chemoresistance. In the kidney, in contrast, the MDR protein is normally expressed at high levels, likely to support the excretion of toxic compounds from the body. Therefore, the protein is present right from the start in RCC and this cancer type displays 'primary resistance'.

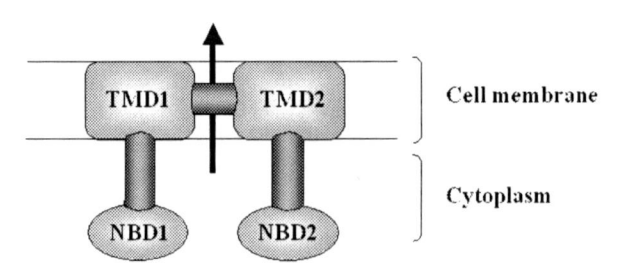

**Figure 15.9** *Structure and function of MDR1 (PGP)*
The transmembrane domains (TMD1, TMD2) form a pore, through which the actual transport proceeds. It is driven by ATP bound by the nucleotide binding domains (NBD1, NBD2).

(3) A crucial change in the development of RCC is decreased apoptosis. Although the underlying mechanisms are not completey clear, they are likely to result in decreased sensitivity towards chemotherapy and radiotherapy (→22.2).

(4) In advanced renal carcinomas, loss of TP53 and PTEN functions are relatively frequent. These losses are in general associated with poor responses to chemotherapy and radiation, by diminishing apoptosis (→7.3) and enhancing tolerance to DNA strand-breaks (→3.3).

Desperation is therefore certainly part of the explanation, why RCC has become one of the favorite objects for immunotherapy. Of course, there are more strictly scientific reasons as well. Most strikingly, 'spontaneous' regression of renal carcinomas has been documented in individual cases. It is commonly attributed to a successful immune response. In accord with these very rare 'miracle' cures, treatment with cytokines that stimulate cytotoxic T-cells has been reported to lead to partial or complete responses (→22.1) in 15-30% of clear cell renal carcinoma patients, although not in other subtypes. Therefore, treatment with interleukin-2 (IL2) and/or interferon $\alpha$ (IFN$\alpha$) is one of the few therapeutic options available for patients with metastatic disease. This treatment occasionally leads to spectacular responses, but it is rarely curative. Moreover, the side effects can be quite intense, comparable to those experienced in a severe case of flu. They would, perhaps, be more acceptable, if one could predict in which patient the treatment is efficacious, but this is not yet possible. So, more experimental approaches to immunotherapy of renal carcinoma are being attempted (→22.5).

In melanoma, another promising target for immunotherapy, immune responses are directed against proteins particular to melanocytes and to cancer-testis antigens ectopically expressed in the cancer (→12.5). In renal carcinoma, oncofetal antigens, i.e. proteins normally expressed only during fetal development and down-regulated in adult kidney, may represent one type of target. More broadly, antigens recognized by immune cells in RCC are derived from proteins as part of the constitutive hypoxia response (Table 15.5), particularly in the conventional type. For instance, a promising cell membrane antigen recognized by the monoclonal antibody G250 is expressed on the surface of essentially every renal carcinoma (of various subtypes), but is not at all detectable in normal kidney. This antigen has turned out to be part of the CA9 carbonic anhydrase, which is induced several-hundred fold in clear cell RCC as a consequence of constitutive HIF1 activation.

## Further reading

Ebele JN et al (eds.) Pathology and Genetics of Tumours of the Urinary System and Male Genital Organs. IARC Press, 2004

Kovacs G et al (1997) The Heidelberg classification of renal cell tumours. J. Pathol. 183, 131-133

Davies JA, Perera AD, Walker CL (1999) Mechanisms of epithelial development and neoplasia in the metanephric kidney. Int. J. Dev. Biol. 43, 473-478

Friedrich CA. (2001) Genotype-phenotype correlation in von Hippel-Lindau syndrome. Hum. Mol. Genet. 10, 763-767

Huebner K, Croce CM (2001) FRA3B and other common fragile sites: the weakest links? Nat. Rev. Cancer 1, 214-221

Meloni-Ehrig AM (2002) Renal cancer: cytogenetic and molecular genetic aspects. Am. J. Med. Genet. 115, 164-172

Bodmer D et al (2002) Understanding familial and non-familial renal cell cancer. Hum. Mol. Genet. 11, 2489-2498

Fromm MF (2002) Genetically determined differences in P-glycoprotein function: implications for disease risk. Toxicology 181-182, 299-303

Kaelin WG (2002) Molecular basis of the VHL hereditary cancer syndrome. Nat. Rev. Cancer 2, 673-682

Pfeifer GP et al (2002) Methylation of the RASSF1A gene in human cancers. Biol. Chem. 383, 907-914

Ambudkar SV et al (2003) P-glycoprotein: from genomics to mechanism. Oncogene 22, 7468-7485

Gitlitz BJ, Figlin RA (2003) Cytokine-based therapy for metastatic renal cell cancer. Urol. Clin. North Am. 30, 589-600

Safran M, Kaelin WG (2003) HIF hydroxylation and the mammalian oxygen-sensing pathway. J. Clin. Invest. 111, 779-783

Linehan WM, Walther MM, Zbar B (2003) The genetic basis of cancer of the kidney. J. Urol. 170, 2163-2172

Mulders P, Bleumer I, Oosterwijk E (2003) Tumor antigens and markers in renal cell carcinoma. Urol. Clin. North Am. 30, 455-465

Pugh CW, Ratcliffe PJ (2003) Regulation of angiogenesis by hypoxia: role of the HIF system. Nat. Med. 9, 677-684

Vieweg J, Dannull J (2003) Tumor vaccines: from gene therapy to dendritic cells: the emerging frontier Urol. Clin. North Am. 30, 633-643

Zbar B, Klausner R, Linehan WM (2003) Studying cancer families to identify kidney cancer genes Annu. Rev. Med. 54, 217-233

# CHAPTER 16

# LIVER CANCER

- Liver cancer is one of the major lethal malignancies worldwide. The main histological subtype is hepatocellular carcinoma, which is derived from hepatocytes, the predominant epithelial cell type in the liver, and often retains biochemical and morphological markers of hepatocyte differentiation.
- Hepatocellular carcinoma develops as a rule in the context of chronic inflammation and liver cirrhosis caused by the hepatitis viruses HBV or HCV, by chronic alcohol abuse, or more rarely by hereditary diseases such as hemochromatosis. Chemical carcinogens such as aflatoxin B1 from the mold Aspergillus flavus act synergistically with the causes of inflammation, in particular with chronic HBV infection. Aflatoxins cause a diagnostic G→T mutation at codon 249 of TP53.
- In addition to disrupting the TP53 network, genetic and epigenetic alterations in hepatocellular carcinoma inactivate cell cycle regulation by RB1 and cause constitutive activity of the WNT signaling pathway, most often by mutations of β-Catenin and more rarely by inactivation of Axin1 and APC. WNT pathway activation may be exacerbated by loss of E-cadherin.
- Several growth factors and their receptors controlling normal hepatocyte proliferation and regeneration are also implicated in the growth and survival of hepatocellular carcinoma, by autocrine and paracrine mechanisms. In addition to activation of HGF/MET and TGFα/ERBB1 circuits, a particular important change may consist in increased stimulation by the insulin-like factor IGF2 through the IGF1R receptor tyrosine kinase, while the scavenger receptor IGFRII may become inactivated.
- The pivotal role of HBV in liver carcinogenesis appears mainly to be due to a continuous hepatocyte destruction by T-cells which attempt to eliminate the infection and their repletion from differentiated cells and eventually liver stem cells. This process may select for genetically altered cells with diminished response to the virus and to apoptotic signals. Increased oxidative stress in the inflamed tissue may contribute to mutagenesis and at the same time increase the selective pressure. Inhibition of TP53 function by the viral HBX protein may aid in down-regulation of apoptosis during early stages carcinogenesis and permit the accumulation of cells with aberrant genomes. Furthermore, integration of viral genomes may promote genomic instability.
- Importantly, while therapeutic options are limited once HCC is established, vaccination against HBV and anti-viral treatment against HCV appear to be efficacious in preventing this cancer.

## 16.1 ETIOLOGY OF LIVER CANCER

Although the liver is a frequent site of metastases from other organs, primary liver cancer is nowadays one of the rarer malignancies in Western industrialized countries. The main histological subtype is hepatocellular carcinoma (HCC), derived from the major cell type in the liver, the epithelial hepatocyte.

A long way into their development, HCC cells remain similar to normal hepatocytes to the extent that early stage cancers can be difficult to distinguish morphologically from normal parenchyma or benign adenomas (Figure 16.1). Some secrete the oncofetal albumin homolog α-fetoprotein (→12.6), which is a useful marker in such cases. Molecular histological techniques such as fluorescence-in-situ-hybridization can be used to detect chromosomal abnormalities and support a diagnosis of cancer.

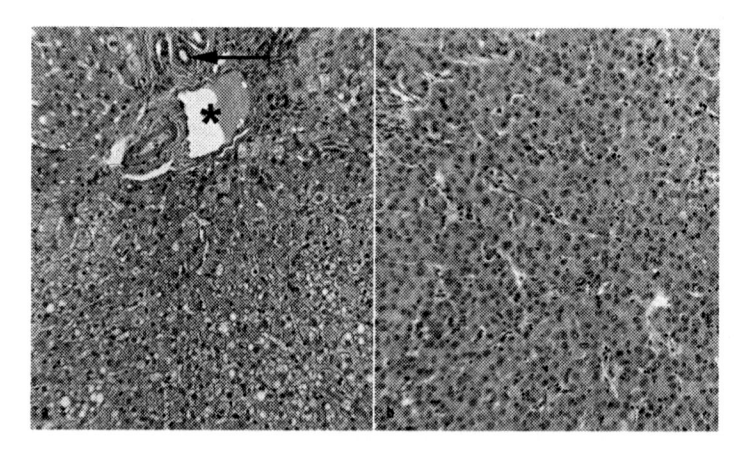

**Figure 16.1** *Histology of hepatocellular carcinoma compared to normal liver*
Aspect of normal liver (left) with a portal field (star) and of hepatocellular carcinoma (right).

Notwithstanding their morphologically differentiated appearance, hepatocellular carcinomas are highly lethal cancers. In spite of progress in surgery, including even the use of liver transplantation, overall survival is <20% two years after diagnosis. As a rule, only small cancers with diameters <2 cm can be cured by surgery. Currently available chemotherapy is unsuccessful.

In many developing countries, the situation is worse. Primary hepatocellular carcinoma is a major health problem in East and Southeast Asia as well as Central and Southern Africa, the more so, as patients tend to be much younger than in the Western world, and even include children. In many countries, the possibilities for treatment are even more limited than in the Western world. As a result, HCC remains the (approximately) fifth-most frequent cause of cancer deaths worldwide.

Independent of geography and socioeconomic factors, HCC arises typically in the context of chronic liver inflammation and liver cirrhosis. For instance, some patients suffering from hereditary hemochromatosis develop chronic liver inflammation leading to cirrhosis, and eventually to HCC in some cases (Figure 16.2).

Hemochromatosis is an iron storage disease caused by missense mutations in the *HFE* gene. Its product is a membrane protein similar to MHC proteins which regulates iron uptake and transport. Inadequate function of HFE leads to iron overload in the liver. Normally, iron is stored in the liver bound to proteins such as ferritin. In hemochromatosis the protein storage capacity is exceeded and free iron ions cause damage to the tissue. Protein-bound ferrous iron ions ($Fe^{2+}$) are relatively innocuous, while $Fe^{2+}$ free in solution or associated with low molecular ligands can be more easily oxidized to the ferrous state ($Fe^{3+}$) in the Fenton reaction (Figure 16.3), which yields highly reactive hydroxyl radicals (→Box 1.1). Hydroxyl radicals damage macromolecules and lipids, causing necrotic cell death and chronic inflammation. They also react with DNA bases and with deoxy-ribose, causing strand-breaks and base mutations.

*HFE* mutations are highly prevalent in Europeans. In this population, 10% may be heterozygous for the 282Cys>Tyr mutation, and up to 20% for the 63His>Asp mutation. Only homozygotes or compound homozygotes (i.e. persons carrying two

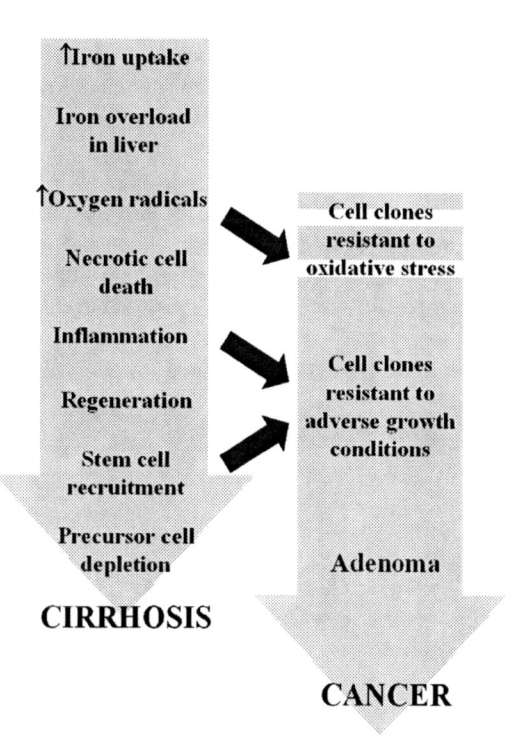

**Figure 16.2** *The pathophysiological chain of HCC development in hemochromatosis*

different mutations in the *HFE* alleles) develop hemochromatosis. In fact, not all do, and only a fraction of these progress towards cirrhosis and hepatocellular cancer (Figure 16.2). This demonstrates that further factors modulate the risk of inflammation and cancer. Nowadays, patients with early signs of hemochromatosis disease can be treated with iron chelators to prevent organ damage, cirrhosis, and cancer.

$$Fe^{2+} + H_2O_2 \rightarrow Fe^{3+} + OH^{\bullet} + OH^{-}$$

**Figure 16.3** *The (standard) Fenton reaction*

Hemochromatosis is a comparatively rare disease, but its pathophysiology is well understood. It provides an example for other, more prevalent, but less well understood factors that cause liver cancer by a basically similar pathway through chronic inflammation and cirrhosis. The most important ones are the hepatitis viruses HBV and HCV, and alcohol abuse (Table 16.1). In a typical Central European population, ≈3% of all HCC cases might be associated with hemochromatosis, and 30% each with HBV, HCV, and alcohol, the remainder with other known or unknown causes. A rising incidence of HCC in industrialized countries over the last decades is mainly caused by an according increased prevalence of HCV infections. Tellingly, the rise in cancer incidence is delayed by ≈15 years compared to that of the viral hepatitis. So, this is the period required for development of cirrhosis and cancer.

HBV persists as a chronic infection in ≈10% and HCV in >50% of infected persons. In chronic viral hepatitis, hepatocytes are continuously destroyed by cytotoxic T-cells attempting to eliminate the virus and need to be replaced. Replacement of hepatocytes is partly achieved by proliferation of differentiated hepatocytes and partly, especially during continuous damage, by recruitment of liver

**Table 16.1.** *Causes of human hepatocellular cancer*

| *Individual factors causing hepatocellular cancer in humans\** |
|---|
| Alcohol |
| Hepatitis virus B |
| Hepatitis virus C |
| Steatosis (fatty liver) |
| Inherited diseases leading to chronic liver damage |
| Hemochromatosis and other diseases leading to iron or copper overload |
| Other causes of liver cirrhosis |
| Aflatoxins |

\* Several factors may interact, often in a synergistic fashion.

stem cells. These are thought to reside in the ducts of Herring at the origin of the bile duct. Their activation can be recognized by spreading of undifferentiated small epithelial-like cells with oval nuclei ('oval cells') into the hepatic parenchymal structure. During chronic alcohol abuse, hepatocyte destruction is initiated by the substance and its metabolites.

During chronic tissue damage and inflammation, nonepithelial cells in the liver become activated and proliferate. While their activation and proliferation initially serves to support the immune response and the regeneration of the tissue, during chronic liver damage and with increasing inflammation their expansion predominates and they gradually replace epithelial structures in the organ. This process eventually manifests as cirrhosis, in which the well-organized parenchymal tissue is displaced by more disorganized and dysfunctional fibrotic tissue.

In this disturbed tissue, hepatocellular cancer may develop from liver stem cells or from more differentiated hepatocytes that are more resistant to the adverse conditions in the organ. Hepatic cancer cells are typically more resistant to viral infection, store less iron, or are less easily triggered to apoptosis by cytotoxic immune cells. So, they may be selected for their ability to survive in the cirrhotic tissue environment. Specifically, their increased resistance to apoptosis, e.g. by the CD95 pathway (→7.2), may be one reason for the primary resistance of HCC to chemotherapy (→22.2).

Hepatocellular carcinoma is preceded by several morphological alterations in the parenchymal epithelium. These were initially identified in experimental animals (Box 16.1), and later confirmed in humans. In some cases, the proliferation of liver stem cells can actually be observed. Otherwise, the first morphological stage are small dysplastic foci, which increase in size to form nodules with progressively aberrant cells. While these precursor stages remain restricted to the epithelial parenchyma, actual HCC invades interstitia and vessels and may form metastases in distant parts of the liver or in other organs. A critical stage, both biologically and clinically, appears to be reached at a tumor volume of $\approx 1$ cm$^3$. Here, angiogenesis by branches of the hepatic artery and capillaries appears to become activated and the tumor cells become more invasive. This transition is quickly followed by spreading of the cancer cells through the liver, and later to other organs.

## 16.2 GENETIC CHANGES IN HEPATOCELLULAR CARCINOMA

Hepatocellular carcinoma cells are aneuploid with a number of consistent chromosomal changes. In short, gains of 1q, 6p, 8q, 11q, 17q and the entire chromosome 7 are prevalent, while losses concern predominantly 4q, 6q, 8p, 13q, 17p and both arms of chromosome 16. The 'cancer pathways' (→6) targeted by these alterations and further genetic and epigenetic changes are the RB1 network regulating the cell cycle, the WNT pathway, the STAT pathway and the TP53 network. In addition, autocrine or paracrine growth factors loops appear to be set up. While these may promote proliferation, they may also contribute to decreased apoptosis together with altered responses to death receptor ligands and overexpression of anti-apoptotic proteins (→7.3).

Regulation of the cell cycle through RB1 is disrupted in most HCC by loss and mutation of *RB1* itself, by hypermethylation, mutation or deletion of *CDKN2A*, and – perhaps more often than in other carcinomas - by overexpression of Cyclin D1 as a consequence of amplification of its gene, *CCND1*, at 11q13. The *MYC* gene at 8q24.1 is also quite often amplified, promoting cell cycle progression and cell growth (→10.3). These alterations leading to deregulation of the cell cycle resemble those in many other carcinomas, e.g. squamous cell carcinoma of the skin (→12.2) and bladder cancers (→14.3).

The WNT pathway is also activated in many hepatocellular carcinomas, although not as regularly as in colorectal cancer (→13.2). In fact, the mode of activation is usually different (Figure 16.4). In colorectal cancer, constitutive activation of the pathway is most often caused by inactivation of both alleles of the *APC* tumor suppressor gene, and in a smaller fraction of cases by point mutations in *CTNNB1* that turn β-Catenin into an oncogenic protein (→13.2). These mutations occur in the part of the protein that is recognized by GSK3β and prohibit the phosphorylation that allows the recognition of β-Catenin by the βTRCP ubiquitin ligase complex.

This type of *CTNNB1* mutation is also prevalent in HCC, at a frequency of 25-30%, whereas inactivation of APC is the exception rather than the rule. Instead, two other alterations promote overactivity of the pathway in some cases. Mutations of Axin1, a protein that helps to assemble the APC/GSK3β/β-Catenin complex (→6.10), are found in ≈5% of HCC. Accordingly, LOH at 16p13, where the *AXIN1* gene is located, is relatively common. In many HCC, loss of the long arm of this chromosome, i.e. 16q, together with mutation and promoter hypermethylation lead to decreased activity of E-Cadherin. Loss of this protein from the cell membrane causes decreased adhesiveness between neighboring epithelial cells and thereby favors invasion and metastasis (→9.2). Moreover, on the cytoplasmic side of the cell membrane, E-Cadherin anchors a network of fibers through the α-Catenin and β-Catenin proteins. In this fashion, in some cell types, E-Cadherin seems to act as a kind of 'buffer' for β-Catenin and to modulate the activity of the intracellular WNT signaling pathway dependent on this protein. In HCC, specifically, the loss of E-Cadherin appears to contribute to enhanced β-Catenin concentrations in the nucleus.

HCC is one of those solid tumors in which enhanced STAT activity has been observed. Among the defects leading to STAT overactivity in these cases is hypermethylation of the *SOCS1* promoter combined with LOH at 16p13 (the same region containing *AXIN1*). SOCS1 is a feedback inhibitor of JAK2 that relays growth and survival signals from cytokines such as IL6 (→6.8). In the absence of SOCS1, signals through the STAT pathway may be prolonged and more strongly promote resistance to apoptosis, in particular. It is important to remember in this context, that although IL6 is labeled a 'cytokine', because it influences the proliferation and function of many cell types of the immune system (→Table 10.1), it also affects the growth and modulates the function of several epithelial cell types, especially of hepatocytes.

TP53 function is disturbed in most HCC, usually by the mechanism most prevalent across all cancers, i.e. point mutations in one allele and loss of the second allele by deletion or recombination (→5.3). In HCC from Europe, the USA, as well as in Japan and Taiwan, these mutations are spread across the central DNA binding domain of the TP53 protein. In HCC from Africa and other Asian countries, a mutational 'hotspot' is observed at codon 249. A transversion in its third base changes AGG to AGT and thus Arg to Ser in the TP53 protein. This particular mutation can be experimentally induced by aflatoxin $B_1$. Food contaminated with the fungus *Aspergillus flavus*, which produces this carcinogen, is mostly consumed

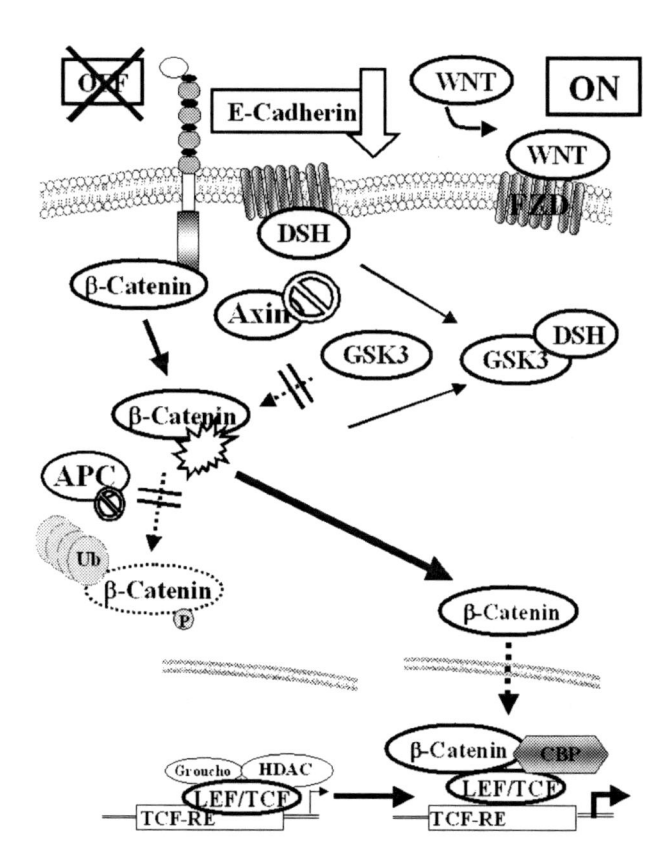

**Figure 16.4** *Mechanisms of WNT pathway activation in HCC*
The WNT pathway is most often activated in hepatocellular carcinoma by oncogenic β-Catenin mutations (exploding star). Alternatively, mutations in Axin and more rarely in APC (prohibition sign) lead to disaggregation of the complex phosphorylating β-Catenin. Constitutive activation of the pathway may often be exacerbated by loss of E-Cadherin which normally retains most β-Catenin at the cell membrane. See also Figures 6.12 and 13.3.

in the hot and humid regions of the world where this mutation is prevalent. So, like the mutations at pyrimidine dimers that are diagnostic for DNA damage by UVB (→12.1), the G→T mutation at codon 249 reveals the influence of a particular carcinogen. Moreover, this mutation is predominantly found in HCC patients with chronic HBV infection, so there is clearly some selection for the mutation from this side.

Throughout all HCC, mutations of TP53 tend to occur together with altered β-Catenin signaling. The mechanism underlying this association is not entirely clear, but it is notable that these changes also concur in colon carcinoma (→13.3). A plausible explanation is that constitutive WNT/β-Catenin signaling activates wild-type TP53 because it induces MYC and Cyclin D1 which increase (directly and indirectly) the transcription of p14$^{ARF1}$ (→6.6). In this fashion, a selective pressure for mutations in TP53 could be exerted. As in many other cancers, loss of TP53 function may contribute in HCC to decreased apoptosis and increased angiogenesis (→6.6). Both may favor cell survival in the environment of a cirrhotic liver.

Among the growth factors and receptors particularly important in HCC are those which also direct normal liver growth and regeneration: TGFα acting through the EGFR (ERBB1), hepatocyte growth factor (HGF) acting through MET and – perhaps more critical than in other cancers – insulin-like growth factor 2 (IGF2) and its receptors (Figure 16.5).

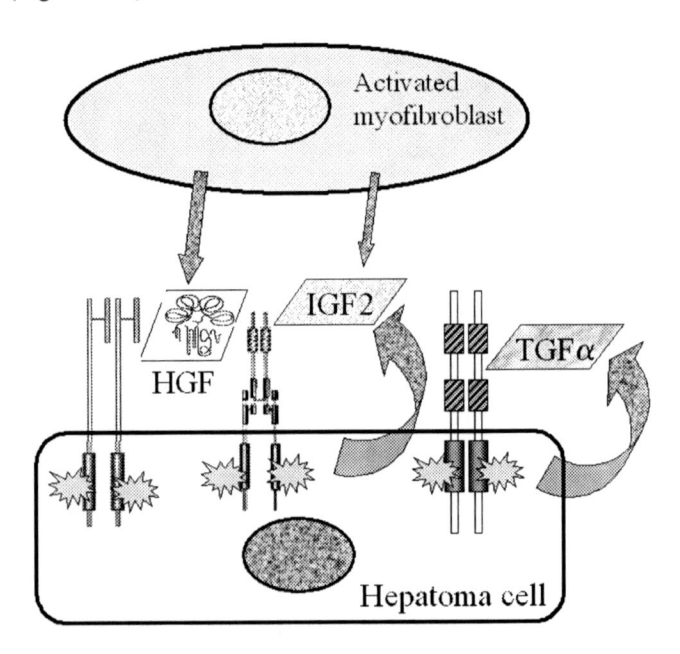

**Figure 16.5** *Growth factors and receptor tyrosine kinases in hepatocellular carcinoma*
See text for further details.

TGFα may be produced by hepatoma cells themselves. If these also express the EGFR, autocrine stimulation of growth and survival may result. In contrast, HGF is produced by nonepithelial cells in the liver. During cirrhosis and development of HCC, mesenchymal cells become activated. In particular, Ito cells which are normally located in the space of Dissé (between endothelial and epithelial cells), change to myoepithelial cells that proliferate themselves, but also secrete growth factors and cytokines acting on other stromal and epithelial cells. They are the likely source of HGF in hepatocellular carcinoma, whose action is enhanced by overexpression of the MET receptor on the hepatoma cells. Production, secretion and maturation of HGF may all be stimulated by factors from the tumor. The genes encoding the EGFR and MET are both located on chromosome 7 which is typically gained in HCC, as in many other carcinomas.

Overexpression of IGF2 is a central event in some cases of the childhood nephroblastoma, Wilms tumor (→11.3), but it is not at all uncommon in carcinomas afflicting older people. Insulin and/or insulin-like growth factors are necessary for the proliferation and function of normal hepatocytes. In HCC, IGF2 can be overproduced as a consequence of various mechanisms. In some cases 'loss of imprinting' (→8.2) is responsible. More complex mechanisms include altered usage of the four promoters of the *IGF2* gene.

IGF2 acts through the receptor tyrosine kinase IGFRI or an alternatively spliced form of the insulin receptor IR1 to promote cell survival and proliferation via the MAPK pathway, but perhaps more importantly via the PI3K pathway (→6.3). In principle, insulin exerts the same effects as IGF2, but it is a relatively weak growth factor, whereas IGF2 affects metabolic functions not as strongly as insulin.

The growth-promoting activity of IGF2 is normally limited by IGFRII. This protein is rightly named a 'receptor', in that it binds IGF2. However, ligand binding does not elicit a proliferation or survival signal. Instead, IGF2 is directed towards the lysosomes for degradation. Therefore, IGFRII is a 'scavenger receptor' that limits the action of insulin-like growth factors. Another limitation to their action is provided by binding proteins like IGFBP3, which is induced by TP53.

So, the cellular reaction to IGF2 is not only dependent on its concentration and expression level of its gene, but also regulated by several other factors. Specifically, the cellular reaction towards IGF2 is dependent on the relative expression levels of the IGFRI and IGFRII receptors. The *IGFR2* gene is located at chromosome 6q27, where LOH is frequent in HCC. Therefore, it is considered as the most likely candidate for a tumor suppressor in this region. There are, however, conflicting findings on this issue. One source of confusion is that Igfr2 is an imprinted gene in mice, and was initially reported to be imprinted in some human tissues as well. Although this idea has now been largely refuted, the notion tends to persist. The importance of insulin-like growth factors in liver cancer, however, is a fact and underlined by occasional point mutations that activate the IGFRI.

## 16.3 VIRUSES IN HCC

The genetic and epigenetic changes that cause HCC develop in a context of chronic tissue inflammation, with infiltration of cytotoxic cells, activation of stromal cells, continuous damage to hepatocytes and according need for regeneration in a tissue that becomes more and more disorganized. Hereditary diseases like hemochromatosis, chronic alcohol abuse, or chronic infection by the hepatitis viruses HBV and HCV can elicit this state each by themselves or together. Chronic inflammation in the liver provides an environment that is per se mutagenic, relieves some of the growth controls in a normal tissue, promotes proliferation of hepatocytic cells and selects for expansion of cells that are resistant to the adverse conditions. Thus, chronic infection by hepatitis viruses by itself would be thought to be sufficient to increase the risk of developing liver cancer. In fact, the risk of HCC in men chronically infected with HBV is increased $\approx$200-fold, which suggests that the virus may contribute to cancer development by additional mechanisms.

In humans, DNA viruses rather than retroviruses are implicated in carcinogenesis, with few exceptions ($\rightarrow$Box 2.1). The best understood human tumor virus is human papilloma virus ($\rightarrow$Box 5.1). Oncogenic strains like HPV16 and HPV18 contain E6 and E7 proteins which bind and inactivate TP53 and RB1, respectively, thereby interfering with two of the most crucial networks that control cell growth, differentiation, and genomic integrity. A causative role of HPV in cervical cancer is established beyond doubt and it is very likely in squamous carcinoma of the head and neck. The evidence is weaker for other cancers. The papovaviruses SV40, JCV, and BKV might act in a similar fashion by impeding the function of TP53 and RB1 through their multifunctional large T proteins. However, there is no conclusive evidence for their active involvement in human cancers, with the arguable exception of mesothelioma and SV40 ($\rightarrow$5.3). A large part of the human population carries JC or BK viruses without obvious adverse effects.

The same is almost true for the herpes virus EBV, although the normally relatively innocuous virus is quite clearly a cofactor in specific lymphatic cancers such as Burkitt lymphoma ($\rightarrow$10.3). As far as known, EBV does not act via TP53 or RB1. Instead, it appears to expand the population of cells amenable to transformation by specific translocations that activate oncogenes like MYC and complements their action by diminishing apoptosis ($\rightarrow$10.3).

HBV is also a DNA virus. However, its life cycle can best be characterized as that of a reverse retrovirus. The virus (Figure 16.6) contains a $\approx$3.3 kb DNA genome in an icosahedrical capsid formed by 240 hepatitis B virus core antigen (HBcAg) proteins. This core is surrounded by a membrane, derived from the host endoplasmatic reticulum, into which several isoforms of the HBV surface antigen (HBsAg) are embedded.

In infectious viruses, the genome is a circular incomplete double-stranded DNA. One strand is complete, but the circle is not closed, because it is covalently linked at its 5'-end to a tyrosine residue in the terminal protein (TP) encoded by the virus. This complete strand is called the minus-strand, because it is used exclusively as the

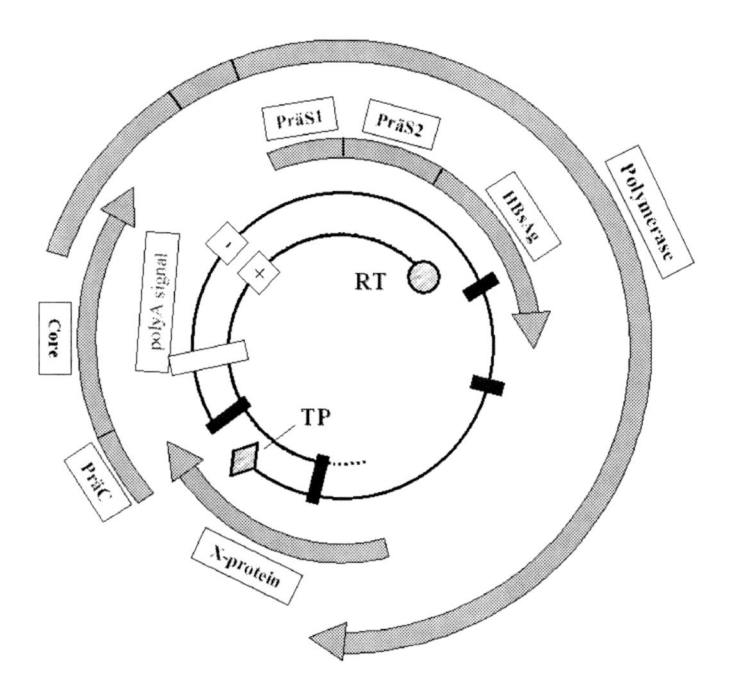

**Figure 16.6** *The genome of HBV*
At the center of the figure the gapped circular double-stranded DNA genome of the virus is depicted with the terminal protein (TP) and reverse transcriptase (RT) bound to it. The black boxes are repeats. The overlapping reading frames for the viral proteins are arranged around the genome. See text for further details.

template for transcription of the viral genome. The plus-strand in the virus is incomplete and of variable length, but always overlaps the gap in the minus-strand (Figure 16.6). At its 5'-end, it contains a conserved short, capped RNA primer and the viral polymerase is bound to its 3' variable end. After infection, the synthesis of this partly double-stranded genome is completed, with removal of the RNA primer. Three different RNAs can be transcribed from the viral genome. They encode the HBcAg, the HBsAg and a shorter secreted isoform of the same protein named HBeAg, the TP protein, and the regulatory HBx protein.

Following uptake into endosomes, the viral capsid escapes into the nucleus, where it is unpacked for replication and transcription to begin (Figure 16.7). As a rule, the DNA genome remains episomal, although it occasionally integrates into the host DNA, usually by recombination involving the single-strand segment (Figure 16.6) of the genome. Replication proceeds through an RNA intermediate. This RNA intermediate, like all other viral RNAs, is synthesized by host RNA polymerases. It is then used by the own polymerase of HBV, a combined reverse transcriptase/DNA polymerase with RNase H activity to generate the minus and plus DNA strands of the genome. The covalently linked polymerase molecule at the 5'-end serves to prime the first strand, and a capped RNA primer is used for the second strand. The genome contains two short direct repeats, which are used for template switching, similar as in retroviruses.

Replication and transcription of HBV take place concurrently. Transcription is stimulated by a liver-specific enhancer in the viral genome, which is responsive to glucocorticoids and – to a lesser extent – to androgens. Once sufficient amounts of viral proteins have become synthesized, the genome is packed into the capsid and the virus assembles in the endoplasmatic reticulum. Complete and incomplete virus particles appear to be secreted via a Golgi pathway.

HBV does not usually lyse cells. Hepatocyte death during infection is mainly caused by cytotoxic lymphocytes (CD8+ T-cells) which appear to act by the FAS/FAS-ligand route (→7.2). During chronic infection, NK cells become as well involved. Acute infections last several weeks and end, when the patient develops a sufficient titer of IgG antibodies against the viral surface and core antigens. Development of immunity is impeded by the secretion of the HBeAg and incomplete viral particles. Young children and other not fully immunocompetent individuals cannot mount an efficient IgG response, if any immune response at all, and the infection becomes chronic, sometimes without symptoms, but often with development of liver cirrhosis and eventually of HCC. Transmission from mother to child is an important route in areas where HBV is endemic.

Cures of HBV infections can sometimes be achieved by interferons $\alpha$ and $\beta$ which activate intracellular responses that prohibit virus replication. However, the best protection against HBV is vaccination using an HBsAg produced in yeast. This vaccine prevents both acute and chronic disease. Importantly, although a few related

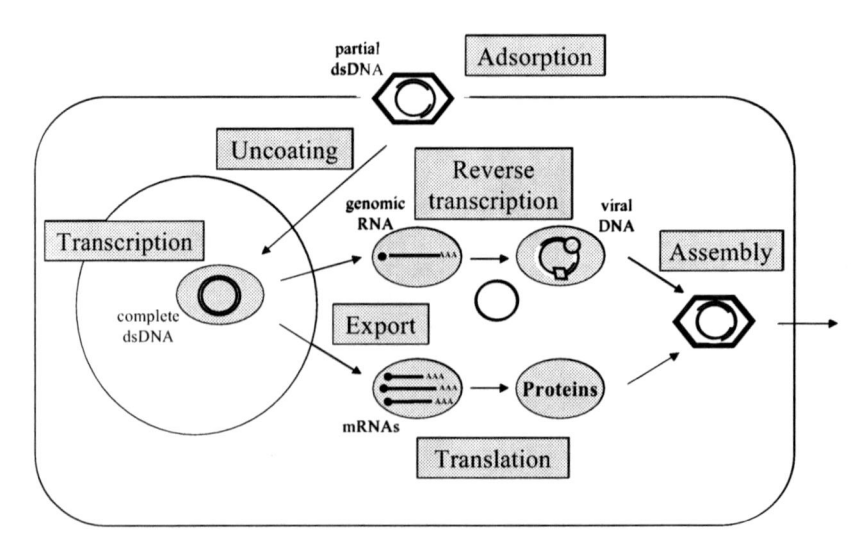

**Figure 16.7** *An outline of the HBV life cycle*
See text for further details.

animal viruses are known, the only known host of HBV are humans and, experimentally, chimpanzees. There is thus a chance to eradicate the virus.

So, what is there in HBV that might specifically promote hepatocarcinogenesis? Obviously, not really much beyond the factors already mentioned, i.e. an increased rate of hepatocyte cell death and regeneration and inflammation with tissue remodeling. Two further factors are debated.

(1) Many HCC arising in HBV-infected patients carry integrated HBV DNA in their genome. These copies are usually incomplete and not replication-competent (even if they could be excised from the genome back into an episomal form). In rare cases, they are integrated into an proto-oncogene such as a Cyclin gene, and may lead to its over-expression through the action of the viral enhancer ($\rightarrow$4.2). In other rare cases, viral integrates may disrupt tumor suppressor genes. There is also evidence that integrates are unstable and favor chromosomal rearrangements, creating a sort of extra 'fragile site' ($\rightarrow$2.2). This effect may pertain to integrates of many different viruses, beyond HBV.

(2) Other than episomal genomes, HBV integrates tend to express substantial amounts of HBx. This is a regulator protein which can transactivate cellular promoters by interaction with cellular transcription factors, even though it does not bind to DNA itself. It is thought that HBx in this manner increases the expression of anti-apoptotic proteins during acute and chronic infection. HBx may also activate cell proliferation by interacting with PKC enzymes ($\rightarrow$6.5). Certainly, HBx binds TP53 and promotes its degradation, like the HPV E6 protein ($\rightarrow$5.3). However, its effect is much weaker than that of E6. In fact, the geographical regions where both HBV and HCC are endemic are those in which co-carcinogens are common. In tropical Africa and Asia, the main co-carcinogen is likely aflatoxin $B_1$ which often causes the very specific G$\rightarrow$T mutation at codon 249 of TP53. It is possible that this specific mutation is not only selected for by the action of the mutagen, but also because it optimally complements the effect of HBx.

## *Further reading*

Knipe DM, Howley PM (eds.) Fields Virology 4$^{th}$ ed. 2 vols. Lippincott Williams & Wilkins 2001
Modrow S, Falke D, Truyen U (2003) Molekulare Virologie 2$^{nd}$ ed. Spektrum

Feitelson MA et al (2001) Genetic mechanisms of hepatocarcinogenesis. Oncogene 21, 2593-2604
Kern MA, Breuhahn K, Schirmacher P (2002) Molecular pathogenesis of human hepatocellular carcinoma. Adv. Cancer Res. 86, 67-112
Tannapfel A, Wittekind C (2002) Genes involved in hepatocellular carcinoma: deregulation in cell cycling and apoptosis. Virchows Arch. 440, 345-352
Thorgeirsson S, Grisham J (2002) Molecular pathogenesis of human hepatocellular carcinoma. Nat. Genet. 31, 339-346
Block TM, Mehta AS, Fimmel CJ, Jordan R (2003) Molecular viral oncology of hepatocellular carcinoma. Oncogene 22, 5093-5107
Llovet JM, Burroughs A, Bruix J (2003) Hepatocellular carcinoma. Lancet 362, 1907-1917
Pollak MN, Schernhammer ES, Hankinson SE (2004) Insulin-like growth factors and neoplasia. Nat. Rev. Cancer 4, 505-518

## Box 16.1 Hepatocellular carcinoma in experimental animals

Nowadays, mouse models dominate cancer research. This is mostly due to the fact that mice can be genetically modified almost at will. In 'transgenic' mice, a putative oncogene can be overexpressed. In 'knockout' animals, one or both copies of a gene can be completely or partially inactivated to study a possible function as a tumor suppressor. Animals harboring different genetic changes can be crossed to analyze gene interactions. Since transgenes and gene knockouts often interfere with fetal development, if the genetic changes are introduced into every cell, more sophisticated techniques have been developed which allow conditional gene activation or inactivation in particular cell types, at particular stages of development, transiently or in only a fraction of cells. In addition, a variety of mouse strains generated by traditional breeding or by random mutations exists which are prone to specific cancers or to cancer in general.

Before mice became so popular in cancer research, rats were more often used, particularly to study chemical carcinogenesis, since rat genetics is less advanced than mouse genetics. Specifically, a variety of chemicals applied in defined regimes elicit hepatomas in rats. Some of these can be transplanted from animal to animal or have given rise to cell lines. These hepatomas range from well-differentiated to anaplastic and provide useful models for molecular and therapeutic studies.

Typically, chemical carcinogens are not only mutagenic, but also elicit widespread death of hepatocytes and other liver cells. Some carcinogens destroy a large fraction of all hepatocytes, inducing fatty livers and/or a state resembling cirrhosis in humans. Precursor cells of hepatocytes and biliary epithelial cells termed 'oval cells' proliferate and replace the damaged parenchyme and ducts. These cells were later also detected in human liver following the clue from rat models.

Other carcinogens cause more subtle changes. First, isoenzyme switches are observed in localized areas of the liver without gross morphological changes ('enzyme-altered foci'), apparently from the expansion of mutated hepatocyte clones. Then, benign nodules develop, from which hepatocellular carcinomas arise. Similar precursor stages were discovered in human livers after they had been seen in animals.

Evidently, histomorphology has benefitted from carcinogenesis research in rats, although molecular biology not as much. Studies of experimental carcinogenesis in rats have contributed surprisingly little to the understanding of genetic changes in human liver cancers. Mutations in p53 or β-catenin were first found in humans and later in carcinogen-treated rats. It is in fact debated whether all of the carcinogens causing liver cancer in rats do so in humans. At least, they do not typically cause cancer in this organ. This difference has led to the recognition that substantial differences exist in the metabolism of carcinogens between rats, humans and other species and from there to the recognition that such differences also exist among humans (→Table 14.1, →21.5). Thus, 'pharmacogenetics' is another field that has been stimulated by experimental hepatocarcinogenesis in rats.

Pitot HC (2002) Principles of oncology, Marcel Dekker, 4$^{th}$ ed.

# CHAPTER 17

# STOMACH CANCER

- Stomach cancer (or gastric cancer) is the second most frequent cause of cancer death worldwide with almost 1 million new cases per year. It poses a serious health problem because of its low cure rate and its severe impact on the quality of life. Fortunately, its incidence has plummeted in many industrialized countries for several decades. The causes of this unique decrease are presumed in improved hygiene, altered diet, and widespread use of antibiotics reducing the prevalence of *Helicobacter pylori* infection.
- *Helicobacter pylori* infection is associated with most cases of stomach cancer. About 50% of the world population carry strains of the bacterium, but <10% develop inflammatory disease and ulcers, and even fewer stomach cancer. The outcome of the infection is determined by genetic variability of the germ and of the host which act in a strongly synergistic fashion. In the bacterium, variations in babA2, cagA, and vacA genes and in the host, polymorphisms in cytokine genes, particularly *IL1B*, influence the risk of chronic inflammation, ulcers, and cancer. Furthermore, carcinogenesis is dependent on cofactors, such as dietary carcinogens and protective ingredients.
- Most stomach cancers arise in the antrum and corpus. The predominant histological subtypes are the intestinal type and the diffuse type. The intestinal type develops from gastric atrophy in areas of intestinal metaplasia. The highly invasive diffuse type consists of small groups of easily scattering undifferentiated cells. Both histological types share some alterations such as the TP53 inactivation.
- Intestinal metaplasia is associated with altered expression of transcription factors that regulate the segmentation of the gut tube during fetal development. From benign metaplasia, intestinal-type gastric cancer develops in a fashion resembling colon carcinoma in some respects.
- The most distinctive alterations in diffuse-type stomach cancers are mutations of the *CDH1* gene encoding E-Cadherin. Rare familial cases of this cancer type are caused by germ-line mutations in E-Cadherin. So, *CDH1* behaves as a classical tumor suppressor gene in this specific cancer type.
- The decreased incidence of 'classical' gastric cancer in industrialized countries is partly offset by a rising incidence of cancers of the esophagus and the upper parts of the stomach. These cancers are associated with alcohol consumption and smoking, and they may be promoted by enhanced acidity of the stomach juice as a consequence of *H. pylori* eradication. They are also typically associated with metaplasia.

## CHAPTER 17

### 17.1 ETIOLOGY OF STOMACH CANCER

Worldwide, about 900,000 persons are diagnosed with stomach cancer (or 'gastric cancer') each year, making it the second most frequent cancer. Since the 5 year survival rate is 3 – 20%, it is also one of the most lethal malignancies. Survival most strongly depends on the tumor stage at diagnosis. Cures can be achieved by complete or partial surgical removal of the stomach. As with many other metastatic carcinomas, chemotherapy is only marginally successful.

While stomach cancer thus clearly remains a major health problem worldwide, the situation is at least improving in many developed countries. For almost 50 years now, the incidence rates have steadily declined in Western Europe and North America (Figure 17.1). This fortunate decline would be admirable, if it had been achieved by conscious human intervention. It was, however, not and only to a small degree caused by improvements in diagnosis and therapy. Instead, the decline in stomach cancer incidence appears to be associated with an altered life-style, in particular with improvements in food quality and with the use of antibiotics. It is

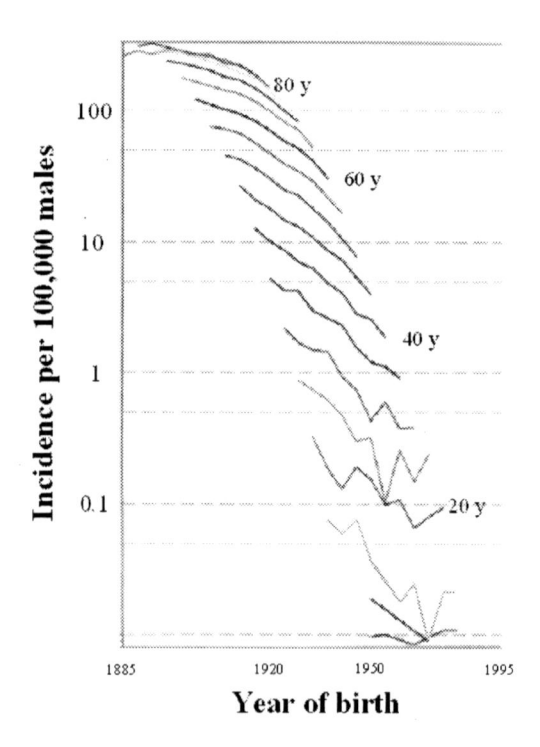

**Figure 17.1** *Decline of stomach cancer incidence in industrialized countries*
Data for the United Kingdom according to the WHO/IARC. This is a birth cohort representation, where the incidence is plotted depending on the year of birth (x-axis) and the age of cancer presentation (in 5 year increments).

important to understand the causes underlying this decrease as thoroughly as possible, not only to further reduce the incidence of stomach cancer in the industrialized countries, but even more to address the problem in those developing countries where high incidence rates persist.

Like liver cancer (→16.1), gastric cancer most often develops in the context of chronic inflammation, named 'chronic gastritis' in the stomach. However, chronic inflammation in this organ is not caused by viruses like HBV and HCV (→16.3) and less strongly by alcoholic drinks. Rather, the crucial agent is a bacterium, *Helicobacter pylori*.

Again as in the liver, complex interactions between the host and the pathogen determine whether chronic inflammation ensues at all and whether it develops further towards cancer. Moreover, some of the co-carcinogens and protective factors are the same, as alcohol abuse and diets low in fresh fruit and vegetables increase the risk of cancers in both organs. Even the role of aflatoxins in the liver may have its complement in the stomach, where dietary nitrosamines are thought to act as mutagens. These factors (Table 17.1) appear to act in a synergistic fashion, although bacterial infection may be the most crucial prerequisite.

Accordingly, the geographical differences in the incidence of gastric cancer can be ascribed to a combination of factors. They appear to result from a synergistic effect of eridication of *Helicobacter pylori* by antibiotic treatment (intentional or incidental) and improvements in food quality. Importantly, stomach cancer is the only clear-cut example of a human cancer caused by bacterial infection. It may not be the only one, though, since bacterial infections causing chronic infections may also lead to an increased risk of cancers in other organs. Another well-documented case of infections causing cancer involves a trematode parasite, *Schistosoma mansoni*, which induces chronic inflammation, metaplasia and a metaplastic cancer, squamous cell carcinoma, in the urinary bladder (→14.1).

With the change in gastric cancer incidence in industrialized countries, the localization of the cancers within the stomach and the relative frequencies of the histological subtypes have also changed. Whereas formerly the vast majority of stomach cancers originated in the lower parts of the stomach, i.e. the corpus (body) and antrum, an increased proportion of cases now originates in the upper sections,

**Table 17.1.** *Causes of stomach cancer cancer*

| *Individual factors causing stomach cancer in humans*\* |
|---|
| *Helicobacter pylori* |
| Nitrosamines |
| Low consumption of fresh fruits and vegetables |
| Alcohol |
| Tobacco smoking |
| Genetic predisposition by high risk gene mutations (*CDH1*) |
| Genetic predisposition by polymorphisms (*IL1B*) |

\* Several factors may interact, often in a synergistic fashion.

the fundus and cardia. In fact, this shift may continue further upwards within the gastrointestinal tract, since the decreased incidence of stomach cancer appears to be accompanied by an increased incidence of carcinoma in the esophagus (Box 17.1). The shift between the histological subtypes takes place, because the decrease in incidence is stronger for the 'intestinal' than the 'diffuse' subtype, which are the major histological subtypes of stomach cancer (Figure 17.2).

The intestinal subtype develops in metaplastic areas. Normally, the surface of the stomach is lined by a simple columnar epithelium in which several different cell types reside. Chief cells produce the protease precursor pepsinogen and enteroendocrine cells secrete peptides and transmitters regulating the function, but also the growth of intestinal and gastric cells. Parietal cells secrete hydrochloric acid to generate the low pH in the stomach. These cells are located in pits, which are protected by mucus produced by epithelial surface cells.

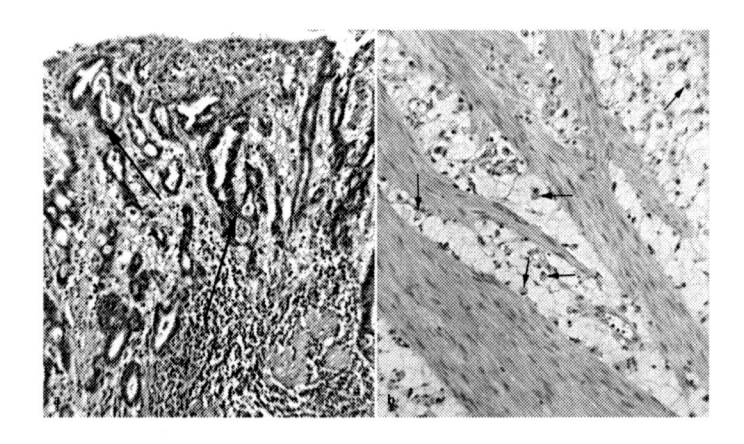

**Figure 17.2** *The major histological subtypes of gastric cancer*
Left: Intestinal-type carcinoma; Right: Diffuse-type carcinoma. The arrows in the right figure point to tumor cell islets.

During intestinal metaplasia, this well-organized structure is replaced by a glandular epithelium which resembles the villous structure of the intestine. Metaplasia of the stomach epithelium precedes the development of cancer. It is one of several clearly defined 'preneoplastic' stages in the development of intestinal type stomach cancer (Figure 17.3). As in the liver (→16.1), most cancers in the stomach arise in a context of chronic inflammation, i.e. gastritis. Chronic gastritis in some cases progresses to an atrophic stage from which intestinal metaplasia develops. Initially, well-differentiated glandular structures are formed, which can develop into benign and increasingly dysplastic adenomas progressing towards invasive carcinomas.

The histological progression of diffuse-type stomach cancer is less well documented. This carcinoma is named for its its characteristic histology. It presents as small groups of loosely attached, highly invasive and undifferentiated tumor cells proliferating into the tissues underlying the gastric mucosa and metastasizing through the peritoneum (Figure 17.2).

Both types of stomach cancer are linked to *Helicobacter pylori* infection, although the relationship to the intestinal subtype is stricter. Patients with the diffuse type cancer are on average younger, whereas intestinal-type stomach cancer usually develops through several preneoplastic stages in older people after decades of gastritis (Figure 17.3).

Normal gastric epithelium
↓
Gastritis
↓
Gastric atrophy
↓
Intestinal metaplasia
↓
Well-differentiated adenoma
↓
Dysplastic Adenoma
↓
Intestinal-type Carcinoma

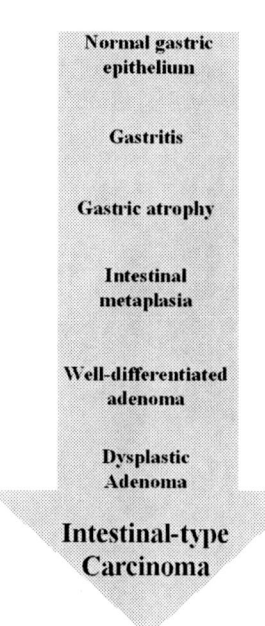

**Figure 17.3** *Stages in the development of intestinal-type stomach cancer*

## 17.2 MOLECULAR MECHANISMS IN GASTRIC CANCER

The two major histological subtypes of stomach cancer share many genetic alterations, e.g. *TP53* mutations. Other alterations differ significantly between them and account for typical characteristics of each subtype.

The crucial molecule in the diffuse type is E-cadherin. This 882 amino acid membrane glycoprotein mediates homotypic adhesion between epithelial cells (→9.2). It connects to the cytoskeleton by way of the catenins, of which β-Catenin doubles as a signaling molecule in the WNT pathway and a transcriptional co-activator (→6.10). In many different types of carcinomas, E-Cadherin is frequently down-regulated during tumor progression. However, it is infrequently mutated. To name just one example from the gastrointestinal tract, down-regulation of E-cadherin and LOH of its gene (*CDH1*) at 16q are frequent in hepatocellular carcinoma (→16.2).

In >50% of all diffuse-type gastric cancers, *CDH1* is mutated, and in almost all others, E-cadherin is strongly down-regulated. Many mutations in the gene are small in-frame deletions or splice site mutations affecting the adhesion domain (Figure 17.4). They lead to a shortened protein that has lost the ability for homotypic interactions. It may, instead, interfere with the function of intact E-cadherin and of related proteins such as N-cadherin (→9.2). The consistent dysfunction of E-Cadherin provides an obvious explanation for the decreased adhesiveness and the 'diffuse' growth pattern of the tumor cells of this cancer. Moreover, loss of E-Cadherin may contribute to WNT-pathway deregulation (→16.2).

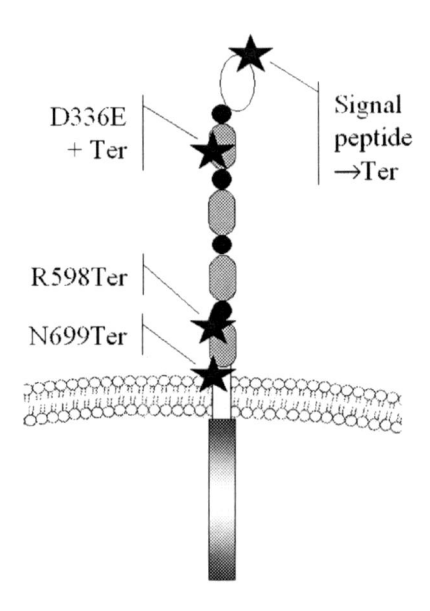

**Figure 17.4** *Mutations of E-Cadherin in diffuse type stomach cancer*
The figure shows the localization of mutations in hereditary cases, which all lead to truncation of the protein or in cases of mutations in the signal peptide to its absence from the cell membrane. Mutations in sporadic cases have the same effects. cf. also Figure 9.3.

The importance of E-Cadherin mutations in sporadic diffuse-type gastric cancer is underlined by the genetic defect in its – very rare – familial form. In a small number of families in which young people succumb to diffuse-type gastric cancer, the disease co-segregates with mutations in *CDH1*. In the cancers that develop, the second, wild-type allele is lost. So, *CDH1* is a classical 'gatekeeper-type' tumor suppressor gene (→5.4) for diffuse-type gastric cancer, with germ-line mutations in familial cancers and biallelic inactivation in sporadic cases. Remarkably, no other cancers appear at elevated rates in the affected families. Thus, there must be something particular about E-Cadherin in the stomach epithelium. Another consistent genetic alteration in diffuse-type gastric cancer are *TP53* mutations, as one might expect in a tumor exhibiting aneuploidy and high invasiveness.

There is no similarly evident genetic predisposition for intestinal-type cancer of the stomach. An increased risk of gastric cancer is observed in HNPCC families (→13.4). A significant proportion of sporadic cases, overall >10%, also exhibit microsatellite instabilities (→3.2). Some mutations in genes found in these cases are similar to those in colon cancers arising by the microsatellite instability pathway (→13.4), e.g. in the *APC, CTNNB1,* and *TGFBRII* genes. This suggests that intestinal type gastric cancer is a sort of colon cancer arising in the wrong organ. Indeed, constitutive activation of the WNT pathway can also be detected in some gastric cancers without an MSI phenotype. It is most often caused by inactivation of *APC*, but other changes like oncogenic mutations of β-Catenin are also found. The changes in the WNT pathway may be compounded through loss of E-Cadherin by LOH and down-regulation (rather than mutation), and by mutations of TP53.

Compared to colon cancer, the TGFβ response (→6.7) may be a more important and probably earlier target of mutations in the stomach. Inactivation of components of the actual pathway does occur, but in gastric cancer a more widespread target may be a 'downstream' transcription factor, RUNX3, which regulates gene expression in gastric cells in response to TGFβ signaling. This factor is inactivated in many gastric cancers by deletion of one *RUNX3* gene copy and promoter hypermethylation of the second, remaining allele.

Metaplasia as such is not a malignancy, although metaplastic tissue is functionally disturbed. Metaplastic intestinal tissue in the stomach is not suited for an environment with a pH as low as 1 and, of course, cannot secrete proteases and peptides like normal gastric pit cells. Similarly, metaplastic squamous epithelium in the urinary tract (→14.1) is not as impermeable to urine components as normal urothelium. In both organs, therefore, the presence of metaplastic tissue tends to aggravate ongoing tissue damage and inflammation.

A straightforward explanation for the development of metaplasia in the context of chronic inflammation is that damage to the tissue over a long period cannot be compensated simply by division and expansion of existing differentiated cells. Rather, chronic tissue damage may necessitate the recruitment of tissue precursor cells. While some of these may differentiate correctly and replenish the damaged tissue, others may be deviated by the adverse conditions in the inflammated organ towards a different cell fate. Conceivably, direct influences from a pathogen like *H. pylori* might also disturb the direction of differentiation (→17.3).

In a sense, intestinal differentiation in the stomach is not exotic, because the stomach and intestine develop from consecutive segments of the embryonal gut tube. During development, a network of transcriptional regulators determines whether segments of the gut assume an anterior, i.e. gastric, or a posterior, i.e. intestinal, identity. There is indeed evidence that intestinal metaplasia in the stomach is associated with and maybe caused by a shift in the pattern of these regulators. The factors involved belong to the HOX and SRY box (SOX) family. The most clearly identified components are the caudal homeobox proteins CDX2 and CDX1 and

SOX2. In addition, specific (canonical) HOX proteins are implicated (Figure 17.5). Expression of CDX2 is normally restricted to the intestine and the protein is not detectable in gastric epithelial cells, whereas SOX2 shows the reverse pattern, being present in cells in the stomach, but not in the intestine. As one might expect, in metaplastic intestinal epithelium in the stomach CDX2 is expressed and SOX2 is down-regulated.

Although they both act as determinants of an intestinal phenotype, CDX2 and CDX1 are actually antagonists. CDX2 promotes the terminal differentiation of intestinal cells, e.g. by inducing $p21^{CIP1}$, whereas CDX1 maintains intestinal cells in a proliferative state. Accordingly, in well-differentiated metaplastic intestinal tissue in the stomach (Figure 17.3), CDX2 is expressed, but CDX1 only weakly. When dysplastic adenomas and carcinomas develop from metaplastic precursor tissue, expression of CDX1 increases and CDX2 becomes down-regulated, in some cases bolstered by hypermethylation of its gene promoter (Figure 17.5).

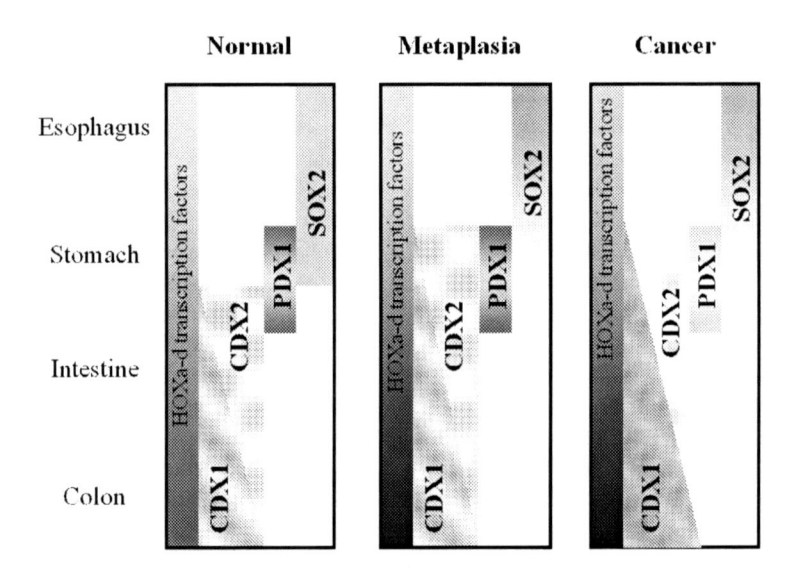

**Figure 17.5** *HOX and SOX factors in stomach metaplasia*
A model of transcription factor changes during development of intestinal metaplasia and intestinal-type cancer in the stomach according to Y. Yuasa. See text for further exposition.

## 17.3 HELICOBACTER PYLORI AND STOMACH CANCER

Since *Helicobacter pylori* was discovered in 1983 (!) and soon after suggested as a cause of gastritis, 'peptic ulcers', and stomach cancer, a large body of evidence has been assembled that supports its classification as a human carcinogen. Unusual in cancer epidemiology, even controlled prospective longitudinal studies have been performed. These demonstrate that infected persons, as indicated by the presence of antibodies against proteins of the bacterium (seropositivity), have a higher risk of developing chronic gastritis as well as stomach cancer and that the risk for stomach cancer increases the longer the bacterium persists in the stomach. The converse approach, intervention, has also been attempted and shown to be successful. Antibiotic treatment that eradicates *Helicobacter* prevents gastritis and cancer. In particular, it appears to prevent in many cases the progression from gastric atrophy to metaplasia and cancer (Figure 17.3).

*Helicobacter pylori* is one of several related bacteria which live in the human gastrointestinal system. Usually they do not cause severe disease, but only an initial gastroenteritis, which can even pass without symptoms. Most people become infected by *Helicobacter pylori* as children and remain so for life, unless the germ is eliminated by antibiotic treatment (usually prescribed for a different infection). The stomach environment is designed to kill or damage bacteria and other infectious agents through its low pH and through the action of the protease pepsin.

*Helicobacter pylori* manages to survive in this environment by a variety of mechanisms developed during the >10,000 years in which it has co-evolved with its host. For instance, it buffers gastric hydrochloric acid by hydrolysing urea to bicarbonate and ammonia through the enzymatic action of urease produced by all *H. pylori* strains. Some strains cling closely to the epithelial surface which is relatively protected by buffering mucins.

Several bacterial proteins mediate the interaction between the bacteria and the epithelial cells of the stomach mucosa.

The babA2 protein, e.g., attaches to the Lewis[b] (blood group) antigen on the surface of gastric epithelial cells.

The bacterial 'cag island' encompasses 31 genes within 40 kb. This 'pathogenicity island', which may move as a unit between bacterial species and strains, encodes a secretion and translocation system that transports the cagA protein into stomach epithelial cells. Within the host cell, the cagA protein becomes tyrosine phosphorylated by SRC tyrosine kinases (→6.5). This phosphorylation elicits changes in cellular cytoskeletal proteins, cell shape and adhesion (Figure 17.6). The tyrosine phosphate in the bacterial protein is recognized by the adaptor proteins SHP2 and GRB2 that activate MAPK and other signaling pathways (→4.4). The resulting change in the epithelial cells may facilitate adhesion of the bacteria and allow them to take cover between and behind the mucosal cells. Introduction of cagA into epithelial cells also induces the expression and secretion of the interleukin IL8. This is a pro-inflammatory cytokine which induces a cellular immune response to the infection. This response is accompanied by increased expression of further cytokines, prominently $IL1\beta$ and $TNF\alpha$ (tumor necrosis factor $\alpha$). On one hand, these cytokines further promote inflammation, on the other hand they decrease the secretion of hydrochloric acid by the parietal cells of the gastric epithelium. This constitutes a second important mechanism by which *Helicobacter pylori* increases the pH in the stomach fluid and can survive in this hostile environment.

The effect of the bacterium on the gastric epithelium is compounded by its vacA protein which induces vacuolization (hence its name) and apoptosis of host cells.

This leads to the release of nutrients and further facilitates penetration of *H. pylori* through the mucosal barrier. The vacA protein is also immunosuppressive, diminishing responses by macrophages and T-cells.

In summary, as the bacterium changes its environment to fit its needs, it induces regenerative proliferation of the gastric epithelium which is stimulated by the peptide gastrin and by cytokines that also contribute to an immune reaction which can lead to tissue reorganization and to chronic inflammation.

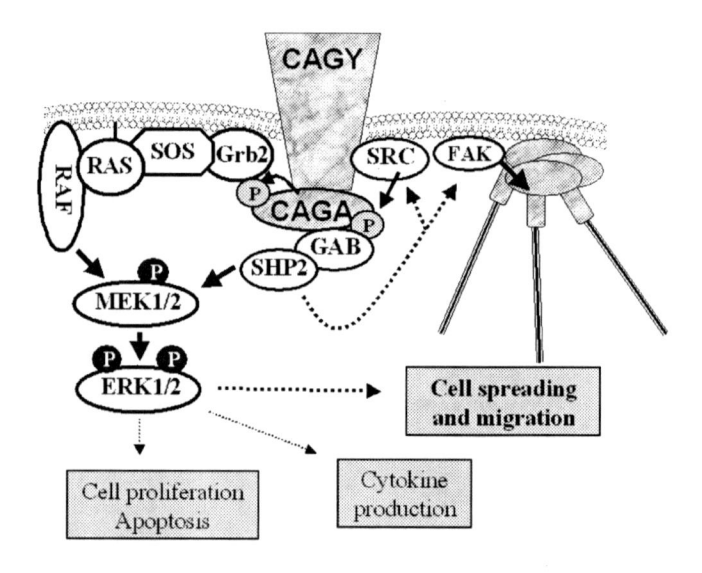

**Figure 17.6** *Effects of the H. pylori cagA protein in stomach mucosa cells* Following its tyrosine phosphorylation by SRC, the protein acts mostly through the two pathways shown to elicit altered cell morphology, interactions, and proliferation.

The outcome of an *H. pylori* infection can vary substantially. Most infected humans do not experience any symptoms over decades. So, one potential, frequent outcome appears to be a stable equilibrium between tissue damage by the bacterium and immune response by the host. In other cases, fulminant inflammation and permanent tissue damage lead to the symptoms of a peptic ulcer and in some individuals, cancer develops. Overall, while between 10% and 65% of individuals in different populations worldwide are infected, only a few percent of them develop chronic gastritis and again a fraction of these eventually come down with stomach cancer. Nevertheless, worldwide, one million cases per year result, with huge differences between different populations. These differences in the relative incidences are caused by an interaction of several factors (Table 17.1).

*H. pylori* strain differences: There are substantial differences between strains of *Helicobacter pylori* in each of the three pathogenetic determinants discussed above, babA2, cagA, and vacA. Strains without functional babA2 genes are much less apt at inducing intestinal-type stomach cancer, probably because fewer bacteria adhere to epithelial cells. The babA2⁺ strains also tend to induce autoimmune reactions much more strongly than others. The cagA island, likewise, is present in most, but not in all strains. Strains lacking cagA induce less inflammation and do not appear to promote cancer development. The presence of actively secreted vacA is also statistically correlated with an increased cancer risk. However, the interpretation of vacA associations is complicated by the peculiar distribution of the vacA genotypes. There are several major polymorphic forms of the gene, designated $s_1a$, $s_1b$, $s_1c$ and $s_2$ (with further subtypes). Each is associated with a particular human population to such an extent that they can be used to follow human migrations. While this association gives good reason to believe that vacA polymorphisms may account for geographic differences in stomach cancer incidence, it complicates studies in those mixed human populations, where people of different origins live under different socioeconomic conditions, and prevents studies of the influence of vacA in homogeneous human populations. Moreover, cagA and vacA genotypes are not completely independent of each other.

Host reaction: There are also substantial differences in the host reaction towards *H. pylori* infection. Some are caused by genetic polymorphisms, while others may be determined by co-infections, e.g. worm parasites. Most of the genetic polymorphisms concern co-determinants of the immune response such as the MHC type or genes encoding cytokines and cytokine receptors. Since stomach cancer is a multifactorial disease, it is not surprising that not all studies agree on the importance of every polymorphism implicated. However, the case for polymorphisms affecting the expression of the cytokine IL1β has been confirmed by several independent studies. This suggests that the effect is large and relevant in different populations.

The *IL1B* gene is polymorphic in man (Figure 17.7). Functional polymorphisms in its promoter influence the expression level of the cytokine, most strongly a C/T

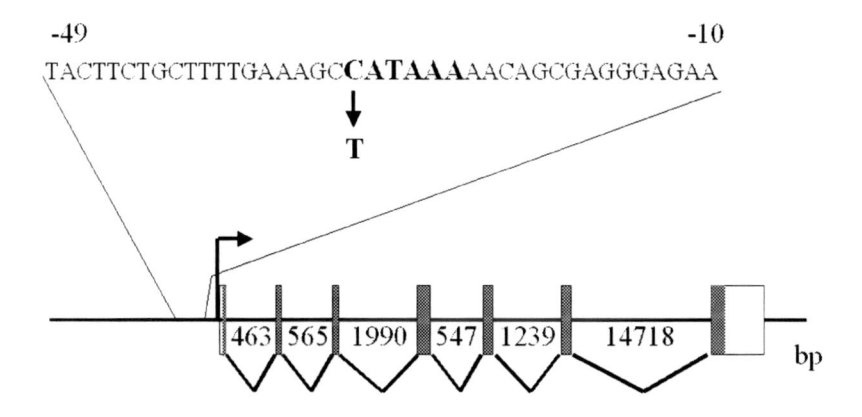

**Figure 17.7** *The IL1B promoter polymorphism related to stomach cancer risk*
The indicated C→T change generates a TATAAA box, substantially increasing the strength of the IL1B gene promoter.

polymorphism at −31 in the TATA box of the gene. In T-alleles, transcription factors bind more strongly than in C-alleles, and expression of IL1B is more strongly inducible. The presence of T-alleles is associated with an ≈10-fold increased risk of gastric ulcers and cancer. Of note, the genes encoding IL1β, IL1α, and their receptor IL1R are all located close to the *ILB* gene. Polymorphisms in these genes, including the *IL1B* -31C/T polymorphisms are often in linkage disequilibrium with others in the gene cluster. This means that a particular form (allele) of the *IL1B* gene is as a rule found together with a particular form (allele) of the *IL1R* gene. Indeed, other studies have found associations of cancer risk with the *IL1R* genotype. As one might suspect, the effect of the *IL1B* polymorphism is synergistic with that of the *Helicobacter pylori* genotype. Across all human populations, the risk of stomach cancer is estimated to be ≈2-3-fold elevated by becoming infected with any strain of *Helicobacter pylori*. By comparison, the risk of *IL1B* −31T carriers infected with cagA$^+$, vacA s$_1$m$_1$ strains is almost 100-fold enhanced.

Dietary carcinogens and irritants: A third important factor in the development of gastric cancer appears to be the presence of co-carcinogens in the diet, specifically alcohol, excessive salt and nitrosamines. Alcohol and salt may act primarily as irritants, aggravating tissue destruction and inflammation. Nitrosamines are implicated as direct mutagens. They are alkylating reagents that react, in particular, with DNA bases and as a rule cause point mutations. They are therefore thought to be responsible for many of the point mutations found in gastric cancers (→17.2), e.g. in *TP53*, *CDH1* (E-Cadherin), and *CTNNB1* (β-Catenin). They are present in the diet, e.g. in pickled foods, or are contained in cigarette smoke. Some may be generated in the acidic milieu of the stomach by reaction of nitric acid with amines contained in food (Figure 17.8). Nitric acid is produced by reduction of dietary nitrates in the oral cavity and during passage through the esophagus. Its concentration is particularly high in spoiling salted or pickled food. Improvements in

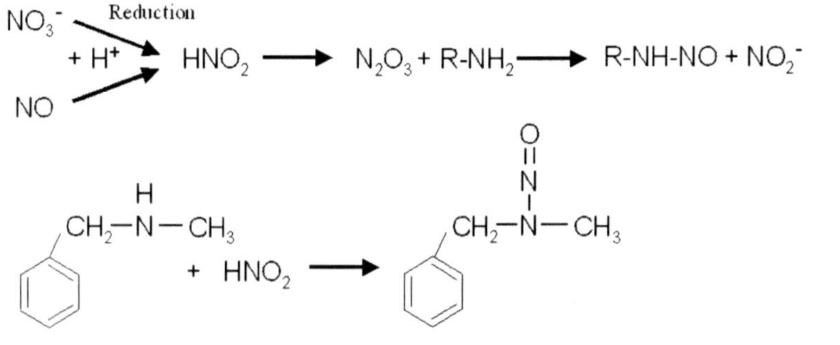

**Figure 17.8** *Nitrosation of amines in an acidic milieu*
Top: General reaction. Bottom: Formation of a nitrosamine (N-nitroso-benzyl-methylamine) from benzyl-methyl-amine.

food quality over the last decades, specifically the replacement of pickling by refrigeration, may have made an equally important contribution to the decline of stomach cancer as the eradication of *Helicobacter pylori* by antibiotics in many individuals.

Fruit and vegetables: Conversely, several other components of the diet appear to protect against stomach cancer. Particularly important may be antioxidants like vitamin C and the methyl group carrier folate, which are present in fresh fruit and vegetables (cf. 20.3). The increasing availability of fresh fruit and vegetables in the Western industrialized countries throughout the year, as opposed to use of pickled, salted and dried food before the advent of global transport and refrigeration, may also have contributed to the decline in stomach cancer.

These compounds may exert their protective effect partly by extracellular and partly by intracellular mechanisms. In the stomach fluid, vitamin C reduces and inactivates nitrosamines. In the inflammated stomach epithelium, antioxidants diminish the effect of reactive oxygen species (→Box 1.2). They may therefore protect against cell damage and decrease the rate of mutations. A sufficient supply of the methyl group carrier folate ensures that thymidine pools are adequate, preventing uridine misincorporation and DNA strand breaks, and ensures an adequate supply of S-adenosylmethionine needed to maintain proper DNA methylation patterns (→8.3, →20.3).

In summary, therefore, the development of stomach cancer as a consequence of infection by *Helicobacter pylori* is not only a multistep, but also a multifactorial process. Several factors synergize (Table 17.1). As in the development of liver cancer following HBV infection (→16.3), chronic inflammation is a crucial prerequisite and is contingent on the genotypes of both pathogen and host. Chronic inflammation as such is likely mutagenic, but most of all it creates an environment in which cells with an altered phenotype are selected for and expand. It may therefore be significant that *TP53* mutations are found already at some frequency in chronic gastritis (Figure 17.3). Again as in liver cancer, both the creation of altered cells and their progression towards cancer cells are accelerated by the presence of mutagens and a lack of protective compounds (→Table 16.1). In stomach cancer, the part of mutagenic aflatoxins in the liver is apparently performed by nitrosamines.

It is likely that each of the above factors may be needed for this cancer to arise at the excessive rate still seen in several parts of the world. The epidemiological data clearly shows that stomach cancer can be largely prevented. Conceivably, each factor involved in gastric carcinogenesis can be addressed for this purpose (→20).

## *Further reading*

Jankowski JA et al (1999) Molecular evolution of the metaplasia-dysplasia-adenocarcinoma sequence in the esophagus. Am. J. Pathol. 154, 965-973

El-Omar EM et al. (2000) Interleukin-1 polymorphisms associated with increased risk of gastric cancer. Nature 404, 398-402

Gonzalez CA, Sala N, Capella G (2002) Genetic susceptibility and gastric cancer risk. Int. J. Cancer 100, 249-260

Hajra KM, Fearon ER (2002) Cadherin and Catenin alterations in human cancer. Genes, Chromosomes Cancer 34, 255-268

Peek RM, Blaser MJ (2002) *Helicobacter pylori* and gastrointestinal tract adenocarcinomas Nat. Rev. Cancer 2, 28-37

Correa P (2003) Bacterial infections as a cause of cancer. J. Natl. Cancer Inst. 95, E3

Gonzalez CA, Sala N, Capella G (2002) Genetic susceptibility and gastric cancer risk. Int. J. Cancer 100, 249-260

Höfler H, Becker KF (2003) Molecular mechanisms of carcinogenesis in gastric cancer. Rec. Res. Cancer Res. 162, 65-72

Yuasa Y (2003) Control of gut differentiation and intestinal-type gastric carcinogenesis. Nat. Rev. Cancer 3, 592-600

Blaser MJ, Atherton JC (2004) *Helicobacter pylori* persistence: biology and disease. J. Clin. Invest. 113, 321-333

Ushijima T, Sasako M (2004) Focus on gastric cancer. Cancer Cell 5, 121-125

### Box 17.1: Barrett esophagus and esophageal cancer

*Helicobacter pylori* is certainly a human carcinogen, but like many pathogens it has co-evolved with its host to provide some beneficial effects. These may include protection against other pathogens, better control of satiety and appetite, and decreased reflux of acidic stomach fluid into the esophagus. So, eradication of the bacterium by antibiotic treatment and better hygiene are expected to diminish the risk of gastritis and stomach cancer, but may increase the risk of obesity and of esophageal inflammation and cancer.

Adenocarcinoma of the esophagus develops usually in the lower part of the organ from a precursor stage named Barrett esophagus, another preneoplastic metaplasia. The esophageal tube is normally lined by a stratified squamous epithelium. In Barrett esophagus, the epithelium near the gastroesophageal sphincter changes into a simple columnar epithelium, typically with goblet cells, as in the intestine. This lesion becomes dysplastic and up to 1% of all affected individuals per year develop adenocarcinoma, which like cancers of the esophagus in general, is frequently fatal.

| Normal epithelium | Barrett metaplasia | Dysplastic metaplasia | Adeno-carcinoma |
|---|---|---|---|

This sequence of events is similar to that in intestinal-type stomach cancer (cf. Fig. 17.3) and so are several etiological factors (Table 17.1), including alcohol, smoking, and a diet low in fresh fruit and vegetables. In addition, Barrett esophagus and adenocarcinoma are closely related to chronic reflux disease (indicated by 'heartburn' symptoms). A key difference between gastric and esophageal cancer is the role of *Helicobacter pylori*, which appears to be opposite in the two diseases. While stomach cancer is promoted by the presence of the bacteria, the risk of esophageal adenocarcinoma increases in its absence.

The reason for this difference may lie in the influence of *H. pylori* on gastric pH. The bacteria produce urease which generates ammonia and bicarbonate that buffer HCl increasing pH. Gastric liquid flowing back into the esophagus is accordingly less acidic and may cause less irritation. Moreover, obesity is more of a risk factor in esophageal cancer and may be counteracted by the influence of *H. pylori* on gastric and intestinal hormones.

Importantly, several genetic and epigenetic changes observed in esophageal adenocarcinomas are already observed in the precursors. They include inactivation of *CDKN2A* or *RB1* and sometimes of *TP53*. RAS mutations and WNT pathway activation may be later changes. In this type of cancer, moreover, epigenetic inactivation of tumor suppressors, specifically of *CDKN2A*/p16$^{INK4A}$, by promoter hypermethylation (→8.3) may be unusually prevalent. Hypermethylation already develops in the Barrett metaplasia.

Wild CP, Hardie LJ (2003) Reflux, Barrett's oesophagus and adenocarcinoma: burning questions. Nat. Rev. Cancer 3, 676-684

# CHAPTER 18

# BREAST CANCER

- Carcinoma of the breast is a major lethal cancer in females in the Western world, on a par with lung and colon cancer. Most cases occur in postmenopausal women, but a significant number of younger women are afflicted, often in families with a hereditary predisposition. Known risk factors include the length of the life-time exposure to estrogens, ionizing radiation, cigarette smoking, and a high-fat diet.
- Between 10% and 20% of breast cancer cases are ascribed to hereditary factors. Inherited mutations inactivating the *BRCA1* and *BRCA2* genes lead to an up to 80% life-time risk of breast and/or ovarian cancers. Breast cancers also occur at an increased frequency in rarer dominantly inherited cancer syndromes including Li-Fraumeni or Cowden disease, affecting *TP53* and *PTEN*, respectively.
- In addition to high-risk mutations in some families, prevalent genetic polymorphisms that confer smaller risk increments to individual women are thought to be significant in the overall population. Examples are variants of genes encoding steroid metabolizing enzymes and certain heterozygous mutations in the *ATM* DNA repair gene.
- The genes *BRCA1* at 17q21 and *BRCA2* at 13q12 encode proteins involved in DNA repair and the control of genomic integrity. They behave as classical tumor suppressors of the 'caretaker' class in many familial cases of breast cancer, with one defective copy inherited and the second copy inactivated by mutation, recombination, deletion or epigenetic inactivation in the tumors. Cancers in BRCA families tend to occur at an earlier age and are more often bilateral. Inactivation of either gene is infrequent in sporadic breast cancers. Since the *BRCA* genes are expressed and function ubiquitously, it is not clear why they predispose mainly to breast and ovarian cancer. Since mutations in these genes are comparatively prevalent, counseling, monitoring and prevention of cancer in potentially affected women are being actively developed, albeit not without controversies.
- Estrogens and progesterone are important factors in the development of breast cancer. Approximately 70% of all breast cancers retain the estrogen receptor α (ERα, encoded by *ESR1*) and the progesterone receptor (PR). A second, distinct estrogen receptor β (ERβ encoded by *ESR2*) is often lost. In the other cases, ERα and PR are down-regulated, usually by epigenetic mechanisms including promoter hypermethylation.
- Like normal breast tissue, many breast cancers remain dependent on estrogens and gestagens for growth and survival and therefore respond to hormone depletion. Hormone depletion in premenopausal women is achieved by surgical removal of the ovaries. In addition, and generally in postmenopausal women

drugs are used which inactivate the estrogen receptor. These act in a tissue-specific manner, with different degrees of side-effects on other organs such as the uterus, the cardiovascular system, and bone. The selective effect of such 'SERMs' (selective estrogen receptor modulators) is mediated by interactions with co-activators and co-repressors expressed in a tissue-specific fashion. SERMs like tamoxifen appear also to be useful in the prevention of breast cancer in women at high risk.

➤ A subset of breast cancers, which are on average more aggressive, do not express estrogen receptors. In many of these, and even in some cancers expressing the ESRs, tumor growth seems to be driven predominantly through receptor tyrosine kinases of the ERBB family. Gene amplification and over-expression of ERBB2, which forms heterodimers with ERBB1 or ERBB3, is prevalent in this group. Accordingly, specific inhibitors of this protein and the specific monoclonal antibody trastuzumab, marketed under the name 'herceptin', are used in selected patients with ERBB2 overexpression, with some success.

➤ The established classification of breast carcinoma, e.g. with respect to steroid hormone receptor status and other molecular markers, is being further refined by gene expression profiling. This method has substantiated presumptions that breast cancers fall into several classes. These may partly reflect their respective cell of origin. Moreover, subgroups with different prognosis and responses to therapy, and specific patterns in patients from families with *BRCA* mutations may be discernible. An individualization of treatment, which is aspired in many major cancers, may therefore be within reach in breast cancer.

## 18.1 BREAST BIOLOGY

In adult humans, structures like the epidermis (→12) and the colon mucosa (→13) undergo a constant turnover, whereas others like the liver parenchyma (→16) and urothelium (→14) only proliferate significantly for the purpose of repair after damage. Breast tissue is different from all of these (Figure 18.1).

First of all, the organ does not develop fully before puberty, so there is one additional growth phase during the second decade of life. During puberty, the immature ducts elongate into the surrounding connective tissue to form 15-20 lobuli. This process involves multiplication of the ductal cells and an expansion of its stem cell population. The connective tissue in the breast likewise expands.

Then, for a period of up to 45 years, the ductular tissue undergoes regular monthly cycles of proliferation and apoptosis. In some women as many as 500 cycles take place, before cessation of ovulation and estrogen production in the ovaries induces menopause.

This regular cycling is interrupted by pregnancies during which the ducts extend further into the underlying connective tissue, where they branch and widen into alveoli. Concomitantly, the secretory cells differentiate and the ducts mature. After parturation, the gland produces substantial amounts of carbohydrates, fat and proteins secreted in the milk, potentially over several years. The secreted proteins provide nutrients, but also include growth factors and immunoprotective proteins.

Weaning induces a partial involution of the gland, again accompanied by apoptosis of glandular cells, particularly the luminal secretory cells in the alveoli. Once again, the tissue, epithelia and stroma alike, is remodeled.

These cycles are controlled by a combination of hormones and locally produced growth factors. The pubertal growth phase is stimulated by estrogens from the ovaries, which become active at this time, and by growth hormone and its 'somatomedin' mediators IGF1 and IGF2. The monthly cycles are controlled mainly by estradiol and progesterone, supported by insulin and further hormones. During the first phase of the monthly cycle, follicle cells secrete mostly estrogens. After release of the oocyte gestagens like progesterone are the major product. Unless fertilization takes place, the follicle degenerates and minimal estrogen and progesterone levels elicit an involution phase in the breast and, more pronounced, in the uterus endometrium. The pregnancy growth phase is stimulated by several hormones, including gestagens, estrogens, growth hormone and insulin, but also glucocorticoids and, of course, prolactin. As in other tissues, proliferation of epithelial cells in the breast is supported by the stroma. Stromal cells and epithelial cells both produce paracrine growth factors, partly in response to steroid hormones.

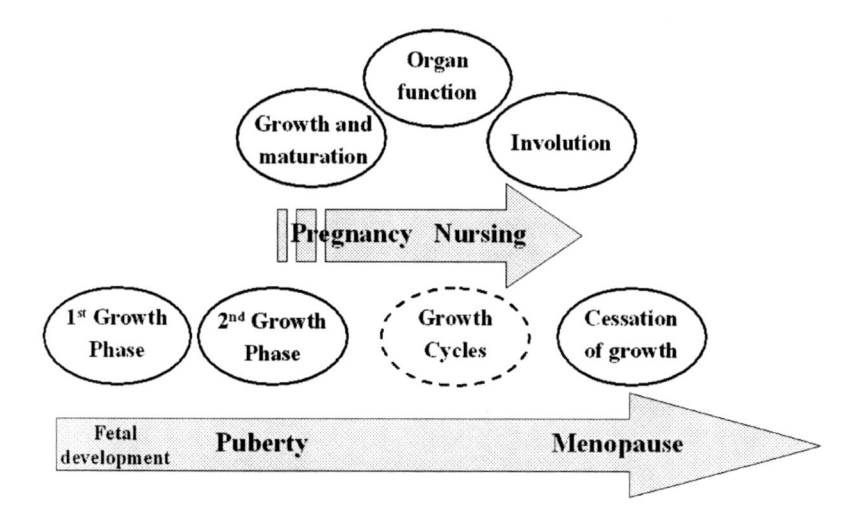

**Figure 18.1** *Growth phases of the breast*

Estrogens represented by estradiol and gestagens represented by progesterone act on cells by binding to specific receptors which belong to the steroid hormone receptor superfamily (Figure 18.2). The superfamily encompasses a large number of DNA-binding proteins with a variety of ligands, e.g. the retinoic acid receptors (→8.5, →10.5). The estrogen receptors and the progesterone receptors belong to a group of more closely related receptors which also includes the androgen receptor (→19.2). The members of this group are ligand-dependent transcription factors that bind as homodimers to specific symmetric binding sites on DNA. These are termed

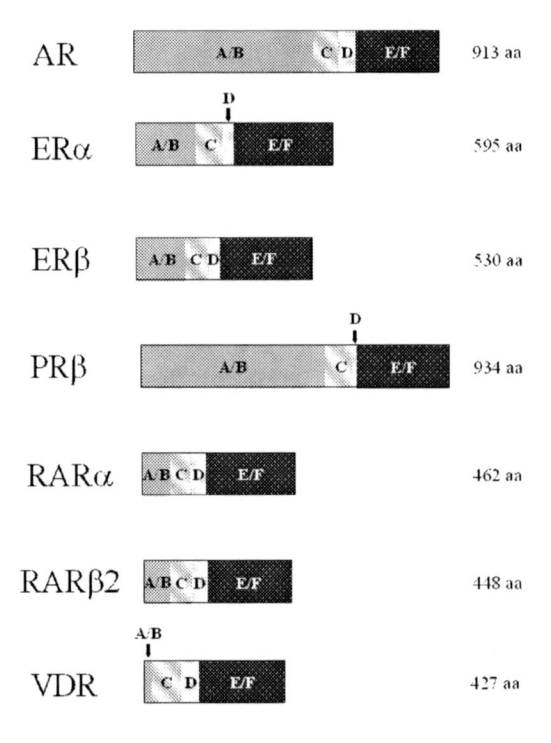

**Figure 18.2** *Some members of the steroid receptor superfamily*
The androgen receptor (AR), estrogen receptors (ERα, ERβ), a progesterone receptor (PRβ), two retinoic acid receptors (RARα, RARβ2), and the vitamin D receptor (VDR), and their subdomains are drawn to scale. Subdomains A/B and E/F are not separated in this representation (compare Figures 18.3 and 19.6).

ERE (estrogen-responsive element), PRE (progesterone-responsive element), ARE (androgen-responsive element), etc.

The estrogen receptor α (ERα) is a typical representative (Figure 18.3). Its DNA-binding domain, which contains two zinc fingers, is located at the center of the primary amino acid sequence. It is flanked on the N-terminal side by a transactivation domain, designated as activation function 1, AF-1. A hinge region on its C-terminal side connects a second transactivation domain, named AF-2. The AF-2 domain binds the ligand and its activity is strongly dependent on ligand-binding. Moreover, protein interactions exerted by this domain control the receptor activity overall.

Inactive receptors are retained in the cytosol and become capable of entering the nucleus only after binding of the ligand and an ensuing conformation change. Some receptors are bound by heat shock proteins like HSP90, others are free, but are shuttled rapidly out of the nucleus, unless occupied by an agonistic ligand. After binding to their specific recognition sites on DNA, the receptor dimers recruit various co-activator proteins through their AF domains. The AF-2 domain of the ERα is known to bind at least five different co-activator proteins specifically. The best characterized of these is the SRC1 (steroid receptor co-activator) protein. Co-activators mediate the interaction between steroid hormone receptors and the general transcription apparatus and the modification of chromatin at the binding site. Binding of co-activators stimulates histone acetylation and the actual initiation of transcription. In addition, they integrate signals from several transcription factors binding to the regulatory regions of the same gene and from signal transduction pathways. For instance, certain co-activators interacting with the estrogen or androgen receptors are regulated by MAPK phosphorylation in response to growth factors of the EGF family. The receptor itself is also phosphorylated. While interactions with co-activators lead to gene activation, interactions with co-repressors can cause gene repression. Repression can be exerted by the ERα, but more pronounced by the progesterone receptor α.

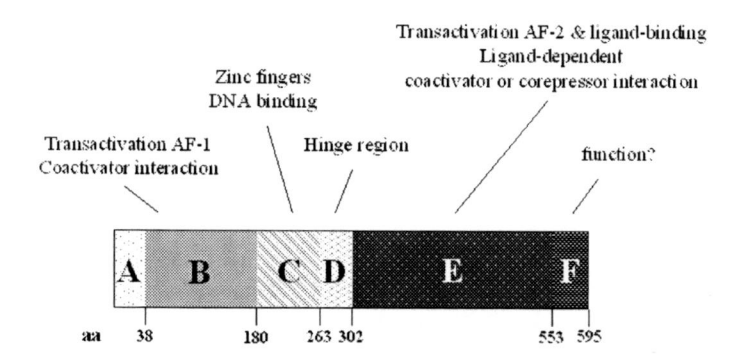

**Figure 18.3** *Structural domains of the estrogen receptor α and their functions*

A second mechanism of ERα action does not require binding of the receptor to DNA. Steroid hormone receptors can interact with several other transcription factors, either sequestering them or modulating their activity while they are bound to DNA. In this fashion, the receptors can regulate the activity of genes that do not possess canonical receptor binding sites. Specifically, the ERα can modulate the activity of AP1 transcription factors. This may be the main mechanism by which estrogens stimulate cell proliferation, i.e. by mimicking activation of MAPK pathways (→6.2) alone or synergistically with growth factors (Figure 18.4). Additionally, estrogens stimulate the production of EGF-like growth factors in breast tissue.

The regulation of breast tissue growth and function by estrogens is in reality more complex than shown in Figure 18.4. Estrogen receptors can also influence gene activity through SP1 and NF$_κ$B sites by direct and indirect interactions. Moreover, ERα activity is modulated by protein-protein interactions with Cyclin D1 (→5.2) and, intriguingly, with BRCA1 (→18.3). In addition, estrogens – like several

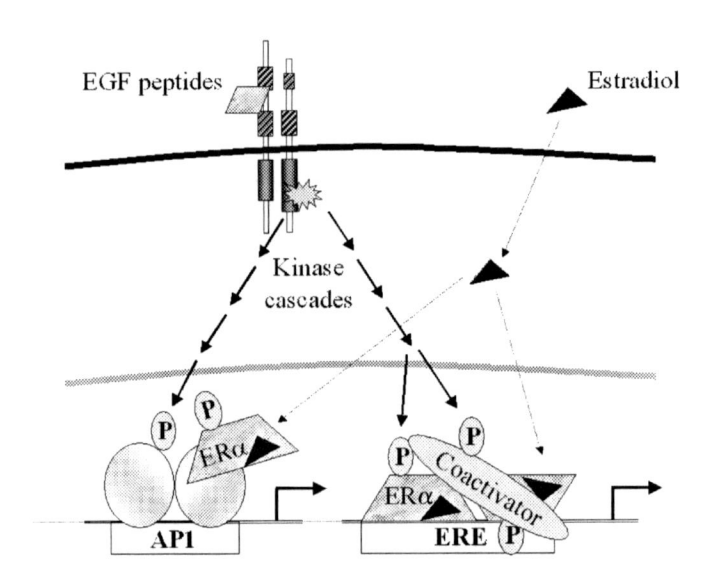

**Figure 18.4** *Interaction of steroid hormones and EGF-like growth factors in the regulation of estrogen receptor α action*

other steroid hormones – elicit very rapid effects at the cell membrane which may not be exerted through their canonical receptors, but through direct activation of ion channels. Most importantly, estrogens act on several different cell types in breast tissue, both epithelial and stromal.

There are two different estrogen receptors in man, estrogen receptor α and estrogen receptor β, which are encoded by two distinct genes, *ESR1* and *ESR2*. They exhibit only 30% homology overall (Figure 18.2), but almost identical DNA-binding domains, both recognizing the 'ERE' sequence AGGTCA NNN TGACCT. ERα mediates most of the proliferative effects in female reproductive tissues. In contrast, ERβ appears to act mostly as an inhibitor of ERα action, and perhaps even of the androgen receptor in males. The two estrogen receptors are expressed in different patterns throughout the human body.

Different subunits of the progesterone receptor (PRA and PRB) are translated from differently spliced mRNAs from the same gene. They can bind interchangably as homodimers or heterodimers. Expression of the PR is induced by ERα. PRA (also called PRα), in particular, appears to act as a feedback inhibitor of ERα. However, it is important to realize that PR action is dependent on gestagens, so its actual effect depends on the relative levels of gestagens.

Even more importantly, while estrogen receptors and progesterone receptors are present in many organs of the female (and even male), the effects of estrogens and gestagens are not the same in each tissue. Estrogens affect e.g. the brain and bone, but they do not stimulate proliferation in these tissues as in breast and endometrium.

Even the effects of estrogens on breast and endometrium differ (like those of gestagens). Therefore, since many organs contain receptors, additional mechanisms ensure the tissue specificity of estrogen action. One possible mechanism is differential expression of the two estrogen receptors. A second mechanism is provided by different expression patterns of co-activators, co-repressors and differences in their regulation. As a consequence, certain compounds with structural similarities to estrogens act as partial antagonists that inhibit estrogen activity to a different extent in different organs. These are designated SERMs, for 'selective estrogen receptor modulators'. Tamoxifen and raloxifene are examples of compounds in this class (Figure 18.5). Tamoxifen, e.g., blocks the stimulation of proliferation by estrogens in the breast, acting as an antagonist, but in the endometrium it rather behaves as an agonist, supporting proliferation. This difference may largely be caused by higher levels of the co-activator SRC1 in the endometrium, which is recruited by the tamoxifen-ERα holocomplex. In contrast, in breast tissue, this complex overwhelmingly recruits co-repressors.

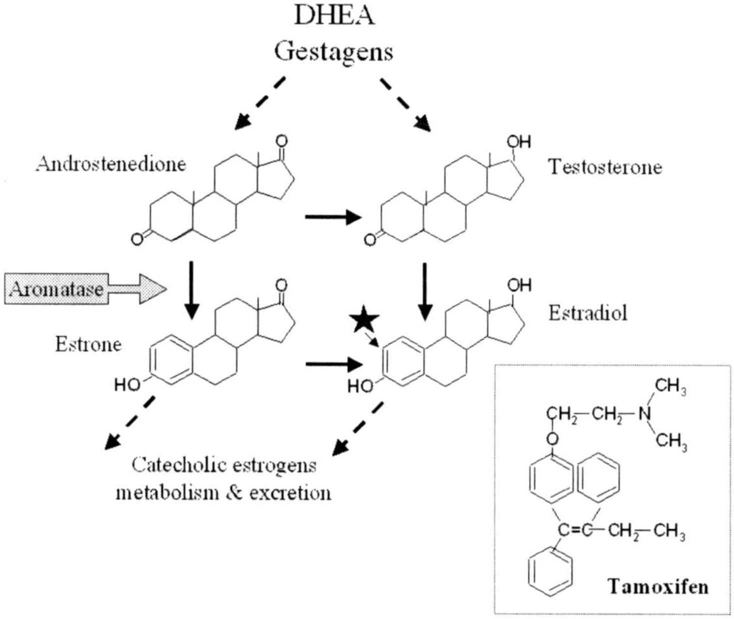

**Figure 18.5** *Structure of estrogens*
Estrone (or oestrone) and estradiol (oestradiol, $E_2$) are the main estrogens in humans formed from the precursors shown by aromatase and other biosynthetic enzymes. The main pathway of metabolism leading to excretion is initiated by hydroxylation at the 2-position in ring A, marked by an asterisk in estradiol. For comparison, the structure of the (partial) anti-estrogen drug tamoxifen is shown in the lower right corner.

## 18.2 ETIOLOGY OF BREAST CANCER

With colon cancer and increasingly lung cancer, breast cancer constitutes the trias of major cancers in females in Western industrialized countries. In these countries, the life-time risk of breast cancer is around 10% for women, and about 30% of them turn out to be lethal. Most breast cancers become apparent in women after their menopause, but a significant fraction is diagnosed earlier. In Western countries, the mean age at menopause is now ≈50 yrs, and menopause takes place in almost all women between 45 and 55. In females aged 40-60 years, breast cancer is the most frequent lethal cancer. In most countries, its incidence is rising, although the increase in mortality has been checked. There are several explanations for this phenomenon, invoking earlier detection and better treatment (→20.4). In any case, it is generally agreed that certain aspects of the Western life style favor the development of breast cancer. There is much less agreement on which aspects these are (Table 18.1).

Estrogen exposure: One risk factor can be summarized as the life-time length of exposure to estrogens. Earlier menarche, later menopause, and fewer (as well as later) pregnancies associated with the Western life-style all lengthen this period. Each change singly is associated with an increased risk of breast cancer. The underlying mechanism could be simply probabilistic. With prolonged exposure to estrogen and a longer period of proliferation cycles, the number of cells that can contract mutations is increased and initiated tumor cells get more time to expand. Specifically, strong signals enforce differentiation of breast ductal cells during pregnancy and others elicit extensive apoptosis of alveolar and ductal cells after weaning. These cycles may help to remove initiated tumor cells and act to 'purify' the tissue. Accordingly, a strong preventive effect is exerted by multiple pregnancies early in life with extended nursing periods.

A second, albeit not mutually exclusive explanation for the effect of estrogen exposure relates to the chemical structure of estrogens. Estrogens and their metabolites are phenolic compounds, after all (Figure 18.5). In particular, diphenolic estrogen metabolites can become partially oxidized to semiquinones which can react with macromolecules in the cell including DNA and induce mutations.

**Table 18.1.** *Potential causes of human breast cancer*

| *Potential factors causing breast cancer in humans\** |
|---|
| Length of exposure to estrogens |
| Ionizing radiation |
| Genetic predisposition (*BRCA1, BRCA2, others*) |
| Sedentary life style |
| High-fat diet |
| Alcohol |
| Tobacco smoking |

\* Several factors may interact, also synergistically.

Semiquinones can, moreover, initiate a process called quinone redox cycling that produces highly reactive oxygen species. So, estrogens may act as chemical carcinogens. If so, the risk of cancers in organs with high estrogen concentrations may strongly depend on the individual ability to metabolize estrogens and to deal with quinone adducts to DNA and redox cycling. Many genes involved in estrogen biosynthesis and metabolism are polymorphic, as are some of the relevant protective enzymes. An interaction between genes and environment is therefore strongly suspected (cf. 18.3).

Exposure to ionizing radiation: Here, too, gene-environment interactions are suspected. They are particularly clear in the case of inherited defects in the repair of DNA damage caused by ionizing radiation. For instance, certain carriers of mutations in the *ATM* gene may be at increased risk for breast cancer (→3.4). The relationship between exposure to radiation and breast cancer risk becomes specifically relevant, when discussing prevention of breast cancer mortality by early detection. There are concerns that population screening by frequent mammography might put susceptible individuals at increased risk (→20).

High-fat diet: The breast cancer incidence in a country is clearly correlated to the content of fat, particularly of saturated animal fats, in the diet. The biological mechanisms that underlie this epidemiologically clear-cut correlation are not really understood. One hypothetical chain of events also invokes estrogens. The adipose tissue of postmenopausal women contains aromatase, a key biosynthetic enzyme for estrogens. So, high-fat diets may lead to overweight with expansion of adipose tissue leading to increased aromatase activity leading to continued production of estrogens after menopause with inappropriate growth stimulation of breast epithelial cells.

Genetic predisposition: The effects of all of the above factors may be dependent on genetic polymorphisms which modulate cancer risk in interaction with environmental factors. Polymorphisms in genes of steroid hormone metabolism, DNA repair (→3.4), cell protection (→3.5), and lipid metabolism, e.g. could all be relevant. While these polymorphisms may exert large effects on the overall incidence of breast cancer, they increase the risk only slightly for each individual woman and they are strongly dependent on other, notably environmental factors (→2.3). In contrast, genetic factors predominate in the smaller fraction of (typically familial) cases that are caused by inherited mutations in one of a limited number of high-risk genes (→18.3).

## 18.3 HEREDITARY BREAST CANCER

At least seven genes are now known, in which inherited mutations in one allele conduce a high risk of breast cancer (Table 18.2). Mutations in most tumor suppressor genes account for a small proportion of all familial cases, certainly less than 10%, while a larger fraction of familial breast cancers originates from mutations inactivating the *BRCA1* and *BRCA2* tumor suppressor genes.

The *PTEN* gene at 10q23.3 encodes a phospholipid and protein phosphatase which controls the activity of the PI3K pathway (→6.3). It is mutated in Cowden syndrome. This rare dominantly inherited disease is characterized by multiple benign hamartomas and by thyroid cancers, but also includes an increased risk for breast cancer.

Mutations in the *TP53* gene (→5.3) are the cause of most cases of the rare Li-Fraumeni syndrome. In line with the general importance of the TP53 protein in the control of genomic stability (→6.6), this is a generalized cancer syndrome which also includes a particularly high risk for breast cancer. A minority of cases might be caused by mutations in genes encoding the checkpoint kinases CHK1 and CHK2 (→3.3).

The more frequent HNPCC syndrome caused by mutations in DNA mismatch repair genes (→3.2) likewise leads to an increased incidence of various cancers, predominantly in the colon and rectum (→13.4). Inherited mutations in the *MLH1* and *MSH2* genes have also been found in breast cancer patients.

Rare mutations in the *LKB1* gene, encoding a serine/threonine kinase that regulates mitosis, likewise cause predominantly cancers of the intestine and colon (→Table 13.1), but also increase the risk of breast cancer.

Up to 50% of familial breast cancers in some populations may be caused by germ-line mutations in the *BRCA1* and *BRCA2* tumor suppressor genes. These mutations predispose almost exclusively to cancers of the breast and of the ovaries, unlike many of those mentioned above. Notably, *BRCA1* and *BRCA2* are considered as 'caretakers' like almost all other tumor suppressor genes in which mutations lead to a strongly increased risk of breast cancer. The obvious exception is *PTEN*.

*BRCA1* and *BRCA2* have actually been named for their role in breast cancer, but they indeed exhibit further similarities.

Table 18.2. *Tumor suppressor genes implicated in hereditary breast cancer*

| Gene | Location | Function | Other cancer sites |
|---|---|---|---|
| *BRCA1* | 17q21 | DNA repair/control of homologous recombination | ovary |
| *BRCA2* | 13q12 | DNA repair/control of homologous recombination | ovary |
| *PTEN* | 10q23.1 | lipid phosphatase/regulation of PI3K pathway | several |
| *TP53* | 17p13.1 | control of genomic integrity and DNA repair | many |
| *MLH1, MSH2* | 3p21, 2p15-16 | DNA mismatch repair | several |
| *LKB1* | 19p13.3 | mitotic control | intestine |

*BRCA1* is a ≈100 kb gene at chromosome 17q21 comprising 24 exons (Figure 18.6). Exon 11 of the gene is unusually large extending over >4.5 kb and accordingly contains a large part of the coding sequence. The BRCA1 protein consists of 1863 amino acids. Several centrally located nuclear localisation signal sequences identify it as a nuclear protein. There are several protein-protein interaction domains, including a RING-finger domain near the N-terminus. Near the C-terminus are two copies of a 110 amino acid sequence called BRCT motif. This

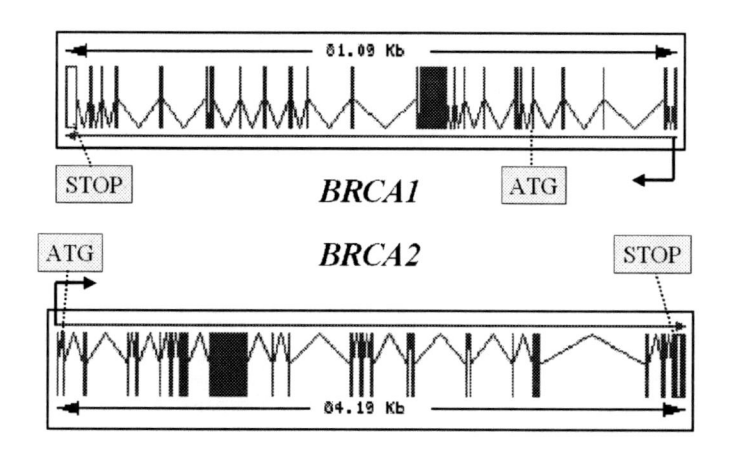

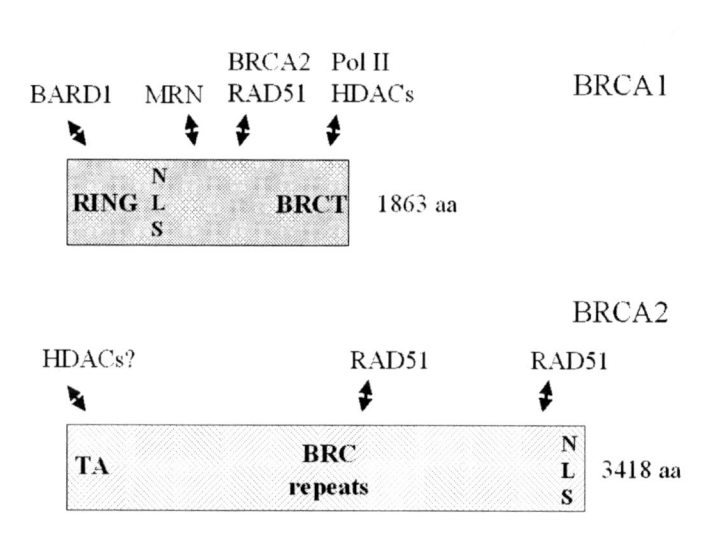

**Figure 18.6** *The BRCA1 and BRCA2 genes and proteins*
A: The genes as shown in the ensembl data base. B: The proteins and some of their interaction partners.

motif was also detected in several other proteins involved in DNA repair and/or cell cycle regulation. BRCA1 also has a transcriptional activation function, unlike BRCA2.

The *BRCA2* gene is located at 13q12 and extends across ≈80 kb. Among 27 exons overall, exon 10 and 11 are – again – unusually large. The BRCA protein comprises 3418 amino acids. Nuclear localization signals are located C-terminally. Multiple protein interactions are supposed or documented. Most importantly, the central part of the molecule contains eight repeats of 'BRC' motifs, each consisting of ≈40 amino acids.

Germ-line mutations in both *BRCA* genes are spread out along almost the entire length of the genes. Most are small deletions or insertions resulting in frame-shift mutations. Nonsense mutations leading to truncated proteins are also prevalent. Splice-site mutations also occur as well as rare missense mutations. In each case, the mutations would be expected to inactivate the function of the proteins or – in some cases – perhaps to exhibit a dominant-negative phenotype.

In breast cancers arising in women with inherited mutations in a *BRCA* gene, the second allele is consistently inactivated as well, by deletion, recombination or point mutation. In this regard, therefore, *BRCA1* and *BRCA2* behave as classical tumor suppressor genes. However, mutations or deletions in these genes are almost never found in sporadic breast cancers. Instead, decreased levels of the proteins are often observed, which in some cases are associated with increased methylation of *BRCA1* regulatory sequences. In this respect, therefore, the BRCAs do not follow the standard scheme for tumor suppressors (→5.4).

The BRCA1 and BRCA2 proteins are expressed in almost all tissues. They are important for DNA repair, prevention of chromosome breaks and checkpoint signaling. BRCA1 and BRCA2 have different, but related functions.

Lack of either protein causes an increase in chromosomal aberrations in proliferating cells, including deletions and translocations, but even fragmentation and formation of multiradial chromosomes. All these aberrations are found in breast cancers with BRCA mutations. They are primarily due to a deficit in homologous recombination repair and secondarily to a failure to activate appropriate cellular checkpoints. Homologous recombination repair (→3.3) is the preferred mechanism for dealing with DNA double strand breaks during the S and G2 phase of the cell cycle in mammalian cells, when homologous DNA strands from sister chromatids are available. In contrast, non-homologous end-joining predominates during G1 (→3.3). This mechanism introduces a larger number of errors than homologous repair at the mended site, including small deletions and insertions. However, these errors do not seem to constitute the major problem in BRCA-deficient cells. Rather, non-homologous end-joining on its own appears insufficient to ensure chromosomal integrity.

The main function of BRCA2 in this regard appears to be control of RAD51 (Figure 18.7). BRCA2 binds and inhibits RAD51 through its BRC repeats, keeping it from binding to DNA as a multiprotein filament. During homologous recombination repair, double-strand breaks must first be processed. Then, RAD51 is released from BRCA2, likely as a result of BRCA2 being phosphorylated, and mediates recombination between the strands that are to be repaired and the intact homologous double helix. This is followed by extension of the single strands by a DNA polymerase, ligation and resolution of the recombinant Holliday junction structure (→3.3). At a later stage of this resolution, RAD51 is removed, apparently by reloading onto BRCA2. BRCA2 interacts with activated FANCD2 protein. This is activated via mono-ubiquitination by the FANC protein complex. It then moves to foci in the nucleus where BRCA2, RAD51 and the MRE11/RAD50/NBS1 (MRN) proteins reside and activates these (→3.3). Homozygous germ-line mutations in

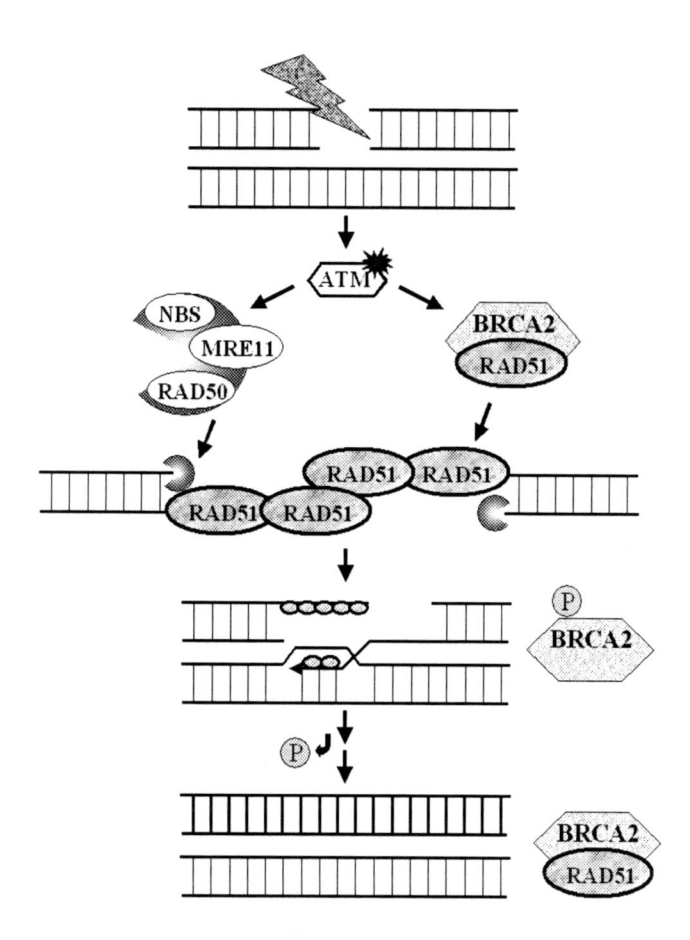

**Figure 18.7** *Interaction between BRCA2 and RAD51 during DNA double-strand repair*
It is assumed that ATM activated by DNA double-strand breaks not only phosphorylates the MRN complex (left) that processes the open ends, but also BRCA2 (right) to release RAD51. RAD51 then mediates recombination between the processed single strand ends and homologous DNA. Rebinding of RAD51 by BRCA2 allows completion of the repair. Several further components are involved as shown in Figure 3.11.

*BRCA2* cause a form of Fanconi anemia and, conversely, breast cancer incidence is enhanced in this recessive cancer syndrome (→3.4).

BRCA1 appears to have a still wider range of functions. In response to DNA double-strand breaks, it is phosphorylated by kinases sensing this damage, such as ATM, ATR or CHK2 (→3.3). It may also be activated by other types of DNA damage, since it interacts e.g. with the MSH2 and MSH6 proteins involved in mismatch repair (→3.1), and perhaps also with the transcription-coupled nucleotide excision repair system (→3.2). Following its activation BRCA1 mediates selected transcriptional responses, such as induction of the GADD45 repair protein, and at the site of damage aids in chromatin remodeling. A crucial function is regulation of the MRN protein complex that resects the DNA ends at double-strand breaks (→3.3). In a similar fashion to BRCA2 directing and limiting the function of RAD51, BRCA1 may control this complex and keep it from overdigesting. An important function of BRCA1 is exerted through its RING finger domain. This domain characterizes substrate recognition proteins of ubiquitin ligases. BRCA1 heterodimerizes with a protein named BARD1. Together they may support mono-ubiquitination of FANCD2 by the FANC protein complex (→3.3), but certainly help to relocate the activated protein to the nuclear foci where BRCA2 resides. In this fashion, BRCA1 mediates repair of DNA crosslinks and stalled replication forks as well as actual strand breaks.

So, both BRCA1 and BRCA2 are involved in regulating or executing homologous recombination stimulated by the FANC proteins and perhaps also in

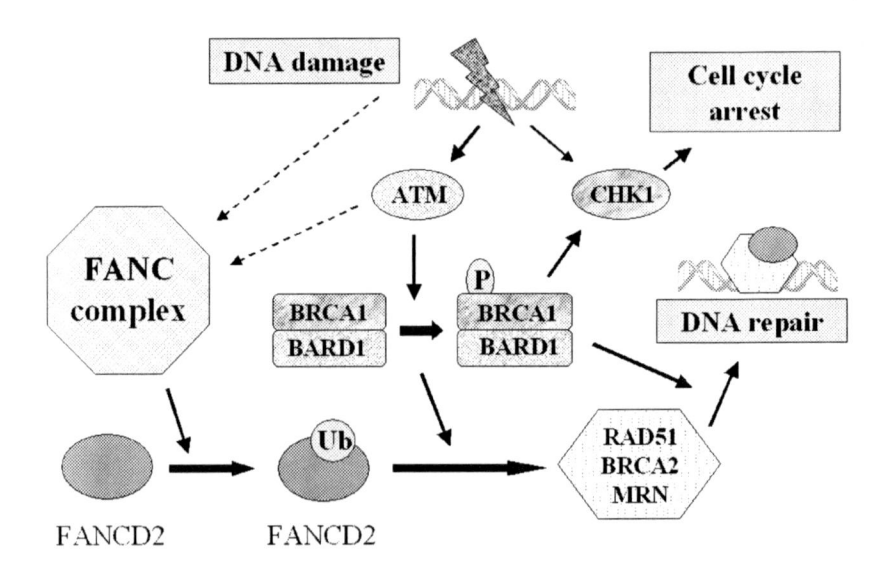

**Figure 18.8** *Interaction of BRCA1 and BRCA2 in the control of DNA repair*
See text for further explanations and cf. also Figures 3.9 and 3.11

regulating the MRN proteins. In a sense, BRCA2 acts 'downstream' of BRCA1 (Figure 18.8).

It is, of course, conspicuous that the functions of so many genes in which mutations convey an increased risk of breast cancer are closely related to DNA repair. On one hand, this has led to the conception of a sort of 'repairosome' that includes the BRCAs as crucial components, but also proteins actually recognizing and removing DNA damage as well as signaling components. The mutations predisposing to breast cancer would have the common effect of impeding the function of this dynamic 'organelle'. Its core structure may correspond to the nuclear foci in which BRCA2 resides.

On the other hand, the question arises why such mutations promote cancer of the breast in particular. In the case of BRCA mutations, the breast and the ovaries are the organs most susceptible to cancers by a large margin. A certain, but much lower increase of risk is observed for cancers of the pancreas, bile duct, stomach, colon, and prostate (in this approximate descending order). To date, the most likely explanation for this specificity seems that during the multiple cycles of growth and involution in the breast there is a high likelihood of incurring chromosomal aberrations unless the DNA damage surveillance and repair system works perfectly. This is plausible, but does not quite account for the increased risk of ovarian cancer as well and, even less, for the increased risk of male breast cancer in carriers of BRCA2 mutations. There are also continuing hints that some of the proteins in the 'repairosome' interact in a more direct fashion with the estrogen response. BRCA1 and TP53, among others, have been shown to interact with the estrogen receptor, mutually regulating each other's activity. So, a second (and perhaps additional) explanation might be that these proteins limit the pro-proliferative and anti-apoptotic action of the activated estrogen receptor in mammary epithelial tissue. Accordingly, BRCA1, in particular, does seem to exert a direct effect on the growth of breast and ovarian cells.

Women with germ-line mutations in *BRCA1* and *BRCA2* have a 40-80% estimated life-time risk of breast cancer compared to the ≈10% risk in the female population in Western industrialized countries at large. The risk of ovarian cancers is roughly 20-fold increased, as they are otherwise less prevalent. The precise increase in risk depends on the particular mutation, on genetic modifiers, and likely on environmental factors. The 4-8-fold increase in life-time breast cancer risk caused by BRCA mutations may seem moderate, but is exacerbated by the typical effects of an inherited tumor suppressor gene mutation. In familial cases, the disease appears earlier and the risk of a cancer in the contralateral organ (breast as well as ovary) is hugely increased over that in sporadic cases.

It is estimated that up to 10% of all breast cancers develop as a consequence of a mutation in a high-risk gene (Figure 18.9). Within this group, mutations in *BRCA1* and *BRCA2* may each underlie approximately one quarter of the cases. Mutations in all other genes listed in Table 18.1 may account for another ≈10%. This leaves >40% of familial cases unaccounted for by mutations in known genes. As during the initial search for the *BRCA1&2* genes in the early 1990's, linkage studies are continuing in families with multiple cases of breast cancer, occuring at an younger than usual age and/or more often bilaterally. Today, such studies are greatly facilitated by the human genome sequence and technical advances allowing high-throughput analyses of genetic markers. It is therefore safe to conclude, even with research ongoing, that a "*BRCA3*" gene with the same impact as *BRCA1* and *BRCA2* will not be discovered. Instead, it is expected that mutations in several more individual genes conferring a very high risk will each be responsible in a small number of families. However, the majority of familial cases may result from mutations that increase the risk of breast cancer only moderately, but sufficiently so to cause an accumulation of cancers in some families that carry them (cf. 19.3). These same mutations will also be responsible for individual cases in other families that would be categorized as 'sporadic'.

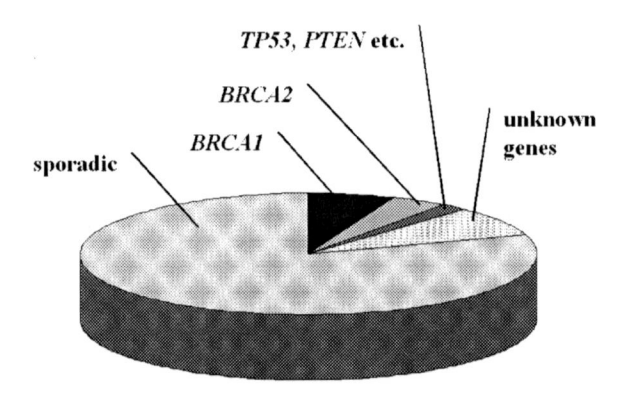

**Figure 18.9** *Contribution of mutations in high risk genes to breast cancer risk*
Estimated proportions of cases contributed by mutations in the indicated genes. Note that the risk in 'sporadic' cases is also modulated by genetic polymorphisms.

This line of thought thus suggests a gradualism in cancer predisposition, at least for breast cancer (Figure 18.9). Mutations in certain 'high-risk' genes may confer such a pronounced increase in risk that they will regularly result in a clustering of cancers within families. They are highly penetrant and to a large degree independent of environmental modulators. Other mutations or polymorphisms in these same genes or mutations in 'low-risk' genes may emerge as a series of cancer cases in some families, but contribute to the development of many more cases categorized as sporadic. At the other end of the spectrum, certain polymorphisms in genes of DNA repair, cell protection, lipid or hormone metabolism etc. may confer only small increments in risk for breast cancer which may be strongly dependent on environmental modulators such as a high-fat diet or exposure to ionizing radiation. Since these polymorphisms are common, in contrast to actual mutations in high-risk genes, their impact on breast cancer may overall be more important. However, these effects are much more difficult to discern than the mutations in high-risk genes (cf. 2.3).

## 18.4 ESTROGEN RECEPTORS AND ERBB PROTEINS IN BREAST CANCER

Estrogens and the estrogen receptors are key regulators of growth in the normal breast, together with a number of other hormones acting through nuclear or membrane receptors. Furthermore, in an incompletely understood fashion, they synergize with growth factors of the EGF family that activate receptor tyrosine kinases of the ERBB family to instruct proliferation and differentiation of the ductal and alveolar epithelia (Figure 18.10). Expression of the growth factors in stromal and epithelial cells and perhaps of the receptors as well is influenced by estrogens. Conversely, active ERBB receptors stimulate the MAPK cascade ($\rightarrow$4.4) which leads to phosphorylation of the ER$\alpha$ and synthesis and phosphorylation of AP1 transcription factors with which the receptor interacts ($\rightarrow$18.1). In breast cancer, these same factors remain relevant, although in a distorted way.

At the time of presentation, in >70% of breast cancers the ER$\alpha$ can be detected by immunohistochemical staining or by biochemical assays such as ELISA. About half of these cancers also express the progesterone receptor. Since the *PR* gene is induced by the estrogen receptor, its expression is an indication that ER$\alpha$ is not only present, but also active. Only <5% of breast cancers express only the PR. The ER$\beta$ is usually down-regulated in breast cancer; quite often the *ESR2* gene is silenced by promoter hypermethylation ($\rightarrow$8.3). This underlines its function as a negative growth regulator that limits proliferative responses to estrogens.

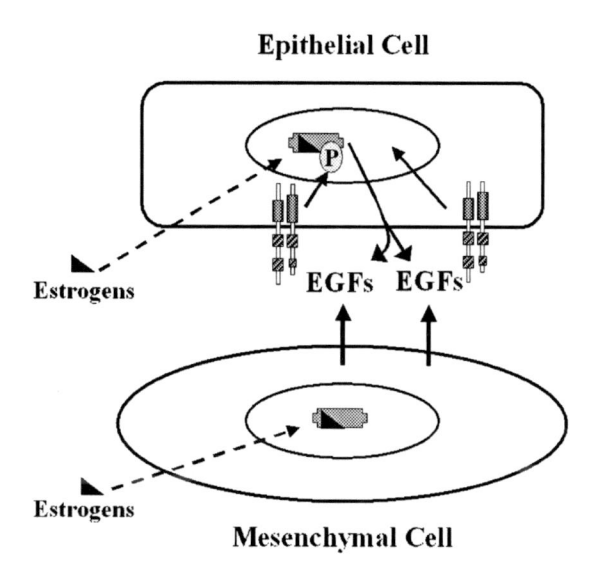

**Figure 18.10** *Presumed interaction of estrogens and EGF-like factors in normal breast tissue* It is thought that estrogen-independent breast cancers lose estrogen receptor $\alpha$, increase production of EGF-like peptides and their responsiveness to them by increased expression of receptors and specifically the co-receptor ERBB2. See text for more details.

The presence or absence of ERα/PR provides the basis for one type of classification of breast cancers. 'ER+' breast cancers are on average better differentiated, grow more slowly, are not as strongly aneuploid, and have a slightly better prognosis than ER- breast cancers. Moreover, as a rule, their growth remains dependent on estrogens. Thus, compounds that block estrogen action or diminish the level of endogenous estrogens are often efficacious against ER+ cancers, but not at all against ER- cancers. Several strategies to specifically treat ER+ cancers are in use or are being explored (cf. 22.3). Compounds such as tamoxifen and raloxifene are partial agonists/antagonists (SERMs) that interfere with estrogen binding and with some interactions of the receptor with co-activators. They are used for chemoprevention, neoadjuvant treatment, adjuvant treatment, or treatment of systemic disease with metastases. Newer full antagonists block the ERα more efficiently and induce its degradation. They are expected to become used mostly in actual tumor treatment rather than in prevention, because of adverse effects on other tissues, such as bones and heart.

In women before menopause, when endogenous estrogens are still produced at high levels, this production must be diminished for efficient treatment. This can be achieved by surgical removal of the ovaries ('oophorectomy') or by treatment with analogues of gonadotropin-releasing hormone (GnRH or LHRH). GnRH is a hypothalamic peptide that stimulates the release of luteinizing hormone (LH) in the hypophysis. It is secreted in a cyclic fashion and regulated by neuronal inputs and by steroid feedback inhibition. Most drugs used in therapy are GnRH receptor agonistic and induce an initial burst of LH. However, since their level during treatment remains steadily high, the GnRH receptors in the pituitary become down-regulated. As a consequence, the production of LH hormone ceases, the ovaries are no longer stimulated by LH and stop to produce estrogens (Figure 18.11). Since in postmenopausal women estrogen synthesis is no longer significantly controlled by LH, a different strategy is required. Usually, inhibitors of the estrogen biosynthesis enzyme aromatase are employed.

By and large, the growth of ER+ breast cancers appears to be promoted by overactivity of those mechanisms that stimulate normal breast epithelial cells to grow during the proliferative phase of the monthly cycle or during pregnancy. Of course, in the cancers, neither the cyclic regression phase nor the terminal differentiation of duct cells take place in an orderly fashion. In contrast, ER- tumors neither depend on nor respond to estrogens. The estrogen receptors are not expressed and the *ESR1* gene as well is quite frequently silenced by hypermethylation. In many ER- breast cancers, a large fraction of the proliferative stimulus appears to be provided through ERBB proteins.

In man, the ERBB family named after their first identified member, the retroviral v-erbB oncogene (→4.1) comprises four structurally related membrane receptors, ERBB1 - ERBB4. The designations HER1 – HER4 are also in use (Table 18.3). Each receptor protein comprises an ≈600 amino acid extracellular ligand-binding domain, a 24 amino acid single-pass helical transmembrane domain and a >500 amino acid cytoplasmatic tyrosine kinase domain with multiple autophosphorylation sites and an autoinhibitory loop (cf. Figure 4.4).

ERBB1, also known as the EGF receptor, of course binds the epidermal growth factor, but also several further peptides that share with EGF a conserved motif with three cystine disulfide bridges, termed the EGF motif. In fact, in most physiological circumstances, the structurally related TGFα, amphiregulin, and heparin-binding EGF (HB-EGF) are probably more relevant than EGF itself. HB-EGF as well as betacellulin (BTC) and epiregulin (EPR) also bind and activate HER4. Neuregulins 1 – 4 (NRGs) are specific for ERBB3 and ERBB4, NRGs 1&2 being predominantly recognized by ERBB3 and NRGs 3&4 by ERBB4. This leaves ERBB2 with no known ligand and there may indeed be none (Table 18.3). Rather, all ERBB receptors form dimers after ligand binding and all prefer to form heterodimers with ERBB2, although homodimers are also active.

Ligand binding induces a conformation change that relieves auto-inhibition of the tyrosine kinase by the pseudosubstrate loop to allow cross-phosphorylation with subsequent docking of adaptor and substrate proteins (→4.4). Signals emanate from the ERBB receptors mainly through the MAPK, PI3K, and STAT pathways (→4.4, 6.3, 6.8).

As ERBB2 is the odd member of the family with respect to ligand binding, ERBB3 is unusual with respect to kinase activity. Crucial residues in its active center are not conserved and ERBB3 may have no kinase activity at all. Thus, its tyrosine kinase activity is provided by its ERBB2 dimerization partner.

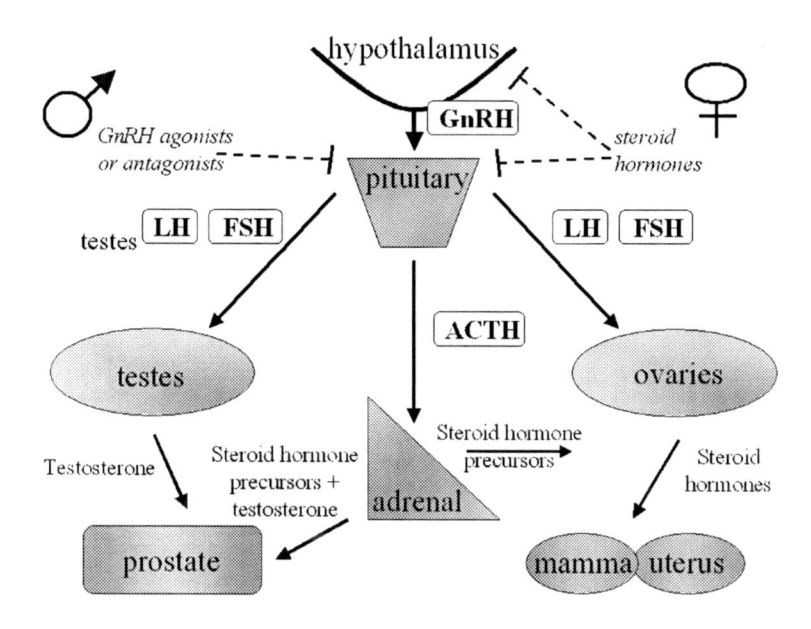

**Figure 18.11** *Regulation of steroid hormone production by pituitary hormones and mode of action of GnRH receptor agonists*

In normal breast epithelial cells, both proliferation and differentiation are influenced by several EGF-related peptides and each member of the ERBB family is involved, at least in a subset of the cells. Production of the peptide growth factors takes place in stromal and epithelial cells and is regulated by estrogens. Both their production and the cellular responses to the growth factors are influenced by cross-talk with other hormones and growth factors. For instance, release of HB-EGF requires proteolytic cleavage by metalloproteinases (→9.3) which are stimulated by endothelin-1 and bombesin acting through G-coupled serpentine receptors. Certain WNT factors also promote release of EGF-like factors. The relevance of WNT pathway activation (→6.10) in breast cancer may therefore have to be considered in this light. At the ERBB receptor step, cross-talk with cytokine receptors may be particularly important in breast tissue. Prolactin and growth hormone receptors induce phosphorylation of ERBB1 by JAK kinases (→6.8).

With multiple growth factors and receptors distributed between several cell types and subject to cross-talk with further pathways, the precise relationships are extremely complex. In addition, they may substantially vary between different phases of breast growth and even segments of the ducts. In summary, it appears that activation of ERBB1, but even more of the ERBB2/ERBB3 heterodimer provides the main stimulus for proliferation. At least in the case of the ERBB2/ERBB3 unit, this stimulation may predominantly be exerted through the PI3K pathway, since ERBB3 after phosphorylation by ERBB2 provides several binding sites for the PI3Kα regulatory subunit (→6.3). In contrast, the ERBB4 receptor which is

**Table 18.3.** *The ERBB family in man*

| Gene | Other names | Chromosomal Localization | Ligands* | Remarks |
| --- | --- | --- | --- | --- |
| *EGFR* | ERBB1/HER1 | 7p12 | EGF, TGFα, amphiregulin, HB-EGF, betacellulin | |
| *ERBB2* | HER2/NEU/p185$^{HER}$ | 17q11.2-q12 | none | heterodimerization partner of other family members |
| *ERBB3* | HER3 | 12q13 | NRG1, NRG2 | lacks tyrosine kinase activity |
| *ERBB4* | HER4 | 2q33.3-q34 | NRG1, NRG2, NRG3, NRG4, betacellulin, HB-EGF | |

*EGF: epidermal growth factor, TGF: transforming growth factor, NRG: neuregulin

activated by heregulins and perhaps by HB-EGF rather promotes the differentiation of mammary epithelial cells.

Approximately 50% of ER- and a smaller proportion of ER+ breast cancers overexpress ERBB2, as detected by immunohistochemistry. In most cases, strong positive staining is caused by amplification of the *ERBB2* gene. Measurements by biochemical assays reveal an inverse quantitative relationship between ERα and ERBB2 protein levels. There is thus a continuum, but in clinical routine, at most four classes are qualitatively distinguished, viz. ER+/ERBB2-, ER+/ERBB2+, ER-/ERBB2+, and ER-/ERBB2-.

Alternatively or in addition to ERBB2, ERBB1 is also overexpressed in some cases and may provide the crucial growth stimulus in some ER-/ERBB2- cancers. ERBB3 is not usually overexpressed, but is always present in ERBB2+ cancers. It is likely essential for the action of ERBB2 by providing the crucial binding sites for PI3Kα and other intracellular mediators. At normal concentrations of the receptors, they would be present as monomers and dimerization of ERBB2 and ERBB3 would be dependent on ligand-binding by ERBB3. Overexpression of ERBB2 may lower the concentration of ligand required or even obliterate it, thereby facilitating constitutive association and activity (Figure 18.12).

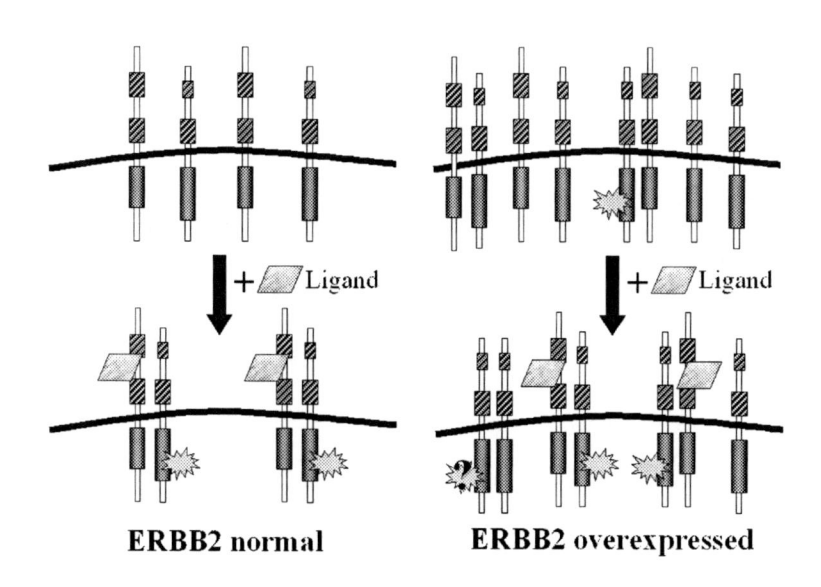

**Figure 18.12** *Effect of ligand upon dimerization of ERBB2 and ERBB3 in normal epithelial cells (left) and cancer cells with ERBB2 overexpression (right)*
ERBB2 is depicted with a shorter extracellular ligand binding domain and a larger intracellular tyrosine kinase domain. The exploding star symbolizes activation. The question mark indicates that it is not certain to which extent ERBB2 overexpression causes ligand-independent activity of its tyrosine kinase.

With increasing expression of ERBB2 or ERBB1, the estrogen requirement for growth of breast epithelial cells appears to be alleviated and eventually lost. Thus, while the normal cells require estrogens and EGF-like growth factors, and ER+ cancers retain these requirements at least to some extent, ER-/ERBB2 breast cancers need only the peptide factors and eventually even less of these. Estrogen action is, of course, partly, but not completely mediated through induction of growth factors and activation of ERBB receptors (→18.1, Figure 18.4). Therefore, in addition to overexpression of ERBB receptors, some kind of 'rewiring' must have occured in ER- cells. This is incompletely understood. Importantly, ER- cancers harbor a large number of further genetic and epigenetic alterations besides loss of the estrogen receptors and amplification of *ERBB2*.

Like the distinction between ER+ and ER- cancers, that between ERBB2+ and ERBB2- breast cancers has prognostic implications, because ERBB2+ cancers as a rule fare worse. It has also consequences for therapy. As for other tyrosine kinases, low molecular weight inhibitors of ERBB2 have been developed and are tested in clinical trials (→22.4). A monoclonal antibody, named trastuzumab, is directed specifically against ERBB2. This antibody was originally generated in mice. To avoid immune reactions when applied to patients, it was humanized. The antigen recognition variable region derived from the mouse was retained, but the constant region was replaced by the human sequence. This biotechnologically produced antibody is now administered to breast cancer patients with metastatic disease, if and only if their cancers stain strongly positive for ERBB2 in immunohistochemistry. *ERBB2* gene amplification is additionally determined. The antibody drug indeed significantly prolongs survival in many of these patients, when administered with standard chemotherapy. In the future, it may also be used in adjuvant and neoadjuvant therapy of ER-/ERBB2+ cancers (→18.5). Determination of ERBB2 gene amplification yields another piece of information. The *ERBB2* amplicon (→2.2) in most cases includes the *TOPO2A* gene encoding the α subunit of topoisomerase II. Its overexpression increases the sensitivity of tumor cells to cytostatic anthracyclins, providing another opening for targeted therapy (→22.2).

## 18.5 CLASSIFICATION OF BREAST CANCERS

Like other carcinomas, breast cancer is treated primarily by surgery. While formerly radical mastectomy with removal of all lymph nodes around the tissue constituted the standard approach, today's philosophy is to keep the surgical intervention as minimal as possible without risking a recurrence.

Breast cancer metastasizes to local lymph nodes as well as to distal organs such as bone, lung, and liver. At the time of surgery, the extent of distal metastasis is often difficult to determine, because many metastases are too small to be discovered by imaging methods and no reliable molecular assays are available for the detection of breast cancer micrometastases. Therefore, if any significant risk of metastasis is assumed, chemotherapy is applied following surgery. Such 'adjuvant' treatment very likely prevents recurrences in some patients, but constitutes an over-treatment for those without metastases. Nevertheless, it is not always efficacious, since some patients develop metastases or locally recurring tumors in the remaining breast tissue in spite of adjuvant treatment. As for other carcinomas, the therapeutic options for established systemic disease are limited.

The most important goal of breast cancer classifications is therefore to provide a basis for choice of therapy in each individual patient. The available choices include the extent of surgery, the administration of adjuvant therapy, and the selection of particular drugs. Several tools are already available for this purpose. 'Classical' histological investigation distinguishes different subtypes such as Paget carcinoma, intraductal carcinoma, lobular carcinoma, and, importantly, precursors like ductal carcinoma-in-situ and benign tumors such as fibroadenoma. Of course, grading is applied and tumor staging is important (→1.4). Specifically, the extent of lymph node involvement is a good indicator of the likelihood of metastases and recurrence.

Molecular markers have improved this classification. Today, determination of ER, PR, and ERBB2 status represents good standard practice. A diagnosis of ER+/PR+/ERBB2-, e.g., indicates a comparatively good prognosis and predicts a good response to anti-estrogenic drugs. By comparison, ERBB2+ cancers are on average more aggressive, but often respond to trastuzumab treatment in conjunction with chemotherapy. Breast cancers in patients with inherited BRCA mutations are special. They are as a rule ER-, and present histological markers of basal duct cells. Moreover, these patients have, of course, a relatively high risk of developing an independent cancer in the same or the other breast (→18.3).

Other molecular markers have been found by studying the correlation between the expression of proteins thought to be involved in tumor invasion and metastasis (→9) and the course of the disease. Several individual proteins constitute useful 'prognostic' markers, e.g. high expression of uPA indicates a high probability of metastasis (→9.3).

Further progress in the classification of breast cancer appears to emerge from gene expression profiles, which identify patterns in the expression of many genes that correlate with biological properties and clinical behavior of a tumor. Most widely, microarrays spotted with cDNA fragments or oligonucleotides corresponding to human genes are employed. Different types of arrays cover thousands to ten-thousands of human genes and their splice variants. RNA extracted from tumor or normal tissues are reverse transcribed and labeled to obtain cDNA mixtures that remain representative of the relative abundancies of mRNAs in the tissues. These are then hybridized to the microarrays. The measured hybridization intensities at each spot give an estimate of the expression level of each mRNA represented on the array.

These techniques allow the comparison of gene expression patterns between different samples, e.g. between tumor and normal tissue or between individual tumors. Because only a small number of measurements are taken on each sample and current microarrays have a limited dynamic range, measurements by microarrays are not very accurate. This means that the expression level of an individual gene in an individual tumor cannot be measured precisely. However, patterns defined by the expression of several hundreds of genes are robust and classify genes as well as samples into groups or 'clusters'. These techniques can therefore on one hand serve to single out individual genes for closer investigation as tumor markers or therapeutic targets, or to identify particular pathways to be active in a certain subtype of cancer. On the other hand, they can provide a method to classify cancers by specific 'signatures', sometimes in unexpected ways.

Investigation of breast cancers by this type of technique has yielded expected as well as unexpected results. ER+ and ER- cancers gave clearly distinct expression profiles. ERBB2+ cancers likewise presented specific signatures. Beyond corroborating current clinical practice, these signatures provide hints at which genes might be induced or down-regulated by estrogens or ERBB2 receptor activation in breast cancers and therefore might provide good targets for novel therapies in these subgroups

Neither unexpectedly, cancers from patients with *BRCA* germ-line mutations formed part of a separate group. This group showed many characteristic markers of basal epithelial cells, such as the cytokeratins 5 and 6. Although lacking the signature of ER+ cancer profiles, these cancers were also clearly distinct from the ERBB2+ group. Thus, there seem to be at least two distinct classes of ER- cancers.

Interestingly, while one group of breast cancers showed expression profiles relating them to basal cells, the ER+ cancers displayed many markers characteristic of the luminal secretory cell phenotype. This finding substantiates previous speculations that different subtypes of breast cancers may derive from different stages of the mammary epithelial lineage, as already shown in other tissues, e.g. the skin (→12.3).

The expression profiles also suggest that ER+ cancers can be divided into further subclasses, which is in line with their quite divergent clinical behavior. These results also underline that the properties of breast cancers are not determined by alterations in any single gene, but are multifactorial.

Expression profiling studies in breast cancer and in other carcinomas have also compared primary cancers and their metastases. In all such studies, individual primary cancers and their metastases proved to be more closely related to each other than primary cancers or metastases among themselves, clustering as separate pairs. This finding bears on an old controversy in cancer biology (Figure 18.13). Metastases could arise from the original tumor by gradual selection of cells that can manage each of the individual steps required for metastasis (→9.1). Alternatively, the ability to metastasize could be determined by the pattern of genetic and epigenetic alterations present in the primary tumor. The close relationship between primary cancers and their metastases argues strongly for this latter hypothesis.

This conclusion is important in clinical practice, since the more the likelihood of metastasis is determined by the properties of the primary tumor, the better it can be predicted by analyzing the tumor tissue at the primary site. Moreover, a comparison of primary tumors that have metastasized to those that have not, may also yield a 'molecular signature' that can help to identify metastatic cancers before overt metastases can be detected. Such a signature has been suggested and, intriguingly, includes not only properties of the tumor cells, but also gene expression patterns in the reactive stroma. This finding underlines current ideas on the crucial importance of both tumor and stromal properties for invasion and metastasis (→9.6).

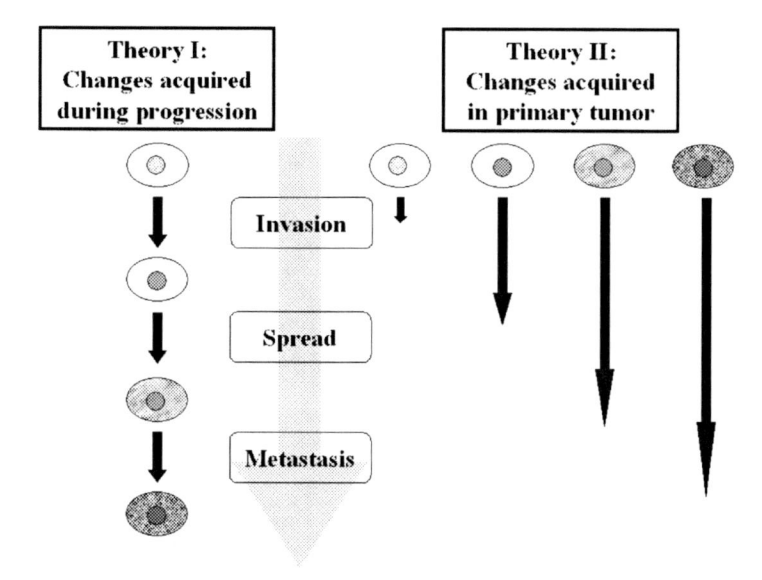

**Figure 18.13** *Two competing models of metastasis*
See text for further exposition.

A specific aim addressed by gene expression profiling in breast cancer is the prediction of prognosis and of response to chemotherapy. Ideally, one would develop a smaller set of markers, detectable by analysis of DNA alterations such as gene amplification, by immunohistochemistry in a routine pathology lab, or perhaps by microarrays containing a smaller set of genes (a 'predictor gene set') which are less expensive and easier to evaluate, but sufficiently robust for prediction. Several approaches have been taken towards this end. A typical approach involves a 'supervised' analysis. Tumor gene expression profiles from patients that were cured by surgery or chemotherapy are compared to profiles from those in which the cancer recurred. The predictor gene set optimally distinguishing these groups is applied to a separate series of tumors for verification. Ideally, it would also classify patients with a favourable outcome or benefitting from a particular type of therapy. Breast cancer-specific microarrays of this type have been developed. It is hoped that they also prove successful in prospective studies.

The developments in expression profiling continue an established trend in breast cancer treatment towards an 'individualized' treatment. This trend can also be observed in many other cancers which show great variation in clinical course and response to treatments (→22.7). An obvious prerequisite is an actual choice of available treatments, as in breast cancer. In fact, the more sophisticated understanding of cancer biology and its molecular basis emerging in breast cancer has initiated a kind of paradigm shift in its therapy that is spreading to other tumor entities. Traditionally, the treatment of a cancer was adjusted according to its histological subtype, extension (as indicated by stage) and apparent biological aggressiveness (as indicated a.o. by grading). In the future, individualized treatment may be gauged by the particular molecular targets provided in a cancer. In breast cancer, this principle is already followed during anti-hormonal therapy contingent on the ER status and trastuzumab therapy contingent on ERBB2 expression. It is expected that identification of further molecular targets in breast cancer, by gene expression profiling or by other techniques, will boost this development.

## Further reading:

Tavassoli FA, Devilee P (eds.) Pathology and genetics of tumours of the breast and female genital organs. IARC Press, 2003

Perou CM et al (2000) Molecular portraits of human breast tumours. Nature 406, 747-752

Scully R, Livingston DM (2000) In search of the tumour-suppressor functions of BRCA1 and BRCA2. Nature 408, 429-432

Katzenellenbogen BS, Katzenellenbogen JA (2002) Defining the „S" in SERMs. Science 295, 2380-2381

van de Vijver et al (2002) A gene-expression signature as a predictor of survival in breast cancer. NEJM 347, 1999-2009

Venkitaraman AR (2000) Cancer susceptibility and the functions of BRCA1 and BRCA2. Cell 108, 171-182

McDonnell DP, Norris JD (2002) Connections and regulation of the human estrogen receptor. Science 296, 1642-1644

Holbro T, Civenni G, Hynes NE (2003) The ErbB receptors and their role in cancer progression. Exp. Cell Res. 284, 99-110

Huang et al (2003) Gene expression predictors of breast cancer outcomes. Lancet 361, 1590-1596

Keen JC, Davidson NE (2003) The biology of breast carcinoma. Cancer 97 Suppl. 825-833

Ramaswamy S et al (2003) A molecular signature of metastasis in primary tumors. Nat. Genet. 33, 49-54

Stern DF (2003) ErbBs in mammary development. Exp. Cell Res. 284, 89-98

Wooster R, Weber BL (2003) Breast and ovarian cancer. NEJM 348, 2339-2347

Jordan VC (2004) Selective estrogen receptor modulation: concept and consequences in the clinic. Cancer Cell 5, 207-213

Venkitaraman AR (2004) Tracing the network connecting BRCA and Fanconi anaemia proteins. Nat. Rev. Cancer 4, 266-276

# CHAPTER 19

# PROSTATE CANCER

- Prostate cancer, called more precisely 'prostate adenocarcinoma', becomes clinically significant in up to 10% of all males in Western industrialized countries. It is rare in younger males, but its incidence increases continuously with age.
- The clinical course of prostate cancer is variable, ranging from clinically insignificant tumors which grow slowly over decades to aggressive cancers which spread locally and metastasize to bone and other organs, killing the patient within a few years.
- The detection of prostate cancer has been improved by modern imaging methods and especially by use of the biochemical serum marker prostate specific antigen (PSA), a moderately specific and very sensitive serum marker. Molecular markers for the classification of prostate cancers are still urgently sought.
- Androgens and the androgen receptor (AR) have been central issues in prostate cancer research for many years and represent major targets for therapeutic intervention. Most prostate cancers respond to treatment by androgen depletion and receptor blockade, apparently by apoptosis of better differentiated tumor cells. Unfortunately, this therapy is palliative and only marginally prolongs survival, because the cancer is repleted from cells with altered responses to androgens that are refractory to anti-androgenic treatment. Several mechanisms contribute, including mutations and amplifications of the *AR* gene as well as changes in signaling pathways acting on the AR.
- The incidence of clinically significant prostate cancer is vastly different between North-Western Europe and South-East Asia. This difference can be partly ascribed to the aging of the population in industrialized countries and improved detection by PSA assays. A difference remains after adjustments for age and detection rate, and points to factors in the 'Western life-style' fostering prostate cancer. One candidate is a diet rich in saturated fat and relatively low in vitamins and micronutrients from fruit and vegetables.
- Genes associated with familial prostate cancer do not behave as classical tumor suppressors. A large number of polymorphisms in genes related to hormone metabolism and action, nucleotide metabolism, and cell protection may modulate the risk for prostate cancer.
- Initial changes in prostate cancer are predominantly chromosome losses and other mechanisms leading to down-regulation of gene expression. In particular, silencing of several genes by promoter hypermethylation coincides with the onset of malignancy. Detection of tumor cells by hypermethylation assays therefore appears particularly promising in prostate cancer.
- Tumor suppressors and oncogenes involved in other cancers such as *TP53*, *PTEN*, *MYC*, and *ERBB1* seem to be predominantly important during the

progression of prostate carcinoma. Initial changes in prostate cancer appear mainly to result in decreased apoptosis and altered interactions with stromal cells.
- As potential tumor suppressors in familial cases, several tumor suppressor candidates in sporadic prostate cancers do not undergo genetic alterations in their second alleles. It is therefore possible that quantitative alterations by decreased gene dosage or epigenetic inactivation rather than qualitative alterations by mutation initiate prostate cancers.
- Epithelial cells in the normal prostate interact tightly with the underlying mesenchyme which supplies growth factors, partly in response to androgens. Throughout much of their development, prostate cancers seem to retain a strong dependency on stromal cells, perhaps more so than other carcinomas. Intense epithelial-mesenchymal interactions are even maintained in bone metastases, where prostate cancer cells appear to find a suitable 'soil' for survival and growth. Here, a vicious cycle may be established, in which the cancer cells stimulate the maturation and activity of osteoblasts and osteoclasts which in turn produce growth factors that promote growth and survival of the cancer cells.

## 19.1 EPIDEMIOLOGY OF PROSTATE CANCER

Prostate cancer develops from the glandular epithelium of the small organ which secretes most of the seminal fluid in males. Most prostate cancers can be clearly identified as adenocarcinomas (Figure 19.1). The prostate is about as big as a

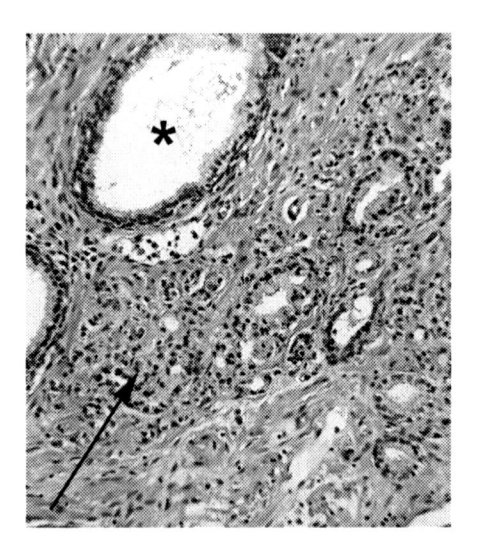

**Figure 19.1** *Histology of prostate adenocarcinoma*
The star marks a normal glandular tubule with clearly distinguishable basal and secretory layers. The arrow points to the carcinoma area with dysmorphic glandular tubules.

chestnut in younger males and almost regularly increases in size after mid-life, mostly by expansion of the mesenchymal stroma. This leads to a benign tumor, 'benign prostate hyperplasia' (BPH). BPH is rarely life-threatening, but by compressing the urethra which passes through the prostate, it can lead to more or less severe problems in passing urine, up to the point of complete obstruction.

Prostate carcinoma is also a disease of elderly men, but often develops in parts of the organ away from the urethra and therefore does not cause urinary obstruction early on. In contrast to BPH, it is a malignant disease. Although BPH and prostate carcinoma are often found in the same organ, the carcinoma is not thought to develop from benign hyperplasia, which is rather characterized by expansion of the mesenchymal component. Precursors of the carcinoma may be dysplastic changes of the glandular ducts such as prostate inflammatory atrophy and prostatic intraepithelial neoplasia (PIN).

Many prostate cancers grow slowly and do not become clinically relevant in the lifetime of an older man. In autopsy studies, up to 40% of males aged over 70 years who died from other causes have been found to harbor cancerous areas in their prostate. While most prostate cancers, thus, may not be 'clinically significant', a significant proportion grow in a more aggressive fashion, expanding locally and metastasizing to lymph nodes, bone, and other organs. Cancers confined to the organ can be cured by removal of the prostate ('radical prostatectomy') or radiation therapy, but metastasized prostate cancer is incurable.

In the mid of the 1990's, the public in the Western industrialized countries realized with some horror that prostate carcinoma was about to become the most lethal cancer in males. This prostate cancer 'epidemy' turned out to have three sources.

(1) Of all cancers, that of the prostate may be the one with the strongest age dependency. Prostate cancer is a rarity below the age of 50, but increases exponentially thereafter (Figure 19.2). With life expectancy increasing, the overall incidence keeps rising and the lifetime risk for clinically relevant prostate cancer is consequentially estimated as around 10%.

(2) The second cause of the apparent prostate cancer boom lay in an improved detection rate. Until the late 1980s, most cases of prostate cancer were detected by the clinical symptoms they caused. Prostate cancer became evident through symptoms caused by bone metastases, more rarely by obstruction of the urethra. Some cancers can be detected by 'digital rectal examination'. However, even these tend to be relatively advanced. So, formerly, the typical patient presented with pain in the lower back, where metastases most often reside, and with advanced, incurable cancer. In principle, the cancer can be detected much earlier, while still confined to the prostate, by histological investigation of biopsies. This is an unpleasant and slightly risky procedure and only performed, if there is some indication for a cancer being present. A main breakthrough in prostate cancer detection was, therefore, achieved by the introduction of the serum marker PSA.

Prostate specific antigen (PSA) is a protease from the kallikrein family (Figure 19.3), which is secreted into the seminal fluid by differentiated secretory prostate epithelial cells. These are separated from the blood by a basal cell layer and a basement membrane. Since PSA is almost exclusively produced in these cells, only

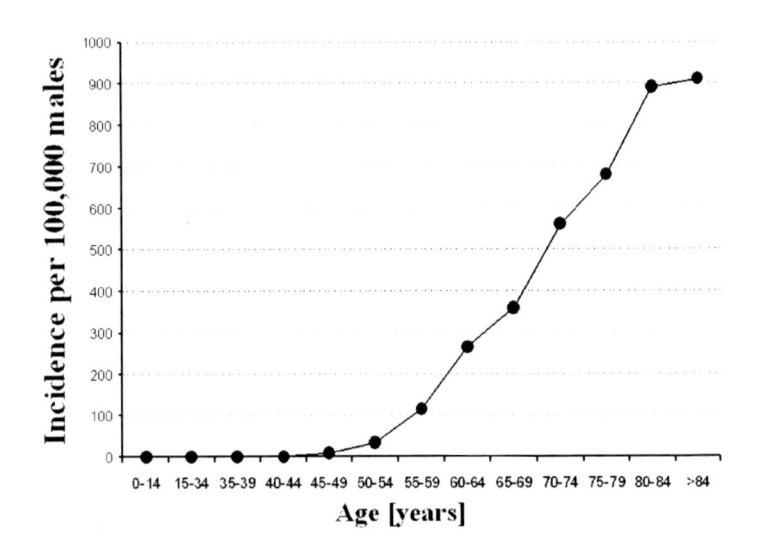

**Figure 19.2** *Age dependency of prostate cancer incidence*
Data are for Germany in 2000 according to the Robert-Koch-Institute.

tiny amounts are normally present in the blood. Most prostate cancers keep producing the protein, although its processing may be altered. Through loss of polarity in the cancer tissue and breakdown of the basement membrane separating the epithelium from the underlying connective tissue and from blood vessels, more PSA appears in serum. Modern ELISAs detect <0.1 ng/ml serum, below the level in healthy middle-aged men. In the serum of prostate cancer patients, concentrations range from 2 ng/ml up to more than 1000 ng/ml (cf. Figure 21.4).

Unfortunately, PSA as a tumor marker is neither completely specific nor perfectly sensitive. On one hand, serum PSA levels can also be increased in benign prostate diseases and they tend to rise with age. This creates a 'grey zone'. Thus, confirmation of the diagnosis by histological investigation of biopsies remains mandatory. Newer assays for PSA exploit its altered processing and binding to serum protease inhibitors to better discriminate between benign and malignant prostatic diseases. On the other hand, some prostate carcinomas secrete little PSA and thereby escape detection.

Although PSA is not a perfect tumor marker, it is good enough to have led to a markedly improved detection rate of prostate carcinomas, even in those countries where no large-scale screening of the male population was attempted (→20.5). Improved detection methods for prostate cancer were paralleled by improvements in the treatment of organ-confined cancers by surgery or radiotherapy and even of metastatic cancers by drugs or radiotherapy. This may also have contributed to earlier detection of prostate cancers by encouraging patients with minor symptoms to follow them up, because a curative therapy was available. If so, one would expect that prostate cancer incidence and mortality would peak somewhere in the late 1990s, because many cancers in years to come were already discovered and cured, which would not have been possible, had they been detected at a more advanced stage. Current preliminary figures suggest that this may indeed be so.

There are, however, two problems with this potential success story. At present, there is no method to reliably distinguish whether an early stage prostate cancer will continue to grow slowly or become aggressive. Thus, some patients identified by PSA assays would not have developed clinically relevant prostate cancer, and may have been over-treated. Conversely, the methods for detection of micrometastases are not perfect either. So, some patients are treated by prostatectomy to no avail, because they will die from metastases (Figure 19.4).

(3) While the apparent prostate cancer epidemic partly reflected the age dependency of the disease and may have been inflated by improvements in its detection, a real rise could be hidden in the statistical figures. Its cause is different to extract. Some pieces of evidence link prostate cancer to 'life-style factors' in Western industrialized countries.

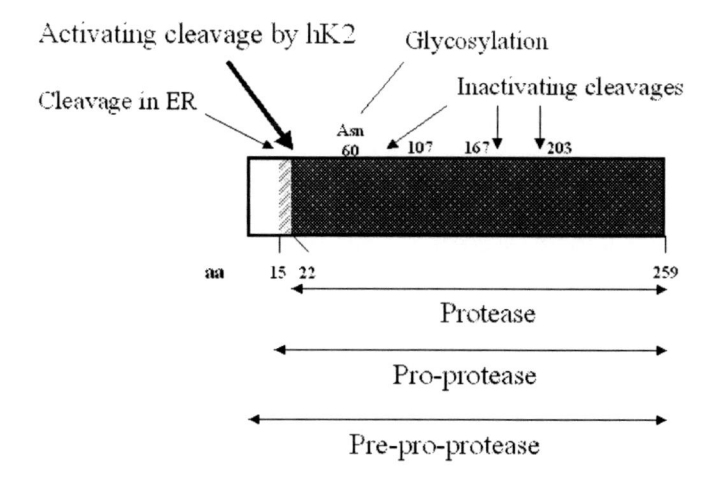

**Figure 19.3** *PSA structure and processing*

The incidence of prostate cancer is very different in these countries compared to that in East Asian countries or in the Mediterranean area, and remains so after reasonable adjustments for age and detection rate (cf. Figure 20.4). More strikingly, second generation immigrants from these countries into the USA show a prostate cancer incidence similar to average Americans. The underlying causes have been much speculated on. The immigrant studies indicate that genetic differences are unlikely to be the dominant factor. Specifically, genetic polymorphisms in hormone metabolism (cf. Table 2.4) have been investigated, but no consistent differences have been identified. Alternatively, different exposures to estrogenic compounds in the environment have been invoked. Plant ingredients called 'phytoestrogens' have

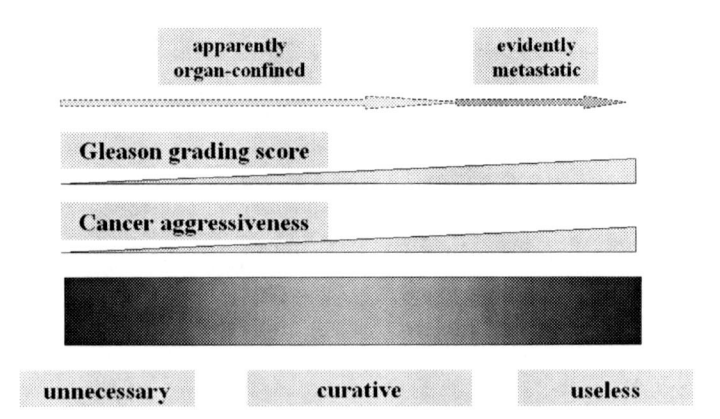

**Figure 19.4** *The prostatectomy dilemma*
See text for further explanation.

been considered as protective, and various synthetic chemicals rather carelessly released into the environment in the 1950s and 1960s have been discussed as 'endocrine disruptors'. Studies on these compounds continue, but are not yet conclusive.

Good evidence points to a relationship between diet and prostate cancer (cf. 20.3). Several individual food items and specifically micronutrients have been convincingly found to be associated with prostate cancer risk. A diet high in saturated fat and in dairy products seems associated with an increased risk of prostate cancer. This same diet may lack protective micronutrients, vitamins, and plant ingredients such as selenium, folate, lycopene (from tomatoes), and genisteine (from soy beans), which have each individually shown to correlate with a decreased risk of prostate cancer. Obviously, it is everything but trivial to ascertain how strong their combined effect may be. Other life-style factors may compound the influence of an unhealthy diet. For once, smoking and alcohol do not seem to be overwhelmingly important, whereas some data suggesting a relationship between prostate cancer risk and lack of exercise look intriguing.

A question for molecular biology is by which mechanisms such life-style factors might affect cancer risk. There are no definitive exogenous carcinogens known in prostate cancer, and the disease may therefore often be caused by endogenous processes. Many individual findings point to an important role of reactive oxygen species (→Box 1.1). They include an association with a high-energy diet and low selenium, which is a crucial constituent of glutathione peroxidase (→3.5). Likewise, weaker genotypes of the OGG1 glycosylase, which removes oxidized guanine from DNA (→3.1) may predispose to prostate cancer. Indeed, the content of oxidized DNA bases in the prostate may increase with age. Moreover, the GSTP1 isoenzyme which inactivates electrophilic metabolites (→3.5) is down-regulated early in prostate carcinogenesis by promoter hypermethylation.

A different sort of link could be mediated by insulin-like growth factors, whose levels depend on life-style, i.e. diet and exercise, as well as on genetic constitution. In a large longitudinal study, the risk of prostate cancer correlated well with high levels of IGF1 and low levels of its inhibitory binding protein IGFBP3 (→16.2) in midlife.

While such findings are intriguing, it is clear that carcinogenesis in the prostate is probably multifactorial and the mechanisms are incompletely understood. Current recommendations for prevention of prostate cancer are therefore relatively unspecific (→20.3).

## 19.2 ANDROGENS IN PROSTATE CANCER

The prostate gland produces part of the seminal fluid. Almost all prostate cancers arise from the glandular epithelium, which consists of a basal and a secretory layer. In the normal gland, basal cells constitute the proliferative fraction. They give rise to intermediate cells which terminally differentiate into secretory cells that line the ducts and produce prostate-specific proteins such as PSA. They turn over slowly and are continuously replaced (Figure 19.5).

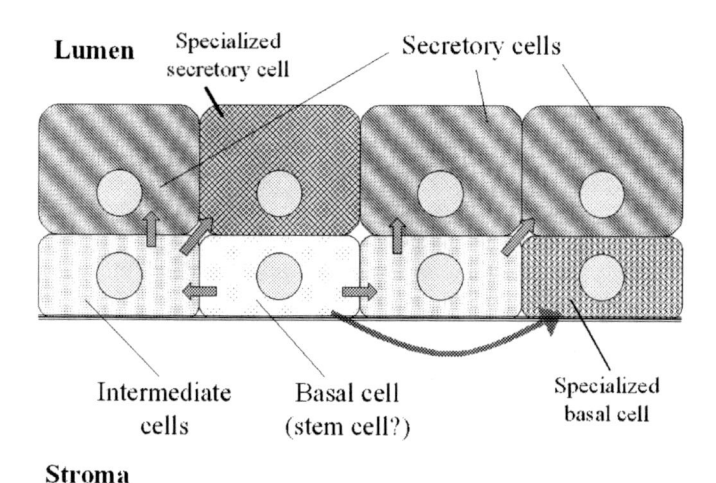

**Figure 19.5** *Organization of cell proliferation and differentiation in the normal prostate*
This hypothetical scheme is accepted in general, although its details are debated: A subset of basal cells is thought to represent progenitors for differentiated cells of the luminal layer by way of intermediate cells. It is not certain to which extent they possess further stem cell properties, in particular, whether they give rise to specialized cells in the basal layer, e.g. neuroendocrine cells. Many cells in the basal layer display an intermediate phenotype and likely represent the transient amplifying population of the tissue. They are supposed to migrate to the secretory layer during terminal differentiation. Stem cells lack the androgen receptor, which is expressed from the intermediate cell stage on.

The secretory cells express high levels of the androgen receptor (AR) and their survival and function is dependent on androgens. The cells in the basal layer contain fewer or no androgen receptors. The androgen receptor (Figure 19.6) is a member of the steroid hormone receptor family and closely related to the estrogen receptors (cf. Fig. 18.2). Its AF-2 domain binds androgens in the same fashion as that of the ESRs bind estrogens.

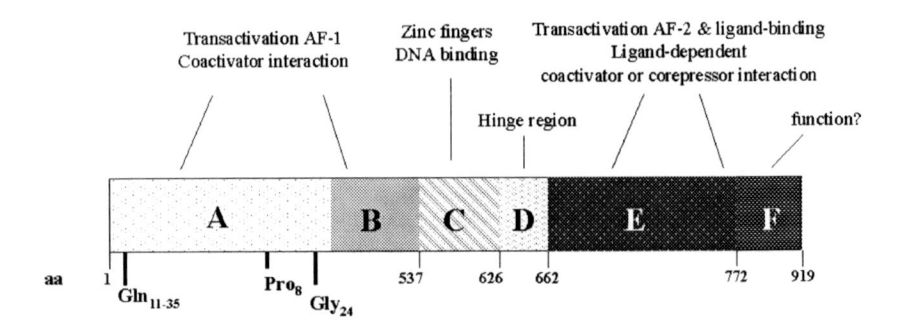

**Figure 19.6** *Structure of the androgen receptor*
The domains are drawn to scale. Note that the aa repeats in domain A are polymorphic.

The main androgen present throughout the body is testosterone. In the prostate, it is locally converted to dihydrotestosterone which binds to the androgen receptor with a ≈10-fold higher affinity. This reaction is catalyzed by 5α-reductase (Figure 19.7). The androgen receptor is also present in mesenchymal stroma cells. These are stimulated by testosterone and dihydrotestosterone to produce FGF and IGF growth factors that regulate the proliferation and survival of the epithelium, particularly of those basal cells, which lack an active androgen receptor.

During fetal development and puberty, the maturation of the gland is also dependent on androgens. Castration before puberty prevents maturation of the prostate, and also prostate cancer. In grown-up men, likewise, removal of androgens causes an involution of the gland. Such observations prompted the idea to treat prostate cancer by androgen depletion. This can be done by surgical removal of the testes in which the cells of Leydig produce >90% of the androgen in the male body. Alternatively, androgen synthesis can be suppressed by interfering with the release of the pituitary hormone LH which stimulates the production of androgens in Leydig cells (cf. Figure 18.11). This is most commonly done by using drugs similar to GnRH which act on receptors in the hypothalamus inducing their down-regulation, as in the treatment of premenopausal ER+ breast cancers. After a short burst of LH (and FSH), no further gonadotropins are produced and the Leydig cells cease to synthesize androgens. Alternatively or concomitantly, binding of androgens to the androgen receptor can be inhibited by antiandrogens which bind to the AF-2 domain of the receptor but do not support its interaction with transcriptional co-activators.

Antiandrogenic treatment is used in the clinic in addition to surgical removal of the prostate (prostatectomy) or for treatment of metastases. Most cancers indeed shrink, with signs of apoptosis in tumor cells. However, unless completely removed by surgery, almost all prostate cancers recur and then respond no longer to continued antiandrogenic treatment. By and large, antiandrogenic treatment reduces the overall tumor mass, often relieving pain and other symptoms, but does not prolong survival by a large margin.

The reason for this is that growth of advanced and certainly of recurrent prostate cancers is dominated by cells that proliferate and survive relatively or completely independent of androgens. It is the subject of a long and continuing debate where these cells come from. According to opinion #1, the androgen-dependent cells in the tumor are differentiated derivatives of the actual tumor stem cells, which are related to each other in a similar fashion as basal and secretory cells in the normal tissue. Under pressure of antiandrogenic treatment, the stem cells would take over. According to opinion #2, initially all prostate cancer cells are androgen-dependent for their survival and perhaps even for their proliferation. The pressure of antiandrogenic treatment then leads to selection of cell clones which manage to grow at very low levels of androgens. These cell clones could be pre-existent in the tumor or develop during treatment. Their emergence might be faciliated by genomic instability in the cancer cells. So, even the nomenclature is debated: Advanced prostate cancers are called 'androgen-independent' by some, but 'androgen-refractory' or even 'androgen-hypersensitive' by others.

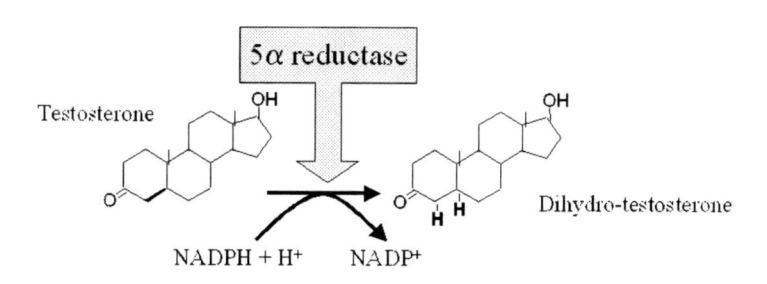

**Figure 19.7** *The 5α-reductase reaction converting testosterone to dihydrotestosterone*
The part of the steroids at which the reaction takes place are drawn in bold.

Indeed, advanced prostate cancers show a bewildering variety of alterations in androgen signaling (Figure 19.8).

(1) Some cancers contain amplifications of the androgen receptor gene at chromosome Xq12. These increase the AR concentration and allow responses to very low levels of testosterone.

(2) Other cancers contain mutations in the AR, typically in the ligand-binding activation domain AF-2 but also in the N-terminal AF-1 domain (Figure 19.6). These mutations increase the affinity of the receptor towards androgens or augment

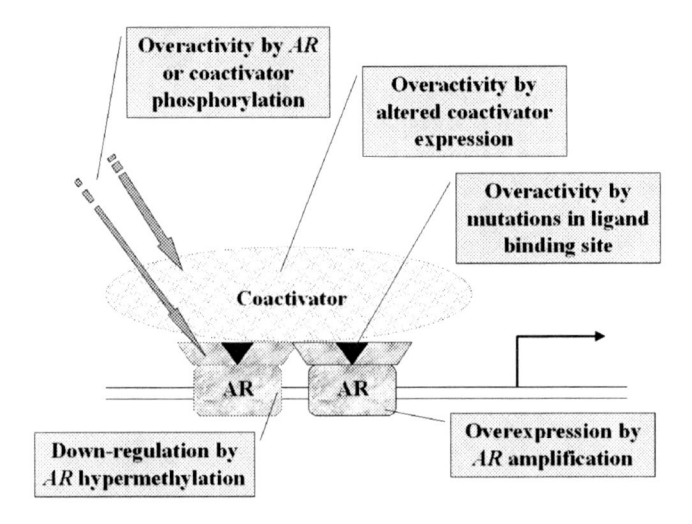

**Figure 19.8** *An overview of androgen signaling alterations in advanced prostate cancer*
See text for further explanations

its transcriptional activation function. Some mutations alter receptor specificity, making it responsive to other steroids in the body such as progesterone, dehydroepiandrosterone or even anti-androgenic drugs.

(3) Like other members of the steroid hormone receptor superfamily (→18.1), the androgen receptor activates transcription through interactions with different co-activator proteins. Some of these show over-expression in prostate cancer, which may imply more efficient androgen signaling with lower requirements for ligand-binding.

(4) Co-activators and the receptor itself are targets of phosphorylation by kinases from several signaling pathways. Androgen receptor activity becomes more dependent on such 'cross-talk' and less on ligand-binding in many prostate carcinomas. In these, growth factors such as TGFα or KGF (a different name for FGF7) augment androgen signaling, likely through MAPK pathways (→6.2). Cytokines may also be involved, especially IL-6 and factors produced by osteoblasts and osteoclasts in bone metastases. The canonical WNT pathway can activate the androgen receptor through β-Catenin.

In some cases, other pathways may completely take over the control of proliferation and survival of prostate carcinoma cells, obliterating the requirement for the AR. Consequently, some prostate cancers lack its expression, and the *AR* promoter can become hypermethylated.

In addition, several genetic changes in advanced prostate carcinomas could relieve the requirement for androgen signaling in cell survival. These include mutations and LOH of *TP53* (→6.6) and loss of *PTEN* (→6.3). These changes may mainly act by diminishing apoptosis. Overexpression of BCL2 (→7.2) is also found in a subset of prostate cancers and appears to presage a more rapid development towards androgen-refractory disease. Interestingly, BCL2 expression in the normal tissue is restricted to basal cells.

In addition, most metastatic prostate carcinomas overexpress the EGFR and MYC, not infrequently as a consequence of gene amplification. These latter changes may go along with an increased rate of proliferation in addition to diminished apoptosis completing the switch from androgen-dependent to androgen-independent growth.

Of course, proponents of opinion #1 point out that with these changes prostate carcinoma cells end up more or less where basal epithelial cells in the organ were in the first place, i.e. with cell growth and survival driven by peptide growth factors rather than androgens. So, why not assume that prostate cancers are derived from these cells? They suggest that prostate cancers initially consist of a small stem cell fraction, which 'lurks' behind the major tumor mass consisting of more differentiated cells resembling the secretory cells of the normal prostate. Unlike their normal counterparts, however, these do not cease to proliferate. Anti-androgenic treatment acts as a palliative by diminishing this more differentiated fraction, which is, however, repleted from the 'lurker cell' fraction (Figure 19.9).

The raging of opinions may seem academic, but in fact has important implications for the development of prostate cancer therapies. Opinion #2 predicts that better antiandrogenic treatment, which takes the possible escape routes into account, will eventually become capable of curing prostate cancer. Opinion #1 predicts that this will not work and genes and instead proteins unrelated to androgen signaling need to be targeted for a curative therapy.

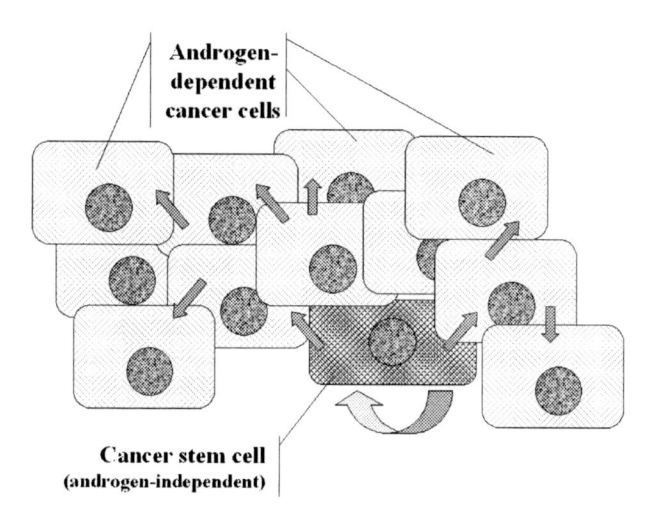

**Figure 19.9** *Prostate cancer as a stem cell disease and the 'lurker cell' hypothesis*
See text for further exposition of the hypothesis.

## 19.3 GENETICS AND EPIGENETICS OF PROSTATE CANCER

Some of the tumor suppressors and oncogenes that are so crucially involved in other common cancers, *TP53*, *PTEN*, *MYC*, *EGFR*, and *BCL2*, contribute also to the progression of prostate cancer towards androgen-independent growth and metastasis. However, they do not appear to be responsible for the initial development of this carcinoma. Likewise, mutations or polymorphisms in these genes certainly do not account for the ≈4-fold increased risk of first-grade relatives of prostate cancer patients to develop the same disease.

In other major cancers, studies of inherited cancer syndromes have helped to identify key genes involved. In prostate cancer, this approach is complicated by several factors. (1) There are very few cases of prostate cancer at a conspicuously early age. Rather, an exponential increase sets in around the age of 50 (Figure 19.2). Specifically, there is no hereditary syndrome with an obvious predisposition towards prostate cancer. (2) With a cancer that appears late in life, it is difficult to investigate several generations within one family. (3) Prostate cancer may be often multifocal, but this is difficult to ascertain as the prostate is a small organ. So, multifocality cannot be used to distinguish familial from sporadic cases. (4) Prostate cancer is so frequent that familial 'clustering' occurs by chance with appreciable frequency.

In spite of these complications, many studies have been performed world-wide. Most have looked at families with several cases of prostate cancer, preferably with one or several at comparatively younger age, by using markers across the entire genome to search for linkage disequilibrium (cf. Box 13.1). Regions in the genome identified in this fashion are then more closely investigated, candidate genes are selected and are screened for mutations which segregate with the disease.

In the case of prostate carcinoma, roughly a dozen different regions in the genome have been implicated in different populations across the world, but few were consistent between several studies (Table 19.1). Most likely, this means that familial clustering of prostate cancer is not caused by the action of one or a few high-risk genes. Rather, different genes may confer some degree of risk, perhaps even by interacting with environmental factors which differ between populations and individuals (Figure 19.10, cf. Fig. 18.9).

In line with this idea, the regions to which candidate hereditary prostate carcinoma genes have been assigned are not those most frequently undergoing LOH in the cancers. Also in accord with this idea, a gene identified in the top candidate region at 1q24, named *HPC1* (hereditary prostate cancer 1), turned out to encode RNaseL, which is involved in regulating cellular responses to infectious agents. Inactivating mutations and certain polymorphisms in this gene appear to go along with an increased risk of prostate cancer. There is little evidence that the gene is a target for 'second hits' in prostate cancer tissue. So, it is probably not a classical tumor suppressor gene (→5.4).

It is, of course, conceivable that how individuals react to infections in the prostate influences their risk to develop a cancer in this organ, as evidenced by the etiology of liver and stomach cancers (→16, →17). However, at this stage, the precise relationship between *HPC1* and prostate cancer is unclear. For several other genes involved in hormone metabolism, DNA repair and nucleotide metabolism, a relationship between polymorphic variants and prostate cancer risk has been observed (Table 19.1). Perhaps, the *HPC* genes are susceptibility genes with a particularly strong influence.

A second useful approach to identifying key genes in human cancers is following up on recurrent chromosomal changes. Indeed, chromosomes 17p and 10q, where the *TP53* and *PTEN* genes are located, often undergo loss in advanced prostate cancers, whereas segments of the chromosomes 7 and 8q, where the *ERBB1* (EGFR) and *MYC* genes reside, are gained or even amplified. In earlier stages of prostate carcinoma, chromosomal losses predominate. The most frequent chromosomal change overall is loss of chromosome 8p, with losses of 13q, 6q, and 16q being relatively prevalent. Mapping of common region of deletion has yielded several regions and various candidate tumor suppressors, but their role is not fully established. The problems encountered are similar to those in the case of the tumor suppressor locus suspected at 9q in bladder cancer (→14.2).

The most convincing candidate is *NKX3.1* (also *NKX3A*) encoding a transcription factor with a homeobox domain. NKX3.1 is largely specific to prostate

Table 19.1. *Gene mutations and polymorphisms predisposing to prostate cancer*

| Gene | Localization | Function | Type of variation* |
|---|---|---|---|
| HPC1 (RNASEL) | 1q24 | ribonuclease | rare mutations in familial PCa frequent polymorphisms |
| ELAC2 | 17p11 | mitotic regulation (?) | rare mutations in familial PCa more frequent polymorphisms |
| AR | Xq11.2-Xq12 | androgen response | polymorphism |
| VDR | 12q12-12q14 | calcitriol receptor | polymorphism |
| SRD5A1 | 5p15 | androgen metabolism | polymorphism |
| GSTM1 | 1p13.3 | detoxification | polymorphisms |
| GSTP1 | 11q13 | cell protection | GSTP1 inactivation in most PCa |
| GSTT1 | 22q11 | | |
| BRCA2 | 13q12.3 | DNA repair | increased risk of PCa by inactivating mutations |
| MTHFR | 1p36.3 | methyl group metabolism | polymorphisms |
| IGFBP3 | 7p13-12 | regulation of growth factor activity | polymorphism serum levels correlate with PCa risk |

*PCa: prostate carcinoma

epithelial cells, is induced by androgens, and is very likely necessary for the proper differentiation of prostate epithelium. Knockout mice engineered to lack the factor fail to develop a functional prostate and hemizygous mice display hyperproliferation of the prostate epithelium. In these respects, *NKX3.1* is as good a tumor suppressor candidate as they come. The problem prohibiting its general acceptance as a tumor suppressor is that in most prostate cancers, including those with LOH or outright deletion of 8p, at least one allele of the gene remains functional and is often expressed at close to expected levels (i.e. 50% of normal). To accept *NKX3.1* as a tumor suppressor, one would have to postulate that diminuation of its expression to half the normal level suffices to inactivate its function. Such 'haploinsufficiency' has also been postulated for other genes (→5.4, →14.2), but is, of course, very difficult to prove in the context of a real human cancer. This is particularly so for prostate carcinoma, which is thought to be unusually heterogeneous and to always contain substantial amounts of stroma. Therefore, the suspicion lingers that the crucial changes in a gene might have escaped detection for technical reasons.

The same sort of problem concerns other established or novel tumor suppressors in prostate cancers. So, *PTEN* is clearly inactivated in some prostate carcinomas by loss of one allele and mutation of the second one; even homozygous deletions have

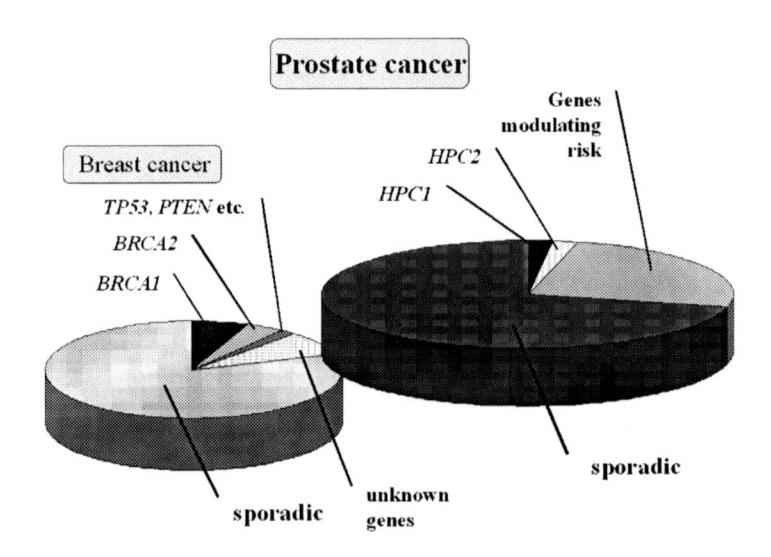

**Figure 19.10** *Genetic predisposition to prostate cancer*
Estimated proportions of cases contributed by mutations in the indicated genes. In prostate cancer (right) true mutations leading to high-risk alleles in *HPC1* and *HPC2* etc. may be extremely rare. However, polymorphisms in these and other genes may strongly modulate the risk conferred by environmental factors (e.g. diet or infections?). Note that the risk in 'sporadic' cases may also be modulated by genetic polymorphisms, albeit more weakly. The situation in breast cancer (left) is in so far different, as high risk alleles in BRCA1, BRCA2 and a number of other genes may account for a substantial proportion of all cases.

been observed. Nevertheless, in many prostate carcinomas, the gene is intact, but strongly down-regulated. It is possible that loss of one *PTEN* allele suffices to promote tumor progression in the prostate, while complete loss of function is a characteristic of very advanced cancers.

Similarly, *RB1* is located at 13q14 telomeric to a region that undergoes LOH and net loss in many prostate cancers, but the remaining *RB1* copy is intact, although its expression level is debated. So, perhaps, the 'classical' concept of tumor suppressors demanding inactivation of both alleles may not apply in prostate carcinomas. There are indeed 'heretical' concepts which attempt to explain cancer development by the overall changes in relative copy numbers, without an absolute requirement for mutational events. Perhaps they may be helpful to understand prostate cancer.

While these ideas are speculative, it is clear that epigenetic mechanisms (→8) are very important in prostate cancers and could account for some of the unexpected findings at the genetic level. Specifically, alterations of DNA methylation (→8.3) are unusually prevalent in prostate cancer. One of the most consistent alteration in prostate cancer is inactivation of the *GSTP1* gene encoding an enzyme protective against electrophilic compounds from exogenous and endogenous sources (→3.5). Loss of expression occurs very early during cancer development and may help to sensitize the tumor cells to further genomic damage. Loss of GSTP1 is almost always caused by epigenetic mechanisms and usually accompanied by dense hypermethylation of the gene promoter (Figure 19.11). More than a dozen genes have now been reported to become hypermethylated in prostate cancer at significant frequencies. Several appear to become coordinately hypermethylated by a sort of 'epigenetic catastrophe' at around the stage of initiation of the carcinoma, perhaps even in advanced prostate intraepithelial neoplasia. Other genes, including those encoding cell adhesion molecules like E-Cadherin and CD44 (→9.2) become hypermethylated later, and still others, including *PTEN*, seem to be down-regulated by epigenetic mechanisms without becoming hypermethylated.

These observations carry interesting prospects for diagnostics and therapy of prostate cancer. Hypermethylation of CpG islands, as in the *GSTP1* gene, can be comparatively easy detected with high sensitivity and specificity, since CpG islands are unmethylated in normal tissues (→8.3). So, detection of *GSTP1* hypermethylation, and that of additional genes, is being developed to assay for the presence of prostate carcinoma (→21.3). Also, since epigenetic gene inactivation is, in principle, reversible, prostate cancer may be a particularly good target for inhibitors of histone deacetylases, histone methylases, and DNA methyltransferases currently under development. In cell and animal models of prostate cancer at least, such compounds are efficacious, usually by induction of apoptosis.

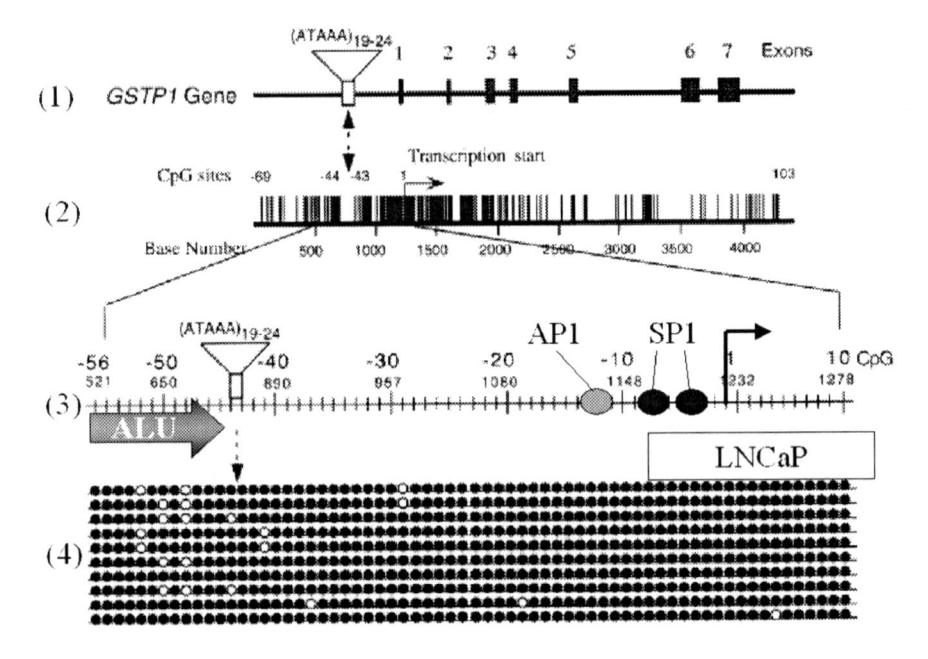

**Figure 19.11** *Hypermethylation of the* GSTP1 *promoter in prostate cancer cells*
(1) Structure of the *GSTP1* gene; (2) Its CpG island, each line indicates a CpG site; (3) 66 CpG sites located within 757 nt around the transcriptional start site. The end of an ALU element and three transcription factor binding sites are indicated; (4) In a prostate carcinoma cell line (LNCaP), almost all of these sites are methylated, while in normal tissues all CpGs 3' of the ATAAA repeat remain unmethylated. Modified from Stirzaker et al, Cancer Res. 64, 3871ff, 2004

## 19.4 TUMOR-STROMA INTERACTIONS IN PROSTATE CANCER

In normal prostate tissue, proliferation and survival of epithelial cells are controlled by growth factors and matrix proteins supplied by the mesenchymal cells which surround the glands as well as on nutrients and oxygen supplied via blood vessels in the mesenchyme. The tissue structure of the prostate, like that of other organs, is dependent on mutual interactions between epithelial and stromal cells (→8.6). As prostate cancers progress, these relationships change, most dramatically in metastases.

In general, tumor progression is associated with increased growth autonomy. In some cancers, this growth autonomy is established by mutations in cell cycle regulators (→6.4) or by inappropriate activation of cancer pathways that control the cell cycle (→6). In many cancers, including prostate cancers, autocrine growth factor loops contribute to growth autonomy. To various extents, however, all cancers remain dependent on interactions with stroma, and this relationship is particularly crucial during metastasis (→9.6). In this respect, prostate cancer may be near one end of a spectrum, as a cancer which retains intense interactions with stromal cells at the primary site as well as in metastases. However, these relationships are really a caricature of those in the normal tissue, because they are distorted by genetic and epigenetic changes in the carcinoma cells, as well as in the stroma.

In the normal prostate, epithelial cells proliferate under the influence of growth factors such as FGFs and EGF-related peptides (Figure 19.12). Receptors for these factors are present mainly at the basal surface of the basal cell layer of the epithelium. The insulin-like growth factor IGF1 may act as a survival factor through the IGFRI receptor (→16.2). The IGF binding protein IGFBP3 may limit the action of IGF1. Luminal secretory cells appear to additionally require androgens for their survival.

Further negative regulator of epithelial cell proliferation are TGFβ factors (→6.7) produced at low levels by the epithelial cells themselves. These levels may be sufficient to support a low level of proliferation of mesenchymal cells necessary for maintenance of the tissue.

Another paracrine factor is the 21 amino acid peptide endothelin-1 (ET-1) which is mainly secreted into the seminal fluid by the secretory epithelial cells, but may also act on the normal prostate mesenchyme. Cells in the prostate stroma express the endothelin receptor A (ETA), which is a G-coupled receptor. It uses $Ca^{2+}$ as a

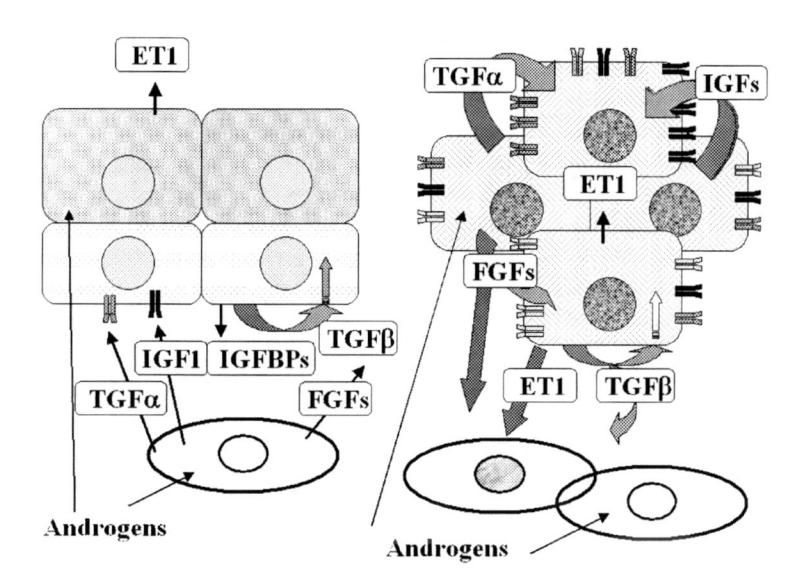

**Figure 19.12** *Paracrine stroma-epithelial interactions in the prostate and in prostate cancer* Left: Normal prostate; right: prostate cancer. A strongly simplified illustration. In particular, increased growth factor secretion by stromal cells is not shown and mechanisms leading to increased stromal growth and angiogenesis are not depicted. See text for more details.

second messenger and is capable of cross-talk with receptor tyrosine kinases (→6.5). Therefore, endothelin and similar peptides can act synergistically with EGF-like growth factors. Normal prostate epithelial cells, however, do not express the ETA, but rather a scavenger receptor, ETB, which binds and removes ET-1.

This orderly organization changes during carcinogenesis (Figure 19.12). Prostate carcinoma cells synthesize FGFs, particularly FGF1 and FGF2, and EGF-like factors, particularly TGFα (cf. Table 18.3). Receptors for FGFs and for EGF-like factors become more strongly expressed and are no longer restricted to basal cells, but are expressed throughout the cancerous epithelium. Insulin-like factors are produced at an increased rate, likely by both carcinoma and stroma. Specifically, IGF2 levels increase. Conversely, IGFBP3 becomes down-regulated at the transcriptional level and by proteolytic cleavage of the secreted protein. Together, these alterations support increased and autonomous proliferation and survival of the carcinoma cells. FGFs, in particular, also stimulate the proliferation of stromal cells and FGF2 is one of the most potent angiogenic factors (→9.4).

Somewhat paradoxically, the synthesis of TGFβ factors increases in prostate carcinoma. However, the receptor subunits, specifically TGFβRII, are down-regulated in carcinoma cells. Therefore, the growth-inhibitory effects on carcinoma cells are diminished, while the proliferation of stromal cells is stimulated. Specifically, TGFβ induces remodeling of the extracellular matrix which is necessary for invasion and angiogenesis (→9.3). Moreover, particularly at later stages of invasion, the immunosuppressive effects of TGFβ may be relevant (→9-5).

Endothelin production is maintained in prostate carcinoma cells. However, compared to normal prostate epithelial cells, ETA is up-regulated and ETB is down-regulated. So, ET-1 now acts on both epithelial and stromal cells.

At some stage of local tumor growth, an HGF/MET autocrine loop is established, as in many other carcinomas (→15.3). In prostate cancers, the expression of both HGF and MET appears to increase gradually. It may be initiated by hypoxia (→Box 9.1). Activation of MET, more than that of other tyrosine receptor kinases, induces scattering of proliferating epithelial cells and a tendency to change their shape towards a mesenchymal cell type. Indeed, a switch of cadherin types may take place during prostate cancer invasion which is typically associated with an epithelial-mesenchymal transition (→9.2). This means that some of the apparent stromal cells in a malignant prostate tumor may in fact be cancer cells. However, prostate cancer metastases typically consist of epithelial cells, which often express E-Cadherin. So, this transition may be transient or may concern only a fraction of the carcinoma cells.

Hypoxia also induces expression of VEGF which synergizes with FGFs and ET-1 to induce angiogenesis. Indeed, an increase in microvessel density and local blood flow is a diagnostic sign of prostate cancers.

Finally, tumor invasion requires an increased expression of proteases which remodel the tissue, aid in tumor cell migration, and liberate latent growth factors from cell surfaces and the extracellular matrix (→9.3). Many such proteases, especially matrix metalloproteinases (MMPs) are produced by stromal cells activated by the carcinoma cells. In the prostate, the carcinoma cells contribute uPA

(→9.3) and, of course, kallikrein proteases like PSA. PSA is only one of several proteases of this class secreted into the seminal fluid, and is systematically designated as hK3. So, hK3 and other members of the family are secreted and activated in prostate carcinoma tissue.

Together, these changes prepare the way for prostate cancer cells to eventually grow beyond the organ and spread through lymph and blood vessels to other parts of the body. Lymph node metastases are usually the first to become established. The most frequent site for hematogenic metastases are bones where prostate cancer cells extravasate in their capillary-rich interior (→9.1). Bones in the spine are the prime target, due to anatomy. However, anatomy is not sufficient to explain the pronounced preference of prostate cancer metastasis for bone tissue. Instead, this affinity is a case in point for the 'seed-and-soil' hypothesis (→9.6).

Bone metastases can be osteolytic or osteoblastic. In osteolytic metastases, the dominating effect of the tumor is dissolution of the bone, whereas in osteoblastic metastases, the dominating effect is stimulation of bone formation. Throughout life, bone is constantly being rebuilt by the combined action of osteoblast cells which deposit trabeculi of calcium apatite and osteoclasts which remove them. Tumor cells in the bone interact with these cells stimulating either cell type or both. Prostate cancer metastases are predominantly osteoblastic.

In bone tissue, prostate carcinoma cells establish a vicious cycle, predominantly with osteoblasts. Autocrine stimulation by FGFs and EGF-like growth factors remains important. Angiogenesis is also promoted in the metastatic tissue by FGFs and VEGF. However, the crucial property that allows prostate carcinoma cells to establish metastases in bones is their adaptation to the tissue by interaction with specific local cell populations, to the point of cellular mimicry (→9.6). So, metastatic prostate carcinoma cells in bone behave in an 'osteomimetic' fashion.

Prostate carcinoma cells secrete TGFβ factors and ET-1, which stimulate the proliferation and maturation of osteoblasts. They also secrete bone morphogenetic proteins (BMPs) acting on these cells. BMPs use similar intracellular pathways for signalling as TGFβ (→6.7), but have distinct receptors. While prostate cancer cells lose their responsiveness to TGFβ, they retain responses to BMPs. So, prostate carcinoma cells in the bone resemble osteoblasts by responding to BMPs and ET-1, thriving in the same microenvironment.

Secretion of proteases by carcinoma cells remains important in bone metastases. Specifically, uPA and kallikrein proteases liberate latent growth factors and allow expansion of the tumor mass. In addition, the protease cleave and inactivate the parathyroid-hormone related peptide (PTHrP). As this is an important growth factor for osteoclasts, its inactivation may be the decisive step that tilts the balance between bone resorption and bone synthesis towards an osteoblastic phenotype. Breast cancer also often metastasizes to bone, but the metastases are more often osteoclastic than osteoblastic. This may be due to secretion of PTHrP by these cancer cells.

In prostate cancer metastases, both osteoclasts and osteoblasts remain active, since activated osteoblasts secrete factors that stimulate the proliferation and maturation of osteoclast precursors, especially the RANK-L protein. The overall effect therefore is an increased turnover, perhaps with osteoclasts allowing the spread of the cancer and liberating factors such as IGF1 which support the cancer cells. These, then, follow, stimulating osteoblasts to produce new bone mineral with a net increase in disorganized bone mass. Systemically, hypocalcemia may develop. Locally, pain ensues.

All these are major problems in patients with advanced prostate cancer, exacerbated by the effects of locally recurrent cancers and metastases in other organs, e.g. the lung. In urine and serum, biochemical markers of increased bone turnover can be detected. These were used to diagnose prostate before the PSA era in patients that accordingly had invariably cancers at an advanced stage and may now be useful to detect metastases. The progression of bone metastases can be slowed and many symptoms can be alleviated by 'bisphosphonates', drugs interfering with bone resorption, or by local irradiation. However, the overall process is too robust to be cured by present therapies.

Prostate carcinoma metastases in bone exemplify how successful establishment of metastases depends on adequate interactions at multiple levels between cancer cells and the tissue at the metastatic site. Breast cancers exhibit a similar preference for bone metastases, but establish a different assortment of interactions. Of course, still other interactions are required for metastasis to tissues such as the liver, lung, and brain which are preferred localizations of metastases in other cancers. Which interactions precisely are involved, is to date poorly understood (cf. 9.6).

## *Further reading*

Chung LWK, Isaacs WB, Simons JW (eds) Prostate cancer: biology, genetics and the new therapeutics. Humana Press, 2002

Ebele JN et al (eds.) Pathology and Genetics of Tumours of the Urinary System and Male Genital Organs. IARC Press, 2004

Koeneman KS, Yeung F, Chung LWK (1999) Osteomimetic properties of prostate cancer cells: a hypothesis supporting the predilection of prostate cancer metastasis and growth in the bone environment. Prostate 39, 246-261

Djakiew D (2000) Dysregulated expression of growth factors and their receptors in the development of prostate cancer. Prostate 42, 150-160

Dong JT (2001) Chromosomal deletions and tumor suppressor genes in prostate cancer. Cancer Metast. Rev. 20, 173-193

Feldman BJ, Feldman D (2001) The development of androgen-independent prostate cancer. Nat. Rev. Cancer 1, 34-45

Isaacs W, de Marzo A, Nelson WG (2002) Focus on prostate cancer. Cancer Cell 2, 113-116

Mundy GR (2002) Metastasis to bone: causes, consequences and therapeutic opportunities. Nat. Rev. Cancer 2, 584-593

Simard J, Dumont M, Soucy P, Labrie F (2002) Prostate cancer susceptibility genes. Endocrinology 143, 2029-2040

Balk SP, Ko YJ, Bubley GJ (2003) Biology of prostate-specific antigen. J. Clin. Oncol. 21, 383-391

Gonzalgo ML, Isaacs WB (2003) Molecular pathways to prostate cancer. J. Urol. 170, 2444-2452

Schulz WA, Burchardt M, Cronauer MV (2003) Molecular biology of prostate cancer. Mol. Hum. Reprod. 9, 437-448

Visakorpi T (2003) The molecular genetics of prostate cancer. Urology 62, Suppl. 1, 3-10

Nakayama M et al (2004) GSTP1 CpG island hypermethylation as a molecular biomarker for prostate cancer. J. Cell. Biochem. 91, 540-552

# PART III

## PREVENTION, DIAGNOSIS, AND THERAPY

# CHAPTER 20

# CANCER PREVENTION

## 20.1 THE IMPORTANCE OF CANCER PREVENTION

Less than 60% of all cancers can be cured by current therapies, even though these are often burdensome with serious side effects. Therefore, the best strategy would seem to prevent cancers in the first place. Moreover, strategies focusing on cancer prevention could circumvent the increasing economic problems in the health systems of Western industrialized countries, where treatment costs are felt to have become exuberant and ressources are strained. In some developing countries, with health care budgets of down to $ 1 per person and year, cancer preventive strategies are plainly the only realistic option.

While these arguments are in principle conceded by everybody, in practice cancer prevention measures are slowly implemented. The scientific literature, too, rarely yields the impression that cancer prevention is high on the list of research priorities. The reasons for this discrepancy are manifold. Political and socioeconomic factors may be dominant. More pertinent to the theme of this book, the science behind prevention is also anything but trivial. Cancer prevention requires a multidisciplinary approach with contributions from many fields, from molecular biology, infectiology, and pharmacology through psychology to health system economy. The scientific basis for successful prevention is not convincing for many cancers. Nevertheless, many programs have already been successfully implemented and further developments are underway. The prime task of molecular biology research in this context is to identify the causes of cancers as precisely as possible and to define the most appropriate stage and optimal means for intervention. Both are different for different cancers. Moreover, in many cases the implementation of cancer prevention hinges on progress in diagnostics and therapy.

Several different types of cancer prevention are distinguished, mostly according to the stage of cancer development at which they are applied (Table 20.1).

**Table 20.1.** *Types of cancer prevention*

| Type of cancer prevention |
|---|
| Primary prevention (avoidance of exposure to carcinogens) |
| Chemoprevention |
| Dietary changes |
| Detection of preneoplastic changes and early cancer stages |
| Prevention of cancer after preneoplastic changes |
| Prevention of recurrences and second cancers |
| Prevention in individuals with inherited high-risk predisposition to cancer |

## 20.2 PRIMARY PREVENTION

Ideally, a cancer can be prevented by eliminating or avoiding the responsible carcinogens (→1.2). These can be chemical compounds such as aromatic amines in bladder cancer (→14.1), physical agents such as UV radiation in skin cancer (→12.1), or biological agents such as the virus HBV in liver cancer (→16.3) and the bacterium *H. pylori* in stomach cancer (→17.3). Obviously, cancers arising overwhelmingly from endogenous processes, which may include many cases of colorectal cancer (→13.5) and prostate carcinoma (→19.1), will require a different approach for prevention (cf. 20.3).

Even in those cancers, in which a responsible carcinogen has unequivocally been identified, the implementation of prevention is not generally straightforward. Arguably, the most tragic case is that of tobacco smoking and lung cancer. A causal relationship has been established at every conceivable level from epidemiological data down to the molecular detail of detecting major carcinogens from cigarette smoke covalently bound to precisely those bases in the *TP53* gene (→5.3) of bronchial epithelial cells, at which mutations are most often observed in lung cancers. Moreover, while addiction to tobacco smoking is difficult to heal in many afflicted persons, tobacco use is not necessary for human life and could principally be avoided. The inability to prevent a large fraction of lung cancers (and others caused by tobacco smoke carcinogens) is hardly due to a lack of scientific insight.

Not all these arguments apply to UV radiation, which is the established major cause of different types of skin cancer (→12.1). The relationship between the carcinogen and the disease can be considered proven. An important piece of molecular evidence is that mutations in *TP53* and other tumor suppressor genes like *PTCH1* in skin cancers bear the signature of induction by UV radiation (→12.1). However, while unreasonable exposure to UV radiation during leisure activities might be avoided, complete avoidance of the carcinogen is not realistic for people with outdoor occupations. Even more importantly, up to a certain dose sunlight with its UV component is beneficial and even necessary for human health.

The critical level, at which danger surpasses benefit, differs between individuals and populations. It depends strongly on genetic polymorphisms that determine the intensity of skin pigmentation (→12.2). In addition, a smaller number of individuals are oversensitive to UV radiation or sunlight, e.g. as a consequence of defects in DNA repair in xeroderma pigmentosum patients (→3.2). Thus, skin cancer prevention requires an individualized approach. It is pursued by a combination of campaigns aiming at the general public and individual counseling by general practitioners and dermatologists.

A different dilemma is posed by chemical carcinogens in the workplace, illustrated by the case of aromatic amines causing bladder cancer in humans (→14.1). Like in the above cases, the relationship is well established and consequences have been drawn. Today, a situation like that encountered by Ludwig Rehn when he went to investigate why so many of his patients developed bladder cancers (→14.1), is - hopefully – not found anywhere anymore. Yet, it was not a century ago, when one third of the employees in another German chemical plant manufacturing benzidine developed bladder cancers. However, after a retreat battle vividly described by Robert Proctor in "Cancer Wars", these experiences have led to the establishment of strict regulations in the production and use of chemicals that are established or likely carcinogens in man.

Unfortunately, many carcinogenic chemicals are necessary and cannot reasonably be eliminated completely. Therefore, it is in the end a political decision which amounts should be produced and which levels should be accepted. Chemical plants and laboratories can be fitted to minimize the risk for those working there, whereas the environment and the population at large cannot. Therefore, the acceptable levels of chemical carcinogens are not only determined from the angle of occupational risk, but also by their presumed impact on the environment and the general population.

In many cases, however, the responsible carcinogens are not known sufficiently precisely for intervention. For instance, epidemiological studies consistently demonstrate an increased risk of bladder cancer in the plastics and rubber industry, but it is not entirely clear which compounds are responsible and which measures could be taken to avoid exposure to them.

Another complication stems from the fact that most cancers have different and potentially interacting causes. Consider the (realistic) case of a bladder cancer patient, who has been employed in a tire factory, has been a long-time smoker, is an amateur painter, and regularly takes (too much) phenacetine against his recurrent headaches. This person has likely been exposed to four different sources of chemical carcinogens that each could cause bladder cancer (→14.1), and it is not unlikely that they synergize. With the exception of smoking, none of these exposures is completely preventable in a real world.

As in the case of UV exposure, there are substantial differences in the way individuals react to chemical carcinogens. In the case of exposure to aromatic amines, several polymorphisms in enzymes responsible for their activation, metabolism, and excretion have been found to modulate a person's risk of cancer and other adverse reactions (→14.1). Their combined effects are not small. Thus, among the employees that developed bladder cancer after occupational exposure to benzidine-related compounds, the vast majority where 'slow acetylators', while few 'rapid acetylators' were affected. These phenotypes are due to polymorphisms in the *NAT2* gene (→14.1).

In the general population, all different combinations of such polymorphisms are represented. It is tempting to select more resistant persons from this heterogeneous group for the inevitable production of arylamines needed for dyes and drugs by testing for their *NAT2* and further genotypes. There are, of course, serious ethical problems associated with this approach, since it might lead to a sort of 'genetic selection'. Proponents point out that this approach is already used in protecting more sensitive persons, e.g. pregnant women are prohibited from working with radioactive materials. Certainly, one would not like to expose an over-sensitive person to a dangerous chemical compound either? On the other hand, a person's genotype can be considered a part of his or her privacy. If it is revealed to employers, who else could be barred from that information? In addition, choosing a vocation is certainly an important constitutent of individual freedom. A general concern is that genetic testing for susceptibility to carcinogens in the workplace might lead to a lowering of the established standards for prevention of occupational carcinogenesis and might eventually endanger the population. Clearly, this issue requires a public consensus as a basis for political decisions. It also requires solid data and excellent counseling for informed choices to be made.

The problem of balancing economic necessities with disease prevention reemerges in the wider context of exposure of the entire population towards chemical carcinogens. Unlike in the workplace, special protective measures are not feasible here. Moreover, the population is even more heterogeneous than the employees in a chemical plant, encompassing not only healthy adults, but also children or people with disabilities or largely increased susceptibilities. Moreover, while the release of synthetic chemical compounds into the environment can often be controlled, many natural sources of carcinogens cannot. It is hardly surprising, therefore, that issues such as the allowable levels of carcinogenic benzene in fuel or of carcinogenic arsenic in drinking water have not only stirred up scientific but also political debates.

These two cases illustrate different kinds of problems that complicate the scientific evaluation of such issues. Benzene causes leukemia after being oxidized to

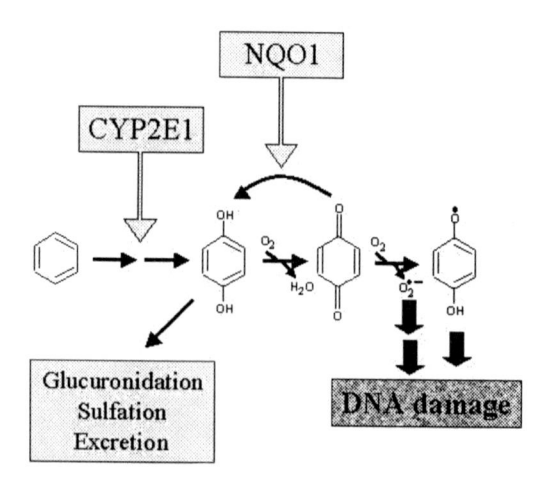

**Figure 20.1** *Metabolism of benzene*
In the bone marrow (and other organs), benzene is oxidized to hydroquinone, predominantly by the cytochrome P450 monooxygenase CYP2E1. This can be excreted following glucuronidation or sulfation. Spontaneous oxidation yields benzoquinone, which upon further reaction with oxygen generates superoxide and semiquinone, both of which are mutagenic. This process can take place in a cyclic manner. NQO1 acts protectively by reducing the quinone to the hydrochinone in a reaction by which two electrons are transferred from NADPH or NADH, with no reactive (semiquinone) intermediate.

phenolic and quinoid metabolites in the bone marrow (Figure 20.1). Its carcinogenicity appears to be limited by the activity of a specific quinone reductase, NAD(P)H:quinone oxidoreductase 1 (NQO1). The *NQO1* gene is polymorphic in man, and <2% of Central Europeans, but >20% in some Asian populations are homozygous for an allele encoding an unstable and essentially inactive enzyme. These homozygotes are likely more susceptible to the carcinogenic effect of benzene. So, which level of sensitivity should be chosen?

Arsenic, too, is established as a human carcinogen by epidemiological data and case studies. It causes primarily cancers of the skin and the bladder. Its mode of action, however, is not at all clear, so all estimates of its impact have a large error margin. It does not induce point mutations, but appears to cause epigenetic changes (→8.3) and perhaps acts by indirect mechanisms as a clastogen, e.g. it favors chromosomal instability (→2.1). Moreover, the uncertainty is exacerbated by evidence pointing to substantial differences in arsenic metabolism between individuals, whose molecular basis is neither elucidated.

While many controversies surround the issue of prevention of chemical carcinogenesis, prevention of carcinogenesis by infectious agents is almost unanimously accepted. Theoretically, elimination of the infectious agents by chemotherapy or vaccination could have major effects. Its potential is illustrated by the geographical differences in the incidence of liver cancer (more precisely: hepatocellular carcinoma) caused by HBV or HCV (→16.1) and the decline of stomach cancer in industrialized countries, which is ascribed largely to the decreasing prevalence of *H. pylori* (→17.3).

Indeed, vaccination programs against HBV appear to be highly effective. For instance, in countries where the virus is endemic, such as Taiwan, even children develop hepatocellular carcinoma. Since the introduction of vaccination for young children, the incidence of such cases has plummeted. Hopefully, this trend will continue in the future. Since HBV appears to act synergistically with aflatoxins from contaminated food (→16.1) in countries with a humid climate, protection against viral infection might be complemented by programs improving the quality of food storage. Like the actual implementation of vaccination programs, however, this may be less of a scientific than of an economic problem.

The RNA virus HCV is likewise implicated as a cause of liver cancer (→16.1) and could be the next target for a vaccination campaign. In Western countries, where this virus is responsible for a slow, but persistent increase in the incidence of hepatocellular carcinoma, efforts are made to advertise a recently developed vaccine against HCV in risk groups.

Several other viruses, mostly DNA viruses, are implicated in human cancer (→Box 5.1, →10.3, →16.3). The relevance of the papovaviruses SV40, BK, and JC (→5.3) for human cancer is controversial, although (or perhaps, because) chronic infection with BKV, in particular, is highly prevalent. The herpes virus EBV is very likely involved in some human cancers (→10.3), but its mode of action is unclear and the majority of the population lives without adverse effects. Apparently, its co-carcinogenic effect is mostly exerted in immuno-comprised persons. This is almost certainly so for another member of the herpes virus family, HHV8/HKSV, which is the causative agent in Kaposi sarcoma (Box 8.1). For both viruses, preventing or improving the immunodeficiency seems the most practical strategy.

Therefore, the current prime candidates for cancer prevention by antiviral vaccination are the oncogenic strains of human papilloma virus HPV (→Box 5.1). Vaccination against HPV is expected to lead to a huge decrease in cervical cancer incidence, but it might also diminish the incidence of other cancers, notably of squamous cell carcinomas of the head and neck.

The most clear-cut case of a bacterium causing cancer in humans is that of *H. pylori* in stomach cancer (→17.3). Quite certainly, the continuous decrease in stomach cancer incidence in most Western industrialized countries is in part due to eradication of the bacterium by antibiotic treatment and increased hygiene. Other factors have very probably contributed. They include lower exposure to nitrosamines from smoked and pickled foods and the increased availability and consumption of protective fruit and vegetables. Since it is considered difficult to eradicate *H. pylori* completely, progress in prevention of this cancer, which worldwide remains one of the most lethal diseases, may also come from improvements in general hygiene and nutrition. There are some concerns that in the absence of accompanying measures, the eradication of *H. pylori* by antibiotic treatment or vaccination might result to some extent in a shift of cancers from the lower parts of the stomach to the cardia and the esophagus (→Box 17.1).

How strongly pathogenic *H. pylori* acts in an individual person depends on genetic variation among the bacterial strains and on genetic polymorphisms in the host (→17.3). Together, these determine whether the bacterium behaves as a symbiote protecting against reflux disease (and perhaps other infections and obesity) or as a carcinogenic agent. In an ideal world with unlimited ressources, one would like to understand these interactions as fully as possible, determine the genotypes of bacterium and host alike and then choose how to proceed for each individual. This is, however, not even practical in a rich industrialized country, least in a $1 per year and person health system. A more realistic strategy is targeting of high-risk strains of *H. pylori* in populations known to carry a high risk overall. Another approach relies on the assumption that carcinogenesis in the stomach can be reversed even at the stage of gastritis and intestinal metaplasia by antibiotic treatment against *H. pylori*. This is in fact the approach that is now pursued, more or less deliberately, in countries with a well-functioning health system. Strictly argued, this type of strategy does not fall into the category of 'primary prevention'. Instead it is a case of 'secondary prevention', because the primary carcinogen is eliminated only after carcinogenesis has begun, but at an early stage.

## 20.3 CANCER PREVENTION AND DIET

In most human cancers, carcinogens are not so clearly defined that they can be eliminated or avoided. Even the carcinogens discussed in the previous section are not the only causes of the respective cancers. Thus, more careful exposure to sunlight will prevent many, but not all skin cancers. Vaccination against HBV (and eventually HCV) is expected to prevent liver cancers resulting from chronic viral infection, but not those resulting from alcohol-induced cirrhosis. Only a fraction of bladder cancers is prevented by strict regulations of arylamine use in the workplace. Nevertheless, primary prevention in these cancers is feasible and is successfully pursued.

Among the four major lethal cancers in Western industrialized countries (→1.1), in only one, lung cancer, a large fraction of the cases is associated with exposure to a clearly defined exogenous agent, i.e. tobacco smoke. As mentioned above (→20.1), primary prevention in lung cancer has proven frustratingly difficult in spite of the evident relationship between carcinogen and cancer. In the other three major cancers, colorectal carcinoma, breast cancer and prostate carcinoma, there is certainly no single major exogenous carcinogen. Therefore, early dectection of localized tumors and preneoplastic lesions is the main option for reducing their mortality (→21.3).

However, in epidemiological studies, all three cancers show associations with a more or less clearly defined 'Western life-style'. While several aspects of this life-style have come under scrutiny, the most convincing arguments point to diet as a major factor. This is not to say that the Western diet is thought to contain a high level of contaminating carcinogens. Rather, diet appears to modulate the risk for these three cancers by affecting endogenous carcinogenic processes. The potential for prevention by changes in the diet is illustrated by the order of magnitude difference in the incidence of prostate cancer between East Asia and North-West Europe (→19.1), and even more impressively, by the same difference emerging in successive generations of Asian immigrants into the USA. The incidences of the three cancers in Southern Europe and in the Near East are typically intermediate, so a 'Mediterranean life-style effect' has been postulated. Of note, diet may be only one of several factors responsible for these differences. The difficulty with prevention at the level of life-style is illustrated by the case of tobacco smoking. If prevention of a cancer caused by a clearly defined recreational drug is frustrating, prevention of cancers by insufficiently defined factors in the everyday diet could prove impossible.

This difficulty may be one reason, why much research on prevention of these major cancers has focussed on single components in the diet. Ideally, one might be able to identify one responsible dietary component that could be added as a supplement to staple foods (e.g. folate to flour) or offered as a drug (e.g. soy extracts). This 'micronutrient' component could either be low in the Western diet or be specifically present in East-Asian foods. In fact, candidates for both types of factors have been proposed (Table 20.2, Figure 20.2).

Compared to 'Westerners', East Asians eat on average fewer calories. This difference in absolute calories is relatively small, whereas the difference in calories provided by animal fat is huge (cf. Figure 20.4). Typical 'Western' diets are reach in red meat, which has a high fat content, and in dairy products. In contrast, typical 'Eastern' diets provide a higher fraction of calories from grains and legumes. There are also differences in the amount and type of vegetables and fruit consumed as well as in the proportion of fish in the diet. These differences are, however, almost as large within the 'East' and 'West' each as between these regions, and do probably not account for the difference in cancer incidence in general. There are more factors of this sort. For instance, selenium deficiency predisposes to several cancer types in some regions of China as well as some in Europe. Selenium deficiency compromises protection against reactive oxygen species, since enzymes such as glutathione peroxidase require this trace element for their catalytic activity.

Among the components discussed to be lacking in the typical Western diet are some that are thought to protect against reactive oxygen species (→Box 1.1). Carotenoids and related compounds are contained in many vegetables and fruits, and their levels in blood correlate with the consumption of yellow/red vegetables. According to such measurements carotenoid supply may indeed have been low in more traditional Western diets. In particular, the main carotenoid of tomatoes, lycopene, has been proposed as a major protective factor in the Mediterranean diet. Several large prospective longitudinal cohort studies, which are among the most reliable tools of epidemiology, have indeed confirmed the presumed relationship between consumption of vegetables and risk for the major cancers. Variously, significant decreases in cancer risk were found between consumption of vegetables overall, of yellow/red or green vegetables. In contrast, intervention studies using pure β-carotene were unsuccessful and in some cases even increased the rates of cancer. Taken together, these results could mean that a combination of compounds rather than a single compound in vegetables and fruit protects against cancer.

Several further compounds contained in fruits and vegetables may synergize with carotenoids to protect against cancer (Table 20.2). Vitamin C (ascorbate) and vitamin E (tocopherol) are also antioxidants. Unsaturated fatty acids contained in vegetable oils and seeds are essential for cell membrane function and the synthesis of paracrine factors regulating tissue homeostasis and function as well as immune responses.

Flavonoids and isoflavonoids (Table 20.2, Figure 20.2) are also antioxidants, but may exert additional effects. Some are weak estrogens and may modulate hormone metabolism. This may influence the development of breast cancer (→18.1) and prostate cancer (→19.1), although the evidence is vague. Some of these compounds are kinase inhibitors and, at least at pharmacological doses, are capable of inhibiting

**Table 20.2.** *Dietary components proposed to influence cancer risk*

| Increasing risk | Decreasing risk |
| --- | --- |
| High fat (saturated animal fat) | Carotenoids (e.g., β-carotene, lycopene) |
| High total calory intake | Vitamin C (ascorbic acid) |
| Alcohol consumption | Vitamin E (tocopherol) |
| Pickled and salted foods | Vitamin $B_{12}$ |
| Smoked and burnt meats | Folic acid |
| Phytoestrogens | Unsaturated fatty acids |
| Mold toxins (e.g., aflatoxins) | Phytoestrogens |
|  | Flavonoids, isoflavonoids (e.g. genistein, resveratrol, epigallocatechin-3-gallate) |
|  | Sulphoraphane |

the proliferation of tumor cells or induce apoptosis. Genisteine from soy beans, resveratol from grapes and wine, and epigallocatechin-3-gallate (EGCG) from green tea belong to this group. Flavonoids and isoflavonoids are present at higher levels in East Asian and Mediterranean diets than in typical Western diets. Accordingly, quantitative, albeit not striking differences in their blood levels are observed between different geographic regions.

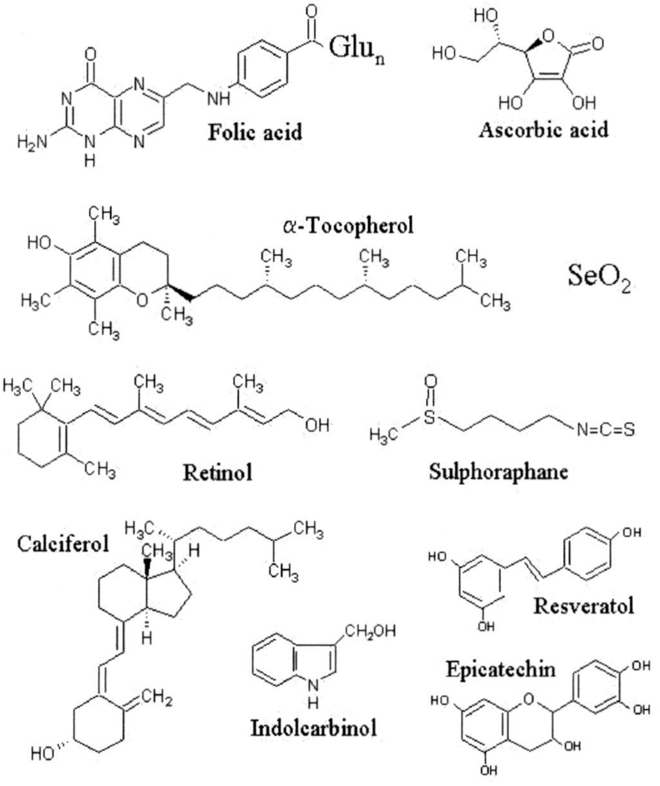

**Figure 20.2** *Chemical structure of some dietary components proposed to prevent cancer*
Folic acid, ascorbic acid, α-tocopherol, retinol, and calciferol are vitamins; selenium is a trace element (mostly present as selenium dioxide or its corresponding acid in the diet). Sulphoraphane, resveratol, indolcarbinol, and epicatechin are each representatives of a huge number of similar plant ingredients (plus meanwhile synthetic analogs).

It is nevertheless unlikely that concentrations of such compounds sufficient to inhibit a receptor tyrosine kinase are reached by consuming typical Eastern or Mediterranean diets. It is also questionable whether consuming these compounds as purified dietary supplements will prevent cancer. More likely, in a well balanced diet they may synergize to impede the early development or progression of cancers.

Notably, these compounds are excellent 'leads' (→22.3) for the development of specific cancer drugs.

Dark green vegetables contain specific compounds, in addition to carotenoids, which could affect cancer development. For instance, sulforaphane contained in several kinds of cabbages, especially in broccoli and brussel sprouts, is a potent inducer of glutathione transferases and other phase II enzymes of xenobiotic metabolism (→3.5). Induction of such enzymes would be thought to accelerate the inactivation of electrophilic carcinogens from exogenous and endogenous sources and prevent mutagenesis.

Dark green vegetables are also an important source of bioavailable folic acid, which is limited in some traditional Western diets and in modern low-quality ('junk') foods. Deficiencies in folic acid are known to synergize with deficiencies in vitamin $B_{12}$, with deficiencies in methyl group donors such as choline and methionine, or with extensive consumption of alcohol to cause cardiovascular disease and developmental defects such as spina bifida. These deficiencies may also advance the development of several human cancers.

Folate deficiency acts by two related pathways (Figure 20.3). Folate is an essential coenzyme for one-carbon metabolism, prominently for the biosynthesis of thymidine and of purine nucleotides. In an initial reaction, the hydroxymethyl-group

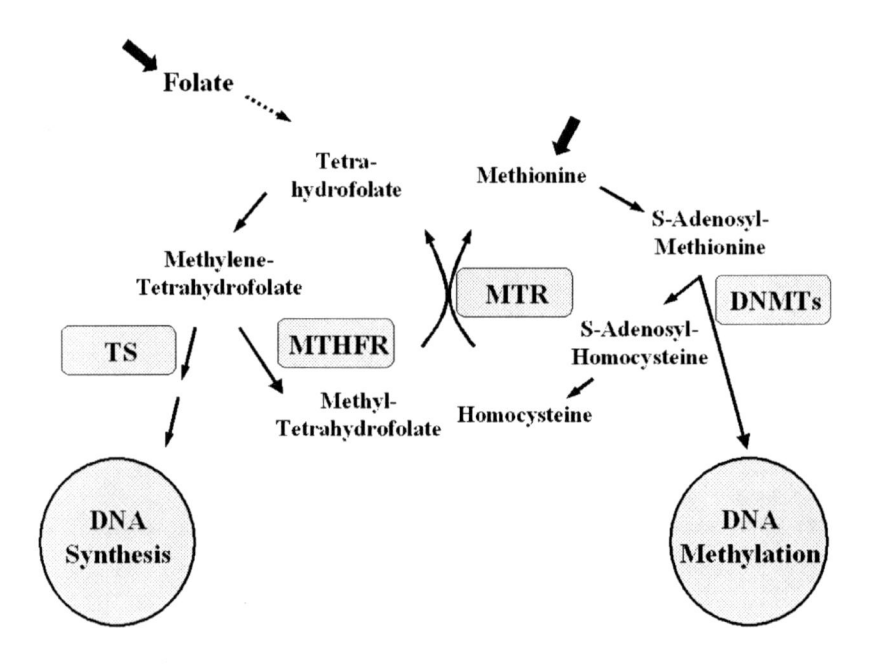

**Figure 20.3** *Functions of folate in DNA synthesis and the 'methyl cycle'*
The enzymes shown are TS: thymidylate synthetase; MTHFR: methylene-tetrahydrofolate reductase; MTR: methionine synthase; DNMTs: DNA methyltransferases

of serine is transferred to tetrahydrofolic acid (THF) yielding $N^5,N^{10}$-methylene-THF. This is used in the thymidylate synthase (TS) reaction to derive dTMP from dUTP. The TS reaction is rate-limiting for dTTP biosynthesis and under some conditions for the synthesis of DNA. This is exploited in chemotherapy by 5-fluorouracil and related compounds (→22.2). In the absence of cytostatic drugs, the thymidylate synthase reaction can be suboptimal, when folate or methyl group donors are insufficiently available. As a result, more dUTP becomes incorporated into newly synthesized DNA. Uracil in DNA is removed by uracil glycohydrolase and replaced by thymine via short-patch repair (→3.1). So, as a consequence of folate deficiency, the number of DNA strand-breaks and the potential of errors during DNA replication increase. An alternative reaction catalyzed by methylene tetrahydrofolate reductase (MTHFR), directs $N^5,N^{10}$-methylene-THF towards the 'methyl cycle'. Its product is $N^5$-methyl-THF. This can be used by methionine synthase to regenerate the otherwise essential amino acid methionine from homocysteine in a reaction that also requires coenzyme $B_{12}$. Following conversion to S-adenosylmethionine (SAM) by SAM synthetase, the methyl group can be used in a variety of methyltransferase reactions, including the methylation of lipids, RNA, proteins and DNA. In each case, S-adenosylhomocysteine (SAH) is formed, which is hydrolyzed to adenosine and homocysteine by SAH hydrolase, thereby completing the 'methyl cycle'. SAH is a product inhibitor of methyltransferase reactions, including the DNA methyltransferases (→8.3). Therefore, the efficiency of DNA methylation depends on the SAM:SAH ratio. Deficiencies in folate, vitamin $B_{12}$, methionine or other methyl group donors can therefore compromise the correct establishment of DNA methylation patterns (→8.3), and perhaps also the methylation of RNA and of proteins.

The extent to which such deficiencies become relevant also depends on genetic factors. Several enzymes involved in one-carbon metabolism and in the 'methyl cycle' are polymorphic in man, most prominently MTHFR. Among several polymorphisms in the *MTHFR* gene, the most prevalent is an exchange of alanine for valine at codon 677. Since the valine variant is less stable, MTHFR activity varies between individuals. Depending on the genotype, limited amounts of methylene tetrahydrofolate are therefore shuttled preferentially towards nucleotide biosynthesis or towards the methyl cycle (Figure 20.3), thereby compromising either DNA synthesis or DNA methylation. Indeed, the *MTHFR* genotype and folate deficiency have been shown to synergize in causing defects in DNA replication and DNA methylation. Moreover, several epidemiological studies have shown cancer risk to increase, e.g. in the colon, most strongly in individuals with specific *MTHFR* genotypes who consume a diet deficient in folate and vitamin $B_{12}$ and consume large amounts of alcohol.

Popular accounts of these studies and others demonstrating similar interactions for cardiovascular disease have labeled this effect 'methyl magic'. There is, in fact, little magic involved. A diet with a good supply of folate and vitamin $B_{12}$ is all that is needed. Current nutritional recommendations such as 'Take 5' (meaning: per day servings of fruit and vegetables) are based on the insights into methyl group metabolism, antioxidant action, and effects of other plant ingredients described above. In some countries, folate is even added to staple foods. Recommendations such as these could lead to a gradual decrease of the incidence of major cancers, even more when they are supported by overall changes in life-style. For instance, immigrants from Southern Europe and the Near East have introduced a greater variety of Mediterrean fruits and vegetable dishes into the Central European diet. Tomatoes, e.g., are now common in the 'Western'diet. Because of the complex interactions involved, the precise causes of changes in cancer incidence that result from such cultural developments will be very difficult to trace.

While individual ingredients of the diet certainly influence cancer risk, and may synergize with each other, the most consistent difference between 'Eastern' and 'Western' diets is the amount of saturated fat consumed, particularly of animal fat. For breast and prostate cancer each, and to a lower degree for colon cancer, the cancer incidence in a country is proportional to the amount of saturated fat consumed (Figure 20.4). The molecular basis of this relationship is not understood. Conspicuously, the risk for these three cancers, among all malignancies, also increases in adipose (obese) persons. Obesity is caused by an interplay of genetic predisposition, cultural factors, overnutrition and lack of exercise that in itself is extremely complex. Even *H. pylori* may feature in this relationship by influencing the levels of hormones secreted by the stomach that regulate satiety. So, the relationship between adipositas and cancer is not expected to be straightforward.

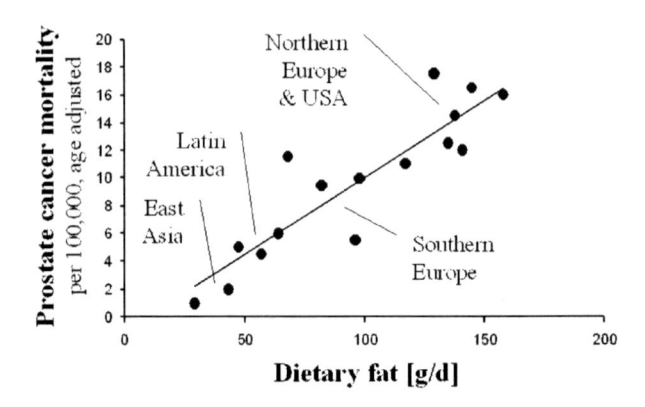

**Figure 20.4** *Relationship between dietary saturated fats and prostate cancer incidence*

An important component in this complex relationship may be insulin-like growth factors. While IGF2 acts mostly locally as a paracrine factor, IGF1 circulates as a hormone in the blood, supporting the growth of several tissues. It also regulates glucose metabolism, but much less so than insulin, which in turn is a weaker growth factor. Increased expression of IGFs is thought to contribute to a wide range of human cancers, including breast and prostate cancers. Normally, IGF1 levels in blood peak during the adolescent growth phase and decline later on. They remain significant, however, and are regulated by nutrition and by exercise, similar to those of insulin. So, in adipose persons, insulin levels are often enhanced, leading to a prediabetic state and often to diabetes, but IGF1 levels too are supraphysiological. Indeed, the level of IGF1 in mid-life has been found to predict the risk of prostate cancer which appears several decades later. The action of IGFs is controlled by binding proteins (IGFBPs). These are likewise subject to regulation by nutrition and by insulin. The levels of IGFBPs in serum also appear to correlate with the risk for breast and prostate cancers. While these relationships are intriguing, they are certainly only part of a complex relationship that needs to be elucidated to define the best approach to addressing obesity as a cause of cancer.

## 20.4 PREVENTION OF CANCERS IN GROUPS AT HIGH RISK

Recommending dietary changes such as 'Take 5!' ($\rightarrow$20.3) is a cancer prevention strategy that addresses the overall population. This strategy is unproblematic. No adverse effects are to be feared and the same changes in diet supposed to decrease the risk of major cancers very likely diminish the risk of cardiovascular disease and diabetes, to a perhaps even greater extent. It is similarly unproblematic to pursue cancer prevention by vaccinating an entire population against HBV or HCV, which likewise has the additional benefit of preventing acute and chronic liver disease. There are risks involved in vaccination, since a few individual show adverse reactions, but they are much lower than the risks associated with actual infections. Moreover, as HBV does not seem to possess an animal reservoir, there is hope that this virus may not only be contained, but eventually be exterminated by vaccination. No foreseeable downside would be associated with its demise.

A bit more problematic is the handling of *H. pylori* ($\rightarrow$17.3), since many strains of the bacterium may actually constitute symbiotes rather than pathogens, even though others present a considerable risk to susceptible humans. Therefore, a vaccination strategy would have to take these considerations into account. As described above ($\rightarrow$20.2), one way out of this dilemma is to restrict anti-bacterial treatment to the susceptible part of the population and/or to those carrying high-risk strains. Ideally, one would identify these persons by molecular diagnosis, but more realistically, patients with gastritis will be identified and treated before ulcers and cancer have a chance to develop.

This is an example of identifying a population at increased risk for a specific cancer. In the case of stomach cancer associated with *H. pylori* infection, there is the added advantage that treatment is available and can usually be admitted without severe side effects and at moderate cost. Moreover, it is not too difficult to identify a precancerous state, viz. gastritis or ulcers, because it is symptomatic. Each of these factors is generally important for the successful prevention of cancers in risk populations, i.e. straightforward delineation of the population at risk, availability of an achievable specific treatment with few or no side effects, and easy identification of preneoplastic or early tumor stages by symptoms or simple and affordable assays (Table 20.3). Unfortunately, instances such as stomach cancer associated with *H. pylori* infection represent the exception rather than the rule.

**Table 20.3.** *Prerequisites for successful cancer prevention in risk groups*

| Criterion |
|---|
| Clear identification of population at risk |
| Sensitive, specific, reliable, and inexpensive assay |
| Detection of significant preneoplastic or early cancer stage |
| Efficient and affordable treatment with acceptable side effects |

Important tasks during cancer prevention in populations at risk are monitoring and – as in the case of stomach cancer – education and information of health professionals and the affected population. This can be very successful, if the population exposed to a specific cancer risk is clearly defined, e.g. in the workplace. In occupations with hazards from radiation or chemicals, regulations aiming at minimizing exposure are put in place and controlled, regular instruction meetings are instituted, and regular examinations are performed to detect early signs of cancers such as aberrant cells in the blood or urine. Monitoring in such instances can be improved by molecular-based techniques for the early detection of cancers or preferably of preneoplastic changes (→21).

Like exposure to defined carcinogens, inheritance of mutations in high-risk genes predisposing to cancer is a risk factor in a specific part of the population (→2.3). Insights from molecular genetics now allow a better definition of this population by identification of those individuals within affected families who carry high-risk mutations. Since monitoring is likewise warranted in these persons, techniques for early detection of cancers are being developed.

One might feel intuitively that preventing inherited cancer today should be a straightforward task. It is unfortunately not, for a variety of reasons.

One complication is encountered in cancer syndromes inherited in a recessive mode (→Table 2.3). These are typically rare diseases caused by mutations in genes involved in DNA repair and become manifest during childhood or during adolescence (→Table 3.2). This means that familial clustering occurs only to a limited extent. In most cases, only siblings of the first child affected (the 'index patient') are also at risk, unless the family belongs to a religious or ethnic group with frequent intermarriage. These diseases often comprise a wider range of symptoms, some of which may precede the manifestation of cancers[15]. This can provide an advantage in so far as monitoring for cancers can be begun after the disease has been identified by other symptoms and has been confirmed by molecular genetic diagnostics.

The major problem then is how to avoid and treat the cancers that develop. In spite of being aware of the risk, and even of understanding the involved mechanisms as precisely as in the hypersensitivity of xeroderma pigmentosum patients towards UV (→3.2), this can be exceedingly difficult. Another problem may arise during therapy, because patients with these afflictions may be hypersensitive to commonly used treatments (→3.3) such as radiation therapy (in ataxia teleangiectasia) or to

---

[15] Still, cancers in these patients may develop much earlier than in the general population (cf. 3.4)

crosslinker cytostatic drugs such as cis-platinum (in Fanconi anemia). So, each case requires a careful, individualized approach. As a practical consequence patients with this sort of diseases are often referred to specialized centers. There, experience can be gathered and therapy can be improved with each case. Some diseases in this group are candidates for gene or stem cell therapy, e.g. Fanconi anemia, where predominantly the hematopoetic system is at risk for cancer (→3.3).

The second, larger group of inherited cancers comprises autosomal-dominantly inherited diseases with an increased incidence of benign and malignant tumors, typically in young adults or in middle age (→Table 2.2). Cancer predisposition in these cases is not necessarily associated with conspicuous symptoms characteristic of a syndrome. Instead, the diseases often run within families, affecting not only siblings of an 'index' patient, but also second and third degree relatives. Almost all are caused by an inherited mutation in a tumor suppressor gene (→Table 5.1). Familial adenomatous polyposis coli (FAP) predisposing to multiple colorectal adenomas and eventually carcinoma (→13.2) and hereditary breast cancer (→18.3) are prominent examples (Table 20.4).

The first hurdle encountered when trying to prevent cancer in affected persons can be awareness. FAP is very distinctive and the emergence of multiple polypous adenomas typically precedes the development of a cancer (→13.2). Therefore, patients are usually identified, even if the familial background is unknown. A familial background in a disease like FAP may seem hard to overlook, but as families become smaller and mobility increases, family history is not always trivial to ascertain.

**Table 20.4.** *A comparison of cancer prevention in adenomatous polyposis coli and hereditary breast cancer*

| Factor | Hereditary colorectal cancer (FAP) | Hereditary breast cancer (BRCA) |
|---|---|---|
| Identification of affected individuals | straightforward | difficult |
| Distinct preneoplastic lesion | multiple polyps | none |
| Penetrance | usually very high | variable |
| Genes involved | one: *APC* | several: *BRCA1*, *BRCA2*, others |
| Distribution of mutations within gene(s) | hotspot regions | hotspots in specific populations |
| Genotype/phenotype correlation | reasonably good | low |
| Prevention by surgery | moderately efficacious | largely efficacious |
| Chemoprevention | possible | likely possible |

This problem is more severe in patients with hereditary breast cancer, since malignancies are generally the first manifestation of the disease, except in patients with more generalized cancer syndromes, e.g. Cowden disease (→18.3). Awareness of hereditary breast cancer may have increased since the discovery of the *BRCA* genes was widely reported in medical journals and in the lay press. Perhaps, the following controversies about how to apply the new knowledge may have increased awareness further. These controversies relate to the most crucial problems in cancer prevention within selected groups mentioned above (Table 20.3), viz. delineation of the population at risk, availability of achievable specific treatments with few or no side effects, and the identification of preneoplastic or early tumor stages.

FAP is not only phenotypically distinctive, but is also essentially a homogeneous disease at the genetic level. Almost all cases are caused by mutations in the *APC* gene (→13.2) and many of these occur in hot spots within the gene. So, molecular genetic confirmation of the disease is feasible at moderate cost. Within a family, the mutation identified in one person can be specifically sought in relatives, or closely linked genetic markers, e.g. microsatellites, can be used. Moreover, a reasonably good correlation exists between the location of the mutation and the severity of the disease (→13.2). So, molecular diagnostics can be helpful to choose an appropriate preventive approach.

Nevertheless, problems remain. The first one is ethical. Diagnosing the disease in one person reveals an information on all relatives, particularly if diagnostics is performed by linkage analysis. In the case of FAP with its very high penetrance, this is mostly a problem concerning children, since asymptomatic adults are rare. Nevertheless, people have a right to to know or not to know. So, adequate counseling is absolutely necessary. Because of the nature of the disease in FAP[16], this issue is not as problematic as in some other diseases, including breast cancer (see below).

In everyday practice, the most serious problem is how to proceed after an *APC* mutation has been verified. One option is preventive surgery by partial colectomy, i.e. removal of the segment of the gut most susceptible to adenomas and carcinoma, usually in young adults before multiple polyps have developed. This diminishes the cancer risk, but does not exclude cancers in other parts of the colon and rectum. Thus, monitoring for carcinomas is regularly performed. Dietary recommendations are, of course, also given (→20.3).

More recently, chemoprevention has been introduced. Non-steroidal antiinflammatory drugs, i.e., inhibitors of COX2 such as celecoxib (→13.6), diminish colon cancer risk not only in patients with inflammatory bowel diseases, but also appear to be efficacious in FAP patients.

Thus, overall early diagnosis of FAP allows to diminish the risk of lethal cancers and extends the life of affected individuals. This is achieved with considerable effort and costs, in life quality of the patient and in economic terms.

Similar, but also additional issues are to be considered in the prevention of hereditary breast cancer. One additional problem is the identification of the

---

[16] i.e. very high penetrance, a symptomatic precursor stage and very high lethality, if treated late

population at risk. Unlike FAP, hereditary breast cancer has no distinctive morphological features and worse, no distinctive precursor stage preceding actual cancer. So, before the advent of molecular techniques, it was not possible to discern whether a familial clustering of breast cancers was accidential or caused by a predisposition inherited in the family.

This distinction can now often be made, but the disease is genetically heterogeneous (Table 18.2). While the majority of hereditary cancers appears to be caused by inherited mutations in *BRCA1* or *BRCA2*, some cases are due to mutations in other known genes. However, a sizeable fraction of familial cases cannot be ascribed to a specific gene. Most likely, no single '*BRCA3*', but several genes are involved (→18.3).

So, verifying a hereditary predisposition toward breast cancer in a family with several 'index' cases can still constitute a considerable problem. Moreover, the *BRCA* genes are pretty large (→18.3) and germ-line mutations are spread throughout the genes. The task of identifying mutations is in some cases facilitated by the occurrence of other cancers in a family. For instance, frequent ovarian carcinomas suggest *BRCA2* as the prime candidate. Also, in some populations founder effects have caused a predominance of specific mutations whose presence or absence can be tested before a larger screening procedure is undertaken. In clinical routine, if a mutation is found in a *BRCA* gene in a member of a family, a hereditary predisposition can be assumed, if none can be found, it cannot be excluded. This is rather unsatisfactory for everybody involved.

If a *BRCA* mutation is detected, the next problem arises from another difference towards FAP. While the penetrance of FAP is generally very high and the exceptional cases with attenuated disease can be defined with some certainty (i.e. by mutations at the 5'-end of the gene, →13.2), considerable differences have been observed in the penetrance of hereditary breast cancer caused by various mutations in BRCA genes. Unfortunately, very few of these differences are consistent enought to allow a prediction of the risk for an individual patient. This means that – by and large – patients with a life-time risk of 20% for development of breast or ovarian cancer have to be treated like those with a life-time risk of 80%. Again, this is not satisfactory for anybody involved, the more so, since the options for cancer prevention are also rather limited.

As in FAP, surgery is the primary option and has been shown to diminish the risk substantially. Oophorectomy not only strongly diminishes the risk of ovarian cancer, but also that of breast cancer, likely because estrogen levels are decreased (→18.1). Mastectomy in various forms can be performed. An important issue is when to perform these operations. Obviously, one would like to delay them as long as possible without incurring the risk of a cancer having metastasized. Likewise, the extent of mastectomy must be balanced between risk of cancer and risks of the procedure, which are physical as well as psychological. Standard recommendations are available, but the actual decision should be made by the patient after counseling.

Monitoring can detect some breast cancers at comparatively early stages, but is not straightforward, especially in younger women. For instance, mammography has been critized as being unreliable in women below the age of 50. There is no good molecular marker for breast cancer yet, which could be used with serum or with nipple aspirates. However, research on this issue is very active.

Brighter prospects emerge from newer developments. Chemoprevention appears to be efficacious in breast cancer as well and specifically in hereditary breast cancer caused by *BRCA* mutations. NSAIDs are also tried for the prevention of breast cancer. In addition, although BRCA-associated cancers normally do not respond well to anti-estrogenic therapies (→18.4), SERMs such as tamoxifen or the newer raloxifene (→18.4) may also be active in their prevention. However, they may delay the manifestation of the disease rather than prevent it. These drugs can have side effects, because they interfere with the normal functions of estrogens in protection of the cardiovascular system and regulation of the balance of resorption and osteogenesis in the bone. Moreover, older SERMs increase the risk of endometrial cancer.

In summary, therefore, prevention of cancers caused by inherited mutations is rarely simple and typically only to some extent efficacious, in spite of considerable efforts and costs. Not least, the quality of life is often compromised. Clearly, the task for molecular research remains to better understand the underlying pathophysiological mechanisms in order to identify more specific targets for prevention, and to improve diagnostic techniques, in particular for the early detection of cancers arising in persons at risk.

## 20.5 PREVENTION OF PROSTATE CANCER BY SCREENING THE AGING MALE POPULATION

There can be little doubt that the considerable efforts to prevent or at least delay cancers is appropriate in small populations at high risk, such as families with inherited cancer syndromes. Likewise, drastic measures such as surgical removal of organs or long-term treatment with anti-hormonal drugs are accepted in persons known to run a high risk of developing a lethal cancer. Such measures would certainly not be acceptable for the whole population, even though one third of the population in a typical Western industrialized country will develop some kind of cancer in their life-time and one fifth will die of it. However, many different cancers in different organs contribute to this morbidity and mortality and no universal prevention scheme is at hand.

Nevertheless, for some cancers, screening of the entire population or a susceptible subpopulation could make sense, if certain requirements are met. They are similar to those defined in high-risk populations (Table 20.3), viz. the part of the population at risk needs to be clearly defined, an affordable and reliable assay for detection of preneoplastic or early tumor stages must be at hand, and an efficacious treatment with acceptable side-effects and – in a large population – affordable costs must be available. There is a continuing debate for which cancers these requirements might be met. Obviously, the four major cancers, lung cancer, colon cancer, breast cancer, and prostate cancer are prime candidates for population screening. While according developments are underway for each of them, they are most advanced in the case of prostate cancer.

Screening for prostate carcinoma (→19) is basically a realistic option. (1) It is very prevalent with a life-time risk of up to 10% in the male population and a considerable mortality (≈3% of the male population). (2) Prostate cancer is extremely rare before the age of 50, increasing exponentially thereafter. In males over 75, the disease often takes a slower course and/or surgical treatment is not advisable. This restricts the population to be tested to males aged between 50 and ≈75. (3) A routine biochemical assay for prostate specific antigen (PSA) in serum (→19.1) can detect most tumors while they are still restricted to the organ. This assay has acceptable sensitivity, although its specificity is only moderate. In most cases, a definite diagnosis of prostate cancer can be achieved by further molecular assays, by imaging and by histological investigation of biopsies. (4) Organ-confined prostate cancer can be successfully treated by surgery or radiotherapy. In advanced cases, progression can be delayed by anti-hormonal treatment.

For these reasons, PSA screening of the male population starting from the age of 50 has been introduced in several countries. In smaller regions within Austria or Canada systematic attempts have been made to screen the entire male population and treat all those with detectable cancers. Indeed, the mortality rates of prostate cancer and meanwhile even its incidence have declined. Not all Western countries have followed suit, however, again for several reasons.

(1) Prostate cancer is biologically heterogeneous (→19.1). While a sizeable fraction of prostate cancers take an aggressive course, many remain relatively indolent, and will not cause clinical symptoms, least death in the age group at risk (Figure 20.5). In contrast, treatment by surgery or radiation can cause considerable morbidity such as incontinence and impotence and carries even a (low) risk of mortality. Moreover, while the introduction of PSA assays has dramatically reduced the fraction of cases that have metastasized at the time of detection, some still are and cannot be cured.

At present, while a combination of biochemical and morphological parameters can predict the prognosis of a specific patient quite well, the distinction between indolent, aggressive, and already metastatic cases can rarely be made with certainty. This means that on one hand a fraction of patients are unnecessarily treated, while on the other hand in some cases treatment is useless (→Figure 19.4). As one might expect, the debate on the relative sizes of the three fractions rages. A study in Scandinavia, where 'watchful waiting' rather than therapeutic intervention is the favored strategy, has suggested that radical prostatectomy prolongs life in only one out of seventeen patients treated. This is likely an underestimate. Nevertheless, it raises doubts about the usefulness of a population-wide screening and intervention approach. Moreover, opponents of screening point out that the incidence and mortality of prostate cancer are also declining in countries where no screening/intervention has been introduced. One way out of the dilemma would be the development of molecular markers that allow to stratify prostate cancers into groups with very high, moderate and low risk (→19.4).

422                                CHAPTER 20

(2) PSA screening is more expensive than it looks at first sight. The PSA assay itself is relatively inexpensive, but is not highly specific. While it yields relatively few false negatives, i.e. misses few cases of prostate cancer being present, it has a relatively high rate of false positives. Specifically, men with benign prostatic diseases such as prostatitis or extensive benign hyperplasia often test positive. In these cases, further diagnostics has to be performed and - although some of these men may indeed have to undergo treatment for their benign diseases - there are considerable costs associated with the diagnostic procedure. Because taking of biopsies is an invasive procedure, a slight risk of serious infections is incurred. All this would not matter much in a small population at risk, but if an entire population of older males is screened, costs are magnified and very slight risks become significant. Consider an assay with 95% specificity (which is an excellent value) used in a population of one million males over 50. It means 50,000 men having to undergo the following more extensive diagnostical procedures unnecessarily. Combined with the uncertainty on the clinical significance of a cancer detected in an individual (argument 1), this leaves room for doubt.

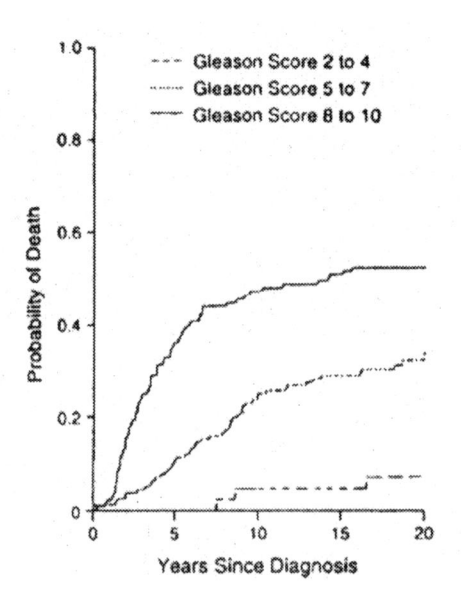

**Figure 20.5** *Natural course of prostate cancer*
Death from untreated prostate cancer depending on Gleason score, a measure of tissue disorganization (increasing with loss of differentiation) used in histopathology. The figure is taken from Schröder FH, Eur. J. Cancer 37, Suppl. 7, p.S129, 2001.

(3) Moreover, if 10% of all males eventually develop clinically significant prostate cancer and are all detected at an early stage, can they all be treated by current methods? In the fictious example population, 100,000 men would be diagnosed with prostate cancer. Currently, many cases are only detected, after the cancer has progressed. Before the advent of PSA assays, most cancers were high-stage, often metastasized and incurable. With optional PSA assays, cases are more often diagnosed at an organ-confined stage at which they can be cured. Not so rarely, they are diagnosed in men older than 75 years or suffering from other - typically cardiovascular - diseases prohibiting surgery. In many of these men, rather slow-growing prostate carcinomas are not limiting life expectation, although the patients may eventually require palliative treatments. If screening was efficiently performed in younger men, e.g. at the age of 60, one could not predict by current methods whether many cancers detected would cause severe clinical symptoms years later or limit life expectancy. Since few males at 60 are nowadays not fit enough for surgery or radiotherapy, most of them would therefore need to be treated right away. This may be feasible in selected regions like Tyrol or Quebec, but in larger populations the ressources would simply not be available. Again, a better definition of the population at risk, more precisely, a classification of the cancers detected by PSA screening and subsequent diagnostic procedures, is required to solve this dilemma.

Evidently, an alternative approach would be chemoprevention rather than definitive therapy in men with suspicious serum levels of PSA. Anti-hormonal treatment is an option, but is problematic because of its side effects. Moreover, there are concerns that applying anti-androgenic treatment early on might spur the early development of androgen-resistant cancer cells ($\rightarrow$19.2), leaving no options if the disease recurs after surgery. Moreover, androgen independence in prostate cancer is typically associated with a more invasive and metastatic phenotype and therefore anti-androgen chemoprevention might prevent the development of less malignant cancers, while promoting the more aggressive ones. A compromise in this situation is being tried by using inhibitors of $5\alpha$-reductase ($\rightarrow$Figure 19.7). This enzyme catalyzes the formation of the more active dihydro-testosterone from testosterone in the prostate. Therefore, its inhibition has fewer side effects than anti-androgenic treatments by androgen receptor antagonists or LHRH agonists ($\rightarrow$19.2). Moreover, the less active androgen testosterone remains present to exert beneficial effects in other tissues. As often in prostate cancer research, acquiring definitive data on the value of this approach will take many years. Intriguingly, preliminary results suggest indeed a decrease in prostate cancer incidence at the expense of a shift towards less differentiated cases.

## 20.6 OTHER TYPES OF PREVENTION

Detection of prostate cancer by PSA assays can, of course, not be considered strictly as cancer prevention. PSA levels in serum increase, after the prostate epithelium has become disorganized and epithelial cells release some of their secretory products into the mesenchyme ($\rightarrow$19.1). This presupposes invasion and therefore a carcinoma by definition. Ideally, however, the cancer is detected at an early stage, while it can still be cured and systemic disease is prevented.

This is the aim of many attempts at detecting cancers early on by molecular markers (→21.3). Ideally, one would detect a preneoplastic state allowing prevention of the actual malignancy. If that is not possible, detection is aspired at a stage where a cure remains possible with minimal harm to the patient.

One step further, viz. following successful therapy of a cancer, e.g. removal by surgery, another type of prevention aims at diminishing the risk of further cancers of the same type. This is prevention of secondary cancers. In contrast, adjuvant therapy by drugs or irradiation aims at eliminating residual cells of the primary cancer that might lead to recurrences.

The distinction between prevention of secondary cancers and preventing recurrences of the first cancer is clear in the prostate, which is completely removed during surgery for prostate cancer. So any cancer appearing later must have been present at the time of surgery, having spread beyond the organ.

In organs that cannot be completely removed this distinction is more difficult to make. The issue is further complicated by 'field cancerization' (→14.1) occurring in many tissues. For instance, carcinomas of the oral mucosa tend to recur. Nevertheless, surgery in this sensitive enviroment must be kept as limited as possible. Therefore, in the standard procedure the cancer is removed with a margin of several cm of morphologically normal tissue around it. Some cancers recur in distant places and indeed can be shown to harbor distinct genetic alterations. Other cancers recur close to the surgical margin in tissue that was morphologically normal at the time of initial surgery. As a rule, they contain the same genetic alterations as the initial cancer.

In such cases, prevention of second cancers and adjuvant therapy tend to overlap. Importantly, in each case, patients need to be closely monitored to detect recurrent as well as second independent tumors.

The main difference between adjuvant therapy and prevention of secondary cancers lies in the type of treatment used. Adjuvant therapy aims at killing tumor cells that have not been eliminated by the primary treatment, i.e. surgery or irradiation. Prevention aims at keeping normal or partially transformed cells from becoming malignant. So, it is certainly prevention to convince a bladder cancer patient treated by local surgery to stop smoking, whereas installing the moderately toxic cytostatic drug mitomycin C into the bladder is really adjuvant therapy.

In other cases, this distinction is even more blurred. For instance, anti-estrogenic therapy after partial mastectomy for breast cancer aims at both residual tumor cells and prevention of independent cancers.

Nevertheless, the distinction between prevention and adjuvant therapy is not only theoretically important. If there is a serious concern that tumor cells are still present after an attempt at definitive therapy, the side-effects of cytostatic drugs will have to be accepted. If the aim is prevention of a second cancer, they will not. For instance, if there is a concern that a bladder cancer has not been completely removed or may have metastasized, systemic therapy based on the more toxic cis-platinum will be used rather than local application of mitomycin C. Likewise, if breast cancers have spread beyond a certain stage, e.g. cancer cells are detected in multiple lymph nodes, cytostatic therapy rather than anti-estrogenic treatment will be considered.

There are several further organs, in which independent second cancers occur frequently, e.g. the skin, particularly in patients with increased susceptibilities, the epithelia of the mouth and throat, particularly in smokers, and the colon, most strongly in patients with inherited susceptibilities. Beyond regular monitoring and general preventive recommendations, such as to avoid smoking and eat a healthy diet, chemoprevention is being developed as an option for these patients. Vitamin mixtures or specific antioxidants such as β-carotene have been tried for several different tumors. For prevention of secondary colon cancers, non-steroidal anti-inflammatory drugs are now the treatment of choice (→13.5) and they are also tried for cancers of other organs. Another group of compounds investigated for the purpose of chemoprevention are retinoids and related compounds (→22.2).

## Further reading

Proctor RN (1995) Cancer wars. BasicBooks

Griffiths K, Denis LJ, Turkes A (2002) Oestrogens, phyto-oestrogens and the pathogenesis of prostatic diseases. Martin Dunitz

Eng C, Hampel H, de la Chapelle A (2000) Genetic testing for cancer predisposition. Annu. Rev. Med. 52, 371-400

Giovannucci E (2002) Epidemiologic studies of folate and colorectal neoplasia: a review. J. Nutr. 132, Suppl. 2350S-2355S

Negm RS, Verma M, Srivastava S (2002) The promise of biomarkers in cancer screening and detection. Trends Mol. Med. 8, 288-293

Freemantle SJ, Spinella MJ, Dmitrovsky E (2003) Retinoids in cancer therapy and chemoprevention: promise meets resistance. Oncogene 22, 7305-7315

Kensler TW et al (2003) Translational strategies for cancer prevention in liver. Nat. Rev. Cancer 3, 321-329

Pollak MN, Foulkes WD (2003) Challenges to cancer control by screening. Nat. Rev. Cancer 3, 297-303

Sabichi AL et al (2003) Frontiers in cancer prevention research. Cancer Res. 63, 5649-5655

Lerman C, Shields AE (2004) Genetic testing for cancer susceptibility: the promise and the pitfalls. Nat. Rev. Cancer 4, 235-241

# CHAPTER 21

# CANCER DIAGNOSIS

## 21.1 THE EVOLVING SCOPE OF MOLECULAR DIAGNOSTICS

There is widespread agreement that insights into the molecular biology of human cancers will make their most rapid impact in the area of cancer diagnosis. As described in the previous chapter (→20), implementation of cancer prevention is not only impeded by our limited knowledge of the complex causes of cancer, but also by a host of socioeconomic factors. As described in the following chapter (→22), the development of molecular-based cancer therapies is also hampered by scientific as well as general factors. In contrast, translation of research results into routine diagnosis is underway and is favored by scientific and general factors.

(1) Other than in cancer therapy (→22.7), the techniques used in research laboratories can be applied in diagnostic laboratories with relatively little additional effort. In general, techniques need to be further standardized and additional controls must be introduced. Often, procedures are developed to make them amenable to automatization. (2) The translation of molecular biology results into diagnostic procedures can built on the existing infrastructures provided by pathology and clinical chemistry laboratories in the clinic. These laboratories are already accustomed to performing biochemical and immunochemical assays and are now adding molecular biology techniques to their repertoire. (3) Similarly, the production of standardized and validated reagents for molecular diagnostic techniques has been taken up by companies that have previously marketed diagnostic kits for immunohistochemical or biochemical assays or by new biotech companies. (4) Molecular biology techniques fit well into an ongoing trend towards individualized therapy, as discussed later in this chapter.

Molecular biology techniques can be applied for a wide range purposes in cancer diagnosis (Table 21.1).

Tumor staging, grading, and differential diagnosis: The first applications that come to mind concern differential diagnosis of tumors before therapy. Here, assays for proteins and nucleic acids in body fluids and tissue samples supplement diagnostic procedures relying on imaging techniques and 'classical' histopathology. Routine histopathology increasingly makes use of molecular markers at the protein, RNA and DNA level. After all, it is not a big leap from traditional staining methods for tissue samples to immunohistochemistry to detect specific protein markers, RNA-in-situ-hybridization to detect specific mRNAs, or fluorescence-in-situ-hybridization to detect chromosomal aberrations. Nevertheless, a new specialty is emerging, labeled 'molecular pathology'.

It is important to be aware that, as developments in molecular biology are changing histopathology, developments in physics and information technology are revolutionizing imaging techniques. Modern computing power, e.g., allows the reconstruction of virtual dynamic 3D images from tomography data to detect and precisely localize very small tumors by non-invasive techniques. The combination of modern imaging techniques with molecular markers, such as labeled antibodies, is a rapidly developing field of applied medical science. This combination may eventually be used to localize even micrometastases consisting of a few tumor cells, to monitor the distribution of therapeutic molecules, and to improve the targeting of radiotherapy.

While modern imaging techniques can excellently circumscribe the extent of a tumor and thereby help to determine its stage (→1.4), they yield little information on its histology and its biological properties, which determine its further clinical behavior. This information is also required for the choice of therapy (→1.5). Traditionally, the tumor subtype was determined by its morphology, usually determined by staining of tissue sections. More recently, immunohistochemistry, e.g. for cell type-specific cytokeratins in carcinomas or CD surface proteins in leukemias, has entered routine practice and improved differential diagnosis. In general, the precise identification of the tumor type already provides a great deal of information on the likely future course of the disease (i.e. the prognosis of the patient) and a basis for the choice of the most appropriate treatment. In routine histopathology, additional information on the aggressiveness of a tumor is obtained by grading, which relies on subjective estimates of the degree of tissue disorganization as well as cellular and nuclear atypia (→1.4).

Like the determination of the tumor type, this estimation of its 'character' has in many cases been improved by the use of antibodies. These can help, e.g., to measure the proliferative fraction of the tumor or to ascertain the intactness of the basement membrane separating a questionable carcinoma in situ from the underlying connective tissue[17]. In spite of these improvements, classification, staging and grading of many tumor types by current methods are far from perfect in predicting

**Table 21.1.** *Applications of molecular diagnostic techniques for cancer detection and classification*

| *Purpose* |
|---|
| Detection of cancer predisposition |
| Detection of preneoplastic changes |
| Cancer detection |
| Tumor staging |
| Tumor grading |
| Differential diagnosis |
| Subclassification |
| Prognosis of spontaneous clinical course |
| Prognosis of response to therapy |

---

[17] disruption of the basement membrane would indicate that the carcinoma is no longer 'in situ', but invasive

prognosis and the response to specific therapies. Molecular diagnostic techniques will thus be helpful for the determination of tumor stage, e.g. by allowing detection of tumor cells in the blood or in the bone marrow.

Cancer subclassification: A major impact of molecular diagnostic techniques is expected in the subclassification of tumors with respect to prognosis and response to therapy. In several cases, particularly in hematological cancers (→21.2), molecular analyses have revealed that a disease which appeared uniform by morphological criteria can be further differentiated. In other instances, e.g. renal cancers (→15.2) or breast cancer (→18.5), molecular classification has corroborated previous suspicions that different subclasses of the disease exist and may need different treatments. Yet another situation is found in some cancers that are morphologically and molecularly similar, but in which progression depends on specific molecular alterations. For instance, mutation of *TP53*, as detected by accumulation of the mutant protein in the tumor cell (→5.3), may predict a worse prognosis in several cancers, including Wilms tumors (→11.4) and bladder cancer (→14.4).

Prediction of response to specific therapies: The advent of cancer therapies directed against specific molecular targets has advanced these developments one step further. Previously, cancer therapy was directed against one type of cancer in general, e.g. against breast cancer, exploiting what was seen as general properties of cancer cells, such as increased DNA synthesis or increased sensitivity to ionizing radiation (→22.2). As therapy increasingly targets specific molecular changes which are present in a subclass of cancers of one histological type (→22.3), diagnosis has to follow suit. So, increasingly, molecular diagnostics is required to determine which cancers express the targets for specific molecular therapies. In this way, the introduction of molecular diagnostics into the clinic strengthens the already established trend towards individualized cancer therapy (→22.7).

Early detection: Although imaging and histopathology have become more sophisticated, they are limited by the size of a cancer. Molecular diagnostics holds the promise of detecting cancers at even earlier stages. This can be applied in several circumstances, particularly in cancer prevention and screening (→20). Another application is in the monitoring of therapy efficacy and of recurrences.

Detection of predispositions: Specifically, molecular diagnostics can be used to identify populations at risk for the development of a cancer (→20.4). This is often not possible by traditional methods. Importantly, the definition of risk conferred by inherited mutations is unique to molecular diagnostics.

## 21.2 MOLECULAR DIAGNOSIS OF HEMATOLOGICAL CANCERS

Arguably, molecular diagnostics is presently best established in hematology, of all subspecialties within oncology. Molecular techniques are used to ascertain the initial diagnosis and to determine the exact subtype among morphologically similar leukemias and lymphomas as a basis for prognosis and therapy selection. They are applied to assay autologous bone marrow and stem cell transplants for residual tumor cells and to match allogenic donor transplants to the recipient patient. During

430                             CHAPTER 21

and after therapy, they are employed to monitor its success and to detect residual tumor cells and recurrences at an early stage.

Since many hematological cancers are characterized by specific cytogenetic aberrations, typically translocations (→10.2), cytogenetic techniques are well suited for differential diagnosis. In addition, detection of surface markers by antibody staining is used, since cells at various stages and sublineages of the hematopoetic lineage express specific cell membrane proteins.

Burkitt lymphoma (BL) invariably carries one of three translocations which bring the *MYC* gene under the control of immunoglobulin gene enhancers (→10.3). These translocations can be detected by karyotyping of tumor cell metaphases, which is best performed on cultured tumor cells. For the direct investigation of tissue sections, interphase two-colour FISH is more convenient (Figure 21.1). Tissue sections are hybridized with two DNA probes, one each from the *MYC* locus at 8q and an immunoglobulin locus (initially for *IGH* at 14q, which is most often involved), labeled by different colours. In normal cells, the loci are on distinct chromosomes and located at some distance from each other in an interphase nucleus.

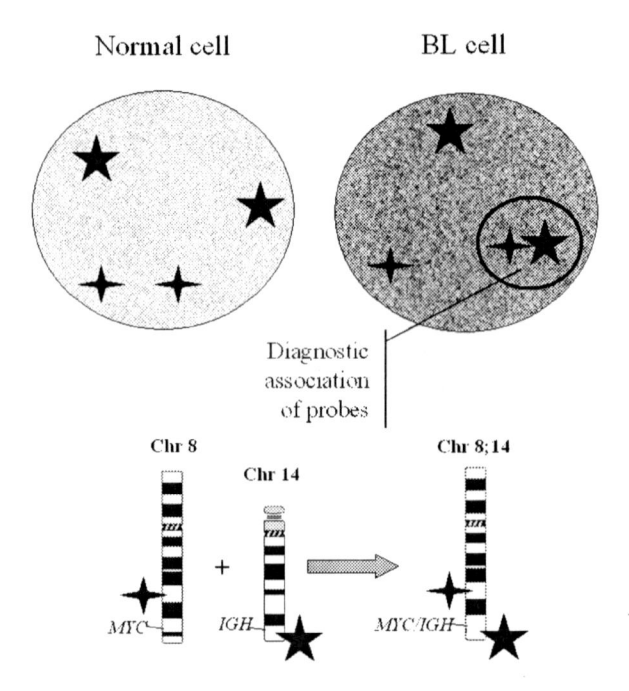

**Figure 21.1** *Detection of specific translocations in Burkitt lymphoma by FISH*
Fluorescence-in-situ-hybridization (FISH) probes for chromosomes 8 (centromeric to the MYC locus) and chromosome 14 (centromeric to the *IGH* locus) are indicated by different type stars. They locate to different regions in most normal interphase nuclei, but consistently close to each other in the nuclei of BL cells.

Even in BL cells, one pair of signals corresponding to the normal chromosomes 8 and 14 remains apart and serves as an internal control. In contrast, the signals from the translocation chromosome appear close to each other. While this apposition may occur by chance in an occasional cell, its regular appearance in a lymphoma tissue proves the presence of this particular translocation chromosome and confirms the diagnosis of Burkitt lymphoma.

This kind of technique can be used in general to detect translocations, including those in chronic myeloid leukemia (CML). However, while the translocations in BL cause the overexpression of MYC mRNA and protein, whose structures are not necessarily altered (→10.3), the characteristic translocation in CML generates a novel fusion gene, *BCR-ABL*, with a transcript and a protein that are unique to tumor cells (→10.4). This provides further opportunities for detection. The rearrangement creating the *BCR-ABL* gene can be detected by Southern blot or PCR analysis. In principle, translocations in BL could also be detected by these methods, but the detection is more straightforward and reliable in CML, since the breakpoints occur within restricted regions in both genes. They therefore create a limited number of different mRNA forms (→Figure 10.8), none of which is present in a cell without a translocation. All can be detected by RT-PCR using a few pairs of PCR primers, one primer each from the *BCR* and *ABL* gene. This detection can be performed qualitatively to verify the diagnosis or quantitatively to estimate the number of tumor cells. Moreover, RT-PCR can be made extremely sensitive by nesting and real-time techniques allowing the detection of one tumor cell in a billion. Therefore, this method is much more sensitive than cytogenetic or morphological detection of tumor cells.

Thus, owing to the development in molecular diagnostics, CML therapy can be monitored at several levels (Figure 21.2) by a set of techniques with different sensitivities. 'Clinical remission' remains an important criterion. The improvement of symptoms in a treated patient and the disappearance of morphologically detectable tumor cells ('hematological remission'), however, are not sufficient to predict whether a recurrence will occur. Indeed, in some cases with clinical improvements and lack of obvious tumor cells, cytogenetic techniques can still detect translocation chromosomes indicative of 'minimal residual disease' (cf. 10.4). Without additional therapy, such patients as a rule relapse. The prognosis is much better, if 'cytogenetic remission' is achieved and no cells with translocation chromosomes can be detected. However, some of these patients still experience recurrences. Indeed, RT-PCR analyses can detect residual tumor cells in the bone marrow and even in blood. In fact, these methods can be made so sensitive as to detect tumor cells in every treated patient and even in some healthy individuals. Therefore, quantitative methods are employed and cut-off values are defined to determine which patients are very unlikely to experience recurrences.

Molecular monitoring in CML is not only important to determine the efficacy of treatment. It is also helpful in the choice of therapy, particularly to decide which patients need stem cell transplantation, and when (→10.4). This treatment can have serious side-effects, such as graft-versus-host disease (→10.4). Therefore, the definitive exclusion of residual disease by molecular diagnostic techniques can

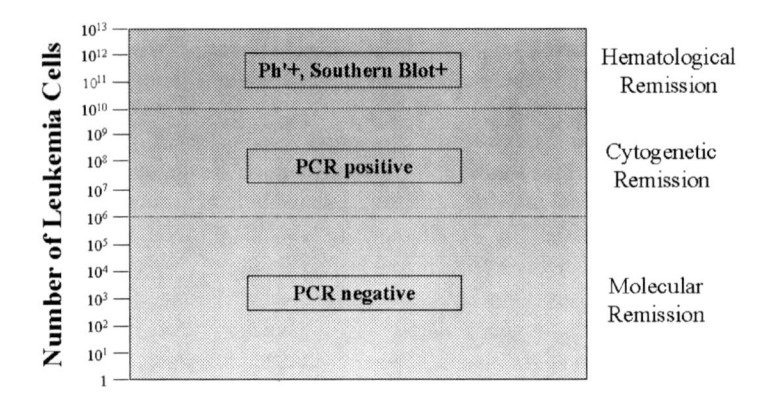

**Figure 21.2** *Molecular detection of CML cells*
See text for details.

identify those patients for whom transplantation can be deferred. Of course, the selection of donors and cells for transplantation is also aided by molecular techniques.

In a similar fashion, molecular diagnostics can help in the choice of acute promyelocytic leukemia (APL) therapy. Cytogenetic or molecular diagnostics can identify patients that carry the t(15;17) (q22;q21) translocation generating the *PML-RARA* fusion gene (→10.5). These can then be treated with all-trans retinoic acid, which has little effect in other acute myeloid leukemias and even in some cases of APL that are not caused by this particular translocation.

Unfortunately, the predominance of *MYC* translocations in BL, of *BCR-ABL* fusion genes in CML, and of *PML-RARA* fusions in APL are the exception rather than the rule in hematological cancers. More often, acute leukemias and lymphomas with similar phenotypes can be caused by several different genetic alterations, i.e. different translocations or other characteristic chromosomal alterations (cf. Figure 10.2). Moreover, not all leukemias and lymphomas are characterized by specific single chromosomal changes or gene mutations. In carcinomas, this is the rule. Phenotypically similar diseases caused by different translocations can exhibit large differences in their spontaneous clinical course and in their response to specific treatments. So, determining the correlation between specific translocations and clinical behavior is a continuing effort in hematological oncology that leads to a steady progress in adapting the treatment to the individual patient's cancer.

The problem is of course exacerbated in those hematological cancers that do not show specific chromosomal aberrations. Such cases represent a significant fraction of acute lymphocytic leukemia (Figure 10.2). Here, advances are expected from 'gene expression profiling' (→18.5). Initial studies using these techniques showed that is feasible to distinguish ALL from AML (acute myeolid leukemias) by analysis of mRNA expression patterns. This is rarely a problem in the clinic, because protein markers for the myeloid and lymphatic lineage (Figure 10.2) can be used, if the morphology of the tumor cells is not already distinctive. Indeed, the mRNAs that showed the most pronounced differences in the expression profiles were those encoding such protein markers. More importantly, the technique can distinguish different subtypes among AMLs lacking conspicuous chromosomal changes.

Another pressing clinical problem is the distinction between different subtypes of follicular lymphomas and large cell lymphoma. They exhibit similar morphological and biochemical markers, but some subtypes are rather indolent, while specific ones are aggressive, and therefore require a different therapy. This distinction can also be made by expression profiling. In other cases, expression profiling has suggested new protein markers for detection of minimal residual disease (→10.4) of specific subtypes.

## 21.3 MOLECULAR DETECTION OF CARCINOMAS

Molecular diagnostics of carcinomas is less straightforward than that of leukemias and lymphomas. The first additional complication lies in obtaining material for analysis. For leukemia diagnostics, blood samples are in many cases sufficient to detect the presence of a leukemia and even establish an initial differential diagnosis. In some cases or later in the diagnostic procedure, bone marrow aspirates or other biopsies are additionally required. Even solid types of lymphomas often present as relatively accessible swellings.

In contrast, samples from carcinoma can often be obtained only through more invasive procedures, which may carry a risk of inadvertently spreading tumor cells, e.g. in the peritoneum. So, often, only limited amounts of useful material from a carcinoma are available until surgery has been performed. This may be one reason why molecular biology of human carcinomas is not as advanced as that of hematological cancers. Another reason is that molecular alterations in carcinomas are often more complex and more heterogeneous than in hematological cancers.

A diagnostic assay applicable in a clinical setting has to fulfil several requirements (Table 21.2). (1) It should require as little material as possible, e.g. from biopsies. This is a strong point of many molecular assays, in particular of PCR techniques. (2) Ideally, an assay should be able to use samples obtained by non-invasive methods. Saliva, sputum, urine, and stool samples can be obtained by completely non-invasive routes. Blood samples also present few problems. In some organs, e.g. the urinary bladder, washes can be performed which yield 'lavage' fluids containing cells and molecules from tumors. (3) PCR techniques have also, in general, made molecular assays more rapid, which is a further requirement in many clinical situations. (4) A more problematic requirement is the stability of the molecule to be assayed. Some proteins and RNA in general are not very stable in tissue samples, unless these are quickly frozen or fixed. Therefore, assays based on DNA or on more stable proteins are generally preferable. RNA instability is even more of a problem, if quantitation is required. This is one factor withstanding the use of expression profiling techniques (→18.5) in clinical routine. (6) Another factor in

**Table 21.2.** *Important criteria in the use of molecular diagnostic assays in the clinic*

| Criterion |
| --- |
| Minimal sample required |
| Applicable with samples obtained by non-invasive or minimally invasive methods |
| Speed of result |
| Stability of molecule assayed |
| Cost effectiveness |
| Reliability |
| Specificity |
| Sensitivity |
| Reproducibility |
| Compatibility with existing expertise, procedures, and equipment |

this case, as in general, is cost-effectiveness. Cost, however, is relative. Using an expensive assay to determine the correct therapy in a small number of patients is a different issue from using an assay for screening a large population. (7) Assay sensitivity and specificity are of course crucial, although again, the required levels depend on the intended use. (8) Last not least, an assay has to be reliable, which means that it measures the parameter it is supposed to measure, not only in a laboratory situation, but also with clinical samples.

Specificity and cost are the major hurdles when a molecular assay is designated for screening larger groups or populations. Such assays also have to be essentially non-invasive, i.e. no more invasive than taking a small blood sample. In the application of molecular diagnostic techniques for carcinomas in the clinic, sensitivity and cost effectiveness most often represent the crucial issues. It is at these points, where the complexity and heterogeneity of molecular alterations in carcinomas become relevant. Nevertheless, a variety of assays have been developed (Table 21.3).

The problem of sensitivity arises mainly, because there are only few instances, in which a single molecular alteration is found in every cancer of one type. This is particularly true for carcinomas. For instance, while almost every colon carcinoma shows constitutive activation of the WNT pathway (→13.2), these are brought about by different genetic changes (Figure 21.3). In most cases, both copies of the *APC* gene are inactivated, while in a smaller fraction activating mutations in *CTNNB1* (β-Catenin) are responsible. In still another, albeit small fraction of colon carcinomas, neither of these genes is affected. This would probably not compromise sensitivity overly. More problematic is that the alterations in *CTNNB1* and *APC* are not homogeneous, either. Mutations that activate β-Catenin are restricted to a small region of the protein (→13.2), but are not uniform. *APC* inactivation is worse, since it occurs by different mechanisms which include small and large deletions as well as point mutations at many different sites in the gene. Since most lead to a truncated protein, alternatively a protein assay could be envisioned that employs an antibody against the carboxy-terminus of the APC protein. This assay could be applied to colon cells obtained by biopsies or from faeces. This method would miss the (rare) cases with missense mutations, in addition to all those not caused by *APC* mutations. More problematic is that this is a 'negative marker' assay. Assays that detect the disappearance of a marker tend to yield lower sensitivities and specificities. Hopes are placed in a microarray technique that detects all possible mutations in *APC* and *CTNNB1* (and perhaps *AXIN1*).

In contrast to loss of a protein, point mutations in DNA represent 'positive' markers that can be detected in spite of a background of DNA from non-tumor cells in samples obtained by non-invasive methods (e.g. stool samples). Therefore, some approaches for the molecular detection of colon cancer have pursued detection of *KRAS* mutations. While these occur only in 40-60% of all colon cancers (→13.3), they are strictly limited to the three codons 12, 13, and 61. These mutations can be highly sensitively detected by appropriate PCR techniques. Of course, as *KRAS* mutations mostly develop during the progression of colon cancers (→13.3), this assay would detect carcinomas rather than adenomas. In theory, then, a sensitivity of 40-60% could be achieved, with a specificity approaching 100%. In practice, neither figure has been reached. The reasons for this disappointing outcome are under investigation. One might speculate that sensitivity is limited by the amount of tumor DNA entering the gut lumen and surviving the passage. There are furthermore indications that specificity is lowered by an unexpectedly high fraction of false positives, i.e. mutations detected in persons without carcinomas. Perhaps, mutations in *KRAS* sometimes occur in normal cells or arise in DNA as it travels through the gut.

A straightforward place to look for molecules originating from a cancer is blood. A number of serum protein markers are employed for the detection and monitoring of specific cancers.

**Table 21.3.** *Some molecular assays applied or considered for detection of carcinomas*

| Assay | Carcinoma | Stage of development |
|---|---|---|
| Immunodetection of PSA in serum | prostate | routine |
| Immunodetection of CEA in serum | colon and others | routine |
| Detection of circulating tumor cells by tissue-specific mRNA | many | trials, some long-term |
| Detection of RAS mutations in stool and serum | colorectal, pancreas | trials |
| Detection of TP53 mutations in body fluids and excretion | many | trials |
| Detection of LOH in serum and urine | many | trials |
| Detection of promoter hypermethylation in serum, urine, lavage fluids, or biopsies | many | trials, some advanced |

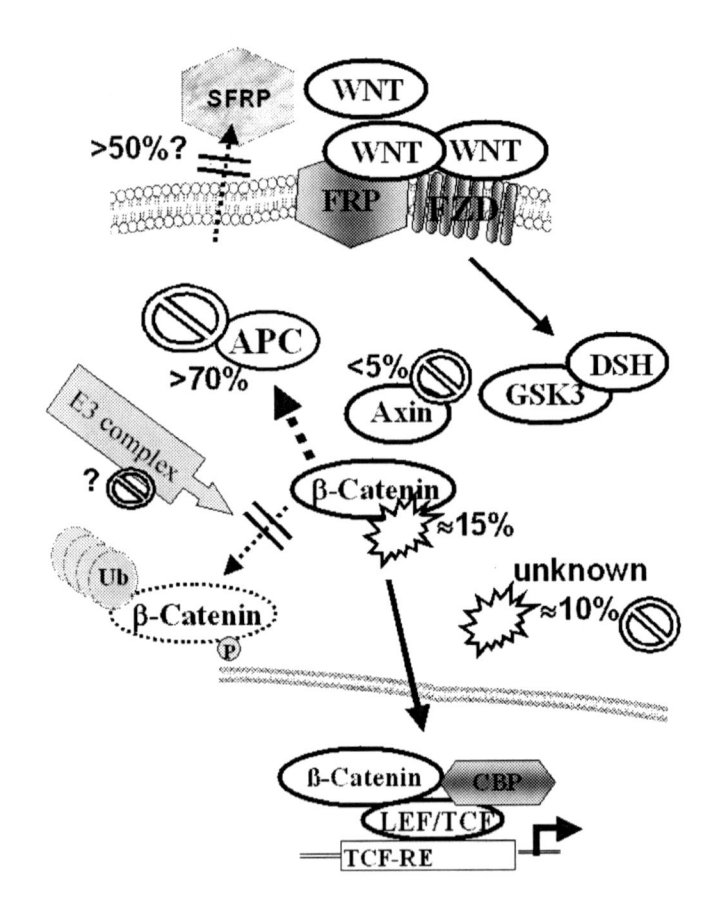

**Figure 21.3** *Variability of genetic changes responsible for activation of the WNT pathway in colon cancer*
cf. Figures 6.12 and 13.3

Some carcinomas, including those of the colon, can be detected by increased levels of carcino-embryonic antigen (CEA) in the blood. This protein marker can be determined very reliably and at moderate cost by a routine immunoassay. Unfortunately, the assay is not very sensitive, since only a fraction of cancers express this oncofetal antigen (→12.5), and not very specific, a.o., because several cancer types express CEA. However, once a cancer has been found to express CEA, the marker can be used to monitor the success of the therapy.

A similar immunoassay for prostate-specific antigen (PSA) is used to monitor prostate cancer therapy (→19.1). Detection of PSA has several advantages that have led to its almost ubiquitous use. (1) Prostate epithelial cells are the only significant source of PSA in males. Therefore, any increase in serum must be due to a process in this tissue. (2) Only few prostate carcinomas cease to synthesize the protein, often only at a very late stage when this is no longer clinically important. (3) The PSA level is roughly related to the tumor volume. Following removal of the prostate harboring the cancer, PSA declines (Figure 21.4). Ideally, the protein should become undetectable (<0.1 ng/ml) and stay so, if the cancer has been cured. A nadir in the ng/ml range typically indicates that the local tumor has not been completely removed. In that case, one would consider adjuvant radiotherapy or anti-androgenic therapy (→19.2). Recurrences caused by micrometastases are usually announced several years in advance by a slow increase of serum PSA from zero levels. In late stage systemic disease, values >1000 ng/ml may be reached.

PSA levels can also be employed in clinical and experimental studies. The presence and size of primary and metastatic tumor masses is an important parameter to determine the efficacy of a new drug or the significance of a molecular marker. However, metastases in prostate cancer can often only be detected with difficulty, least their size be measured. So, in prostate cancer the success of a new therapy is often monitored via the PSA level, which is used as a 'surrogate parameter' of cancer extension. In general, a 50% decrease is taken as evidence of remission. While PSA assays are thus helpful, they do have downsides, on the scientific as well as on the psychological side. On the scientific side, PSA synthesis is induced by androgens. Therefore, treatments that influence androgen signaling may decrease PSA levels much more than they affect tumor growth. On the psychological side, as PSA plays such a dominant role in the treatment of prostate cancer, not only patients, but also doctors and scientists are tempted to ascribe more value to a PSA level than it is worth.

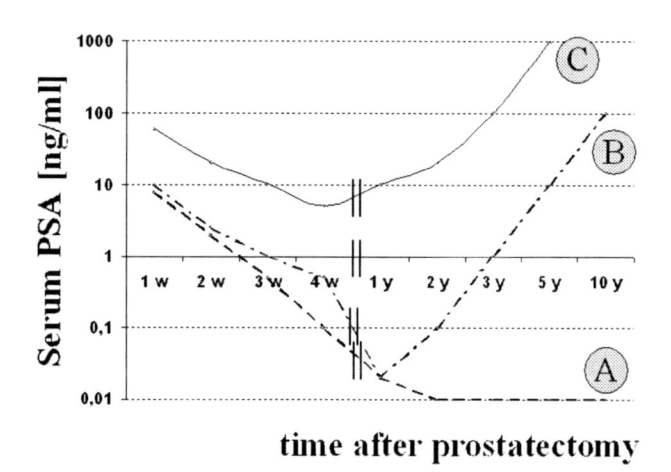

**Figure 21.4** *Use of PSA serum levels for monitoring of prostate carcinoma therapy*
Three idealized time courses. A: Complete resection of the cancer without recurrence; B: Recurrence after several years, presumably due to micrometastases; C: Incomplete removal; in this case, adjuvant radiotherapy would be considered. Note that the time scale changes and that the PSA value axis is logarithmic.

While PSA is an excellent marker for monitoring of prostate cancer, its specificity for the initial detection of the carcinoma is limited (→19.1). Here, additional molecular diagnostic assays might be helpful, which detect cancer cells in prostate biopsies or blood, or cancer-specific markers in blood, urine, or ejaculate.

As in colon cancer, genetic alterations in prostate cancer are heterogeneous, if anything, much more so (→19.3). In some patients, tumor cells can be detected in the blood, e.g. by an RT-PCR assay for prostate-specific mRNAs such as PSA mRNA. Detection of PSA mRNA is a more specific indication of cancer than detection of PSA protein, because the protein is secreted, whereas its mRNA can only be dectected if prostate epithelial cells are present in the circulation. These are in most cases cancer cells. For a while, this method was even thought to be useful for the distinction of metastatic from organ-confined cancers, i.e. for 'molecular staging'. However, this may not prove to be true.

Other assays are based on DNA alterations in tumor cells. In many tumor patients, increased levels of DNA are found in blood plasma, which is cell-free. Its source is not entirely clear, but much of it appears to be derived from tumor cells, perhaps released from cells dying in the circulation (→9.1). Typical mutations present in cancer tissues can also be detected in this circulating DNA, e.g. *KRAS* mutations from a colon or pancreatic carcinoma.

In prostate cancer, specifically, no characteristic mutations occur regularly enough to be exploited for this kind of assay. However, alterations of DNA methylation are highly prevalent, in particular hypermethylation of the *GSTP1* gene, which may occur in >80% of all prostate carcinomas (→19.3). Since DNA hypermethylation affects CpG-islands which are unmethylated in normal tissues (→8.3), its detection is very specific and sensitive because of a negligible background from normal tissues. Moreover, while hypermethylation of some genes is also found in aging tissues or early preneoplastic lesions, the hypermethylation of *GSTP1* appears to be cancer-specific, or at least restricted to late preneoplastic lesions in addition. So, detection of *GSTP1* hypermethylation in blood or prostate fluid may provide a valuable technique to supplement PSA assays.

DNA hypermethylation is also found in many other cancers (→8.3) and similar assays are being developed for their detection. Unlike in the case of prostate cancer and *GSTP1*, in most cancers, hypermethylation assays must be performed for several genes, since each is hypermethylated in only a fraction of cancers (cf. Table 8.2). Moreover, hypermethylation of *GSTP1* is relatively specific to prostate cancer. It is otherwise only found in a smaller fraction of renal carcinomas and hepatomas. These cancers are relatively straightforwardly excluded in prostate cancer patients. In contrast, another gene, *RASSF1A* (→6.2), is hypermethylated in prostate cancer, but also in many others and even some precursors. This property could make *RASSF1A* hypermethylation useful as a general sensitive tumor marker, but limit its specificity.

Tumor DNA in plasma can also be assayed for mutations or for allelic imbalances. As in general, the sensitivity of these assays is limited by the heterogeneity of genetic alterations within the tumor type. Some current research therefore aims at developing assays which detect every mutation in one gene, e.g. in

*TP53*. These assays would, however, still miss alternative pathways of TP53 inactivation, such as MDM2 over-expression (→6.6). The sensitivity of assays for allelic imbalances is in addition limited by the extent to which tumor DNA is diluted in the plasma by DNA originating from normal cells.

A hotly debated question in the use of plasma DNA for tumor detection is how its overall presence and amount relate to tumor stage. A similar question pertains to the issue of tumor cells found to circulate in the blood. If DNA from a tumor is present in blood, it appears to have gained access to the circulation, which is almost certain, if tumor cells are encountered in blood outside the tumor tissue. So, while early stages of tumor development can perhaps not be discovered by assays using blood samples, the detection of tumor-specific DNA alterations or even of tumor cells may yield information on how far the cancer has advanced. This approach may therefore yield a chance to achieve a 'molecular staging'. Alternatively, the type of mutations and allelic imbalances might be used for 'molecular staging', if they can be assigned to specific stages of tumor development.

## 21.4 MOLECULAR CLASSIFICATION OF CARCINOMAS

Detection of a cancer is only the first step towards therapy. Whether and which therapy is administered, depends on a further, more precise classification of the cancer. Traditionally, cancers were assigned to different histological subtypes by their morphology and further classified by stage and grade by imaging techniques and histopathological examination (→1.4). Experience and carefully collected observational data were used to assess the prognosis of the cancer and select the most appropriate therapy.

The criteria underlying tumor staging, in particular, are not arbitrary, but have usually been chosen to correspond to those steps in tumor progression, at which the prognosis and accordingly, the most appropriate treatment changes. Such steps may be evident, such as beginning invasion of outer layers in a tissue or growth beyond an organ, or they may have to be determined by careful follow-up of large numbers of patients. For instance, renal carcinomas rarely metastasize unless they exceed a certain volume, but are as a rule incurable, once metastases have developed (→15.1). Accordingly, the criteria for staging of renal carcinomas have been changed back and forth based on observational data relating the course of disease to the diameter of the tumor. In this case, prognosis additionally depends on the specific histological subtype, with some histological subtypes metastasizing more readily than others (→15.1). Therefore, knowledge on the histological type of the tumor is important for the determination of prognosis. More recently, identification of characteristic chromosomal alterations for each subtype of renal carcinomas (→15.2) has been introduced to aid in the classification of ambiguous cases.

There are many carcinomas, in which the 'classical' trias of histology, staging, and grading does not consistently yield sufficient information for optimal selection of therapy. It is not at all exceptional for carcinomas with identical stage, grade, and histology to take divergent clinical courses. The most severe problem in general is that micrometastases escape detection by current imaging methods. Therefore, staging is not really precise and the extension of the primary tumor as well as its histology and grade only yield an estimate on the presence of micrometastases. These are responsible for recurrences and patient death, even if the primary tumor can be completely removed or destroyed. Thus, the decisions on how to deal with the primary tumor and whether to apply an adjuvant therapy (and which, if there is a choice), are based on probabilities rather than definitive information. A large set of empirical data has been collected for each cancer type which can be used in these decisions. In some cancers, algorithms and nomograms taking all known relevant parameters into account have been introduced as a help for patients and doctors (Figure 21.5). Nevertheless, the overall situation is far from satisfactory.

In a sense, progress in therapy has aggravated this dilemma. Since a larger choice of treatments has become available, criteria are needed to determine which patients will respond to which therapy. In addition, therapies differ with respect to their side-effects and their costs, which can often not be neglected.

For these reasons, improvements in the classification of carcinomas are a major goal of current molecular research. Breast cancer may represent a major carcinoma, where this type of research is advanced and is translated into the clinic at a rapid

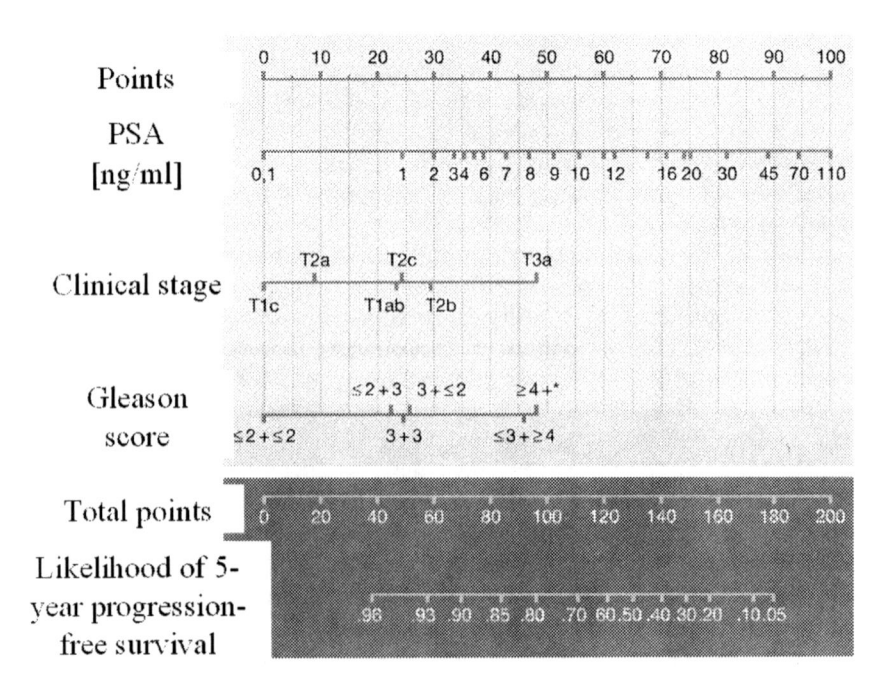

**Figure 21.5** *A nomogram used to determine the prognosis of prostate carcinoma patients*
Points are accorded for PSA serum level, clinical stage (i.e. as apparent prior to surgery), and Gleason score of biopsies (a histopathological measure of tissue disorganization). Total points are related to the likelihood of not experiencing a tumor recurrence after prostatectomy.

pace (→18.5). Molecular assays are increasingly used in this cancer as supplements to staging, grading and histology.

As in other carcinomas, prognosis in breast cancer depends on tumor size, tumor grade, and the histological subtype. Neither unusually, breast cancers in young patients tend to be more aggressive. More specific to breast cancer, prognosis and therapy differ before and after menopause (→18.2). The single most significant prognostic factor in this cancer is the extent of lymph node involvement. Less than 30% of cancers without detectable tumor cells in the lymph nodes recur, whereas >75% of cancers with several positive lymph nodes have progressed to systemic disease and will recur, if only the primary tumor is destroyed. So, almost all patients with lymph node involvement receive adjuvant therapy. Moreover, since recurrent breast cancer is rarely curable, chemotherapy or anti-estrogenic therapy is also administered to most of the patients with no positive lymph nodes as well, unless additional favorable factors are found, such as small tumor size or low grade, or a rarer less aggressive histological subtype is present. This means that overall ≈60% of breast cancer patients unnecessarily receive an unpleasant and toxic therapy, but it is difficult to determine, whether this is so for each individual patient.

Some molecular markers are already in routine use to help with this decision. Low concentrations of the plasminogen activator uPA and its inhibitor PAI-1 (→9.3) indicate a favorable prognosis, at least in cancers that are well or moderately differentiated. Patients with cancers designated ER+/PR+ that express both the estrogen and progesterone receptors (→18.4) also fare better, as a rule. More recently, determination of the ERBB2 status (→18.5) by immunohistochemistry and FISH analysis has been introduced. In general, breast cancers with ERBB2 overexpression caused by gene amplification are more likely to have metastasized. However, determinations of the ER/PR and ERBB2 status are employed primarily for the choice of therapy rather than for the determination of prognosis. Cancers that are ER+/PR+ tend to respond well to anti-estrogenic therapy, whereas cancers that do not express the steroid hormone receptors or that overexpress ERBB2 do not. Instead, some cancers with ERBB2 over-expression can be treated more successfully by a combination of chemotherapy, including anthracyclins, and an antibody directed against ERBB2, trastuzumab (→18.4).

Overall, >100 individual molecular markers for the prognosis of breast cancer have been suggested over the last years. Promising candidates that may make it into clinical routine are the proliferation markers Ki67 and PCNA, which are also useful in several other cancers (→1.4). They can be detected in a semi-quantitative manner by immunohistochemistry. Similarly, the cell cycle regulator Cyclin E is overexpressed in more aggressive breast cancers, and $p27^{KIP1}$ is accordingly down-regulated (→6.4).

It is generally presumed that no single clinical or molecular parameter may be sufficient for an optimal prognosis of breast cancer. Rather, several are already in use and further ones are being added. Considering appropriately all the known factors and their complex relationships becomes increasingly difficult. Therefore, algorithms and nomograms have been introduced that help to take into account the relative importance of each information obtained, including histology, staging, grading, patient age, and various molecular markers, and their relationship to each other. These can help in the choice of therapy, which ultimately rests with the doctor and the patient. As more and more interacting factors become known and can be assayed, computer programs are being developed that calculate the risks associated with each treatment strategy. Learning algorithms and neuronal networks are particularly suited to this task, as they can improve with their 'own' experience and integrate information from new clinical studies.

In a sense, therefore, expression profiling using microarrays (→18.5) is a logical continuation of a development that is already underway in breast cancer diagnosis. Analysis of the expression levels of a large number of genes indeed allows to classify breast cancers into ER+ and ER- types. It distinguishes previously unrecognized luminal cell-like and basal cell-like subtypes, with ERBB2+ cancers representing a distinct subclass within the basal-cell-like subtype. Cancers arising in patients with inherited mutations in the BRCA genes (→18.3) also exhibit characteristic profiles. Most importantly, metastatic cancers appear to show distinct expression patterns from those still growing locally.

The next step therefore will be to verify these profiles in larger groups of patients in prospective studies. A large population study of this kind is underway in the Netherlands. If these studies are successful, the technical and cost problems that currently prohibit routine use of expression profiling in the clinic are likely to not present serious obstacles in the long run. In parallel with these studies, much smaller sets of genes that are decisive for the distinctions provided by microarray data will be assayed for their potential as prognostic markers. For prostate cancer, e.g., a set of just four genes has been proposed for this purpose. Expression of all four can be followed by immunohistochemistry, facilitating the introduction of these markers into routine laboratories.

## 21.5 PROSPECTS OF MOLECULAR DIAGNOSTICS IN THE AGE OF INDIVIDUALIZED THERAPY

The examples of breast cancer and leukemia described in the previous sections illustrate a general development in cancer diagnosis and therapy. Cancer therapy has moved towards individualization. This development began actually quite independently of the availability of adequate molecular markers and any in-depth understanding of the molecular basis of cancer pathophysiology. Already, surgery, chemotherapy and radiotherapy are administered contingent on histopathological parameters and on the patient's general state of health and psychosocial circumstances. A vast amount of empirical data can serve as a basis for the decision in each individual case ('evidence-based medicine'). This individualization helps to achieve optimal therapeutic results, to minimize suffering not only from the cancer, but also from the treatment, and to avoid unnecessary expenses. In this situation, molecular markers come in handy to continue an ongoing development in all subdisciplines of oncology.

However, the potential of molecular markers in the diagnosis of cancer goes beyond providing better distinctions between subclasses of one cancer and serving as prognostic markers. It is captured in the novel notions of 'pharmacogenetics' and 'pharmacogenomics'.

In a sense, pharmacogenetics is also a continuation of an existing trend, since it has been known for quite a while that individual patients can react very different to some drugs (Table 21.4).

*NAT2*: 'Slow' and 'fast' acetylators, e.g. not only metabolize carcinogenic arylamines at a different rate (→14.1), but also a variety of drugs in medical use. Therefore, this phenotype influences susceptibility to cancer as well as the response to therapy. The 'slow' and 'fast' acetylator phenotypes are due to polymorphisms in genes encoding N-acetyltransferases, mostly in *NAT2*.

*UGTA1*: Individuals with Gilbert syndrome[18], like slow acetylators, are otherwise asymptomatic, but excrete certain drugs more slowly than others. These can therefore accumulate to dangerous levels. The molecular basis of Gilbert syndrome is a polymorphism in the promoter of the *UGTA1* gene encoding UDP-glucuronyl-transferase A1. The promoter contains a variable number of TA repeats. The allele *UGT1A1\*1* with six repeats yields maximum expression, whereas a frequent polymorphic allele containing seven repeats, *UGT1A1\*28*, is associated with diminished expression. The UGTA1 enzyme transfers glucuronic acid to hydroxyl groups of endogenous or exogenous compounds, which increased their solubility in aquous solution and facilitates their excretion. In cancer therapy, this polymorphism is most relevant for the metabolism of irinotecan, a topoisomerase inhibitor (→22.2) used a.o. in the treatment of colorectal and lung cancers. Severe toxicity of the compound is observed predominantly in persons homozygous for the *UGT1A1\*28* allele.

*TS*: The promoter of the *TS* gene encoding thymidylate synthetase is likewise polymorphic. It contains either two or three repeats of a 28 bp tandem repeat. The respective alleles are designated *TSER\*2* and *TSER\*3*. The *TSER\*3* alleles lead to increased expression of the enzyme that is crucial for the synthesis of dTTP required for DNA replication (cf. 20.3 and 22.2). Increased expression of TS diminishes the

Table 21.4. *Examples of individual genetic differences in the reaction to anti-cancer drugs*

| Gene | Polymorphism | Effects on |
|---|---|---|
| NAT2 | multiple alleles with missense changes | amines |
| UGTA1 | number of repeats in promoter | irinotecan, others |
| TS | number of repeats in promoter | 5-FU and other nucleoside drugs |
| TPMT | limited number of alleles, mostly Y240C and A154T | thiotepa and other thiopurines |
| ATM | multiple alleles with missense and nonsense changes | radiotherapy, drugs causing DNA strand-breaks |

---

[18] also 'Morbus Meulengracht' in parts of Europe

effect of inhibitors, such as 5-fluoro-uracil (→22.2), which are employed in the treatment of many different cancers.

*TMPT*: Thiopurine methyltransferase (TPMT) metabolizes and detoxifies purines containing a thiol group, especially the drug azathioprine, which is used in the chemotherapy of leukemias. About 1% of all individuals lack enzyme activity due to a genetic polymorphism. In these persons, administration of thiopurine drugs at the standard dose can be lethal. Heterozygotes for the *TPMT* polymorphism tolerate intermediate doses.

*ATM*: Sensitivity to radiotherapy, likewise, is influenced by genetic polymorphisms, as exemplified by individuals heterozygous for mutations in the *ATM* gene. This gene encodes a protein kinase controlling the cellular response to DNA damage by ionizing radiation, but also to other forms of DNA damage (→3.3).

Pharmacogenetics, thus, is the systematic study of all genetic variation that determine the individual response to drugs. Of course, pharmacogenetics is not restricted to drugs used in cancer therapy, and as indicated by the case of *ATM*, genetic polymorphisms are also relevant to therapies other than drugs.

In current clinical practice, pharmacogenetics is rarely used explicitly. Rather, it is implicit in the way drugs are administered. Patients are asked about known hypersensitivities and are observed for adverse reactions known to occur with specific compounds. New drugs are monitored for side effects while being developed and are continued to be monitored for side effects after their introduction to the general market. This latter monitoring is important, since some polymorphisms are present in only a few individuals or are only prevalent in specific subpopulations. Drug trials can never be comprehensive in this respect. While these general procedures are well established, specific genetic analysis for polymorphisms influencing drug sensitivity are currently only used in selected circumstances. For instance, some institutions have an *TPMT* assay set up routinely.

The reasons for this are practicability and cost to a much greater extent than a lack in understanding of individual variabilities in the reaction to drugs. For many drugs in current use, the mechanisms underlying different toxicities are in fact well elucidated. However, the genetic polymorphisms underlying individual variabilities are often complex and assaying them is currently more expensive than relying on careful observation in a trial-and-error fashion.

The 'slow' and 'rapid' acetylator phenotypes (Table 21.4), e.g., are brought about by quantitative interactions between >10 different *NAT2* alleles. To predict the phenotype from the genotype, a range of polymorphisms must be tested (Figure 21.6). So, observing the patients or monitoring the excretion of a test drug dose, is more practical. Similarly, individuals with low UDPGTA1 expression can usually be identified by a slight elevation of serum bilirubin in the absence of other signs of liver disease, as determined by routine clinical chemistry. A molecular genetic assay is preferred for assaying *TPMT*, because the genotype/phenotype relationship is straightforward and the polymorphisms are relevant for a clearly defined class of drugs administered for selected diseases.

This situation may change radically in the future. One factor driving the change is the continuing automatization of techniques for molecular genetic analysis which reduces expenses. Many in the field envision a chip-based assay, which at one stroke detects all polymorphisms relevant for drug metabolism in an individual. This analysis could be performed once and for all for each person and drugs could be prescribed accordingly. However, not everybody is enthusiastic about this scenario, for diverse reasons. One concern is safeguarding of individual genetic data. It is one thing to determine the genetic properties of an individual responsible for the sensitivity to one particular drug administered against a deadly disease, but it is another to record many polymorphisms influencing the response to a variety of medical and other drugs. There are also purely scientific concerns. As the example of the acetylator phenotype shows, the relationship between genotype and phenotype is often not straightforward. The individual reaction to most drugs is determined by several factors, of which only some are determined by the DNA sequence assayed in molecular tests. Even if genetic factors predominate, several genes may be involved and their interactions can be complex.

The issue is different, if not adverse reactions, but positive response to a drug or treatment are at stake. A substantial number of otherwise efficacious drugs have failed in early clinical trials (i.e. in phase I or phase II) or have not even proceeded to being tested in humans (i.e. to phase I) because they display intolerable side-effects. Others seem efficacious in a subset of patients too small to warrant their further development for the general market. If one could predict which patients tolerate or respond, respectively, to such drugs, they could still be used in selected patients. This would extend the range of individualisation of therapy.

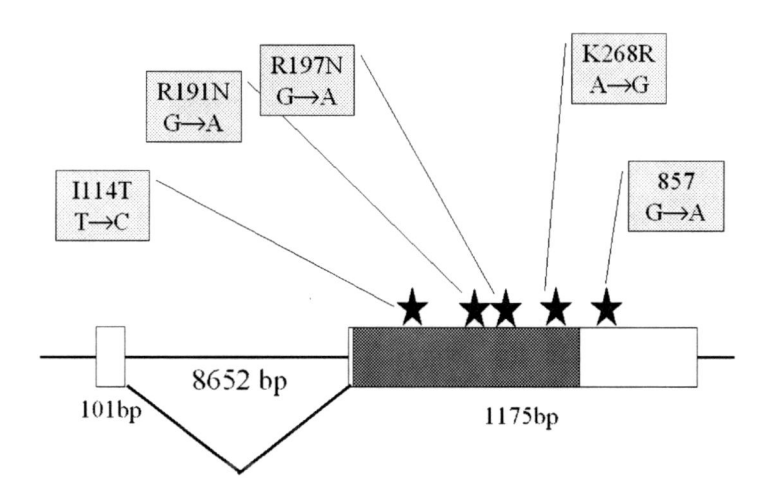

**Figure 21.6** *Alleles of the NAT2 gene influencing the acetylator phenotype*
The two exons of the gene are shown with the coding region in grey. The asterisks denote the location of major polymorphisms responsible for the slow vs. rapid acetylator phenotype.

The notions of 'pharmacogenetics' and 'pharmacogenomics' are still pretty fresh and are often used rather loosely, so some confusion has arisen. In fact, prediction of positive responses to therapy is one area in which pharmacogenetics and pharmacogenomics meet and overlap. The precise distinction should be that pharmacogenetics deals with the patient's reactions to specific therapies, whereas pharmacogenomics considers the therapeutic targets specific for a disease. A broad definition of pharmacogenomics would therefore encompass almost the entire molecular biology of human cancers. In everyday use, the term 'pharmacogenomics' more specifically denotes the investigation of drug targets in specific cancers.

A good illustration of the purpose of pharmacogenomics is the case of ERBB2. In a specific subclass of breast cancers, ERBB2 is overexpressed, typically as a result of gene amplification (→18.4). Determination of ERBB2 expression and of amplification of its gene gives a very good indication of whether cancers will respond to an antibody directed against the protein. This may sound like an issue for pharmacogenetics, but it is a genetic property of the particular tumor and not the individual patient that provides the basis of the treatment.

As more and more therapeutic agents become available that are targeted to specific molecules in specific tumors, pharmacogenomic testing will become more important.

Moreover, other than in pharmacogenetics, prediction of the response of a cancer to a drug typically relies on an molecular assay. Obviously, the response of a breast cancer patient to an anti-ERBB2 antibody cannot be predicted from her previous experience with everyday drugs. Similarly, the simple strategy of applying the drug and monitoring the response is inefficient and costly, since the treatment is expensive and is efficacious in only a fraction of the patients. So, the molecular assay has to precede the application of the drug. The necessity of pharmacogenomic tests is underlined by failed efforts to extend the use of ERBB2-targeted therapy to other cancers. Amplification and over-expression of ERBB2 are rare in prostate and bladder cancers and clinical trials with the antibody directed against the protein were by and large unsuccessful.

In contrast, imatinib, an inhibitor of the BCR-ABL protein kinase, was found to induce remissions not only in chronic myeloid leukemia (→10.4), which is driven by this fusion protein, but also in other cancers. In gastrointestinal stromal tumors, its effect could be related to the inhibition of the KIT receptor tyrosine kinase, already known as an alternative target of the drug from in vitro assays. However, among these, only those cancers in which KIT was activated by a mutation responded (→22.4). This experience makes a strong argument in favor of pharmacogenomical approaches to therapy. However, further cancer types, without KIT activation, have responded to imatinib. In these, the effect of the drug is ascribed to inhibition of the receptor tyrosine kinase PDGFR (platelet-derived growth factor receptor).

Pharmacogenetics and pharmacogenomics are two specific areas which may be representative for the overall direction that molecular diagnostics in oncology is taking. It is expected that molecular diagnostic techniques will supplement the established methods of tumor diagnosis, rather than replace them. The main impact of innovation will be to better tailor therapy to each individual patient and to each individual cancer.

## *Further reading*

Golub TR (2001) Genomic approaches to the pathogenesis of hematologic malignancies. Curr. Opin. Hematol. 81, 252-261

Sorlie T et al (2001) Gene expression patterns of breast carcinomas distinguish tumor subclasses with clinical implications. PNAS 98, 10869-10874

Bok RA, Small EJ (2002) Bloodborne biomolecular markers in prostate cancer: development and progression. Nat. Rev. Cancer 2, 918-926

Chung CH, Bernard PS, Perou CM (2002) Molecular portraits and the family tree of cancer. Nat. Genet. 32 Suppl. 533-540

van't Veer et al (2002) Gene expression profiling predicts clinical outcome of breast cancer. Nature 415, 530-536

Watters JW, McLeod HL (2002) Recent advances in the pharmacogenetics of cancer chemotherapy. Curr. Opin. Mol. Therapeut. 4, 565-571

Widschwendter M, Jones PA (2002) The potential prognostic, predictive, and therapeutic values of DNA methylation in cancer. Clin. Cancer Res. 8, 17-21

Etzioni R et al (2003) The case for early detection. Nat. Rev. Cancer 3, 243-252

Goldstein DB, Tate SK, Sisodiya SM (2003) Pharmacogenetics goes genomic. Nat. Rev. Genet. 4, 937-947

Laird PW (2003) The power and the promise of DNA methylation markers. Nat. Rev. Cancer 3, 253-266

Liu ET (2003) Classification of cancers by expression profiling. Curr. Opin. Genet. Devel. 13, 97-103

Russo G, Zegar C, Giordano A (2003) Advantages and limitations of microarray technology in human cancer. Oncogene 22. 6497-6507

Tribut O et al (2003) Pharmacogenomics. Med. Sci. Monit. 8, RA152-163

van der Velde VHJ et al (2003) Detection of minimal residual disease in hematological malignancies by real-time quantitative PCR: principles, approaches, and laboratory aspects. Leukemia 17, 1013-1034

# CHAPTER 22

# CANCER THERAPY

## 22.1 LIMITATIONS OF CURRENT CANCER THERAPIES

Today, surgery, radiotherapy, and chemotherapy, i.e. 'scalpel, ray, and pill', remain the standard tumor therapies. Novel therapies like gene therapy or immunotherapy are only administered in the setting of experimental studies. In addition, outside 'school' medicine, a variety of 'alternative' treatments are sought by patients and their families who do not have full confidence in standard therapies.

This scepticism is, unfortunately, not entirely unfounded. With current therapies, between 30% and 40% of all cancer patients die of their disease (Figure 22.1). This average percentage conceals huge differences. Basal cell carcinoma and squamous carcinoma of the skin, e.g., are almost always detected before they metastasize and can be cured by local surgery, radiotherapy or drug application (→12.1). A few formerly generally lethal tumors like testicular cancers, Wilms tumors (→11.1) and certain hematological cancers (→10) respond excellently to current therapies. The routine establishment of stem cell transplantation has further improved the prospects of patients with hematological cancers, most impressively of children.

On the other hand, cure rates are still abysmal for some cancers that spread early and do not respond well to chemotherapy. Mortalities exceed 95% for pancreatic carcinoma, and 85% for lung cancers beyond very early stages and some acute leukemias in adults. The rates for other common cancers fall between these extremes. As a rule, carcinomas can be cured by surgery or by radiotherapy as long as they are confined to the organ where they originate (cf. 13.1, 14.1, 15.1, 18.2, 19.1). In contrast, cure rates for metastasized carcinomas have generally remained dismal, although present-day drug and radiation therapies often alleviate symptoms and prolong survival.

Regarding this state of things, two conclusions are evident.

(1) Cancer prevention is superior to cancer treatment (cf. 20.1). Therefore, prevention ought to be consigned a high priority. At the least, if cancers cannot be prevented completely, they ought be detected at an early stage, while cures are still feasible.

(2) Better therapies are required, most urgently for advanced stage and/or metastatic carcinomas.

There is a general feeling that the current therapeutic concepts are approaching their limits. With the therapeutic strategies pursued till the 1990's, quantitative improvements in the therapy of major cancers may still be achieved, but no 'breakthroughs' are expected. In a sense, therefore, the present trend towards individualization of therapy (→21.5) recognizes the limits of current therapies, while attempting to exploit their full potential.

450                                         CHAPTER 22

Breakthroughs in cancer therapy are instead anticipated from novel 'molecular' approaches based on the emerging insights into the molecular biology of human cancers. It is doubtful whether such breakthroughs have already been achieved. So far, only few 'new-age' drugs developed 'rationally' on the basis of molecular biology insights have entered clinical routine (cf. 22.4). There are, however, good reasons to believe that this situation may change. After all, the molecular biology of cancer described in parts I and II of this book is largely a product of the late 1980's and 1990's and the establishment of a new cancer therapy requires at least a decade of development and testing. Moreover, in several cases in which initial attempts at molecular-based cancer therapy have not been successful, advances in the knowledge of cancer biology allow to understand why these failures have occurred and to improve on them. In fact, this knowledge has also elucidated the causes why current cancer chemotherapy is only efficacious up to a certain point.

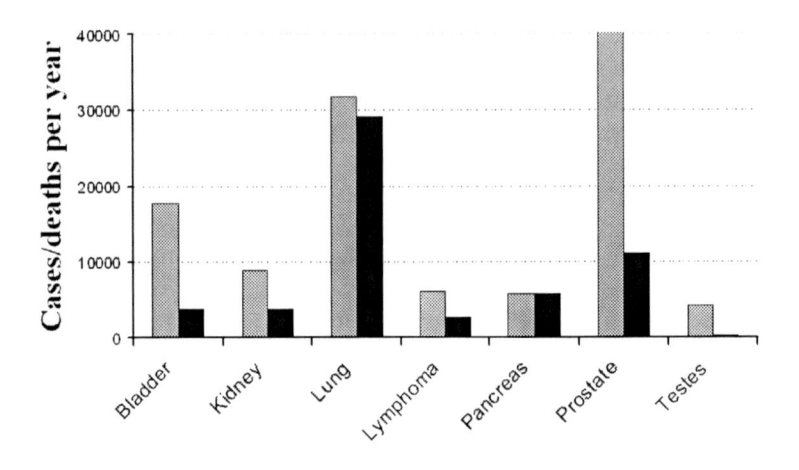

**Figure 22.1** *Mortalities of selected cancers*
Incidence (grey) and mortality (black) of selected cancers in males in Germany in 2000. Data are from the Robert-Koch-Institute.


## 22.2 MOLECULAR MECHANISMS OF CANCER CHEMOTHERAPY

The cancer drugs that are most widely used today were basically developed in the 1950's to 1970's (Table 22.1). A few novel classes of compounds have since been added, such as the taxols, and many older compounds have been chemically modified or replaced by related substances to increase their efficacy and diminish their toxicity. Most of the currently used anti-cancer drugs have been found empirically by screening synthetic and natural compounds for their effect on tumor cells, while others were designed against specific targets. Target-directed drug development, therefore, is not a wholly new invention. However, prior to the recent developments in the understanding of cancer biology, different targets in cancers cells were regarded as important than today. Specifically, increased cell proliferation was regarded as the central property of cancers. Therefore, many 'older' cancer drugs (Table 22.1) are directed against DNA, DNA replication, or mitosis.

<u>Drugs binding to DNA</u>: A large class of anticancer drugs react directly with DNA. For instance, cis-platinum reacts with DNA bases, causing intra-strand and inter-strand crosslinks which block DNA replication and cause cell death, unless repaired. Cis-platinum is the crucial component in many drug regimes used to treat common carcinomas. It is the single most important compound in the combination of drugs that has revolutionized the treatment of testicular cancers, where cure rates of >95% can be achieved.

<u>Nucleoside analogs</u>: Following conversion to nucleotides in the cell, nucleoside analogues interfere directly with DNA replication, impede it indirectly by limiting the synthesis of deoxy-nucleotide triphosphate precursors, or cause strand breaks after incorporation into DNA. 5-fluorouracil (5-FU) is a widely employed member of this class, which acts mainly by inhibition of thymidylate synthase (Figure 22.2). Methotrexate is not a nucleoside analogue, strictly spoken, but also interferes with deoxy-nucleotide biosynthesis by inhibiting dihydrofolate reductase. Thus, both compounds diminish the level of dTTP, the nucleotide precursor specifically needed for DNA replication. A special case is 5'-aza-deoxy-cytidine. It is incorporated into DNA, where it reacts with DNA methyltransferases ($\rightarrow$8.3) attaching them covalently to DNA. This reaction has a double effect. It depletes the methyltransferases causing a genome-wide decrease in DNA methylation and often reactivation of genes inactivated by DNA hypermethylation.

**Table 22.1.** *Some contemporary anti-cancer drugs in frequent use*

| Class of drugs | Examples |
|---|---|
| Drugs binding to DNA | cis-platinum, mitomycin C, adriamycin (doxorubicin), bleomycin, actinomycin, alkylating agents |
| Nucleoside analogues or nucleotide biosynthesis inhibitors | 5-fluoro-uracil, cytosine arabinoside, 5-aza-deoxy-cytidine*, 5-fluoro-cytosine and prodrugs, thiopurines, methotrexate, hydroxy-urea, difluoro-methyl-ornithine** |
| Topoisomerase inhibitors | irinotecan, topotecan, etoposide, also several intercalating drugs like doxorubicin |
| Microtubule-binding | vinblastine, vincristine, taxols |
| Biological agents | antihormones (tamoxifen, raloxifen, antiandrogens), GnRH antagonists and agonists, estrogens, progesteron, retinoic acids and analogues, interferons, interleukins |

\* is thought to act predominantly by blocking DNA methyltransferases
\*\* blocks biosynthesis of polyamines

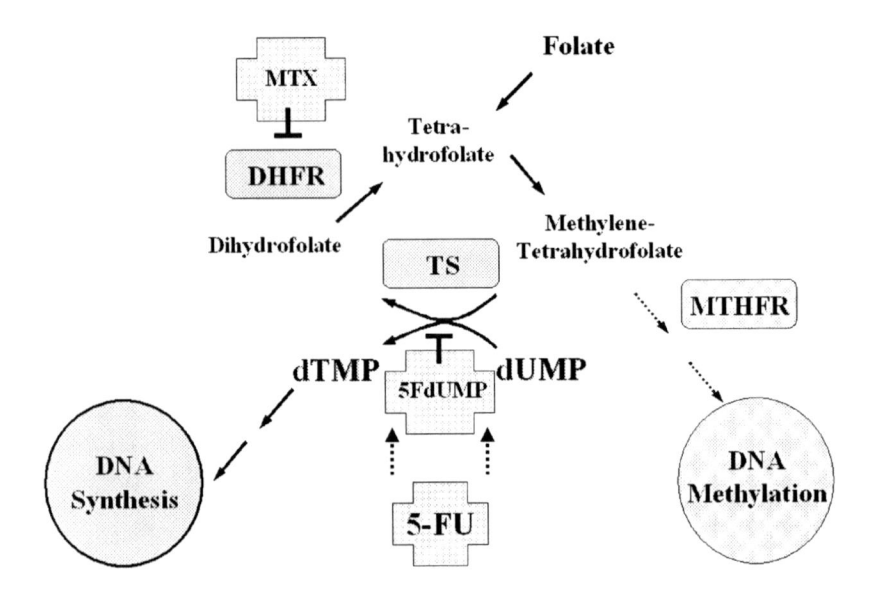

**Figure 22.2** *Mechanisms of action of 5-FU and methothrexate*
cf. also Figure 20.3

The protein-DNA complex may also interfere with DNA replication, unless removed by bulky adduct repair.

Topoisomerase inhibitors: Etoposide exemplifies a third class of compounds which bind and inhibit enzymes involved in DNA replication. Etoposide specifically binds to topoisomerase II and blocks the enzyme at a critical stage. Topoisomerases are necessary for DNA replication (as well as for transcription), since they relax the torsional stress that is caused by the unwinding of the DNA helix. Topoisomerase I enzymes reversibly insert a single-strand break, allow the DNA strands to swivel around each other, and re-ligate the strand-break. Inhibitors of topoisomerase I used in cancer chemotherapy comprise innotecan, irinotecan and topotecan. Topoisomerase II enzymes catalyze a more dramatic reaction, in which a double strand break is reversibly introduced and another DNA helix (or a distant part of the same helix) is passed through, before the ends are resealed by the enzyme (Figure 22.3). This is a more fundamental reaction, which in addition to relaxing torsional stress allows the untangling of DNA knots and loops. Etoposide inhibits type II topoisomerases at a crucial stage of this reaction, i.e. after the helix has been cleaved, but not yet been resealed. In this fashion, DNA replication is inhibited and DNA is fragmented, more efficiently than by topoisomerase I inhibitors.

Microtubule-binding compounds: Taxoles are perhaps the best-known among different compounds reacting with microtubules, while vinblastine or vincristine are used for specific diseases. Some drugs of this class block the assembly of or disrupt existing microtubuli, while others block the turnover of these dynamical structures. Either way, cellular functions depending on microtubules are compromised or inhibited. The most important process affected by interference with microtubule function is mitosis, but intracellular vesicle transport and cell migration are also inhibited.

Biological agents: The designation 'biological agents' is sometimes used as a summary designation for a diverse group of compounds that do not directly interfere with basic cellular functions, such as DNA replication and mitosis. Rather, they act on signaling pathways controlling cell proliferation and differentiation. By activating or inhibiting receptor molecules, they redirect cancer cells in a more subtle fashion towards normal behavior. Hormones and antihormones used in the treatment of breast cancer ($\rightarrow$18.4) and of prostate cancer ($\rightarrow$19.2) as well as inducers of differentiation such as retinoic acid used in the therapy of acute promyelocytic leukemia ($\rightarrow$10.5) can be assigned to this category. They act selectively on certain cancers since they activate or inhibit receptor proteins that are specifically required for their growth and survival. This does not automatically imply that such compounds do not have adverse side effects. However, these are typically not caused by toxicity. Rather, effects of these drugs on the proliferation, differentiation, or function of normal cells are mediated by the same receptor(s) as in cancer cells. For instance, anti-estrogens favor osteoporosis and cardiovascular disease, because they block the beneficial effects of estrogens on bone and heart tissue.

Biological agents act through specific receptors present only in certain cells,

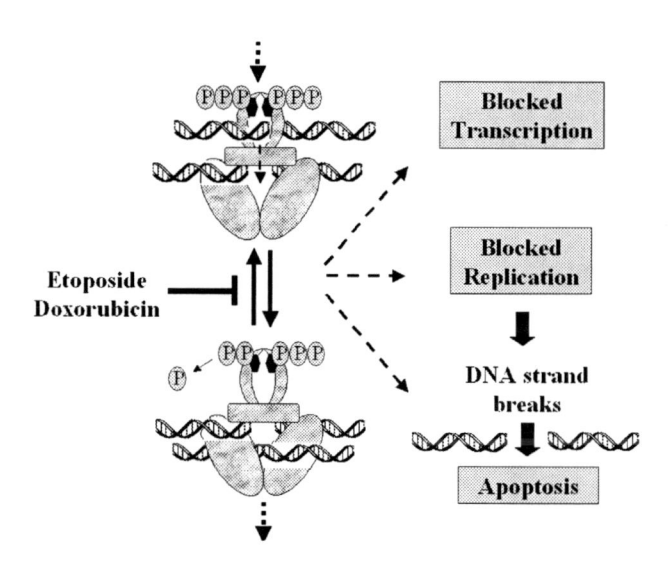

**Figure 22.3** *Effects of topoisomerase II inhibition by chemotherapeutic drugs*
See text for further explanation

which explains their specificity. But how can drugs directed against DNA replication and mitosis, i.e. basic processes essential in many different cells of the body, or drugs reacting with DNA itself act selectively on cancer cells at all? Three major reasons have been recognized (Table 22.2).

Differences in proliferation: Many cancers contain a higher proliferative fraction than normal tissues, and many cancer cells replicate faster than most normal cells. These differences constituted the main rationale in the early years of cancer chemotherapy development. Unfortunately, many normal tissues, too, contain fast-replicating compartments. Accordingly, treatments that aim purely at rapidly replicating cells cause damage to such tissues as well. For this reason, side effects of chemotherapy are common in organs with a rapid turnover, primarily the hematopoetic system, gut and skin. Adverse side effects of chemotherapy include leukopenia (low numbers of leukocytes), diarrhoea (as a consequence of damage to the gut mucosa), and alopecia (hair loss). These side effects can be severe and limit the dose of cytostatic drugs that can be applied. The severity of adverse effects in clinical trials and routine use of drugs is categorized from I-V for each type of effect. Grade III or IV side effects will be cause for concern and may represent the reason for termination of the treatment or for dose reduction, and grade V means a fatal outcome of the treatment.

Adverse side effects are a general problem with cancer therapy, but worse, many cancers do not confirm to the fast-replication stereotype. In many carcinomas, in particular, relatively few cells are actively proliferating at one point in time and those that are do not traverse the cell cycle very rapidly. Therefore, while cycling cells in the cancer may indeed be killed by the drug, they will later be replaced by other cells from the tumor that were not in a critical phase of the cell cycle or not cycling at all, when the drug was present. Chemotherapy of prostate carcinoma ($\rightarrow$19.1) and renal cancers ($\rightarrow$15.6), e.g., is vexed by this effect. It represents, however, a wider problem and constitutes a second, specific limit to the efficacy of therapy directed at DNA and DNA replication, in addition to the first general limit provided by the toxicity of the therapy. An interesting new approach to circumvent the low proliferation problem is 'metronomic therapy'. In this type of chemotherapy, cytostatic drugs are applied regularly at relatively low doses over longer periods than in standard regimes. Compared to standard therapy, the aim of this treatment is no longer to cure the cancer, but to slow its growth and prolong survival with minimal harm to the patient.

**Table 22.2.** *Mechanisms responsible for the selectivity of anti-cancer drugs*

| *Mechanism responsible for differential response of tumor vs. normal cells* |
|---|
| Increased proliferative fraction and shortened cell cycle |
| Inefficiency or inactivation of cellular checkpoints |
| Defects in DNA repair |
| Altered apoptosis |
| Dependence on 'cancer pathways' |

CANCER THERAPY 455

Defects in cellular checkpoints and DNA repair: The second set of reasons why cancers are more sensitive to cytotoxic chemotherapy than normal cells was unknown when the first generation of active drugs was developed. Many cancer cells are defective in DNA damage checkpoints (→3.4), e.g. as a consequence of mutations in the TP53 pathway (→6.6). So, following the covalent reaction of a drug like cis-platinum with DNA, cancer cells may continue to replicate and attempt to divide in spite of the damage, with catastrophic consequences (Figure 22.4).

In addition, some cancers are defective in the repair of specific types of DNA damage, due to the inactivation of particular repair systems. For instance, some cancers lack the MGMT enzyme which removes alkyl groups from guanine (→3.1), and for this reason are hypersensitive to drugs alkylating DNA at guanines. Similarly, colon carcinomas with a microsatellite instability phenotype respond on average better to chemotherapy than those with a chromosomal instability phenotype (→13.5). This difference may be due to the inactivation of the mismatch repair system in cancers with microsatellite instability (→13.4). In selected cancers, e.g. of the ovary, DNA crosslink repair may be compromised by epigenetic inactivation of FANC genes (→3.3). Encouraged by such examples, one current line of applied

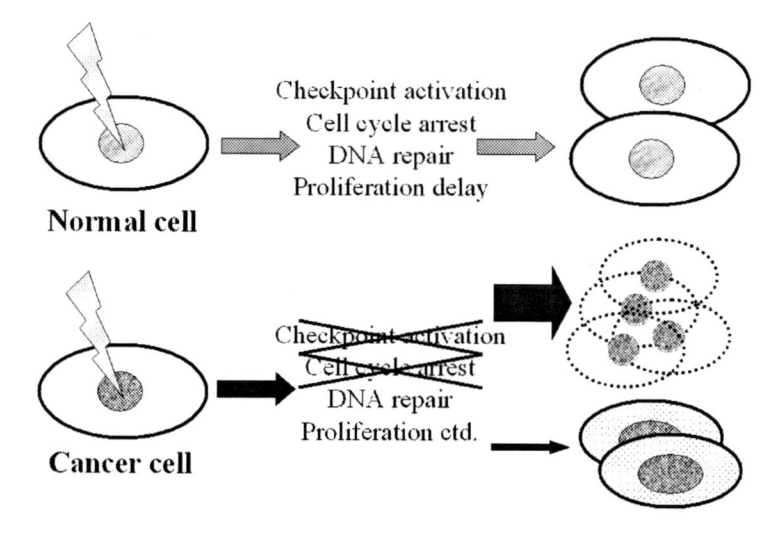

**Figure 22.4** *Exploiting deficiencies in checkpoint control for cancer therapy*
Normal cells react to DNA damage during drug or radiation therapy by checkpoint activation, with cell cycle arrest and resume proliferation (if at all) only after DNA repair is completed. Because of defective checkpoint control, cancer cells proceed through mitosis and G1 irrespective of DNA damage. This can lead to mitotic catastrophes, mitotic arrest, or to persistent double-strand breaks that elicit apoptosis during the next round of replication. However, while most cancer cells die or arrest, a few may escape with severely damaged genomes and increased genomic instability (bottom right). These are responsible for remissions and are usually resistant to treatment.

cancer research aims at identifying further DNA repair defects in specific cancers and exploit them for selective therapy.

Another novel approach consists in using drugs that aggravate the checkpoint deficiencies in cancer cells, e.g. by blocking kinases involved in the control of the G2→M checkpoint. In normal cells several mechanisms ensure checkpoint control. In cancer cells, some or all of these may be deficient, rendering them more sensitive to their inhibition. Combination treatment together with compounds that interfere with DNA replication would then lead to checkpoint arrest in normal cells, but to a mitotic catastrophe in cancer cells (Figure 22.4). One compound acting in this fashion is caffeine, although at millimolar concentrations not tolerated in a human person.

<u>Altered apoptosis:</u> Somewhat counterintuitively, another explanation for the selectivity of cytostatic drugs towards cancers is related to altered apoptosis (→7.3). This sounds paradoxical, as apoptosis is often impeded in cancer cells. Indeed, defects in apoptotic signaling and execution can contribute to resistance against chemotherapy. However, many cancer cells can be considered as being 'poised' for apoptosis. Inappropriate growth control, genomic instability, and nucleotide imbalances generate pro-apoptotic signals, which do not elicit apoptosis because anti-apoptotic signals prevail in cancer cells. In this critical constellation, drug treatment may add further signals that 'tip the balance' towards apoptosis (Figure 22.5).

For instance, cancer drugs like cis-platinum, 5-FU, and etoposide lead to the induction and activation of death receptors like FAS (→7.2). Others, including methotrexate, activate the intrinsic, mitochondrial pathway of apoptosis (→7.2). The reaction of a cancer to drug treatment therefore depends on which defects precisely are responsible for decreased apoptosis. If the block to apoptosis is very efficient, it will protect the cell against drug-induced apoptosis as well. For instance, strong overexpression of IAP type proteins like survivin which inhibit caspases or strong overexpression of BCL2 which prohibits activation of the intrinsic pathway (→7.2) can also cause resistance to chemotherapy.

Loss of TP53 function also influences the response to chemotherapy in many cancers, but its effect is complex. TP53 is important for checkpoint signaling following DNA damage (→5.3). So, cancer cells with TP53 loss of function tolerate more DNA damage than normal cells and continue to proliferate in spite of it. This is a questionable advantage, since they run a higher risk of mitotic catastrophes or incurring damage to essential genetic material. On the other hand, loss of TP53 certainly impedes the induction of apoptosis (→7.2). Overall, therefore, cancers with loss of TP53 function tend to respond less well to chemotherapy, but there are exceptions to this rule.

From these arguments, it can be deduced why certain cancers respond well to chemotherapy, while others do not. For instance, testicular cancers respond excellently to chemotherapy and are particularly sensitive to cis-platinum. They usually contain a high proliferative fraction with rapidly replicating cells. Checkpoints in testicular cancer cells appear to be not fully functional and nucleotide excision repair, in particular, is poorly efficient. Finally, TP53 is usually not mutated and can be induced by cytostatic drugs to support the induction of apoptosis. So, in this cancer type, all pertinent factors favor therapeutic success.

Unfortunately, carcinomas in general rather resemble renal cell carcinoma (→15.6). In this cancer, a low proliferative fraction, a slow, but relentless growth, and the presence of strong anti-apoptotic signals, with loss of TP53 in some cases, tilt the balance against the success of chemotherapy (as well as radiotherapy), even though checkpoints may not be fully intact and TP53 may remain functional in a subset of the cases.

These general factors that counteract successful therapy are exacerbated by specific mechanisms of chemoresistance. In renal carcinoma, the expression of the multidrug resistance protein Pgp/MDR1 and other protective proteins (→15.6) further limits the impact of chemotherapy, contributing to 'primary resistance'. Expression of MDR1 is also found in other cancers, even if expression in the corresponding normal tissue is not as strong as in the kidney. In some cases, the protein becomes expressed only during therapy in resistant cancer clones, eliciting 'secondary resistance'. High levels of the multidrug resistance protein confer

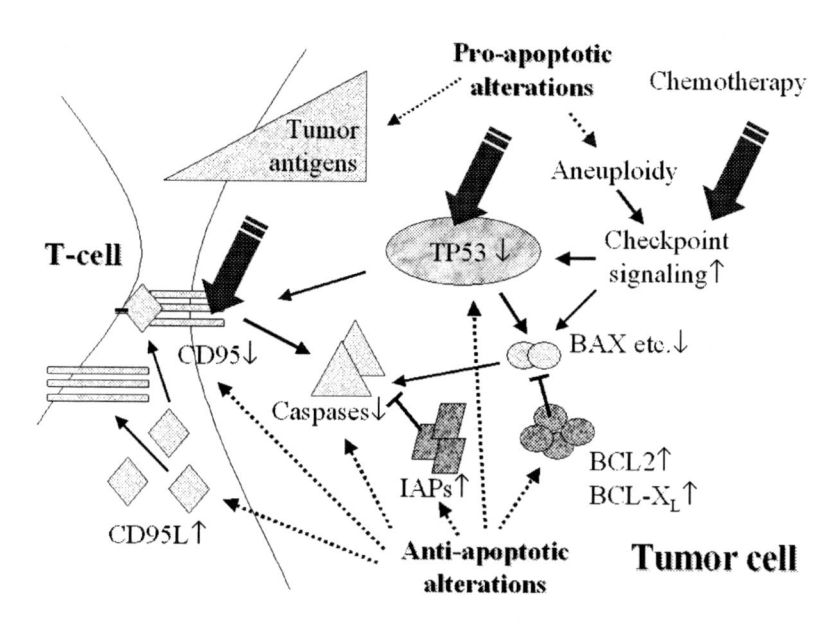

**Figure 22.5** *Roles of altered apoptosis in cancer cells in resistance and hypersensitivity to therapy*

Pro-apoptotic changes in cancers cells (top) are held in check by anti-apoptotic alterations (bottom), which often also confer resistance to chemotherapy. In other cases, chemotherapy (stippled arrows) increases pro-apoptotic signals sufficiently to tip the balance towards apoptosis, if only slightly. Some of the changes present in tumor cells such as partial caspase activation and CD95 ligand expression may then serve to amplify this incremental change.

resistance to hydrophobic drugs by transporting them out of the cell. Multidrug resistance ensues, since many anti-cancer drugs are hydrophobic compounds.

Multidrug resistance by over-expression of MDR1 is a general mechanism of drug resistance. Similarly, increased levels of protective proteins such as glutathione transferases (→3.5) diminish the sensitivity of a cancer to a range of drugs. Activation of antiapoptotic pathways, specifically of the PI3K (→6.3) and NFκB (→6.9) pathways, also contributes to decreased sensitivity against a broad range of chemical and physical therapies. Additional mechanisms confer resistance to individual drugs.

Resistance to cis-platinum, e.g., can be caused by overexpression of metallothioneines which protect cells from the toxic effects of metal ions in general. Resistance to topoisomerase inhibitors may be due to overexpression of the target enzyme as a consequence of gene amplification.

Resistance to 5-FU and related compounds targeting thymidylate synthetase (Figure 22.2) is particularly complex. The response to the drug depends on properties of the tumor as well as on the genetic constitution of the patient (cf. 21.5). Resistence can be caused by altered metabolism of the drug, by mutations and amplifications in the *TS* gene, and can be favored by genetic polymorphisms in this and other genes (cf. 21.5).

Most advanced cancers are characterized by genomic instability, which can be associated with increased rates of chromosomal gains and losses, gene amplification, deregulation of gene expression, and/or point mutations. These mechanisms are not only relevant for the development of the cancer as such, but also open a variety of escape routes during drug therapy. For instance, drug targets can be rendered insensitive to inhibitors by point mutations or become less sensitive by amplification of the gene encoding the target.

In an advanced cancer, which is genetically heterogeneous, a fraction of its cells may carry an alteration leading to decreased sensitivity towards a cytotoxic drug. Administration of the drug will then select these cells from all others leading to the emergence of a new cell clone with altered properties (Figure 22.6). As a rule, this cell clone will not only be resistant to the specific drug and often to others, but also be more genetically unstable than the overall tumor before treatment (cf. Figure 22.4).

In some cancers, the cells more responsive to drug therapy correspond to the more differentiated fraction. Induction of apoptosis and/or growth arrest in this population therefore may expose a more malignant fraction of cancer cells and sometimes release restraints on these. Such 'lurker' cells are suspected to be responsible for the recurrent growth of breast and prostate cancer following anti-hormonal therapy (→18.4, →19.2). However, this phenomenon is not restricted to anti-hormonal treatment.

Worse, cancer chemotherapy may in some cases directly promote genomic instability, if it causes damage to the genome without actually killing cells or arresting their growth irreversibly. In such cases, the treatment effectively acts as a mutagen that induces a resistant cell clone, often with further genetic alterations

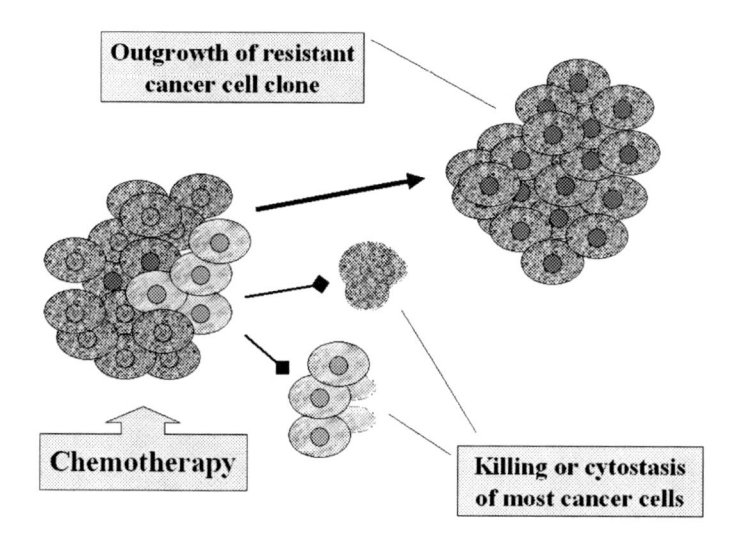

**Figure 22.6** *Selection of resistant cancer cells during drug therapy*
See text for further elaboration.

(Figure 22.6). Outgrowth of a more malignant cancer is therefore a common observation after failed cytostatic drug chemotherapy.

## 22.3 PRINCIPLES OF TARGETED DRUG THERAPY

One strategy to circumvent the problems associated with conventional chemotherapy (→22.2) is to develop drugs against more specific targets in the cancer. This is not a fundamentally novel idea. Many drugs in current use interact with highly specific targets such as microtubular proteins or topoisomerase enzymes (→22.2). Their targets are not specific to cancers, though. Those drugs called 'biological agents' in the previous section come closer to the ideal, since they act on specific receptor proteins which may occur preferentially in certain tissues, but, more importantly, are essential for the growth of specific cancers. These, then, are the forerunners of a novel drug generation.

All-trans retinoic acid, e.g., binds to receptors that are more or less ubiquitous in the body (→8.5). Indeed, synthetic analogues of this tissue hormone are also used for the treatment of benign skin diseases like acne, because retinoic acid promotes cell differerentiation in the skin, as in many other epithelia. Retinoids therefore have been tried as anticancer drugs in almost every type of cancer, usually with detectable, but limited effects on tumor growth. In contrast, retinoic acid is highly active in most cases of acute promyeolocytic leukemia (→10.5), out of all acute leukemias. What makes the difference towards all other cancers is that in this particular type of leukemia the causative genetic change involves the retinoic acid receptor α, whereas in other cancers changes in the response to retinoids may well occur, but are non-essential for their growth and survival.

This case, then, comes close to the ideal of target-oriented cancer therapy. Elucidation of crucial events that drive cancer growth should provide targets for therapy. Targets for rational therapy ought to be at least specific to the tumor, but better essential for its growth and survival. Elucidating these crucial events and identifying suitable target molecules are however no simple tasks.

In many hematological cancers, the presence of a characteristic chromosomal translocation points to an essential genetic event (→10.2). Yet, even in leukemias and lymphomas developing a therapy from that knowledge can be difficult. Acute promyeolocytic leukemia (APL) is exceptional in so far as the fusion protein formed by the causative chromosomal translocation (→10.5) contains a receptor protein whose ligands are well characterized. Developing a therapy for chronic myelocytic leukemia (CML) by targeting the causative BCR-ABL fusion protein was still quite straightforward, since it contains an essential protein kinase activity (→10.4).

Unfortunately, not all fusion proteins display functions that lend themselves to inhibition or activation by small molecule drugs. In the jargon of pharmaceutical research, they are not easily 'drugable'. Furthermore, APL and CML are untypical in constituting essentially homogeneous diseases, whereas other hematological cancers may be caused by a variety of different translocations and gene fusions (cf. Figure 10.2).

As carcinomas are characterized by multistep development with accumulation of a larger number of various genetic and epigenetic alterations (cf. 13.3), it is generally even less clear which targets are optimal. Optimists assume that many of these alterations are essential for the growth and survival of the cancer and conclude that the multitude of changes in carcinomas offer a wide choice of targets for therapy. Pessimists point out that the more alterations have already occurred, the higher the chance that some of them may be passenger alterations. Worse, a cancer with many genomic alterations is likely to develop further ones allowing escape from therapy. Likewise, optimists suggest that cancer pathways activated in specific cancers (cf. 6) present excellent targets, as they are crucial for driving tumor growth. Pessimists point out that these same pathways are also important for normal cells which might mean a narrow 'therapeutic window'.

In practice, potential targets for cancer-specific therapy are defined based on a variety of considerations (Table 22.3).

**Table 22.3.** *Molecular targets for cancer therapies*

| Type of target | Examples |
| --- | --- |
| Ectopic proteins | viral proteins, cancer-testis antigens |
| Overexpressed proteins | oncogene products, particularly receptor tyrosine kinases |
| Altered proteins | mutated products of oncogenes |
| Cancer pathway components | MAP kinases, CDKs |

Ectopic targets: An ideal drug target in a cancer would never occur in a normal tissue. Since cancer cells are derived from normal cells, one would think that such targets might be rare, but some do exist. (1) In cancers induced by viruses or cancers harboring viruses, viral proteins can be targeted. The E6 and E7 proteins of HPV (→Box 5.1) are involved in carcinogenesis in several tissues. Other viruses like EBV and HBV, while not necessarily driving tumor growth, are at least present in many Burkitt lymphomas (→10.3) and hepatocellular carcinomas (→16.3), respectively. (2) Fusion proteins in leukemias and lymphomas are composed of proteins which are also present in normal tissues, but always separately. Their fusion confers novel properties which can be exploited to target them selectively. (3) Many cancers express proteins (→12.5) which are otherwise only found in fetal tissues ('oncofetal proteins') or in a very small range of other tissues, e.g. in the testes ('cancer testis antigens'). These are often not essential for the growth of the cancer, but they can be used for the targeting of toxins or for immunotherapy.

Overexpressed proteins: A second class of targets is provided by proteins overexpressed in cancer cells. Some oncofetal proteins and cancer testis antigens actually belong to this class, because they are expressed at very low levels in normal tissue. The most important group, however, of such proteins are the products of oncogenes that have become activated by overexpression, e.g. as a consequence of gene amplification.

Several strategies have been developed, e.g., to exploit the overexpression of the EGFR or ERBB2 receptor tyrosine kinases in many advanced carcinomas for therapy. An evident disadvantage of using such proteins as drug targets, though, is the very fact that they are overexpressed. This point is illustrated by the amplification of the androgen receptor gene in prostate carcinomas which have become unresponsive to anti-androgenic treatment (→19.2). Accordingly, targeting an amplified protein kinase by an inhibitory drug may invite further amplification as a mechanism of resistance. Nevertheless, as the case of trastuzumab shows, overexpressed cell-surface receptor tyrosine kinases can be used for targeting by antibodies (→18.4). Antibodies can also be directed at proteins that are not as essential for the growth and survival of the tumor cell as ERBB2 is for many breast cancers. Modern high-throughput proteomics and expression profiling approaches are excellently suited for the identification of proteins overexpressed in cancer cells. There is therefore no shortage of candidates for this approach.

Proteins with altered structures: The third class of targets are proteins whose structure is altered in cancer cells. Fusion proteins could also be assigned to this class. They are excellent targets, because their structure is different in tumors compared to normal cells and they are essential for cancer growth. The same is true for oncogenic proteins activated by point mutations, such as KRAS in colon cancer (→13.3) or β-Catenin in the same cancer and more often in hepatocellular carcinoma (→16.2).

Even tumor suppressor proteins inactivated by point mutations deserve consideration. They could provide targets for immunotherapy, but drug therapy is not inconceivable. A favorite candidate in this respect is TP53, since it is more often inactivated by missense mutations than by deletions or promoter hypermethylation.

Most missense mutations in TP53 appear to interfere with the conformational activation of the protein, which strongly accumulates in cancer cells because the mutated protein is more slowly degraded (→5.3). A drug pushing the protein into an active state would therefore activate a comparatively huge amount of TP53 protein and likely elicit apoptosis.

In addition to these clear-cut cases of mutated oncoproteins, there is some evidence that cancer cells in general may harbor a larger proportion of misfolded and altered proteins than normal cells. It is not quite clear what causes this defect, but it may make cancer cells more susceptible to inhibition of chaperones and of proteasomal degradation. Inhibition of either sort of target may overload the cell with misfolded proteins, like during a heat-shock. Indeed, inhibitors of heat-shock proteins acting as chaperones have emerged as surprisingly good inhibitors of cancer growth, with few side effects. Likewise, inhibitors of proteasome function, e.g. of threonine proteases, have turned out to be surprisingly specific for cancer cells.

Cancer pathway signaling: The fourth category of targets comprises molecules that regulate 'cancer pathways' (→6). The proliferation and survival of cancer cells depend on a relatively restricted number of signal transduction pathways. In different cancers, one or the other of these are overactive or inactive. Inhibition of overactive pathways or restoration of inactivated pathways is a major goal of many current drug development.

However, the designation 'cancer pathways' is in so far imprecise, as the same pathways also control the proliferation, differentiation, survival and function of normal tissues. So, differences between normal and cancer cells are expected to be quantitative rather than qualitative. Hope that these differences may still be sufficient to allow improved cancer therapy is based on observations and ideas summarized by the 'addiction hypothesis'.

Compared to normal cells, signaling pathways in cancers are thought to be 'rewired'. For instance, overactivity of pro-proliferative signals relayed through the canonical (ERK) MAPK pathway would in normal cells be counteracted by increased apoptosis (→6.4). In cancer cells, this increase in apoptosis is impeded by over-activity of other pathways such as the PI3K or the NFκB pathway (→6.4) or by overexpression of anti-apoptotic proteins (→7.3). Therefore, the survival of cancer cells is much more dependent on these anti-apoptotic activities than that of normal cells, which display more moderate and transient activities of MAPK pathways (Figure 22.7). In other words, the cancer cells have become 'addicted' to the activity of the anti-apoptotic cancer pathway.

Of note, the addiction hypothesis predicts that inhibiting the MAPK pathway which actually drives proliferation could be less efficient than inhibiting the PI3K pathway that influences proliferation only indirectly but allows cell survival. This hypothesis may also provide an alternative explanation why cancer cells react more sensitively to inhibitors of heat-shock proteins. They may have become dependent on mutant proteins that could not be assembled in the absence of these molecular chaperones.

Targets defined by such considerations can be exploited by different kinds of therapy. Development of pharmacological therapy with small molecules is the most obvious approach. It is best suited, but not restricted to proteins with enzymatic activities. This strategy has the important practical advantage that it can build on established procedures. Nowadays, pharmaceutical companies possess compound libraries comprising ten thousands of synthetic and natural chemicals that can be screened by high-throughput methods for activating or inhibitory activity against a specific target enzyme. Even protein-protein or protein-DNA interactions can be influenced. A molecule with activity is considered a 'lead' compound. Lead compounds can be chemically modified by a host of well established techniques and procedures to achieve increased specifity and better general pharmacological properties. Determination of the structure of the target protein by modern biophysical methods and drug design using sophisticated computer methods have further facilitated this strategy.

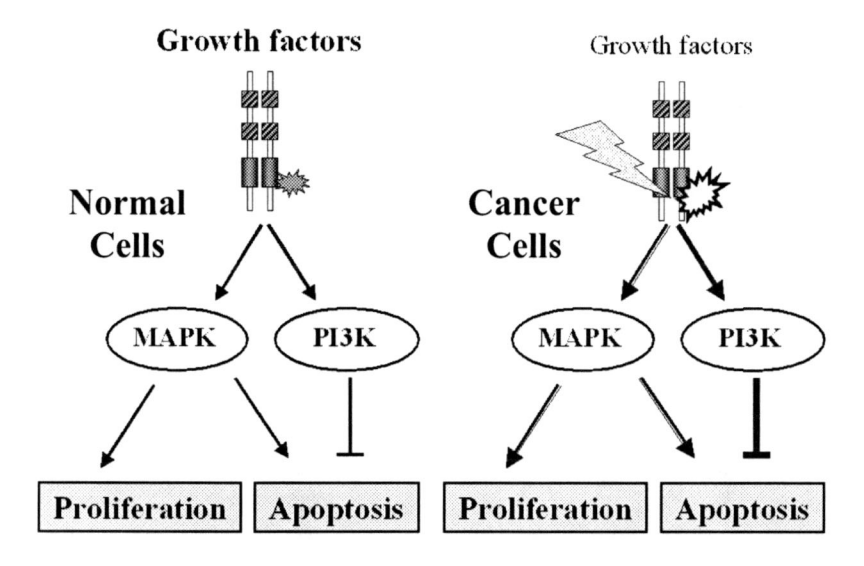

**Figure 22.7** *The 'addiction hypothesis' illustrated by the MAPK and PI3K pathways*
The width of the arrows indicates the activity of the pathways. The lightning symbolizes an upstream mutation activating primarily the MAPK pathway, such as a receptor tyrosine kinase or a RAS mutation. See text for a detailed exposition of the hypothesis.

However, small molecule drugs are not the only option anymore. In fact, some of the most successful 'novel' cancer drugs are antibodies against growth factor receptors. Their development, likewise, has benefited from the availability of a wide range of sophisticated molecular biology methods. For instance, therapeutic antibodies can now be detected and optimized not only in animals, but also in bacteria and phages. If an antibody is initially developed in an animal, it can be 'humanized', i.e. the constant chains can be replaced by a human immunoglobulin sequences using standard methods of genetic engineering. This adaptation impedes the development of an immune response towards the therapeutic antibody in the patient. Without humanization, antibodies become inactive upon repeated administration or may even cause serious adverse reactions such as an allergic shock.

Application of antibodies can be regarded as a type of immune therapy, although antibodies can also be used in a similar fashion as small molecule drugs. In contrast, a suitable target molecule can be also used as the basis for a true cancer vaccine. Some modern approaches at cancer immunotherapy do indeed use defined targets (→22.5).

Defined targets in cancer can also be exploited for gene therapy. In theory, gene therapy is a more straightforward approach than drug or immune therapy. However, the development of new drugs and vaccines can be pursued on a strong fundament of established procedures and long-term experience, whereas in gene therapy almost everything has to be developed from scratch (→22.6). Of all the strategies contemplated and tried in gene therapy, the use of antisense oligonucleotides or siRNA against overexpressed or altered proteins in cancer most closely resembles the approach in the development of drugs. It therefore runs a good chance of becoming established in the clinic first among gene therapy approaches. There are, however, strategies that are unique to gene therapy and may in the long run prove superior. For instance, gene therapy can be used to re-introduce a tumor suppressor that has been inactivated in cancer cells or to exploit the presence of an oncogenic change to allow the replication of a cytolytic virus (→22.6).

Last not least, combinations of novel therapies, drug, immune, and gene therapy, are pursued, and either type of 'novel' therapy can be used in conjunction with established chemotherapies or radiotherapies directed at less cancer-specific targets.

## 22.4 EXAMPLES OF NEW TARGET-DIRECTED DRUG THERAPIES

Drug development based on target-oriented approaches is now routine. Still, the development of a new drug can take a decade from the discovery of a 'lead compound' to routine clinical use. So, in oncology relatively few 'novel' drug are already being used in everyday practice, although many clinical studies are underway (Table 22.4). The experience obtained from the use of these novel drugs in clinical routine most clearly illustrates both the potential and limitations of target-oriented therapies in oncology.

**Table 22.4.** *Selected targeted drugs employed in the clinic or clinical trials*

| *Drug(s)* | *Target(s)* | *Stage of development* |
|---|---|---|
| Imatinib | BCR-ABL, KIT, PDGFR(?) | routine use |
| Farnesyl transferase inhibitors | RAS (RHO?, RAC?) | clinical trials |
| Trastuzumab | ERBB2 | routine use |
| Gefitinib | ERBB1 | routine use starting |

Imatinib (alias STI571, alias Gleevec or Glivec) was developed as an inhibitor of the BCR-ABL protein kinase, which is a fusion protein resulting from the characteristic chromosomal translocation in chronic myelogenous leukemia (CML) and is causative for this cancer (→10.4). The inhibitor blocks the tyrosine kinase activity in the ABL domain of the protein that is essential for its oncogenic function. The drug is thus directed against a target largely specific to this cancer, since the normal ABL protein is not essential for growth and survival in normal somatic cells, although it is important for the control of cellular responses to DNA damage (→3.3). Imatinib is now used for the therapy of CML in its chronic phase alternatively to interferon α with cytogenetical remissions in ≈70% of the patients (compared to ≈10% with former drugs). It even induces clinical remissions in many patients in which the disease has progressed into the terminal blast crisis. For these patients, no treatment was previously available (→10.4).

Imatinib is relatively nontoxic, as one would hope for a targeted drug. When applied in the chronic phase of the disease, it may achieve complete cures, as indicated by molecular remission (→21.2). Whether this is really so, will have to be ascertained by long-term follow-up of the treated patients. Ideally, the drug may help to spare many patients from stem cell transplantation. In blast crisis patients, however, remissions are as a rule temporary. Many cancers develop resistance against the drug. One mechanism of resistance involves amplification of the *BCR-ABL* gene. In other cases, mutations render the BCR-ABL protein less sensitive to the inhibitor. Typically, these mutations lead to changes in amino acids at the ATP binding site of the kinase where imatinib binds. So, while the drug is much more efficacious than previous treatments, it is not immune to the development of resistance. Tellingly, resistance develops more regularly in the accelerated phase or blast crisis of CML characterized by a high level of genomic instability.

Somewhat unexpectedly, imatinib was also found to be highly active against a different cancer. Gastrointestinal stromal tumor (GIST) is a relatively rare type of sarcoma, for which few therapeutic options beyond surgery had previously been available. A subset of these cancers responds very well to imatinib. Of note, this shows that these cancers are genetically more heterogeneous than they appear morphologically. In some cases, cures are achieved, e.g., because cancer regresses to such an extent that the primary cancer and isolated metastases can be surgically removed. In others, the progression of the disease is significantly delayed. All cancers showing remissions under imatinib treatment display activating mutations of the KIT receptor tyrosine kinase. Indeed, the drug *in vitro* also inhibits the KIT and PDGFR kinases with high affinity, in addition to the BCR-ABL kinase. In fact, the efficacy of imatinib in GIST even depends on the precise mutation in the *KIT* gene. The frequent mutations in exon 11 presage a good response, whereas a specific mutation in exon 17, D816H, is associated with a lack of response, i.e. the cancers display primary resistance. Mutations like the latter one are also found in cancers regrowing under imatinib therapy following an initial remission, i.e. displaying secondary resistance. So, clearly, this is a good case in point for pharmacogenomics (→21.5).

During treatment of GIST patients with imatinib, a number of interesting observations were made which suggest that responses to novel cancer drugs may show quite different characteristics from those to standard cytotoxic chemotherapy.

(1) Responses were often slow, at least when the size of the tumor was taken as a parameter. Rather than being killed, tumor cells appeared to be arrested and even to terminally differentiate. Such changes cannot be detected by many methods routinely employed in the monitoring of chemotherapy. Instead, their detection requires imaging methods based on the metabolic activity of the cancer, e.g. positron emission tomography. (2) Tumor endothelia were often severely damaged, probably as a consequence of a decreased supply of growth factors from the cancer as well as by direct inhibition of the PDGFR, which is important for endothelial cell growth (→9.4). This damage may lead to bleeding, which may contribute to the observed initial increase in tumor size after the start of treatment. (3) As the typical adverse effects of chemotherapy were not pronounced with imatinib, it appears that the clinical complications during the use of such novel drugs may be quite different from those during current cytotoxic chemotherapy. In fact, this finding is precedented by the experience with the treatment of acute promyelocytic leukemia using all-trans retinoic acid (→10.5). Here, complications can arise when a large number of cancer cells apoptose at one stroke and/or differentiate into almost normal cells (mostly granulocytes) which exhibit a range of biological activities including the secretion of cytokines (→10.5).

Like protein kinases, mutated RAS proteins constitute a promising target for cancer therapy. They are overexpressed in some cancers, but more importantly, they carry mutations at very specific sites that lead to their constitutive activation in about 30% of human cancers overall (→4.3). In different cancer types either HRAS or KRAS are mutated in an almost exclusive fashion, providing another potential level to achieve specificity. Even normal RAS proteins may exert an oncogenic action in some cancers by transmitting signals essential for cell growth and survival from oncogenic receptors (→4.4).

The problem with targeting mutated RAS proteins by small-molecule drugs lies in the precise mechanism causing their oncogenic activation (→4.4). RAS proteins are over-active in human cancers because their intrinsic GTPase activity is decreased. Mutations at very specific sites block the interaction with GTPase activating proteins (GAP), thereby prolonging the state in which RAS proteins can stimulate downstream effectors such as the kinases RAF or PI3K (→6.2). So, it would probably be relatively simple to find drugs that inhibit RAS GTPase activity. However, it is more difficult to find compounds that promote GTP hydrolysis.

To circumvent this dilemma, another strategy was conceived to target oncogenic RAS proteins. RAS proteins are tethered to the inner face of the membrane by post-translational modifications that make them more hydrophobic (Figure 22.8). RAS proteins and their small GTP-binding protein relatives like RHO and RAC end in the amino acid sequence CAAX, where C is cysteine, A is an aliphatic amino acid (like alanine), and X is serine, methionine, glutamine, or cysteine. This carboxy-terminal tetrapeptide is recognized by protein farnesyltransferases that transfer a farnesyl residue to the cysteine thiol side chain. The terminal three amino acids are subsequently cleaved off by a specific protease and the new terminal carboxyl group is methylated by a protein methyltransferase. Additionally, a palmitoyl chain can be added to a penulmitate cysteine.

Related proteins and a small fraction of RAS proteins are alternatively modified through geranylation by the enzyme geranyl-geranyl-transferase (GGT1). The substrates for these reactions, farnesyl-pyrophosphate, and geranyl-pyrophosphate, are ubiquitous intermediates of the cholesterol biosynthetic pathway. As a consequence of these modifications, the C-terminus of the protein becomes sufficiently hydrophobic to stick to the membrane, whereas the unmodified protein is cytosolic and, importantly, inactive.

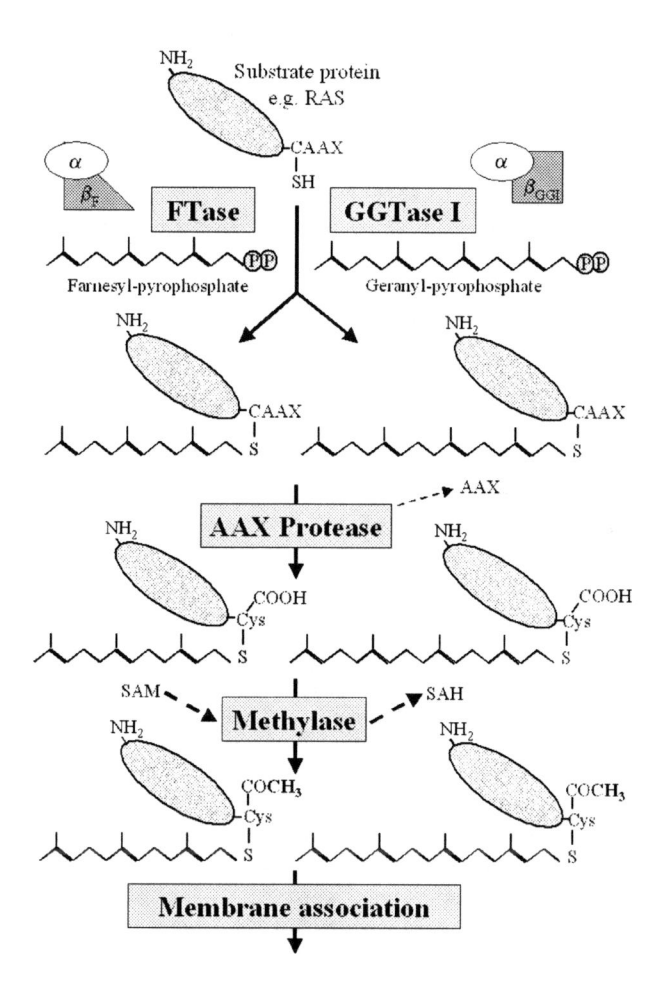

**Figure 22.8** *Posttranslational modification of RAS and other proteins*
See text for further explanations. Note that -CAAX stands for -Cys-Ala-Ala-any amino acid.

This modification, then, can be targeted by drugs. Actually, blocking cholesterol biosynthesis by statins at the level of hydroxymethyl-glutaryl-coenzyme A reductase, which is a common treatment for hypercholesterolemia, may have some effect on this modification as well. A more specific target for interfering with RAS function is, of course, the enzyme farnesyl transferase. A weak point of this strategy could be that it is not specific for oncogenic RAS, because normal RAS also depends on the same modification for its function. This argument can be turned around by arguing that 'upstream' oncogenic alterations such as overactivity of receptor tyrosine kinase may depend on non-mutated RAS proteins and might be blocked by the same drugs. So, the initial idea of targeting an altered protein has in reality transformed into another approach based on the 'addiction hypothesis' (cf. Figure 22.7), because the succes of the strategy hinges on the question whether cancer cells are significantly more dependent on RAS functions than normal cells.

Two basic types of farnesyl transferase inhibitors (FTIs) are available that were obtained by the two principal strategies now used in drug development, i.e. rational design based on the known structure of a target protein and its substrates and random screening of libraries of synthetic and rational compounds for inhibition (in this case) of the target activity. Rational design of FTIs was based on the structure of the tetrapeptide known to be essential for substrate recognition. This tetrapeptide was modified until optimal inhibitory specificity was obtained. Additional modifications were in this case necessary to improve solubility and uptake. In the second strategy, screening of natural compounds libraries yielded non-peptide compounds (Figure 22.9), which served as 'leads', i.e. their basic structure was varied until pharmacological requirements were reasonably met.

Several such compounds have been tested in clinical trials. They were found to be moderately active against some acute leukemias and some carcinomas. Efficacy has been seen against pancreatic carcinoma, in which KRAS mutations may be most prevalent among all human cancers, but in general no correlation was evident between response to FTIs and the presence of a mutated RAS protein. Moreover, some FTIs display severe side effects, e.g. in the gastrointestinal tract, which are not too different from those of traditional cytotoxic chemotherapy.

Since the idea of using farnesyl transferases as therapeutic targets was conceived, many other proteins besides RAS and related small GTPases have been shown to become farnesylated. The list of such proteins even includes major structural proteins of the cell such as lamins. It is therefore not at all clear that the anti-cancer action of FTIs results from their interference with RAS function. In some cancers, they induce apoptosis, which could be due to decreased activity of the PI3K pathway as a consequence of RAS inhibition (→6.3). So, perhaps, cancers in which RAS is important for PI3K activation may be treated with these compounds. In most cancers, however, cells arrest at the G2→M border or in prometaphase, because they cannot form mitotic spindles. It is not understood, how this effect might be caused by RAS inhibition, and so it is considered to more likely result from the inhibition of a distinct protein. It is also not clear, whether the limiting toxicity of the FTIs is caused by inhibition of the normal functions of RAS proteins or by interference with that of other farnesylated proteins.

# CANCER THERAPY

The greatest efforts in the development of novel anti-cancer drugs so far have been directed at receptor tyrosine kinases (RTKs, Table 22.4). Overexpression or mutation of membrane proteins from this class contribute to the growth of many human cancers (→4.3). Specifically, members of the ERBB family are overexpressed in a wide range of metastatic carcinomas, whose treatment constitutes one of the major unsolved problems in cancer therapy (→22.1). The RTK superfamily also comprises proteins such as IGFRI, MET, and FGFR3 implicated in the causation of, e.g., cancers of the liver (→16.2), the kidney (→15.3), and the urinary bladder (→14.3). In addition, the PDGFR and receptors for VEGF such as FLT1 are essential for endothelial cell growth and angiogenesis (→9.4) in a wide range of cancers from different tissues.

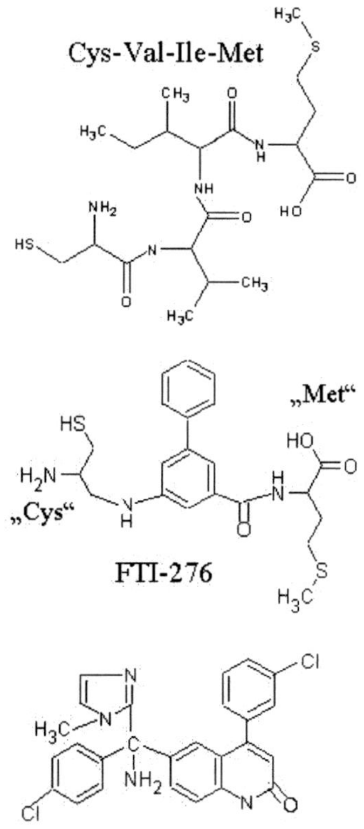

**Figure 22.9** *Peptide and non-peptide inhibitors of farnesyltransferases*
Inhibitors of farnesyltransferase were modelled after the tetrapeptide shown at the top, which resembles the CAAX sequence at the N-terminus of RAS proteins. A designed and optimized synthetic peptide analog is shown at the center and an unrelated heterocyclic compound identified as an inibitor by chemical library screening is depicted at the bottom.

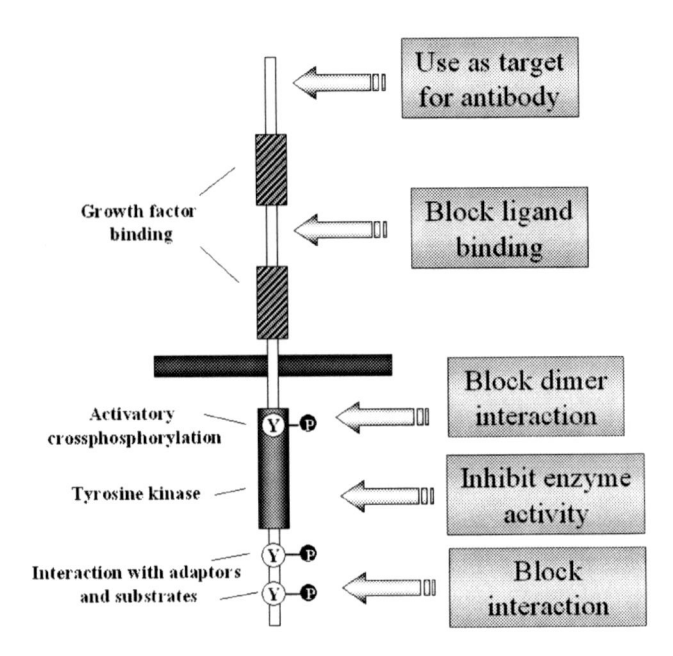

**Figure 22.10** *Drugable activities of receptor tyrosine kinases*

All RTKs are located at the cell membrane, making them accessible not only to drugs that can penetrate into the cell, but also to those acting on the outside of the cell and even to antibodies (Figure 22.10). Indeed, antibodies directed at the extracellular domains of several receptor tyrosine kinases appear to provide valuable drugs. The function of RTKs, with very few exceptions such as ERBB3 (→18.4), depends on their intracellular tyrosine protein kinase activity. This activity is well drugable, and tyrosine kinase inhibitors in general are designated as tyrphostins.

In contrast, the ligand binding activity of the receptors located in their extracellular domain has proven a more difficult target. Antibodies are themselves proteins which recognize epitopes on other proteins consisting of several amino acids and their modifications. Binding of an antibody or a growth factor ligand to a receptor tyrosine kinase is a protein-protein interaction. This kind of interaction typically involves a large number of comparatively weak interactions (often hydrophobic or van der Waal interactions) across a relatively large surface. The binding of an enzyme substrate or inhibitor depends instead on a small number of stronger and individually more specific binding interactions (often ionic bonds or hydrogen bonding in addition to hydrophobic interactions, or even covalent bonds). In general, protein-protein interactions are therefore relatively difficult to inhibit by small molecules. For this reason, antibodies are superior to small-molecule drugs for inhibition of ligand binding to growth factor receptors. These structural requirements also provide a plausible explanation why almost all inhibitors of receptor tyrosine kinases found so far by screening approaches inhibit binding of ATP, but not of protein substrates.

While several antibodies to RTKs have proceeded to advanced stages of clinical development, the only one already widely employed in clinical routine is trastuzumab (alias 'herceptin'), a recombinant, humanized antibody directed against ERBB2. Although it has been tested in several other malignancies, it is routinely used mainly in the treatment of breast cancer (→18.5). Like imatinib in its application for the treatment of CML and GIST, trastuzumab is paradigmatic, because the rationale for its application differs from that of traditional cytostatic drugs. Herceptin is only prescribed against a specific subset of breast cancers, which is defined by a molecular marker, viz. overexpression of ERBB2 with amplification of the gene. So, while the administration of cytotoxic chemotherapy was contingent on histopathological parameters, tumor grading and staging, the application of trastuzumab therapy is dependent on the classification of the cancer as ERBB2+ (and ER-). This principle is likely to be extended to many of the target-directed novel therapies currently under development.

Administration of trastuzumab is clearly beneficial for the group of breast cancer patients with ERBB2+/ER- metastatic cancers, whose prognosis is in general dire. However, while trastuzumab extends survival in these patients and improves their quality of life, it is not a miracle drug that might lead to a cure. Administered usually in combination with a cytotoxic drug such as adriamycin, it induces apoptosis and growth arrest in many cancer cells, but does not stop their growth entirely. Moreover, some patients with ERBB2+ cancers do not show remissions or even stable disease. An important area of research is therefore to determine the precise mode of action of this therapeutic antibody in human patients as well as the mechanisms underlying therapeutic failures.

Since in the age of robotized high-throughput assays it is fairly straightforward to screen 10,000s of compounds for their ability to inhibit a tyrosine kinase, a multitude of inhibitors for receptor tyrosine kinases are now available for research purposes. However, very few are presently used in the clinic. This is, of course, because use of a drug in human requires careful testing (Table 22.5).

Many tyrosine kinase inhibitors are now investigated in phase I or phase II studies. In phase I studies, the dose tolerated without adverse effects is established in a small number of individuals. In addition, it is determined whether the drug actually reaches levels that are sufficient to inhibit the intended target. For instance, its level can be quantitated in serum or in leukocytes and, if possible, target enzyme activities or the state of their substrates are measured. If phase I studies are performed in patients, which is the rule with cytotoxic drugs, some indication of efficacy can be gained.

However, determination of efficacy is really the formal aim of phase II studies. These involve a larger number of patients. Therefore, they provide information on whether the drug induces complete or partial remissions or at least delays the further progress of the cancer ('stable disease'). They also reveal a fuller range of the side effects to be expected. Due to differences in drug metabolism and general constitution, susceptibilities for adverse effects vary widely and in some cases side effects become only apparent after a large number of patients have been treated (→21.5). For instance, a substantial number of novel drugs, including tyrosine kinase inhibitors, could not be further developed because they interfere with a specific cardiac ion channel, causing a state called 'long Q-T'. This particular complication is, as a consequence, now routinely evaluated in the preclinical phase (or latest in phase I) to avoid later disappointments. Of course, novel kinds of difficulties may arise with every novel drug.

If phase II studies have been successful, with reasonable efficacy and safety, phase III studies are initiated with large numbers of patients, in a double-blinded set-up. Depending on the tumor type, these studies in particular can take a long time. For instance, the period needed to determine whether a new drug aimed at preventing the development of androgen-refractory prostate cancer (→19.2) indeed does so, is estimated as 5 – 10 years.

A drug can be introduced into general use following phase III. However, its efficacy and safety continues to be monitored for quite a while. This is considered phase IV. Its purpose is to detect adverse reactions in smaller subsets of the population, e.g. due to genetic polymorphisms in drug metabolism (→21.5) and to define even better which patients exactly benefit from use of the drug.

Before cancer drugs can be entered into these phases of 'clinical development', they have to be optimized in a 'preclinical' phase. Part of the task is biochemical characterization. The specificity of an inhibitor has to be determined by measuring its $K_i$ for a variety of kinases, including non-tyrosine kinases. As a rule of thumb, $K_i$s of suitable inhibitors are in the nanomolar range for the target kinase, while other

**Table 22.5.** *Stages in drug development*

| Stage | Study subjects | Purpose |
| --- | --- | --- |
| Preclinical | none (cell and animal models) | activity against target and disease, optimization of pharmacological properties, prediction of side effects |
| Phase I | tens | establishment of tolerated dose and dose efficacious against target |
| Phase II | tens to hundreds | establishment of efficacy (often response), observation of frequent adverse effects |
| Phase III | tens to hundreds (comparative and double-blinded, often long-term) | establishment of efficacy (often long-term effect), observation of frequent adverse effects |
| Phase IV | populations (and/or selected subgroups) | further optimization of administration mode (and sometimes dose), monitoring for rare adverse events and interactions with other therapies |

kinases are inhibited by micromolar or higher concentrations. Very few tyrosine kinase inhibitors are specific for one or a few kinases, those in clinical use like imatinib (STI571, Gleevec), gefitinib (Iressa, ZD1839), or OSI-774 are among them. The biochemical explanation for this problem is that they all bind to the ATP binding site, which is relatively conserved between many enzymes and even more so within the superfamily of receptor tyrosine kinases.

Another step in preclinical development is optimization of the pharmacological properties of the inhibitor drug. Medicinal chemists wield a large repertoire of modifications of a small molecule affecting its solubility, its ability to pass through cell membranes, its stability against metabolic degradation, its binding to carrier proteins, and its half-life in the patient and in the tumor overall (Figure 22.11). The drug entered into a phase I trial may look quite different from the lead compound emerged from a high-throughput kinase inhibitor screen and even from the compound that is used in laboratory research to inhibit a particular kinase.

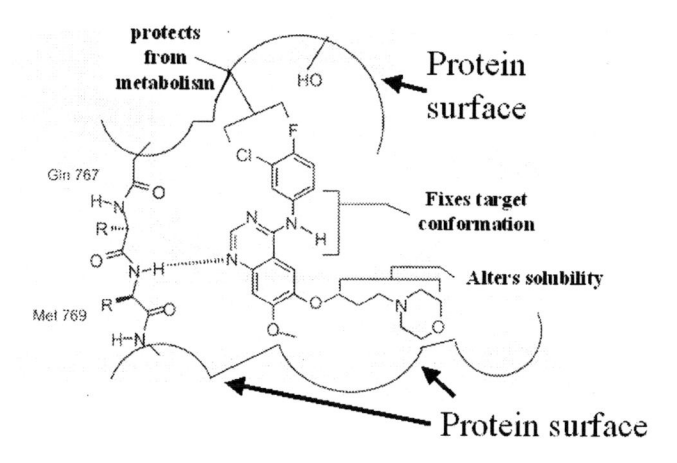

**Figure 22.11** *Optimization of the pharmacological properties of a tyrosine kinase inhibitor* Gefitinib in the ATP binding pocket of the EGFR tyrosine kinase. The purposes of several modifications made on the original lead compound are indicated.

Preclinical development of cancer drugs further comprises their testing in cell culture and animal models. Cell culture models include established cancer cell lines[19] and increasingly often primary cultures of cancer and normal cells. Typical animal models are xenografts of human cancers or cancer cell lines in mice, but also transgenic or gene knockout animals. For some cancers, animal models can be used in which a cancer similar to that in humans arises spontaneously or can be induced by a carcinogen. For instance, bladder cancers that are similar to those in humans (→14.1) can be induced by chemical carcinogens in rats or dogs. Liver cancers, likewise, can be induced by chemical carcinogens in rats or mice. However, while

---

[19] The NCI at Bethesda, MD, USA, maintains a standard set of such cell lines.

these are similar in some respects to human cancer, they do not mimic well the etiology of human liver cancers through the stage of cirrhosis (→ Box 16.1).

Cell culture and animal models are absolutely necessary for the development of new therapies. However, in many cases, novel therapies looked extremely promising in preclinical development, but did not live up to that promise in the clinic. This is also true to a certain degree for tyrosine kinase inhibitors. This criticism extends even to immunotherapy and gene therapy approaches. It is a very important aim of current research to understand the causes of such discrepancies and to establish better models and criteria for prediction of therapeutic efficacy in humans.

This is a complex matter, but some factors are evident. (1) The growth fraction of many 'real' metastatic carcinomas in humans is small, whereas that of model cell lines and xenografts is larger. This can hardly be avoided, since one would not want a preclinical experiment to extend over several years. Nevertheless, this difference is at least partly responsible for the differential effect of novel target-directed drugs in models vs. in patients, as it was for the same difference seen with 'classical' cytotoxic drugs. (2) Human cancers are very heterogeneous and the cell lines and xenografts used as models are at best representative of a subset of each cancer type. In some cancers, e.g. prostate carcinoma, they may even be exceptional, as most cancers do not grow in culture or as xenografts. (3) Animal models may not reflect certain aspects of human cancers. This may be particularly true for rodent tumors, which are most widely used at present. Only some of the differences are understood at the molecular level, such as the differences in the regulation of cellular senescence, telomeres and the *CDKN2A* locus (→7.4).

Gefitinib (Figure 22.11) or ZD1839, which is marketed as Iressa, is a relatively specific inhibitor of the EGFR. It was developed for application in metastatic carcinomas, exhibiting excellent efficacy in preclinical models. Indeed, the compound has shown significant activity in several cancers that so far had defied all attempts at treatment, viz. several types of recurrent and metastatic carcinomas. Adverse side effects were often moderate and rarely exceeded grade II. Not unexpectedly, they occurred in the skin and gut, where the EGFR is thought to mediate signals for tissue renewal. Indeed, in skin samples from treated patients, autophosphorylation of the EGFR was largely blocked and likely downstream effectors of EGFR activation followed suit. AKT phosphorylation was likewise decreased and $p27^{KIP1}$ became induced (→6.3).

Still, the therapeutic benefits of the drug were moderate. Responses were seen in only a fraction of the patients, and the treatment usually resulted in stable disease rather than remissions. In general, overall survival was only slightly improved, if at all. So, contrary to hopes the drug has proved to be at best palliative rather than curative. Most unexpectedly, neither the presence nor the absence of responses to gefitinib were found to correlate with the expression or activity level of the EGFR in a particular cancer. Present evidence suggests that the cancers responding to the drug are those in which the EGFR is activated by mutations rather than by overexpression. If confirmed, this observation would allow a pre-selection of the patients to be treated with gefitinib.

There are several potential explanations for the unexpectedly moderate success of this novel drug, which are being explored. The most likely one is that human cancers are even more heterogeneous than previously assumed. The most worrying explanation is that the well-established over-expression of the EGFR in many metastatic cancers does not really have the presumed significance, and is not as essential for growth and survival of the cancer cells as one might have hoped.

In conclusion then, the novel cancer drugs developed against carefully selected targets are still largely at the beginning of their application in the clinic. Limited successes have been achieved, but with the exception of imatinib in chronic phase CML and some GISTs, they have not been greeted unanimously as breakthroughs. Experience with the drugs already used in the clinic and those currently or previously tested in phase I and phase II studies will certainly be helpful to generate better drugs by this relatively new approach. After all, the first generation of 'novel' drugs was based on the understanding of cancer molecular biology of the early 1990's which today in retrospect many would consider as quite naive. Also on the positive side, drugs directed against specific targets in cancer cells have in general shown lower toxicity than the previous generation of cytotoxic drugs. What pharmacologists label ADMET parameters (for absorption, distribution, metabolism, excretion, and toxicity) remain challenges for the new drug generation as well. Most importantly, this first generation of novel target-oriented drugs has made it very clear that a thorough understanding of the molecular biology of cancers is a prerequisite to generate further therapeutic drugs that are at least as successful as imatinib.

## 22.5 NEW CONCEPTS IN CANCER THERAPY: IMMUNOTHERAPY

Immunotherapy is arguably the most elegant concept in cancer therapy. After all, its central idea is to harness the body's own ressources against a cancer. Furthermore, diverse evidence indicates that immunotherapy might represent an extension of anti-cancer immunity that prevents many cancers from developing in the first place. (1) Various lymphomas, leukemias, sarcomas, skin carcinomas and cervical carcinomas occur with increased frequencies in patients with immunodeficiency diseases, e.g. in AIDS patients (→Box 8.1). (2) Stimulation of immune reactions by interleukins and interferons is routine in the treatment of several cancers, including chronic phase CML (→10.4) and metastatic renal carcinoma (→15.6). (3) Stem cell therapy by transplantation of hematopoetic precursor cells is thought to be active against leukemias (and even some solid tumors) through a graft-versus-tumor reaction. (4) There are a few documented cases of cancers which have spontaneously disappeared, and these are usually ascribed to a successful immune response.

Indeed, cancers as a rule do elicit an immune reaction (→9.5). Antibodies to cancer-specific antigens (→12.5) and T-cells directed against cancer cells are found in many cancer patients. The detection of such antibodies can even be exploited for cancer diagnostics. Tumor tissue contain almost invariably infiltrating immune cells, macrophages, neutrophils and even cytotoxic T-cells that can be shown to exhibit specificity against tumor cell antigens.

However, the spontaneous immune response can evidently not contain all cancers. In late stage cancers, the patient's immune system may simply become overwhelmed by the tumor mass spreading through the body. In fact, it may cave in completely and become as well incapable of coping with unrelated bacterial and viral infections.

Even at earlier stages of cancer progression, however, the patient's immune response to the cancer often appears inefficient or muted. Cancer cells evade the immune response by a variety of mechanims (→Table 9.4), e.g. secretion of TGFß1 (→6.7), down-regulation of FAS (→7.3), or the diminished presentation of antigen peptides and co-activator proteins at the cancer cell surface (→9.5).

Worse, in many cancers, the effect of the immune response is ambiguous. While cancer cells are being attacked, the destruction of the tissue structure associated with the immune response and the ensuing inflammation facilitate invasion and metastasis (→9.5). Some cancers may even respond to cytokines and chemokines secreted by immune cells by enhanced proliferation or migration. The CXCR4/CXCL12 system, e.g., may be involved in directing metastases to the bone (→9.6).

So, if immunotherapy is to be successful, it has to overcome the mechanisms that prohibit an efficient anti-tumor response by the patient's immune system (Table 22.6). This is likely what happens in those CML and renal carcinoma patients in which immunotherapy using IFNα or IL2, respectively, causes remissions. These cytokines 'awake' cytotoxic T-cells directed against tumor cells. Typically, the immune cells are already present at the site of the cancer, but are relatively inert. Cytokine treatment then may activate them, stimulating their proliferation and their ability to kill tumor cells.

Of note, not all patients respond to these treatments. This is very typical of immune therapy. In many immunotherapy trials, including those of novel therapies, some patients respond excellently, to the point of 'miracle cures', whereas others show no response at all, or only adverse reactions to the treatment. So, while the

Table 22.6. *Some established and novel immunotherapies*

| *Immunotherapy* | *Rationale* |
| --- | --- |
| Administration of cytokines (e.g., IL-2, IFNα) | Activation of cytotoxic T-cells |
| Ex-vivo-treatment or transfection of T-cells with cytokines or cytokine genes | Activation of cyotoxic T-cells |
| Adjuvant treatment (e.g., BCG) | Stimulation of endogenous immune response |
| Vaccination against tumor antigens | Generation or expansion of anti-tumor T-cells or B-cells |
| Ex-vivo generation or treatment of dendritic cells presenting tumor antigens | Activation of T-cells |
| Vaccination against tumor viruses | Prevention of cancers or elimination at early stages |

results of such trials are often encouraging, they get stuck at this point, because it is incompletely understood what is responsible for the variable responses. 'Proximate' parameters such as killing efficiency of CD8+ T-cells or the levels of cytokines in the tumor and serum can, of course, be ascertained, but why such cells become activated in one patient, but not in another, remains largely unclear.

If the lack of response can, perhaps, not be amended, it would at least be helpful to be able to predict which patients respond to a treatment. For instance, 15-30% of metastatic renal carcinomas show partial remissions after administration of a combined cytokine/cytostatic drug cocktail, while many more develop adverse effects resembling the symptoms of a severe flu. It would therefore be helpful to have some indication of which patients might benefit from the treatment, but this is not yet possible.

An alternative approach towards activating cytotoxic T-cells is to isolate them, activate them in vitro, and reintroduce them into the patient or directly into a tumor. This can be done in various ways. T-cells can be treated with cytokines. They can be incubated with professional antigen-presenting cells challenged with tumor extracts or specific antigens, which activate the T-cells by direct contact and by cytokines. In a combination of immunotherapy and gene therapy, they can be transfected or transduced with a gene encoding a cytokine, such as IL-2, which stimulates the activation and proliferation of cytotoxic T-cells. In fact, one of the first experiments in human gene therapy was exactly this one.

Meanwhile, professional antigen-presenting cells, dendritic cells, have become much better characterized and their crucial function in the activation of T-cells has become recognized. Specifically, their insufficient function in tumors was identified as another weak point in the spontaneous immune response to cancers (Figure 22.12). Moreover, dendritic cells can now be differentiated from monocyte precursors and manipulated in vitro. So, several newer attempts at immunotherapy have used dendritic cells isolated from the tumor or differentiated from blood precursors in vitro.

These dendritic cells can be exposed to tumor antigens in several different fashions. They can be exposed to specific antigens of the tumor (→12.5) or transfected with expression constructs for tumor antigens. They can be incubated with tumor lysates or with lysates from primary cultures of tumor cells. In still another approach, they can be fused with tumor cells. In each case, they are reintroduced into the patient, either alone or in combination with T-cells.

The procedure of treating dendritic cells in vitro with specific tumor antigens or tumor protein lysates resembles the process occurring in vivo during a vaccination. So, this is also considered a sort of vaccination. It is only one step from there towards using tumor lysates or antigens as vaccines in vivo. These vaccines must be supported by adjuvants to elicit a significant immune reaction. In fact, one such adjuvant, BCG, consisting of *Mycobacterium tuberculosis*, is used to prevent recurrences of localized bladder cancers (→14.1).

Vaccination against tumor antigens could also be considered for the prevention of cancers. The prime candidates in this regard are viral proteins in cancers caused by or associated with viruses. Vaccination against HBV appears indeed to be

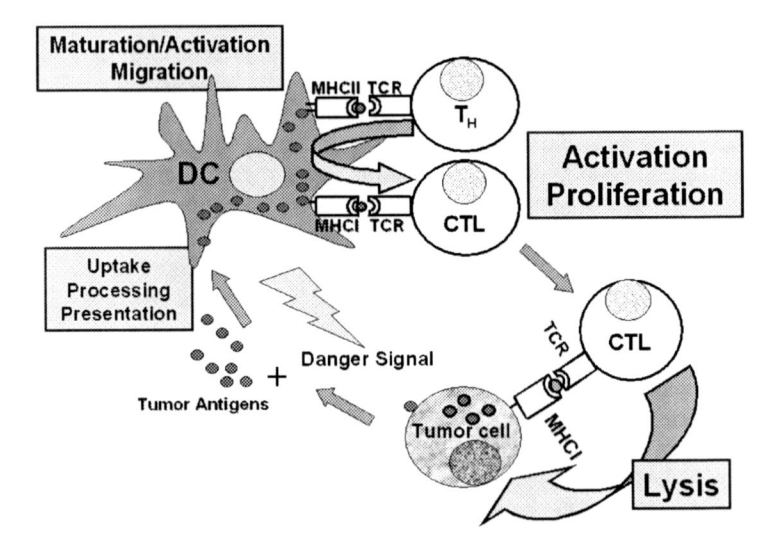

**Figure 22.12** *Function of dendritic cells (DC) in the immune response against cancers*
CTL: cytotoxic T-cell originating from a naive precursor ($T_H$); TCR: T-cell receptor; MHC: major histocompatibility complex (presenting antigens). Courtesy: Dr. R. Sorg

effective for the prevention of hepatocellular carcinoma (→20.2). Likewise, vaccination against HPV is expected to diminish the incidence of cervical and other cancers (→20.2). It is uncertain whether such vaccines would still be beneficial once cancer are established. In the case of HBV, in particular, immunotherapy might do more harm than good (→16.3).

In spite of occasional successes, overall, 'novel' approaches at immunotherapy have not made a major impact in the clinic, yet, compared to the more established (but also not veteran) immunotherapies using cytokines or stem cell transplantation. This may have several causes that cover the range from understandable to worrysome. Novel therapies are typically tried in patients with advanced stage cancers for whom no other treatments are available. These may be the very patients, in which the immune system is on the verge of collapsing, and the chances of succeeding are therefore minimal. At the end of the range of arguments, it remains possible that some cancers manage to turn even immune therapy to their advantage, in a similar fashion as they likely distort spontaneous immune responses (→9.5). An improved understanding of these interactions is certainly required, particularly in those cases in which immune therapy fails.

## 22.6 NEW CONCEPTS IN CANCER THERAPY: GENE THERAPY

If one considers cancer primarily as a genetic disease, it is consequent to treat it by therapy of the genes that cause cancer. Moreover, the fact that genetic alterations are essentially specific to the cancer cells ought to obliterate the problem of therapeutic selectivity that complicates other therapies, at least in theory. Since genetic alterations in cancer comprise the increased or misdirected activity of oncogenes as well as the lack or insufficiency of tumor suppressors, gene therapy can aim at inhibiting oncogenes or at restoring oncogene function. With ≈250 genes implicated as oncogenes or tumor suppressor genes in human cancers, there is no shortage of targets.

In fact, the range of gene therapy approaches is even broader than that (Table 22.7). Gene therapy does not have to be addressed directly at oncogenes or tumor suppressor genes, but can exploit their altered activities indirectly. The lack of the function of a tumor suppressor may be exploited to allow the replication of a cytotoxic virus in tumor cells or the increased activity of an oncogene may permit the selective expression of a toxin gene.

Genes that are not oncogenes or tumor suppressors themselves, but which oppose or mediate their effects, can also be employed. Genes that modulate the interaction between tumor cells and immune cells can be introduced into either cell type. Genes that protect sensitive populations like hematopoetic cells from toxic therapy can be introduced, thereby allowing the administration of increased doses of drugs or radiation.

So, ideas are plenty and even experiments, to the stage of human trials. In fact, the majority of gene therapy trials conducted in humans so far have been attempts at cancer therapy. In most studies the goals were to establish the safety of the procedure and to determine whether the therapeutic gene reached the target tissues, while proving the efficacy of the treatment was not an official aim. These

**Table 22.7.** *Gene therapies suggested for treatment of human cancers*

| Gene therapy | Example |
|---|---|
| Resubstitution therapy | reintroduction of *TP53* by adenoviral vectors into cancers with mutations in the gene |
| Oncolytic virus | infection of cancers with RB1 or TP53 inactivation by adenoviruses lacking E1A and/or E1B |
| Prodrug activation | transfection or transduction with Herpes simplex thymidine kinase followed by treatment with gancyclovir |
| Toxin expression by selective promoters | hypoxia-responsive promoters directing expression of diphteria toxin |
| Antisense oligonucleotides | thiophosphonate oligonucleotides directed against BCL2 mRNA |
| siRNA | double-stranded RNA directed against MYC mRNA |

experiments were thus comparable to phase I studies in drug development, and usually were labeled as such. In more than a hundred trials involving thousands of patients, few, if any have been cured. It has, however, become clearer which requirements gene therapy has to meet if it is to provide the intended breakthrough in cancer therapy and what the obstacles to its clinical application are.

The limited successes in clinical trials stand in striking contrast to the results obtained in gene therapy experiments with preclinical models, such as cell cultures and animals. There, the full panoply of gene therapy approaches has been tried, with often impressive results.

Resubstitution therapy is particularly effective in such models. Introducing tumor suppressor genes that are inactivated in a specific tumor cell line or in a tumor xenograft causes growth arrest or apoptosis in all tumor cells that receive the gene. To reach as many tumor cells as possible, viral vectors are typically employed in this type of experiment. An alternative is gene transfer by liposomes or dendrimers.

The most efficacious gene for the purpose of resubstitution may be TP53. (1) It often elicits apoptosis in tumor cells with chromosomal instability. (2) It is functionally inactivated in many different types of human cancer (→5.3). (3) It augments the effects of cytostatic drugs and radiation (→22.2). In clinical trials using this approach, adenoviral vectors have been employed, in which a TP53 expression cassette replaces a non-essential viral gene (Figure 22.13).

Trials by resubstitution therapy in humans have been little successful, for several related reasons. The central problem is that essentially only those tumor cells receiving the therapeutic gene can undergo growth arrest or apoptosis. It is difficult to administer sufficiently high amounts of the therapeutic virus, for technical as well

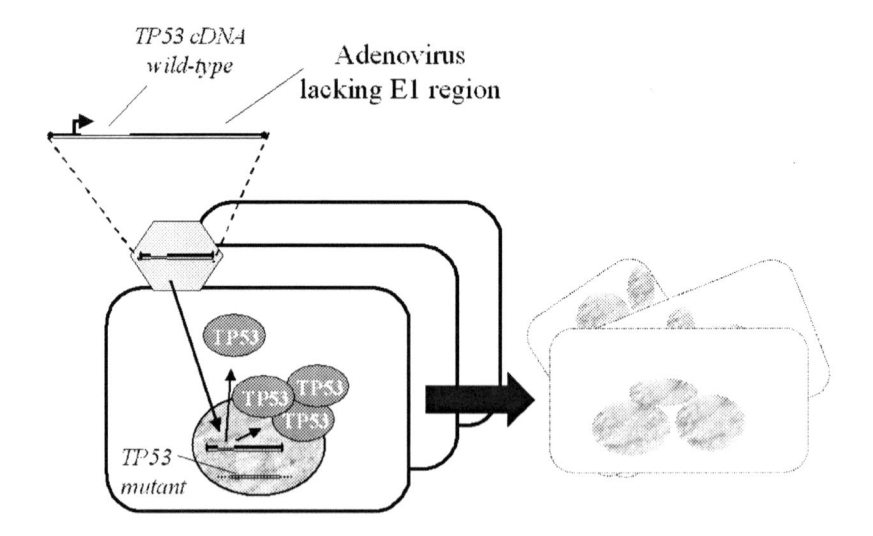

**Figure 22.13** *Resubstitution gene therapy of human cancers carrying mutant TP53 by reexpression of wild-type TP53 using an adenoviral vector*

as safety reasons. The concentration of the recombinant virus (the 'titer') that can be produced in cell cultures is limited, and administration of too high titers can elicit an lethal immune reaction in the patient.

Systemic administration of therapeutic viruses, especially of adenoviruses, is also complicated by the ability of the liver to filter and remove viral particles. Up to 90% of all adenoviruses administered are removed during a single pass through the liver. Getting viral particles into the right place is further hampered by mechanical factors due to the disturbed anatomy of solid cancers. As a rule, they have a suboptimal vessel system (→9.4), which is barely sufficient to provide nutrients, growth factors, and oxygen, and which limits the penetration of therapeutic agents. Even small molecule drugs may not always get optimally distributed into a tumor, least viral particles. The pressure ('turgor') within a tumor can be higher than in the surrounding tissue, further impeding the entry of larger particles.

Finally, the repeated administration of therapeutic viruses can lead to immune reactions that decrease efficacy and/or cause adverse immunological reactions in the patient. This effect may limit the applicability of adenoviruses. These are good vectors capable of accomodating quite large genes, infect many human cell types, and are otherwise relatively safe. The unmodified viruses cause at most a light, common cold-like disease when they naturally infect epithelial cells in the airways. Almost everybody has been exposed, and most humans harbor memory B-cells and T-cells directed against adenoviral antigens. These are activated when the viruses are administered as a therapeutic agent and become a more severe problem with each repeated application or upon adminstration of high doses.

For such reasons, the use of recombinant adenoviruses carrying tumor suppressor genes is limited to local administration, for the time being. This could still be useful in those cases, where local tumor growth presents the main problem, but surgery or irradiation are impossible, too risky, or disfiguring. For instance, aggressive brain tumors, glioblastomas, spread within the brain tissue and are often impossible to remove without serious damage to the brain. However, since they metastasize relatively late, local treatment by gene therapy is an option. Similarly, carcinomas recurring in the mouth and throat are often difficult to excise without compromising breathing, swallowing, or speaking. There are thus useful and important applications for this type of gene therapy, but its use in the treatment of systemic disease remains a remote possibility, unless fundamental innovations are made.

Some of these problems are circumvented by a related approach using 'oncolytic viruses'. Tumor suppressor proteins like RB1 and TP53 not only prevent the carcinogenic effects of tumor viruses, but also impede their replication. Specifically, the replication of DNA viruses, including papovaviruses (SV40), papillomaviruses (HPV), and adenoviruses is inhibited by these tumor suppressor proteins. Therefore, these viruses contain proteins that in turn block the function of RB1 and TP53 (→5.3). In SV40, the large T antigen is responsible for both, while HPV and adenoviruses each contain two separate proteins that inactivate RB1 and TP53, viz. E6 and E7 in HPV (→Box 5.1), and E1A and E1B in adenoviruses.

The large T protein of SV40 is indispensible because it also functions directly in viral transcription and replication. In contrast, adenoviruses lacking E1B can still replicate and remain cytolytic. However, their replication is blocked in cells that contain functional TP53. Inactivation of TP53 function is common in human cancers (→6.6), expecially those metastatic carcinomas which represent a major problem in cancer therapy. So, the cells of these tumors ought to allow the replication of adenoviruses that lack the E1B protein directed at the tumor suppressors (Figure 22.14).

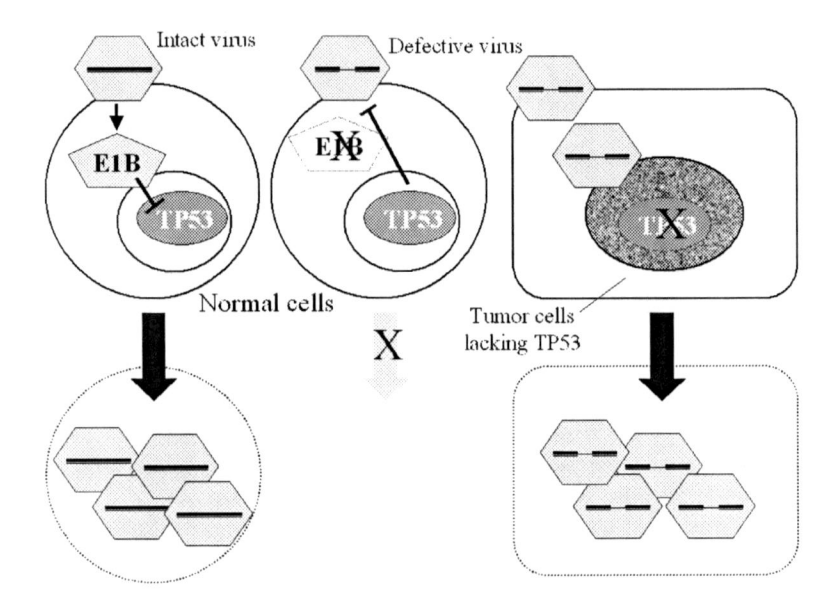

**Figure 22.14** *Selective replication of an adenovirus lacking E1B in cancer cells with TP53 inactivation*

As such viruses remain competent for replication, they are supposed to spread within a cancer, lysing the tumor cells, but they should not be capable of replicating and lysing normal epithelial cells. Since replication-competent viruses are used in this approach, some problems arising in resubstitution therapy are circumvented, viz. the limited efficiency of gene transfer and the requirement for unrealistically high and dangerous virus titers. Other problems remain, such as an eventual immune reaction by the host. However, in this case, there is a chance that the immune reaction might actually be helpful. When it becomes effective some time after administration of the virus, it might be directed not only at the therapeutic virus, but also against the tumor cells in which it replicates. So, this approach could turn out to represent an immunotherapy in disguise.

In clinical trials, however, this elegant concept has not proven as efficacious as one might have hoped, although promising results have been seen. The precise reasons limiting the efficacy of this approach are under investigation. One factor limiting the effect of oncolytic as well as other therapeutic viruses is that cancer cells often express low amounts of receptor proteins used for viral attachment and entry, such as a membrane folate transporter protein (CAR) employed by adenoviruses. Non-viral methods for gene transfer suffer from similar or even worse problems.

Gene therapy strategies that depend on getting a therapeutic gene into each and every cell of a cancer may be generally unrealistic. It appears that a gene therapy strategy aiming at a cure rather than palliation must evoke a 'bystander' effect, i.e. tumor cells that do not receive the therapeutic gene must also be affected. Eliciting an immune reaction against tumor cells by the replication of a cytolytic virus is one example of a bystander effect. In this case, one hopes that the immune response would be directed at a wider range of antigens, not only those provided by the virus, but also some specific to the tumor cells.

There are more explicit strategies to achieve bystander effects. In one approach employed in several variations, a gene encoding an enzyme that activates a prodrug is introduced into cancer cells. The prodrug is then activated in successfully transfected cancer cells, from which it diffuses to neighboring cells in the tumor. In this fashion, it would kill cancer cells and stromal cells needed for tumor growth, such as endothelial cells. The thymidine kinase of the herpes virus HSV has been used in early trials of this approach. Other than cellular thymidine kinases, this enzyme phosphorylates the nucleoside drug gancyclovir which thereby becomes capable of inhibiting DNA replication (Figure 22.15). Neighboring cells are affected, if they are linked to the successfully transfected cancer cell by gap junctions which allow the passage of gancyclovir nucleotides. Such cells would

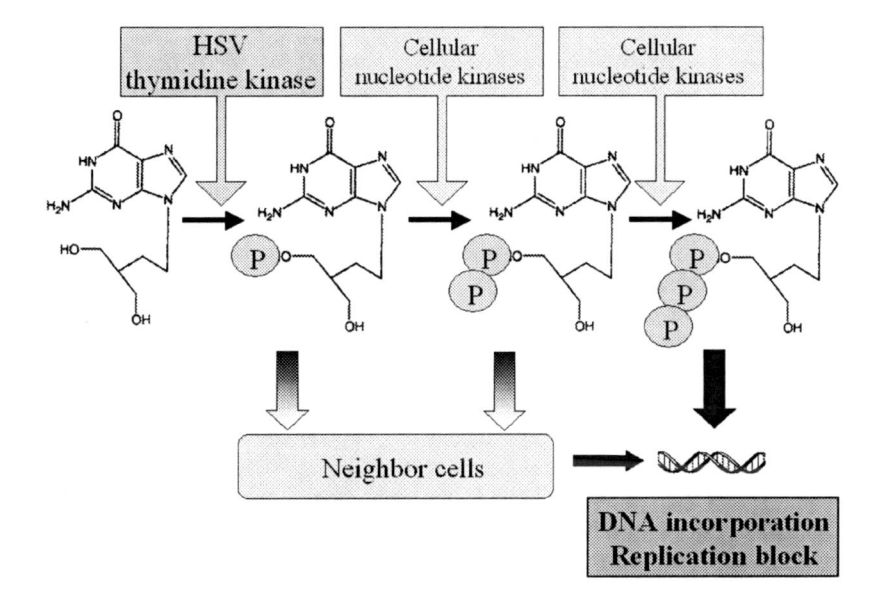

**Figure 22.15** *Activation of the prodrug gancyclovir by the HSV thymidine kinase*

most likely be also cancer cells, because gap junctions are typically formed in a homotypic fashion. However, in most cancers, gap junctional communication is down-regulated (→9.2), which limits the distribution of the active drug. Newer trials therefore employ enzymes that yield better diffusible and more active drugs, e.g. cytidine deaminase, which catalyzes one step in the activation of the prodrug capecitabine to 5-FU.

In this approach, the selectivity of the treatment depends on delivery of the therapeutic toxic gene to the correct cell or on its selective expression there. Again, several strategies are being explored. For instance, the expression of tumor antigens on the cell membrane can be used to target accordingly engineered recombinant viruses. Within a cancer cell, selectivity of expression of a toxic gene can be achieved by using promoters that respond to specific alterations in the tumor cell or to conditions prevailing in the cancer. For instance, promoters responsive to hypoxia, i.e. to HIF factors (→15.4), have been designed. These may be particularly useful in clear-cell renal carcinomas in which these factors are constitutively active (→15.4), but they are certainly not limited to this tumor type.

Other selective promoters exploit the lack of a tumor suppressor that represses them or respond to the oncogenic activation of a cancer pathway in a specific cancer. This is one way to exploit the activation of oncogenes for selectivity in gene therapy. For instance, the activation of β-Catenin in primary hepatocellular carcinomas (→16.2) leads to an increased activity of TCF-dependent promoters. So, such promoters could be employed in gene therapy of liver carcinomas, but also of colon cancers with the same genetic change. Increased promoter activity would also be expected in the large proportion of colorectal cancers in which activation of the WNT pathway is caused by loss of the APC tumor suppressor function rather than oncogenic activation of β-Catenin (→13.2). A concern in this application is, whether tissue stem cells dependent on WNT signals for their maintenance (→8.6) are also targeted and might become depleted. The promoter of the catalytic subunit of telomerase, hTERT, is also considered (→7.4), and ought to be applicable in an even wider range of cancers, albeit with the same caveat.

Gene therapy can be targeted at oncogenes and their protein products in a more fashion. For instance, a variety of techniques have become available to down-regulate the expression of specific genes. They are well established in laboratory research and are being developed for suppressing oncogene action in cancer patients. The expression of an oncogenic protein can be blocked at several steps from transcription to translation.

The *MYC* gene, e.g., has been targeted by several methods, although none of them has seriously moved beyond the preclinical stage. It is an attractive target, because it is causally involved in several lymphomas, including Burkitt lymphoma (→10.5), and at least important, if not essential for the growth of many solid tumors (→4.3). The *MYC* promoter region contains DNA sequences, e.g. purine-pyrimidine tracts and G-rich regions, which have the potential to form unusual DNA structures such as triplex or quadruplex DNA. So, this gene might be targeted specifically by oligonucleotides designed to stabilize these unusual structures and block *MYC* transcription[20].

In other genes, where no such unusual structures are available, transcription might be blocked specifically by attaching an intercalating compound like the anthracyclins adriamycin or daunomycin as an obstacle to transcription to an oligonucleotide specific for the oncogene sequence.

Several techniques are available to target the mRNAs of oncogenes by prohibiting their processing or translation, or by promoting their degradation. Antisense oligonucleotides directed against several oncogenes have proceeded to advanced clinical trial stages. They usually contain a modified backbone to protect them from metabolism and increase the stability of the duplex they form with a specific mRNA. Phosphothionates are frequently employed. Nevertheless, huge amounts of oligonucleotides are required to achieve active concentrations in humans, and they have to be highly pure to avoid side effects, especially immune reactions. Thus, their application in humans is not only limited by concerns about their efficacy.

It is hoped that treatment costs may be lower with siRNAs, which are moreover expected to be more efficacious. Antisense oligonucleotides appear to work mainly by blocking the processing and translation of the mRNA to which they bind, although additional effects may contribute. In contrast, siRNAs induce the degradation of their corresponding mRNA. This mechanism increases their efficiency. Furthermore, in theory and indeed in some cells, siRNAs elicit an amplification effect by which a relatively low concentration of siRNA induces an RNase-containing complex that degrades the corresponding mRNAs in the cell. Its activity may be maintained over a period of days to weeks. Thus, less siRNA at larger intervals may be required than in the case of anti-sense oligonucleotides. The concept of using siRNA for gene regulation in mammalian cells, least for therapy in humans, is still very new. It remains to be seen how straightforwardly it can be translated from the laboratory into the clinic.

Anti-sense constructs and siRNA directed at oncogenes are expected to be applied systemically. This raises two critical issues. (1) How can specificity for cancer cells be achieved? (2) Will the down-regulation of one oncogene suffice to inhibit the growth of a cancer? In essence, the answers hinge on how far the 'addiction theory' discussed previously in this chapter (→22.3) is correct. In other words: There is no doubt that the activation of oncogenes is important in the development of many cancers, but are cancer cells so much more dependent on these proteins than normal cells that there is a useful therapeutic window? And do advanced cancers remain dependent on the activity of oncogenic alterations, or can they evade therapy, e.g., by switching to a different cancer pathway?

In this regard, the example of imatinib use against CML (→22.4) can be interpreted as a pro or as a contra argument. A drug directed at an oncogene thought to be crucial for this cancer indeed induced lasting remissions in the chronic phase of the disease and even transient responses in many patients at the advanced blast

---

[20] In fact, this may be achieved also by small molecule drugs.

crisis stage. This proves that the oncogene product is essential in the chronic phase and remains at least important in some tumor that have progressed. Moreover, the tumor cells recurring after treatment with the drug often harbor genetic changes which alter the structure or expression level of this particular target oncogene protein. However, not all patients respond in the first place, especially among those with advanced disease, and not all cases of drug resistance can be explained by such specific mechanisms.

It seems therefore safe to argue that selection of oncogene targets for antisense and siRNA therapies, too, will require a very thorough understanding of the 'rewiring' of cancer pathways in a particular malignant tumor.

## 22.7 THE FUTURE OF CANCER THERAPY

The first wave of optimism which accompanied the development of novel drugs, immunotherapy and gene therapy for the treatment of human cancers has passed. There is a general feeling that the high hopes pinned on their development have not been fulfilled. A more guarded optimism now prevails and goals are set more realistic (as far as one can tell). In looking back on the 1990's, one realizes how naive some of the novel approaches were. However, this is the benefit of hindsight. After all, human oncogenes and tumor suppressors as such were only discovered during the 1980's and many were identified only in the 1990's. Many of their actions and interactions are not fully understood to this day. So, of course, if one wanted to act at all on the emerging knowledge, it would have to be on an incomplete basis. When dealing with a lethal disease of major importance, it is probably unavoidable that researchers, doctors and companies rush for solutions, once they appear on the horizon. One criticism that may be made with some right is that too often the difficulties ahead have been downplayed and the incompleteness of our understanding has been obfuscated in the rush for success. This may be the reason why now many involved feel that we have been taught a lesson on the complexity and diversity of human cancers which we are to heed in our future efforts at curing this disease.

Some consequences appear to have been drawn. Prevention and early diagnosis are seen as more important than they were in the prime of the enthusiasm associated with the development of novel therapies. Novel therapeutic approaches will be based on a more thorough understanding of the molecular biology of human cancers, acknowledging their diversity and complexity at an earlier stage of the development of therapies. The qualities of the established therapeutic methods, surgery, irradiation and cytostatic drugs, are probably better appreciated, which should faciliate the integration and combination of different types of therapies. For instance, gene therapy with oncolytic viruses has been combined with cytotoxic chemotherapy. Indeed, both 'traditional' and 'novel' therapy approaches appear to converge towards the concept of individualized therapy, which aims at taking the properties of each cancer and the constitution and wishes of each patient into account.

The implementation of individualized therapy will constitute new challenges for cancer diagnostics and classification by molecular techniques, as will any extension of cancer prevention and early detection programs. Education of clinicians and scientists will have to evolve to facilitate translational research and its application in the clinic. New options for the therapy, diagnostics and prevention of cancer will have implications for the organization of health systems, in the industrialized Western world as well as in the developing countries of the South. This will require political action and consensus decisions based on public understanding of complex issues. Providing an adequate and solid scientific basis for these developments remains a task that is as challenging as it is important.

## Further reading

Friedman T (ed.) The development of human gene therapy. Cold Spring Harbor Laboratory Press, 1999
Templeton NS, Lasic DD (eds.) Gene therapy: therapeutic mechanisms and strategies. Marcel Dekker, 2000
Stuhler G, Walden P (eds.) Cancer Immune Therapy: Current and Future Strategies. Wiley VCH, 2002
Prendergast GC (ed.) Molecular Cancer Therapeutics: Strategies for Drug Discovery and Development. Wiley, 2004

Bange J, Zwick E, Ullrich A (2001) Molecular targets for breast cancer therapy and prevention. Nat. Med. 7, 548-552
Rosenberg SA (2001) Progress in human tumour immunology and immunotherapy. Nature 411, 380-384
Vogelstein B, Kinzler KW (2001) Achilles' heel of cancer? (*i.e. p53 inactivation*) Nature 412, 865-866
Caponigro F (2002) Farnesyl transferase inhibitors: a major breakthrough in anticancer therapy? Anticancer Drugs 13, 891-897
Darnell JE (2002) Transcription factors as targets for cancer therapy. Nat. Rev. Cancer 2, 740-749
Hurley LH (2002) DNA and its associated processes as targets for cancer therapy. Nat. Rev. Cancer 2, 188-200
Johnstone RW, Ruefli AA, Lowe SW (2002) Apoptosis: a link between cancer genetics and chemotherapy. Cell 108, 153-164
Arteaga CL (2003) ErbB-targeted therapeutic approaches in human cancer. Exp. Cell Res. 284, 122-130
Blackledge G (2003) Growth factor receptor tyrosine kinase inhibitors: clinical development and potential for prostate cancer therapy. J. Urol. 170, S77-S83
Deininger MW, Druker BJ (2003) Specific targeted therapy of chronic myelogenous leukemia with imatinib. Pharmacol. Rev. 55, 401-423
Fei P, el-Deiry WS (2003) P53 and radiation responses. Oncogene 22, 5774-5783
Felscher DW (2003) Cancer revoked: oncogenes as therapeutic targets. Nat. Rev. Cancer 3, 375-379
Grünwald V, Hidalgo M (2003) Developing inhibitors of the Epidermal Growth Factor Receptor for cancer treatment. J. Natl. Cancer Inst. 95, 851-867
Millar AW, Lynch KP (2003) Rethinking clinical trials for cytostatic drugs. Nat. Rev. Cancer 3, 540-544
Nagrasubramanian R, Innocenti F, Ratain MJ (2003) Pharmacogenetics in cancer treatment. Annu. Rev. Med. 54, 437-452
Nevins J et al (2003) Towards integrated clinico-genomic models for personalized medicine: combining gene expression signatures and clinical features in breast cancer outcomes prediction. Hum. Mol. Genet. 12. Spec. No. 2, R153-157
Rothenberg ML, Carbone DP, Johnson DH (2003) Improving the evaluation of new cancer treatments: challenges and opportunities. Nat. Rev. Cancer 3, 303-309
Sebti SM (2003) Blocked pathways: FTIs shut down oncogenic signals. Oncologist 8 Suppl. 3, 30-38
Vlahovic G, Crawford J (2003) Activation of tyrosine kinases in cancer. Oncologist 8, 531-538.
Waxman DJ, Schwartz PS (2003) Harnessing apoptosis for improved anticancer gene therapy. Cancer Res. 63, 8563-8572
Eskens FALM (2004) Angiogenesis inhibitors in clinical development; where are we now and where are we going. Brit. J. Cancer 90, 1-7
Harris M (2004) Monoclonal antibodies as therapeutic agents for cancer. Lancet Oncol. 5, 292-302
Jordan VC (2004) Selective estrogen receptor modulation: concept and consequences in the clinic. Cancer Cell 5, 207-213
Wang S, El-Deiry WS (2004) The p53 pathway: targets for the development of novel cancer therapeutics. Cancer Treat. Res. 119, 175-187

# KEYWORD INDEX

B: box; Ch: chapter; F: figure; T: table,
page numbers in bold: definitions or closer treatment
In case of entries with several designations, the one most widely used in the literature is preferred; if an abbreviation is more commonly used than the full designation, this should be the main entry.
Note that gene and organism names are not italicized in the keyword index. Official gene names (as in the OMIM database) are used, with few exceptions.

## 1ff
3T3 assay 81, 118
14-3-3 103F, 104, 122, 130f
chromosome
  1 246T, 246, **247F**, 252, 311, 314, 331, 394
  2 82, 227, 311
  3 119, 231, **261**, 268, 299, 311, 315ff, 315F, 322f, 323F
  4 331
  5 274ff, 288B, 299
  6 224, 246, 268, 299f, 300F, 311, 331, 335, 395
  7 183, 246, 299, 303, 311, 313f, 321, 331, 335, 395
  8 81, 227, 246, 299, 311, 331, 331, 395, 430
  9 106F, 110, 121, 233, 260, 263f, 267, **301f**, 306B, 322, 395
  10 33, **34F**, 122, 268, 311, 322, 395
  11 83, 170ff, 202, 237, 246, 250ff, 251F, 299, 301, 331
  12 82, 246, 264, 301, 311
  13 95, 224, 246, 261, 299, 311, 331, 368, 397
  14 227, 311, 430
  15 237
  16 121, 137, 246, 250, 299, 311, 331f, 346, 395
  17 81f, 101, 106f, 231, 237, 246, 250, 267, 281, 299, 311, 315, 321, 322, 331, 367, 395
  18 281
  19 250
  20 299, 311
  22 182, 227, 232f, 246, 252

## A
A1 152T
abasic sites 51f
Abelson leukemia virus 233
ABL 66, 74T, 132, 224T, 233f, 236, 431, 465
acetylator phenotype 293, 405, 443ff
achondroplasia 304
actinic keratosis 260
acute lymphoblastic leukemia *see* ALL
acute myeloic leukemia *see* AML
adaptor protein 85, 128
'addiction hypothesis' 462, 463F, 468, 485
adenine nucleotide translocator 151
adenocarcinoma 13T, 206, 272ff, 383ff
adenoma 13T, 272ff, 328, 344, 348, 418
adenovirus 46, 108, 480ff, 480F, 482F
adherens junction 197f, 198F
adjuvant treatment 21, 272, 374, 378, 424f, 440
ADMET 475
ADPKD 311f, 312T
adrenal cancer 97, 316
adriamycin 471, 485
adverse reactions 444f, **454**, 464, 466, 474, 476ff, 481
aflatoxin 6, 6T, 56, 260, 333f, 339, 343, 353, 407
AFP 65, 269, 328
age 1, 9
aging 58, 65, 163, **165B**, 420
  syndrome 162, 165B
  theories 165B
Ah receptor 318
AIDS 192B, 227, 475
AIF (apoptosis-inducing factor) 151
AKT 74T, 88, 116T, 120F, **120ff**, 124f, 124F, 127, 130, 159, 474
alcohol 40T, 330ff, 336, 343, 352, 355B, 388, 409, 412f
aldehydes 9
algorithms (in therapy) 440ff
alkylation 52, 455
ALL 224, 224T, 225F, 432f
alopecia 454
alpha-fetoprotein 12
ALT (alternative mechanism of telomere stabilization) 162
ALV (avian leukosis virus) 72, 73F
Alzheimer disease 141
amino acids (aa) 206
AML 61, 224T, 225f, 236ff, 241, 432f
AML-EVI 236
ammonia 206, 349
amplicon 34f,
anaplastic 245f, 252
androgen 338, 363F, 389ff, 399, 423, 437
androgen receptor 40T, 173, 359, 360F, 362, **390ff**, 390F, 395T, 423, 461
anemia 217
anergy 211f
aneuploidy 15, 28T, 33, 106, 226, 236, 275, 281, 284f, 296, 331, 347

489

angiogenesis 17f, 65, 104f, 131, 186f, 186F, 196, **206ff**, 212, 217B, 310, 319f, 331, 334, 398ff, 469
angiogenic switch 207
angioma 316
angiomyolipoma 309
angiopoetin 208, 208T
angiostatin 208, 208T, 210
animal model 340B, 473, 480
anoikis 129, 200
anthracyclin 378, 441, 484
antibiotics 343, 349, 353, 355B, 408
antibody 210, 229, 325, 338, 378, 428, 430, 441, 446, 461, 463, 470f, 475ff
  humanized 378, 463
antigen 223, 483
antigen presentation 211
anti-hormonal (therapy) 21, 187, 207f, 208T, 217B, 382, **390ff**, 420, 420ff, 424, 437, 441, 453, 458, 461
anti-oncogene 109 see tumor suppressor
antioxidants 10, 353, 410, 413, 425
$\alpha_2$-anti-plasmin 203
$\alpha_1$-anti-protease 204, 206F
antisense 464, 485
AP1 86F, 124, 142, 184, 185F, 204, 361, 373
APAF 105, 131, 151F, 151ff, 158, 178T
APC 38T, 93T, 96, 109, 116T, 138F, 140f, **274ff**, 274T, 275F, 277F, 285, 288B, 298, 332, 347, 419T, 420, 434f, 484
AP (apurinic/apyrimidinic) endonuclease 49T, 52
APE1 52
APEXL2 52
APL (acute promyelocytic leukemia) 21, 182, 185, 223f, 224T, **237ff**, 432, 453, 459, 466
apoptosis 9f, 12, 18, 21, 63, 68, 80T, 88, 98, 104f, 117ff, 125, 135ff, **145ff**(Ch), 189, 202, 211, 217B, 223, 226, 231, 233, 238f, 258, 279, 281, 286, 320, 322, 358, 364, 391, 393, 456f, 457F, 462, 468, 480ff
  extrinsic pathway 150, **153ff**, 153F, 157
  intrinsic pathway 105, **150ff**, 151F, 158, 456
  resistance 331
apoptosome 151F, 152ff
apurinic site 50F
apyrimidinic site 50F
AR see androgen receptor
arachidonic acid 286, 286F
arbitrary PCR 26
aromatase 363F, 365, 374
ARNT 209, 318
aromatic amines see arylamines
arsenic 6, 6T, 8F, 292, 406f
arylamines 6, 6T, 8F, 9, 40T, 291ff, 292F, 404ff, 443

asbestos 8, 108
ascorbate see vitamin C
asparagine hydroxylation 319
aspirin 137, 286
assay requirements 433ff, 434T
asymetrical division 187, 279
ataxia telengiectasia 37, 39T, **65ff**, 67T, 417
ATM 39T, 40T, 42, **64ff**, 67T, 103, 103F, 108, 116T, 129, 131, 159, 161, 235, 365, 370, 370F, 444
ATP binding site 465, 471, 473
ATR 64, **66**, 103, 103F, 370
ATRA see retinoids
attenuation 76, 228
autocrine (factor, loop) 78, 139, 264, 321, 331, 398ff
autoimmunity 154
Axin 116T, 138F, 140f, 275ff, 277F, 332, 435
5'-aza-deoxy-cytidine 451f
azathioprine 444

# B
bab2 349ff
bacteria (carcinogenesis by) 9, 348ff
BAD 120F, 122, 152T, 158
BAK 150ff, 152T
BARD1 370
Barrett esophagus 355f
basal cell carcinoma of the skin see BCC
base adduct 50F, 56F, 365
base exchanges 27
base excision repair 48ff
basement membrane 16, 195, 199f, 202, 207, 272, 290, 386, 428
base misincorporation 48, 283
'battle of the sexes' hypothesis 170
BAX 103F, 105, 116T, 131, 150ff, 151F, 152T, 158, 284, 320
B-cells 135, 168, 210ff, 221ff, 226ff, 481
B-cell lymphoma 231
BCC 58, 97, 141, 143, 150, 190, 257ff, 258H, **262ff**, 280, 298, 306, 449
BCG 295, 477
BCL1 see CCND1
BCL2 80T, 83, 88, 105, 130, 150ff, 151F, 152T, 155, 158, 224, 224T, 393, 456
BCL2 family 150ff, 152T
BCL6 224T, 231
BCLW 152T
BCLXL 130, 135, 150, 152T, 155, 158
BCR 224T, 233, 431
BCR-ABL 33, 134, 225f, **232ff**, 234F, 235F, 431ff, 446, 460, 464T, 465
Beckwith-Wiedemann syndrome 171f, 246T, 250ff
benzene 8F, 40T, 406f, 406F
benzopyrene 6, 6T, 8F, 41, 56, 259,

BH3 domain 150ff
BHD 38T, 312T, 315
bHLH proteins 183f
bicarbonate 206
BID 151F, 152T, 153F, 155
BIK 152T
biological agents (in cancer therapy) 21, 451T, 453f, 459f
biotechnology 21, 378, 463
BIR domain 152
bisphosphonates 402
BK virus 46
bladder cancer 1f, 6f, 9, 19F, 20F, 44F, 85, 95, 110, 288B, **289ff**(Ch), 291H, 332, 395, 404ff, 424, 429, 446, 469, 473, 477
blast crisis 232ff, 465, 485f
blastema 244f
BLK 152T
bleomycin 62
BLM (helicase) 39T, 49T, 64, 239
Bloom syndrome 37, 39T
BMF 152T
BMI1 182, 188
BMP 131ff, 401
BOK 152T
bombesin 376
bone 17, 137, 187, 196, 214, 304, 362, 378, 385, 420, 453, 476
bone marrow 189, 199, 221ff, 233, 401, 407, 429
bone morphogenetic proteins *see* BMP
boundary element 31, 171
BPH 382f, 422
BRAF 79T, 83, 87, 116T, 119, 268
brain 46, 362, 402
brain cancer 1, 7, 42, 46, 61, 79T, 264,
BRCA 64, 417T
BRCA1 38T, 61, 66, 67T, 93T, 96, 361, **365ff**, 366T, 367F, 370F, 419F
BRCA2 38T, 39T, 40T, 49T, 61, 67T, 93T, 96, **365ff**, 366T, 367F, 369F, 370F, 395T, 419f
breakage-fusion-bridge cycles 35F, 36, 162f
breast 338ff
breast cancer 1, 21, 37, 82, 95, 158, 182, 197, 205, 213, 239, 278, 288B, **357ff**(Ch), 401f, 409ff, 429, 440, 446, 453, 458, 461, 471
  hereditary 38T, 61, 67T, 93T, 93, 96, 122, 365ff, 417ff, 417T
bulky adduct 49T, 452 *see also* base adduct
Burkitt lymphoma 33, 43, 46, 224ff, **226ff**, 227H, 336, 430f, 430F, 461, 484
Burt-Hogg-Dubè syndrome *see* BHD
bypass repair 49T, 57, 260
bystander effect 483ff

C
cachexia 14, 16
CAD (caspase-activated DNase) 155
cadherins 128, 198f, 201T, 214, 400
cadherin switch 198
cadmium 6, 6T, 70
caffeine 456
cagA 349ff, 350F
CAK (CDK activating kinase) 98
calcitonin 14
calcium (ions) 14, 127, 151, 198, 399, 402
calcium-dependent protein kinase 141
CAM 201T,
cAMP 199, 266
cancer
  age dependency 39, 41, 44F, 67, 244, 273, 385ff, 386F, 421
  diagnosis 17ff, 197, 226, 328, 342, 397, 403, 415ff, **427ff**(Ch), 475, 487
  incidence **1ff**, 44F, 94, 165, 257, 273, 291, 308, 328, 342f, 342F, 364, 385, 407f, 414, 421, 449, 450F
  mortality **1ff**, 328, 364, 409, 421, 449, 450F
  prevention 3, 5, 137, 353, **403ff**(Ch), 449, 477f, 486
  susceptibility 395
  therapy 378, 393, 396, 403, 427ff, 439ff, 442ff, **449ff**(Ch)
cancer gene 43
cancer of head and neck *see* SCC
cancer pathways 83, 110, **113ff**(Ch), 190, 380, 453, 460, 462
cancer predisposition 37, 43f, 66ff, **67T**, 179, 266, 347, 365, 372, 395T, 416ff, 429
cancer syndrome 92ff, 137
cancer testis antigens 12, 177, 269, 325, 461
capecitabine 484
capillary 196, 207ff, 310
CAR 483
carbonic anhydrase (CA9/12) 320, 322, 325
carcinoembryonic antigen *see* CEA
carcinogenesis 5ff, 62, 68ff, 157, 262ff, 278, 328ff, 340B, 352f, 400
carcinogens 68ff, 256f, 299, 352f, 404ff, 473
  biological 5ff, 336ff, 348ff, 407f
  chemical 5ff, 62, 291ff, 333f, 340B, 404ff, 409
  complete 5, 259
  endogenous 5, 9, 364, 404, 409
  physical 5ff, 256ff
carcinoma **13T**, 79fT, 131ff, 194ff, 260ff, 262ff, 272ff, 308ff, 327ff, 341ff, 357ff, 433ff, 439ff, 449
  in situ 19F, 296, 428
  metastatic 4, 474

CARD domain 152
caretaker 43f, 93T, 110, 284, 366
carotene (carotenoids) 69, 388, 410, 412, 425
caspase 151F, 151ff, 153F, 211
  substrates 156T, 157ff
catecholamines 316
catenins 198F, 198f, 199F, 331
β-Catenin 116T, 122, 128, 138f, 140f, **275ff**, 277F, 298, 317f, 331ff, 333F, 340, 345, 392, 461, 484
CBF (CSL) 142
CBL 85, 183, 235
CBP 239
C/BPA 224T, 225
CCND1 80T, 83, 88, 224T, 278, 311, 332ff *see also* Cyclin D
CCND2 80T, 88 *see also* Cyclin D
CDC2 (CDK1) 98, 99F
CDC25 98, 124F, 124f
CDC42 117
CD9 201T, 202
CD19/20 226
CD34 221
CD44 201T, 202, 397
CD95 *see* FAS
CDH1 38T, 96, 178T, 200, 299, 352 *see also* E-Cadherin
CDK2 97ff, 98F, 99F, 103, 104F, 124F, 124f
CDK4 80T, 82f, 88, 97ff, 98F, 99F, 124F, 124f, 267, 297, 318
CDK6 99F
CDK7 *see* CAK
CDK (protein) inhibitors **98ff**, 100T, 116T, 148, 159, 162ff, 188, 281
CDKN1A/B 98, 100T
CDKN1C 98ff, 100T, 171f, 172F, 178T, 246T, 299
CDKN2A 38T, 98ff, 100T, 104, 106F, 112B, 119, 125, 163f, 177, 178T, 226, 236, 260, 267f, 285, 298, 301, 302F, 322, 331, 355B, 474
CDKN2B 98, 100T, 178T, 226
CDKN2C/D 98, 100T
CDX1/2 347f
CEA 12, 20, 269, 436
C/EBPα 224T, 225
C/EBPε 239
celexocib 286, 418
cell adhesion 78, 117f, 126F, 135, 155, 159, 197ff, 230, 234, 278, 286, 313, 322, 397
cell-cell interaction 197ff, 244, 268
cell clone (in tumor) 15, 88, 173, 295, 299, 304, 458, 459F
cell culture 473f, 480
cell cycle 12, 87, 148, 230, 261, 265, 454ff
cell cycle arrest 104, 119, 143, 162f, 238

cell cycle regulation 80T, **97ff**, 98F, 99F, 109, **123ff**, 129ff, 239, 297ff, 318, 322, 331, 369ff, 398ff
cell death 10, 12, 148ff, 259
cell differentiation 10ff, 19, 137ff, 168f, **182ff**, 244, 279, 299, 347f, 364, 453, 459
cell fate 137, 300, 347
cell growth 14, 87, 230, 262, 280, 331
cell-matrix interaction 128, 197ff, 281
cell migration (motility) 87, 118, 127f, 198ff, 204ff, 207, 268, 295, 313, 452
cell proliferation 11ff, 19, 87f, 109, 116ff, 129ff, 135, 137ff, 146ff, 155f, 184, 202, 208, 223, 226, 230f, 241, 244, 262, 286, 299, 304, 315, 358ff, 361, 399, 451ff
cell protection 68ff, 165B, 365
cell shape 118
cell surface protein 201f, 201T, 221ff, 226, 428, 430, 476
cell survival 86, 88, 122, 280
cell turnover 286
cerebellum 316
cervical cancer 1, 7, 85, 304, 336, 408, 475, 478
cesium 7
chaperone 463
checkpoint 43, 50, 62ff, 68, 129ff, 161, 258, 299, 369, 455f, 455F
chemokine 135, 186, 210, **212ff**, 213F, 304, 476
chemokine receptors 212ff, 213F, 268, 476
chemoprevention 374, 418ff, 423ff
chemotherapy 3ff, 15, 21f, 51f, 61f, 68ff, 108, 129, 217B, 227, 232ff, 238, 240, 245f, 252, 272, 285, 296, 310, 324f, 328, 342, 378ff, 407, 413, 441ff, 449f, **450ff**, 465, 486
  resistance 70, 157, 159, 217B, 226, 236, 324, 456ff, 461, 486
CHK1/2 66, 103, 103F, 130f, 366, 370, 370F
chlorinated alkene 41
cholesterol 14, 141, 467f
choline 412
chromatin 97, 169ff, **179ff**, 228, 237, 239, 300, 361
  remodeling 181f, 239, 370
chromosomal aberrations 26, 28T, 62, 67f, 173, 217, 223, 236, 252, 282, 288, 310ff, 311T, 328, 368, 371, 427, 439
  numerical 28T, 34
  structural 28T, 34, 36
chromosomal instability 45, 65, 161f, 177, 284, 299, 311, 407, 455
chromosomal inversions 28T, 33, 250
chromosomal translocation 28T, 32, 62, **82f**, 88, 223ff, 224T, 226T, 227f, 228F, 233, 250, 430ff, 460

# KEYWORD INDEX

chromosomes 15
chronic myeloid (myelogenous) leukemia see CML
CIMP+ 285
CIN see chromosomal instability
CIP see CDK (protein) inhibitors
circulating tumor cells 438
cis-platinum see platinum
CKII (casein kinase) 103F, 104F, 130,
classification of cancers 17, 20, 252, 305, 310ff, 374, 378ff, 422ff, 428ff, 439ff, 471, 487
clastogen 40T, 407
claudin 198F
clear-cell renal carcinoma see renal carcinoma
clinical stage 17
clinical trial (study) 204, 464, 464T, 471ff, 472T, 480ff, 485
CLL 224T
clonality see cell clones
CML 16, 132, 190, 199, 223ff, **232ff**, 431f, 432F, 446, 460, 465, 471, 475f, 485f
co-activator (transcription) 169, 180, 183, 239f, 319, 345, 361, 363, 374, 390ff
co-carcinogen 5, 46, 72, 339, 343, 352f
Cockayne syndrome 58f, 59F, 67T
colitis ulcerosa 285f
collagen 198F, 202ff, 204T, 208
colon cancer (colorectal carcinoma) 1, 9, 12, 53ff, 79fT, 93, 96, 110, 122, 130, 132, 140, 150, 190, **271ff**(Ch), 273H, 293, 296, 297, 317, 332, 404, 409ff, 434ff, 455, 461, 484
computer tomography 17, 309, 427f
conductin 275
congenital 42, 244
connexin 199, 201T
co-repressor (transcription) 180, 239f, 361, 363
counseling 276
counterattack 137, **158**, 212
Cowden syndrome 37, 38T, 92, 93T, **122**, 274, 365f, 418
COX 136, 208, 212, 286f, 286F, 418
CpG dinucleotide 174ff
CpG-island 175ff, 175F, 438
CREB 266
crosslink (repair) 39T, 48T, 49T, **55ff**, 60F, 62, 370, 451
crosstalk 117, 128, 141, 376, 392
cryptic splice sites 30
CSA, CSB 49T, 56ff, 67T
CTCF 170F, 171
CTMP 116T, 120F, 121
CTNNB1 80T, 83, 109, 276ff, 285, 332ff, 347, 352, 434f see also β-Catenin
cullin 125, 317

curcumin 287
CXCL12 212, 214, 476
CXCR4 212, 214, 476
Cyclin A 99F, 339
Cyclin B 99F
Cyclin D 88, **97ff**, 98F, 99F, 116T, 122, 124f, 124f, 135, 262, 297, 303, 331, 361 see also CCND
Cyclin E **97ff**, 98F, 99F, 124f, 318, 441
Cyclin H 56, 98,
cyclin-dependent kinase see CDK
cyclooxygenase see COX
cyclophosphamide 6, 292
CYLD 116T, 137
cylindrimatosis 137
CYP 40T, 292F, 293, 294T, 406F
cytidine deaminase 484
cytochrome C 151F, 151ff
cytochrome oxidase 10B
cytogenetics 310ff
cytogenetic techniques 20, 26, 31, 33, 83, 235, 238, 430f, 430F
cytokeratin 20, 198F, 199, 214, 257F, 262, 291, 298, 428
cytokine 9, 21, 61, 78, 90B, 117, 119, 132ff, 135ff, 152, 153, 185, 205, 208, 211f, **222T**, 223, 259, 286, 325, 332, 335, 341ff, 466, 476
cytokine receptor 85, 90B, 126, 153ff, 268, 342ff, 376
cytosine 51F, 174ff
cytosine deamination 50F, 51, 176, 260
cytoskeleton 78, 80T, 86, 118, 120, 123, 127ff, 141, 155, 198ff, 198F, 234, 280, 345

## D

dam methylation 53
DAXX 239
DBCCR1 306B
death domain 154F, 155
death receptor 105, 153ff, 157, 456
decoy receptor 154f
DEK 300
deletion 27, 28T, 31ff, 96F, 302, 317
demethylation 174F, 178f, 181
dendritic cells 211, 257, 477, 478F
de-novo-methylation 174F, 178
Denys-Drash-syndrome 246T, 250
deoxynucleotide 14
deoxynucleotide-triphosphate 49ff
Desert hedgehog (DHH) 141
desmoplakin 198
detection (early) 364, 386, 396, 409, 416ff, 423ff, 429, 433ff, 449, 486
detoxification 9
developing countries 1, 328, 343, 403, 487

development *see* fetal development
DHEAST 294
diabetes 414f
diacylglycerol 127
diarrhoea 454
dicentric chromosome 35, 161
diepoxybutane 61
diet 9, 40T, 70, 179, 273, 343, 352f, 355B, 365, 372, 388f, **408ff**, 418, 425 *see also* food
differentiation therapy 237ff
dihydrofolate reductase 451, 452F
diploid 15
DISC (death-inducing signaling complex) 153F, 155, 157
DLG 275, 275F
D-loop 64, 159
DNA 451, 451ff
DNA bases (chemical modification) 9
DNA crosslinks see crosslink
DNA damage 9, **47ff**(Ch), 109, 156ff, 258, 281, 444, 456, 465
DNA demethylase
DNA ligase 49T, 52, 57, 62ff
DNA methylation 15, 53, 168, 171f, **174ff**, 174F, 175F, 181, 188, 226, 285, 397, 413
  inhibitor 177, 285, 397, 451
DNA methyltransferase 172, 174, 174F, 178f, 181, 412F, 413, 451
DNA mismatch repair 43
DNA-PK (DNA-dependent protein kinase) 64, 103, 103F, 129, 155, 233
DNA polymerase ($\alpha$, $\beta$, $\delta$, $\epsilon$, $\gamma$, $\eta$, or $\iota$) 48, 49T, 52, 57f, 160, 178, 260
DNA repair 10, 16, 31, 37, 40T, 43f, **47ff**(Ch), 104f, 155, 159f, 165B, 233, 235, 239, 365, 368ff, 404, 416f, 455f
DNA replication 6T, 9, 48ff, 53ff, 59, 62, 160, 174, 181, 311, 395, 413, 443, 451ff
  end-replication problem 160
  fidelity 48
  stalling 64, 370
DNA strand break 31, 35F, 49T, 50F, **62ff**, 258, 324, 329, 353, 413
  double-strand 22, 31, 35F, 49T, 50F, 62ff, 63F, 102ff, 150, 159, 161, 230, 368ff
  single-strand 48
  strand break repair 62ff
DNA synthesis 12, 14, 21, 122, 179, 217B, 429
DNA virus 31, 46, 72, 149, 336, 407f
DNase 155
DNMT *see* DNA methyltransferase
dominant-negative 66, 106, 225, 241, 250, 264, 369
double minutes 35
double-strand break *see* DNA strand break
  double-strand break repair 39T

DP1 98F
Drosophila 141
'drugable' 460, 470, 470F
DSH (disheveled) 138F, 276, 277F
dsRNA 40T
Ductus-Bellini carcinoma 309
dyskeratosis congenita 162
dysplasia 11, 19F, 272, 296, 355B, 385

E
E1A/B 46, 108, 481f, 482F
E2F **97ff**, 98F, 99F, 103, 125, 162, 182, 184, 231, 300
E6 108, **112B**, 143, 318, 339, 461, 481
E7 108, **112B**, 143, 461, 481
early-response genes 87
EB1 275, 275F
E-box 86, 183f, 184F
EBV 6T, 8, 46, 90B, 157f, 227, 231, 232F, 336, 407, 461
E-Cadherin 128, 198F, **198ff**, 199F, 201T, 278, 332, 333F, **345ff**, 346F, 397, 400
ECM *see* extracellular matrix
ectopic expression 12
ectopic structures 244
EGCG 411
EGF 132
EGF-like growth factors 202, 204, 208, 262, 303f, 321, 361, 362F, **373ff**, 375f, 376T, 399ff
EGFR 77, **84ff**, 127, 133, 183, 262, 303, 320ff, 334, 376, 393, 461, 474f, 473F *see also* ERBB1
EGLN family 318
EGR 249
EIF4 121, 239
ELAC2 395T
electromagnetic radiation 7
ELISA 386, 436
ELK1 86F, 87, 124
elongin 317
EMT *see* epithelial-mesenchymal transition
endocrine cancers 43, 214
endocrine disruptors 388
endogenous retroelement 31
endometrial cancer 282, 420
endometrium 207, 359, 362f
endonuclease 52, 56f, 64
endostatin 208, 208T, 210
endothelial cell 105, 187, 189, 207ff, 212ff, 286, 466, 469
endothelin (ET1) 319, 376, 399ff
endothelin receptors (ETA/B) 399ff
energy demand 14
enhancer 31, 228
ephrin 280
epicatechin 411

epidermis 142f, 148, 256ff, 257F, 290H, 322, 358
epigenetic (alterations, mechanisms) 27, 52, 68, 110, **167ff**(Ch), 200ff, 214, 250ff, 283, 300, 394ff, 407
epithelial-mesenchymal transition 200, 206F, 310
epithelium 16
  squamous 290H
  transitional 290, 290H
Epstein-Barr-virus *see* EBV
ERBA 74T
ERBB 74T, 373ff, 376T, 469
ERBB1 77, 77F, 79T, 80ff, 299, 375f, 376T
ERBB2 79T, 82f, 85, 321, **375ff**, 376T, 377F, 379ff, 441f, 446, 461, 471
ERBB3 83, 375ff, 376T, 377F
ERBB4 83, 375, 376T
ERCC 39T, 49T, 56ff, 66, 67T
ERK 86F, 87f, 115ff, 121, 123ff, 130, 132, 204
erythroblastosis 75
erythrocyte 75, 148, 221
erythropoesis 217B, 221
erythropoetin 22, 132, 222T, 319f
esophagus, esophageal cancer 1, 344, 352, 355B, 408
ESR1/2 see estrogen receptors
ESS1 261
estrogen 358ff, 362F, 363F, 364f, 371, 387, 410, 419, 453
estrogen receptors 178T 359ff, 360F, 361F, 371, 373ff, 379ff, 390, 441f
ethical issues 405ff, 418, 445
etoposide 452, 453F, 456
ETS 124
evidence-based medicine 442
execution genes 43
EXO1 53
exonuclease 53, 102, 161, 165B
expression profiling 42, 197, 206, 379ff, 432f, 442
extracellular matrix 130ff, 139, 147, 186, 195ff, 200ff, **202ff**, 208, 211, 221, 322, 398ff
extravasation 17, 196, 213F, 401
eyes 93ff, 148, 158
EZH2 182

**F**
FADD 153F, 155, 157
FAK 128, 129F, 155, 200, 233f
familial adenomatous polyposis coli *see* FAP
familial cancers 37, 93ff, 94F, 200, 245f, 250f, 288B, 371ff, 416ff
FANC (genes, proteins) 39T, 49T, 60ff, 66, 67T, 369f, 370F, 455

Fanconi anemia 37, 39T, 60ff, 67T, 370, 417
FAP 38T, 93T, 93, **272ff**, 274T, 275F, 286, 417ff, 417T
farnesylation, farnesyltransferase (inhibitors) 85, 464T, 466ff, 467F, 469F
FAS 105, 135, 154F, **154ff**, 157f, 201T, 211f, 322, 331, 338, 456, 476
FAS ligand 14, 135, 154ff, 157f, 211f
fat 365, 388, 409ff, 414F, 414f
fatty acids 14
fatty acid synthetase 14
FEN1 49T, 52, 63
Fenton reactions 10, 329, 330F
ferritin 329
fetal development 9, 115, 137ff, 147, 170, 176, 181f, 200, 221, 244, 248, 264, 266, 303, 308, 313, 340B, 347, 358, 390
α-fetoprotein *see* AFP
FGF 79T, 84, 185f, 187, 202, 204, 208, 303f, 390, 399ff
FGFR 84, 210, 303f, 303F, 469
FH (fumarate hydratase) 38T, 306B, 312T, 314
FHIT 315, 315F, 323
fibrin 203
fibrinogen 207
fibroblast 202, 205, 212ff, 286
fibronectin 128, 198F, 199, 204T
fibrosis 331
field cancerization 294, 424f
fingerprints 11
first-line therapy 21, 236
FISH 235, 328, 427, 430, 430F, 441
FKHRL1 120F, 122, 159
flavonoids 410f, 411F
FLIP 135, 153F, 155, 157, 211
flow cytometry 20
FLT3 241
fluoro-uracil *see* 5-FU
fms 74T
focal adhesion (contacts, points) 78, 128
focus formation assay *see* 3T3 assay
folate 40T, 51, 179, 353, 388, 409ff, 411F, 412F
follicular (B-cell) lymphoma 80T, 83, 88, 150, 156, 224, 433
food 2, 6, 352, 407 *see also* diet
forkhead 122, 159, 262
FOS 74T, 86F, 87, 124, 184
founder effect 419
FOXM1 262
fragile site 311, 315, 315F, 323, 339
frame-shift mutation 31, 284, 318, 346, 368
Frasier syndrome 246T, 250
Fringe 140F, 142
frizzled 138F, 140, 276
fruit 343, 353, 361B, 408ff

5-FU 413, 444, 451, 452F, 456, 458, 484
Fused 139F, 262
fusion gene 32F, 234F, 300
fusion protein 32F, 33, 134, 225ff, 233ff, 234F, 238F, 461
FZD *see* frizzled

## G
GADD45 103F, 105, 370
gag-fusion protein 72f
gain-of-function 250
gancyclovir 483
GAP (GTPase activator protein) 85, 86F, 109, 118, 466
gap junction 197f, 198F, 483
Gardner syndrome 273ff, 274T, 275F
GAS element 133
gastric cancer *see* stomach cancer
gastrin 350, 355B
gastrointestinal stromal tumors *see* GIST
gatekeeper 43, 93T, **110**, 264, 276, 288B, 346
G-CSF 222
GEF (guanine nucleotide exchange factor) 85ff, 118, 120
gefitinib 473, 473F, 474f
gelsolin 155
gene amplification 28T, 34f, 81f, 88, 100, 107, 183, 217B, 297, **300f**, 377, 381, 391, 446, 458, 461, 465
gene dosage 42, 321, 397
gene regulatory sequences 31
gene silencing 169, 177, 179f, 183
gene therapy 22, 78, 417, 449, 464, 474, 477, 479ff, 479F, 486
genisteine 388, 411
genome stability 248
  control of 40T
genomic instability 11, 15f, 26ff, 45, 88, 125, 159, 162, 197, 210, 217B, 226, 236, 260, 281, **284f**, 297ff, 304, 391, 456, 458, 465
genomic integrity 109, 297
genotype-phenotype relationship 276, 318, 418ff
geranylation (geranyl-transferase) 467f, 467F
germ cells 62, 176f
germ cell cancer 150, 171, 188
germ-line 26, 43, 95ff, 100, 149, 160, 176, 187, 200, 248, 267, 275F, 346, 368, 371f, 419
gestagen *see* progesterone
Gilbert syndrome 443
GIST 79T, 446, 465f, 471, 475
gleevec *see* imatinib
GLI (1-3) 82, 116T, 139F, 141, 262ff, 263F
glio(blasto)ma 88, 95, 107, 264, 278, 481
glivec *see* imatinib
glucocorticoids 338, 359

glucose (metabolism) 14, 119, 206, 250, 320, 414f
GLUT1 320
glutathione 68f, 69T
glutathione peroxidase 66f, 388, 410
glutathione reductase 68f, 69T
glutathione transferase *see* GST
glycogen 317, 320
glycogen synthase kinase *see* GSK
glycosylase 51
glycolytic enzymes 317, 320
GM-CSF 22, 132, 186, 185F, 222T
GnRH 374, 375F, 390, 423
gonads 248
Gorlin syndrome 37, 38T, 93T, 263f
GPCR (G-protein coupled receptor) 79T, 119, 126f, 266, 376, 399
grading (of tumor) 19, 302, 382, 427f, 439f, 471
graft-versus-host-disease *see* GVHD
graft-versus tumor reaction 475
granulocyte 16, 210, 212, 221, 237f, 286, 466
GRB2 85, 86F, 128, 234, 349
Groucho 138F, 278
growth autonomy 11
growth factor 9, 21, 43, 74T, 77, 79T, 87, 100T, 115ff, 125, 131ff, 138ff, 171, 189, 202ff, 221ff, 222T, 230, 261, 279f, 320, 334f, 334F, 358ff, 393, 398ff, 463, 466
  latent 202
growth fraction 474
growth hormone 359, 376
growth regulation 11
growth signals 18
GSK 122, 124f, 124F, 138F, 140f, 275ff, 277F, 331
GST 40, 43, 69f, 412, 458
GSTM1 40f, 40T, 293, 294T, 395T
GSTP1 69, 83, 178T, 294T, 388, 395T, 397f, 398F, 438
GSTT1 40T, 41, 294T, 395T
G-T mismatch glycosylase 51ff
GTP-binding protein 74T, 79T, 85ff, 117ff, 466
guan(os)ine (nucleotide) 49ff, 56F, 102, 260, 455
gut 139, 148, 189, 272ff, 299, 347f, 355, 358, 454, 474
GVHD 233

## H
H2AX 66
H19 170f, 170F, 172F, 250ff, 251F
hair follicle 137, 265
hallmarks of cancer 18B
hamartin *see* TSC2
hamartoma 121, 274T

# Keyword Index

haploinsufficiency **97**, 110, 264, 302, 306B, 396f
Ha-Ras 73, 74T, 81
HAT 98F, 180f
HB-EGF 303, 375f, 376T
HBV 6T, 8, 32, 46, 78, 158, 161F, 330f, 334, **336ff**, 337F, 338F, 343, 353, 407, 415, 461, 477f
HBx 337ff
HCV 8, 46, 330, 336, 343, 407, 415
HDAC 98F, 181, 240
    inhibitors 181, 240, 397
HDM2 see MDM2
heart 453
heat-shock protein see HSP
Hedgehog pathway see SHH pathway
helicase 165
helicobacter pylori 6T, 9, 40T, 343ff, **348ff**, 355, 404, 407f, 414, 415
helper virus 73
hemangioblast 207
hemangioblastoma 316
hematological cancers 20, 32, 82f, 132ff, **221ff**, 300, 429ff, 449, 459f
hematopoetic lineage 199, 221ff, 232, 430
hematopoetic stem cells 16, 182, 190, 207, 221ff, 232ff, 475
hematopoetic system 22, 60f, 132, 221, 222T, 417, 454
hematopoesis 239
hemidesmosomes 198F, 199
hemizygosity 248, 298
hemochromatosis 329ff, 336
hepatitis virus B see HBV
hepatitis virus C see HCV
hepatoblastoma 172
hepatocellular carcinoma see liver cancer
hepatocyte 148, 184, 189, 313, 328ff
hepatocyte growth factor see HGF
hepatoma see liver cancer
HER1-4 see ERBB1-4
herceptin see trastuzumab
hereditary cancer 92ff, 311ff, 394ff, 416ff
hereditary gastric cancer 38T, 346
hereditary leiomyoma renal cell carcinoma (HLRCC) 38T, 306B, 314, 312T
hereditary nonpolyposis colorectal cancer see HNPCC
hereditary papillary renal cancer (HPRC) 38T, 43, 93, 312ff, 312T
heregulin 365ff, 366T
herpes virus 8
heterochromation 169, 172, 181
HFE 329f
HGF 202, 313f, 314F, 321f, 334f, 400
HHR23 56

HHV8 8, 46, 157, 192B, 407
HIF (hypoxia-induced factor) 187, 210f, 217B, **318ff**, 319F, 321T, 325, 484
HIP1 139F, 262f, 263F
high-risk mutation (allele, gene) 39, 268, 365ff, 394, 396F, 415ff
histological subtype 17, 308ff, 379f, 439
histone acetylase see HAT
histone acetylation 169, 172, 180, 239, 361
histone deacetylase see HDAC
histone methylation 169, 172, 180
histone methyltransferase see HMT
histone modification 179, 180F
histone phosphorylation 169, 180
histopathology 17, 427ff, 439ff
HIV 8, 46, 76, 161F, 192B, 422F
HLA-typing 233
HMG-CoA reductase 14, 468
HMT 181f, 397
HNF 184
HNPCC 38T, 54f, 67T, 93T, 274T, 281, 282ff, 299, 347, 366, 366T
Hodgkin lymphoma 46
Holliday junction 64, 369
homocysteine 413
homogenously staining regions (HSR) 35
homologous recombination repair 49T, 61, 63ff, 65F, 159f, 162, 368ff, 369F
homozygous deletion 31, 100, 298, 306B, 396
hormone response 40T
hormones 6, 21, 43, 355B, 359ff, 414, 453
hotspot 333
HOX 239, 347f, 395
HP1 172, 181
HPC1 40T, 97, 394f, 395T
HPV 6T, 7, 46, 108, 112B, 143, 269, 336, 339, 408, 461, 478, 481
HRAS 79T, 81, 261, 303
HRE 319
hSNF5/INI4 182
HSP 360, 462
HSV 483
hTERC 160 see telomerase
hTERT 160, 484 see telomerase
HTLV1 8, 46, 72, 74T, 90B
human-herpesvirus-8 see HHV8
human papilloma virus see HPV
human T-cell leukemia virus see HTLV1
Hutchison-Gilford syndrome 165B
hyaluronic acid 202
hydrogen peroxide 10, 68
hydroxycarbamates 203
8-hydroxy-guanine 49, 52, 388
hydroxyl radical 10, 69, 258, 329f
hydroxy-methyl-cytosine 51
hydroxymethyl-thymidine 51

hygiene 2, 355, 408
hypermethylation 55, 95, 96F, 100, 119, 131, 134, 158, **176ff**, 177F, 178T, 182, 185, 200, 205, 284, 298, 301, 306B, 321f, 345, 347f, 355B, 373, 388, 392, 397, 438, 451
hyperplasia 11, 272
hypertrophy 207
hypomethylation 176ff, 177F, 246, 269
hypoxanthin 51
hypoxia 65, 129, 148, 187, 207ff, 217B, 318ff, 325, 400

## I

IκB 135ff, 136F
IAP 130, 151F, 152ff, 158f, 322, 456f
ICE 152
ID proteins 183, 184F
IG *see* immunoglobulin
IGF 88, 121ff, 125, 130, 359, 389f
IGF1 88, 399ff, 414f
IGF2 79T, 88, **170f**, 170F, 172F, 250ff, 251F, 334f, 414
IGFBP 130, 335, 389, 395T, 399f, 415
IGFR1 79T, 85, 335, 399f, 469
IGFR2 335
IKK 135, 136F
IL1 185f, 185F, 222T
IL1B 40T, 349, 351f, 351F
IL1R 352
IL2 211, 222T, 325, 476
IL3 222T
IL6 90B, 132ff, 222T, 332, 392
IL6R 133f, 134F
IL8 208, 222T, 349
ILK 128, 129F, 200
imaging techniques 17, 427f, 439
imatinib 236, 446, 464T, 465ff, 471, 473, 475, 485f
imiquimod 259
immigrant studies 387, 409
immortalization 16, 108, 149, 230
immune cells 135, 147ff, 186, 189, **210ff**, 269, 281, 331, 475, 479
immune defense 6, 230
immune-privileged sites 158
immune response 21, 40T, 210ff, 229, 233, 257, 259, 269, 295, 325, 331, 349, 410, 475ff, 481f
immune surveillance 233, 236, 259
immune system 8, 14, 43, 46, 131, 153, 156, 195f, 202, **210ff**, 257
immunodeficiency 46, 407, 475
immunoglobulin (gene) 62, 150, 168, 224, 224T, 227ff, 430, 463
immunohistochemistry 20, 297, 373, 377, 381, 427, 441
immunosuppression 16, 90B, 400

immunotherapy 22, 269, 310, 324f, 449, 461, 464, 474, **475ff**, 476T, 486
imprinting 100, 168, 170ff, 176, 250ff
imprinting center (IC) 170F, 171f, 172F, 250ff
Indian hedgehog (IHH) 141
individualized therapy 379ff, 427, 429, 442ff, 449, 486
indol carbinol 411
industrialized countries 1, 273, 293, 308, 328, 343, 353, 364, 371, 385ff, 403, 407ff, 420, 487
inflammation 6, 6T, 9, 16, 46, 135ff, 148, 153ff, 186, **195ff**, 207, 213F, 259, 285ff, 294, 328ff, 336ff, 343ff, 350ff, 408, 476
initiating agents 5
innotecan 452
iNOS 136, 319
insertion 27, 28T, 31
retroviral 31, 72f, 78, 141
insulin (receptor) 121, 335, 359, 414
insulin-like growth factor *see* IGF
integrase 62
integrin 128, 129F, 198F, 199, 201T, 203, 203F, 235, 262
interferons 22, 132ff, 232ff, 325, 338, 465, 475ff
interleukins 22, 132, 204, 475 *see also* IL
inter-strand crosslink 50F
intervention 349, 403ff, 410, 421
intravasation 196
invasion 11, 14, 16ff, 43, 131, 148, 157, **193ff**(Ch), 262, 268, 280ff, 293, 304, 321, 400, 476
invasion program 201
involucrin 260, 291,
ionizing radiation 6T, 40T, 62, 65ff, 102ff, 232, 258, 365, 372, 429, 444
IP3 etc *see* phosphatidylinositol
Iressa *see* gefitinib
irinotecan 443, 452
iron 6, 319, 329f
isochromosome 246, 247F
isoflavonoids 410f, 411F
ISRE 134
Ito cells 335

## J

Jagged (JGD) 140F, 142
JAK (janus kinase) 133, 134F, 234, 332, 376
JC virus 46
JNK 103, 104F, 115ff, 132
JNK (MAPK) pathway, 115ff, 116F, 130, 141, 186, 204, 239, 286
JUN 74T, 86F, 87, 124, 184, 186
juvenile polyposis coli 274T
juxtacentromeric 174, 246

## K

KAI1 201T, 202
kallikrein 385f, 401
Kaposi sarcoma 8, 46, 192B
karyotype 34F, 223, 310ff
KCNQ1 171, 172F, 250f, 251F
keratin 258F, 260
keratinocyte 142f, 185f, 257ff, 257F
  lineage 264f
KGF 185f, 185F, 392
Ki67 19, 226, 303, 441
kidney 139, 184, 196, 244ff
  cancer 1, 20, 44F, 79T, 196, 278,
  **307ff**(Ch), 429, 469 *see also* renal cell
  carcinoma
kidney cysts 9, 311f, 316
KIP *see* CDK (protein) inhibitors
KIR 211
Ki-ras 73, 74T, 81
KIT 74T, 79T, 85, 189, 446, 465
knockout mice 101, 106, 340B, 396, 473
Knudson model **94ff**, 94F, 110, 246, 264, 274f,
  302, 306B, 397
KRAS 79T, 81, 261, 280ff, 435, 438, 461, 468
KSHV *see* HHV8
KU70/KU80 49T, 63ff, 160

## L

lactate 4, 206
lamin 165B, 468
laminin 198F, 199
Langerhans cells 257, 259
large cell lymphoma 433
large T-antigen 46, 481
lead (in drug development) 412, 463, 468,
  473F
leiomyoma 314
leukemia 1, 4, 8, 13T, 20ff, 26, 42, 65, 72,
  74T, 80F, 100, 134, 137, **219ff**, 406, 429,
  444, 460, 468, 475
  acute 182, 449
  crisis 34
leukopenia 454
LFA1 230
LH 374, 375F, 390
LHRH *see* GnRH
life-style 364, 387ff, 409ff
Li-Fraumeni sydrome 37, 38T, 42, 92, 93T,
  101, 268, 366
LINE 31, 161F, 174
linkage analysis 261, 288B, 306B, 371f, 394,
  418
lipid 14, 320, 329, 365
lipid phosphatase 122
liver 158, 196, 202, 272, 358, 378, 402, 415,
  481

liver cancer 1, 6, 8, 12, 46, 79fT, 130, 140,
  196, 260, 278, 282, 311, 314, 317,
  **327ff**(Ch), 328H, 343ff, 353, 394, 404,
  407f, 438, 461, 469, 473f, 477, 484
liver cirrhosis 9, 328ff
LKB1/STK11 274T, 366, 366T
LOH 36, 36F, 96F, **96**, 101, 106f, 202, 231,
  246f, 263f, 276, 281, 297f, 301f, 306B,
  331f, 335, 346, 393f, 396
LOI *see* loss of imprinting
long-patch repair 52, 53F, 62
long Q-T 472
loss of heterozygosity
  *see* LOH
loss of imprinting 171, 251f, 335
LRP 140, 277F, 278
LTR (long terminal repeat) 73F, 75, 75F
lung 196, 272, 378, 402
lung cancer 1f, 12, 14, 41, 141, 143, 261, 264,
  297, 315, 404, 409, 449
'lurker cell' hypothesis 393, 458
lymphangiogenesis 208f
lymph node 18, 196, 226, 272, 378ff, 385, 401,
  441
lymphocyte 16, 202, 210ff
lymphoid cells (lineage) 135, 221, 433
lymphoma 1, 4, 13T, 21, 62, 65, 72, 74T, 80,
  90B, 137, **219ff**, 309, 429ff, 460, 475, 484

## M

β-macroglobulin 204f, 206F
macrophages 10, 16, 135, 153, 156, 205, 210,
  212, 221, 226, 286, 350, 475
MAD 83
MAGE 269
magnetic resonance 17
'major four' 1, 420
mammography 365, 419
MAPK (cascade, pathway) 87f, 100, **115ff**,
  116T, 122ff, 123F, 126F, 134f, 234, 261,
  313, 335, 349, 361, 373, 375, 462
MARCKS 127
marker chromosomes 36
marker (tumor, biomarker, molecular) 20, 158,
  223, 290, 296, 328, 379ff, 386, 402, 420f,
  424, 427ff, 442, 471
matrix metalloproteinase *see* MMP
MAX 83, 86F
MBD2/3 181
MBD4 49T, 54, 176, 283
MC1R 40T, 266, 267F
MCL1 152T
M-CSF 222T, 241
MDM2 66, 80T, 82f, **102ff**, 116T, 122, 129ff,
  231, 264, 298, 317f, 439
MDR1 324, 324F, 457f

# Keyword Index

McCP2 181
medulloblastoma 264, 278
MEK 86F, 87, 115ff
MEKK 86F, 87, 115ff, 202
melanin 266, 269
melanocyte 40T, 257ff, 266ff
melanoma 1, 12, 58, 79T, 83, 119, 207, 214, 257ff, 258H, 266ff, 325
   hereditary 38T, 93T, 100
memory cells 160
MEN *see* multiple endocrine neoplasia
mesenchyme 16, 185f, 185F, 398ff *see also* stroma, fibroblast
mesenchymal-epithelial transition 244, 248f
mesomeric forms 49
mesothelium 248
mesothelioma 8, 46, 143
MET 38T, 43, 79T, 85, 93T, 97, 312T, 313f, 313F, 321f, 400, 469
metabolism 11, 14, 17, 119, 123, 217B, 317, 320
   of carcinogens 40T, 292f, 292F, 294T, 340B, 412
   of drugs 40f, 445, 458, 471
   of hormones 365, 387, 395, 410f
metallothioneines 69f, 69T, 458
metaplasia 12, 291, 298, 343ff, 347f, 355B, 408
metastasis 14, 16, 18, 21, 43, 131, 137, 148, 157, 187, 193ff(Ch), 245, 257, 262, 268, 272, 280ff, 296, 304, 309, 314, 331, 374, 378ff, 385ff, 392, 398ff, 421ff, 449, 465, 476
   hematogenic 17, 196, 401
   lymphogenic 17, 196
   osteoblastic 401
   osteolytic 401
   suppressor gene 110, 201f
   theories of 197, 214, 389, 381f, 401
   transcoelomic 196, 345
methionine 412f
methotrexate 51, 451, 452F, 456
methylation 52, 85, 413f, 467, 467F
'methylator phenotype' 285
methyl cycle 179, 412F, 412f
methylcytosine 51, 51F, 171, 174ff
methylcytosine-binding protein *see* MBD and McCP2
metronomic therapy 454
MGMT (methyl-guanine methyltransferase) 49T, 52, 178T, 455
MHC 201T, 210f, 329
MICA/B 201T, 211
micrometastasis 17, 196, 207, 378ff, 387, 428, 437, 439ff
micronutrient 388, 409ff

microsatellite 31, 36, 39, 45, 49, 96, 282, 418
   analysis 173, 282F, 295
   contraction 49, 55
   expansion 49, 55
   instability *see* MSI
microtubule 119, 262, 275, 451T, 452f, 459
mimicry 214, 401
minimal residual disease 235, 238, 431f
mismatch repair *see* MMR
missense mutation 27
mitochondrial permeability transition 151
mitogenic (signal, cascade) 87, 115
mitomycin 56, 61, 295, 424
mitosis 12, 19, 21, 97, 119, 122, 177, 275, 278, 366, 451ff, 468
mitotic non-disjunction 35F, 36, 95f, 177, 285
MKK 117F
MKP1 87
MLH1 38T, 49T, **53ff**, 54F, 67T, 93T, 178T, 274T, 283, 366, 366T
MLV (murine leukemia virus) 73F
MMP 201T, 202ff, 203F, 204T, 206F, 207, 268, 298, 376, 400f
MMR 49, 53ff, 67, 283ff, 370, 455
model systems 321, 397
molecular diagnostics 415ff, 427ff, 429ff, 442ff, 446, 487
molecular epidemiology 260
'molecular pathology' 427
'molecular staging' 438f
monitoring 235f, 295, 304, 416ff, 419ff, 428ff, 436ff, 466
monocyte 153, 205, 221, 286, 475, 477
monosomy 301
morbidity 296
mouse 163
MOZ 224T
MRE11 49T, 63ff
MRN complex 63f, 63F, 65F, 160, 369ff
αMSH 266, 267F, 268
MSH2 38T, 49T, 53f, 54F, 67T, 93T, 274T, 283, 366, 366T, 370
MSH3 53f, 54F, 283
MSH6 49T, 53f, 54F, 283, 370
MSI 55, 68, 282ff, 299, 347, 455
MTHFR 40T, 395F, 412F, 413
mTOR 120F, 121f
MTR (methionine synthase) 412F, 413
mullerian inhibitory factor 131
multiple endocrine neoplasia 1 38T
multiple endocrine neoplasia 2 38T, 93T, 97
multiple myeloma 223
multistep 194ff, 280ff, 340ff, 340B, 345, 353, 355B
mutagen 5, 102, 286
mutation probability 95
mutator phenotype 45, 285

MXI1 83
MYB 74T, 87
MYC 33, 43, 62, 74T, 75ff, 76F, 80T, 80ff, 86F, 87, 116T, 122, 123F, 124f, 130, 135, 160, 163, 183f, 184F, 189, 224ff, 224T, 227ff, 236, 239, 262, 278, 298f, 303, 318, 321, 331ff, 393, 430, 484f
   effects on cell functions 230T
MYCL 80T, 82f
MYCN 80T, 82f
myelodysplastic syndrome 61
myeloid lineage 221, 241, 433
MYH 49T, 54, 283
myoblast differentiation 183f, 184F
MYOD 183f
myristylation 85, 128, 466

**N**
N-acetyltransferase *see* NAT
nasopharyngeal carcinoma 46
NAT1/2 40T, 292F, 293f, 294T, 405, 443f, 443T, 445F
NBS 39T, 49T, 63ff, 67T
NCoR/SMRT 239
necrosis 10, 146ff, 207, 329
negative markers 435
negative regulatory element 75
neo-adjuvant treatment 21, 374
neoangiogenesis *see* angiogenesis
neoplasia 13T
nephroblastoma *see* Wilms tumor
nephrogenic rest 244ff, 252, 308
NER *see* nucleotide excision repair
NEU *see* ERBB2
neuroblastoma 82
neurofibromatosis-1 (NF1) 38T, 109
neurofibromatosis-2 (NF2) 38T
neuronal network 442
neuronal precursor 142
neurons 148
nevoid basal cell carcinoma syndrome *see* Gorlin syndrome
NFκB 142, 361
NFκB pathway 116T, 130, 135ff, 136F, 155, 158f, 211, 286, 458, 462
nibrin *see* NBS
nickel 6, 6T, 70
nicotine 5, 259,
Nijmegen breakage-syndrome 37, 39T, 64, 67T *see also* NBS
NIK 135, 136F
NIP3 320
NIX 152T, 320
nitroaromates 292
nitrogen mustards 6

nitrosamine 6, 6T, 8F, 9, 292, 343, 352, 352F, 408
nitrosation 51
nitrous oxide 10
NK cells 210f, 338
NKX3.1 110, 395f
NNK 8F
NO *see* nitrous oxide
nomogram 440, 440F
non-coding RNA 171f
non-Hodgkin lymphoma 46
non-homologous DNA end joining (NHEJ) 49T, 63ff, 63F, 368
nonsense-mediated decay (NMD) 29
nonsense mutation 29, 107, 276, 368
NORE1 118F
NOTCH 116T, 140F, 142f, 224T
NOTCH pathway 116T, 138, 140F, 142f, 263
NOXA 103F, 105, 150, 152F, 158
NPM 237
NQO1 40T, 294, 406F, 407
NRAS 81, 268,
NRF1 152T
NSAID 137, 286f, 418ff, 425
nuclear bodies 239f
nucleolus 14, 303
nucleoside drugs 451, 451T
nucleosomal ladder 147, 155
nucleotide
   biosynthesis 40T, 49, 412f, 451f, 452F
   imbalance 129f, 281, 456
   metabolism 395
   precursor pools 49
nucleotide excision repair 39T, 48ff, 55ff, 57F, 62, 258, 260, 456
   global genome 56f, 59F
   transcription-coupled 56f, 59F, 370
null-allele 40, 293
NUMA 237
nutrient 17

**O**
obesity 355B, 365, 408, 414f
occludin 198f
OCT3/4 188
oesophagus *see* esophagus
OGG 40T, 49T, 52, 388
oncocytoma 309ff, 309H
oncogene 37, 43ff, 46, **71ff**(Ch), 97, 102, 109ff, 141ff, 227ff, 300, 313, 394, 461, 479ff, 486
   cooperation 73f, 78, 88
   families 73f, 83
oncofetal (marker, antigen) 12, 269, 325, 328, 436, 461
OSI-774

osteoblast (osteoclast) 137, 187, 214, 392, 401f
osteomimetic 187, 401f
osteosarcoma 4, 74T
oval cells 331
ovarian cancer 1, 119, 196, 419, 455
  hereditary 38T, 61, 93T, 96, 282, 366ff
     germ cell 150
ovaries 189
oxidative stress 10, 66
oxo-guanine see hydroxy-guanine
oxygen 6f, 9, 10B, 17, 187, 206ff, 217B, 318ff

**P**
p14$^{ARF}$ **103ff**, 108, 125, 130, 163, 231, 267, 281, 298, 302, 304, 334
p15$^{INK4B}$ **98ff**, 99F, 100T, 116T, 267f
p16$^{INK4A}$ **98ff**, 99F, 100T, **103ff**, 108, 112B, 116T, 125, 148, 162ff, 182, 188, 260, 267f, 297ff, 302, 304
p18$^{INK4C}$ 98, 100T
p19$^{INK4D}$ 98, 100T
p19$^{Arf}$ 104
p21$^{CIP1}$ 66, **98ff**, 99F, 100T, 103F, 104, 125, 130, 142f, 148, 162f, 239, 297, 348
p27$^{KIP1}$ **98ff**, 99F, 100T, 122, 124F, **124f**, 297, 318, 441, 474
p38$^{MAPK}$ 103, 104F, 117F, 186, 204, 286
p53 340 see TP53
p53AIP 105
p57$^{KIP2}$ **98ff**, 100T, 116T, 162f, 250ff, 297
p63α 30
p70$^{S6K}$ 121
p73α 30
p90$^{RSK1}$ 87, 124
P450 monooxygenase see CYP
PAI1 203, 319, 441
PAK2 155
palliative therapy 22
pancreas 184
pancreatic cancer 1, 38F, 93T, 100, 196, 264, 268, 297, 371, 438, 449, 468
papillary tumors 293, 299, 302ff,
papovavirus 7, 46, 336, 407
paracrine (factor, interaction) 78, 139, 185ff, 185F, 214, 223, 244, 262, 276f, 286, 320, 331, 398ff, 399F, 410
paraneoplastic symptoms 15, 320
PARP 155
'passenger' alteration 299ff
pathological stage 17
PAX6 246T, 247
paxillin 234
PCNA 19, 52f, 57, 178, 226, 303, 441
PCR 31, 431, 433ff
PDGF 208, 320,
PDGFR 210, 446, 465f, 469
pericentromeric 174, 177

pericytes 207
peroxynitrite 10
Peutz-Jeghers syndrome 274T
PGP see MDR1
pH 320, 322, 344, 347, 349, 355B
phagocytosis 147, 159
pharmacogenetics 340B, 443ff
pharmacogenomics 443, 446ff, 465
phenacetine 405
pheochromocytoma 316, 318
Philadelphia chromosome 232ff
phorbol esters 127
phosphatidylinositol 119ff, 121F, 127
  phospholipase C see PLC
phospholipids 14, 156
phosphorylation 97ff, 117ff, 125, 133, 183, 229, 277, 349, 373
phosphothionates 485
phosphotyrosine 85
photodynamic therapy 22, 217B
photoproducts 55, 55F, 258, 260
phytoestrogens 387
pigmentation 58, 257ff, 266ff, 404
PI3K (phosphatidylinositol-3'-kinase) 79T, 85, 116T, 118, 121F, 234, 306B, 376, 466
PI3K pathway 86, 88, 109, 116T, **119ff**, 120F, 123F, 126F, 128, 130, 134, 159, 211, 261, 280, 304, 312f, 320, 335, 365f, 375, 458, 462
PKA 139F, 262, 266, 267F
PKC 87, 103, 104F, 121, 127f, 129F, 141, 183, 339
PKD (protein kinase D) 120f, 120F, 127
PKD1/2 (genes in ADPKD) 312T
PKR 103, 104F
plakoglobin 198
plasma DNA 438f
plasmin 202f, 203F
plasminogen 202f, 203F
plasminogen activator see uPA
plasminogen activator inhibitor see PAI1
platelets 221
platinum 6, 56, 61, 417, 424, 451, 455ff
PLC 85, 127f
pleckstrin homology (PH) domain 120f
pluripotent 137, 187ff, 221
plutonium 7
PLZF 224T, 237, 240
PLZF-RARα 237, 240, 241F
PMA 127
PML 224T, 237ff, 238F
PML-RARα 33, 237ff, 238F, 241F, 432
PMS1 49T, 53f, 54F, 283
PMS2 67T, 274T, 283
pocket domain (protein) 95, 182
podocyte 249

# KEYWORD INDEX

point mutation 15, **26ff**, 67, 85ff, 95ff, 102F, 106f, 176, 200f, 225f, 229, 247, 251, 259, 263f, 280ff, 288B, 297f, 301, 303f, 306B, 313f, 329f, 333f, 339, 391, 407, 435, 439, 446, 458, 461f, 465, 474
polyamines 68, 69T
polyaromatic hydrocarbon 40T, 292 see also benzopyrene
polycomb proteins 181f, 188
polymorphism **39ff**, 52, 67ff, 179, 266, 288, 293, 294T, 351f, 351F, 365f, 372, 387f, 394ff, 395F, 404ff, 413, 443ff, 445F, 458, 472
polyp 272ff, 273H, 417f
polyploidy 15, 33f, 148, 168
porphyrins 10, 14, 22
positional cloning 274, 288B, 299
positive markers 435
positron emission tomography 17, 466
posttranslational modification 85, 466
POUF5 see OCT3/4
pp110 see RB1
PRA/B see progesterone receptor
PRAD1 see CCND1
pre-clinical phase 472, 474
precursor cells 137ff, 187ff, see also stem cells
preneoplastic (precursor) stages 272, 294, 344f, 355B, 409, 415ff, 424, 438
presenilin see γ-secretase
prevention see cancer prevention
primary cancer 21
primary prevention 404ff
primary resistance 324, 457f, 465
pro-angiogenic factors 207, 208T, 400
pro-apoptotic 205
procaspase see caspase
prodrug 483f, 483F
progesterone 359ff, 360F, 392
progesterone receptor 359, 360F, 361f, 373ff, 379ff, 441f
prognosis 223, 272, 298, 378ff, 428ff, 439ff
prolactin 359, 376
proliferation index 304
proliferative fraction 19, 146, 389, 428, 454
proline hydroxylase 217B, 318
promoting agents 5
promoter 31, 76, 484
promotor hypermethylation see hypermethylation
promyelocytic leukemia see APL
proof-reading 48
pro-oxidants 10
prostaglandin 208, 212
prostate 384ff
prostate cancer 3, 21, 44F, 69, 97, 110, 132, 156, 158, 182, 187, 200, 207, 213, 239, 278, 288B, **383ff**(Ch), 384H, 404, 409ff, 420ff, 422F, 436ff, 442, 446, 453f, 458, 472, 474
prostate-specific antigen see PSA
protease 9, 200, 202ff, 211, 286, 344, 347, 385f, 387F
   cysteine 152
   inhibitor 203ff, 386
   threonine 462
proteasome 102, 211, 229, 462
protective factors 343
protein degradation 125, 229
protein kinase 72ff, 88, 202, 268, 392, 444, 460, 466, 472f
protein kinase C see PKC
protein phosphatase 87, 98, 122, 124f, 134
protein (bio)synthesis 14, 86, 102, 119ff, 281
proteoglycan 202ff, 204T
PSA 385ff, 387F, 400, 421ff, 436ff, 437F
pseudo-triploid 34, 34F
proto-oncogene 75ff, 83, 109, 264, 339
PTCH 38T, 93T, 97, 109, 116T, 139F, 141, 261, 262ff, 262F, 298, 306, 404
PTC/RET 33F
PTEN 38T, 93T, 97, 109, 116T, **119ff**, 120F, 121F, 159, 214, 274T, 322, 324, 365f, 366T, 393, 396f
PTHrP 401f
PUMA 103F, 105, 150, 151F, 152F
pyrimidine dimer 50F

## Q

quality of life 21f, 296, 420, 471
quinones 6, 9, 40T, 365, 406F, 407

## R

RAC 117, 120, 128, 129F, 466
RAD9 66
RAD17 66
RAD50 49T, 63ff
RAD51 49T, 368f, 369F
RAD52 49T, 64
radiation (α, β, γ) 7, 22, 40T, 150
radioactive
   cesium 292
   iodine 22, 33
   isotopes 7
radiotherapy 3ff, 21f, 108, 217B, 324, 385f, 416, 421, 428, 437, 442, 449, 481, 486
RAF 74T, 83, 86F, **86f**, 115ff, 127, 157, 268, 466
RANKL 401
RAL 118
raloxifene 363, 374, 420
RARα 184, 224T, 237ff, 238F, 359, 360F, 459f

# Keyword Index

RARβ 178T, 184f, 239, 261, 323, 359, 360F
RARγ 184, 239, 322
RARE 184, 239, 322
RAS 29, 73ff, 81, **85ff**, 97, 103, 109, **115ff**, 116T, 118F, 125, 127ff, 130, 149, 159, 163, 208, 226, 234, 264, 286, 321, 355B, 466ff
RASSF1 118F, 119, 178T, 261, 323, 438
rat 340B
RB1 38T, 43, 93T, **94ff**, 104, 112B, 116T, 119, 124F, **125**, 129f, 155, 161ff, 178T, 182, 184, 188, 226, 236, 248f, 248F, 261, 264, 267, 285, 297ff, 304, 331, 336, 355B, 396, 481
RBX1 317
reactive oxygen species 6, 6T, 9f, 10B, 22, 51f, 68ff, 135, 165B, 212, 258, 265, 286, 313f, 353, 365, 388, 410
rearrangements 27
receptor 21, 459
    degradation 85
    internalization 85
receptor tyrosine kinase **85ff**, 86F, 97, 115ff, 125ff, 134, 159, 189, 208, 241, 304f, 321, 373ff, 400, 411, 446, 461, 469ff, 470F
recombinase 61, 62, 64, 230
recombination 39T, 96F, 161, 337
    homologous 369ff
    illegitimate 36, 96, 177, 247
    meiotic 62, 176
    repair 39T, 61, 63ff, 65ff
rectum cancer *see* colon cancer
redox cycling 40T, 407
5α-reductase 390, 391F, 395T, 423
REF (rodent embryo fibroblast) assay 81, 125
reflux 355B, 408
REL 74T, 80T, 116T, **135ff**, 136F
remission 432F
    clinical 431, 465
    cytogenetic 232, 235, 431, 465
    hematological 232, 235, 238, 431
    molecular 236, 465
renal (cell) carcinoma 21f, 96, 187, 210, 311T, 438, 439, 454, 457, 475ff
    chromophobic 309ff, 309H
    clear cell 119, 261, 308ff, 309H, **321ff**, 312T, 484
    papillary 97, **308ff**, 309H, 312ff, 321
replicative potential 18
replicative senescence *see* senescence
resubstitution therapy 480
resveratrol 411, 411F
RET 33, 33F, 38T, 43, 79T, 85, 93T, 97
retina 273, 316
retinoblastoma 37, 38T, 42f, 93T, 94ff, 245f
retinoblast 94f, 98, 248
retinoic acid receptor *see* RAR

retinoids (retinoic acid) 21, 184, 238, 261, 411F, 425, 432, 453, 459f, 466
retrotransposon 31, 65, 174, 177
retroviral insertions 31
retrovirus 73F
    acute transforming 72ff
    defective 73
    human 8
    insertion 31, 75F
    oncogenic 43, 46, 72ff, 127, 229
    slow-acting 75ff
reverse transcriptase 73, 160f, 161F, 337f, 337F
RFC 57
rhabdomyosarcoma 245
RHO 118, 129F, 141, 466
ribosomal proteins 14
RING-finger domain 240, 367, 379
risk modulating gene 43
RNA editing 249
RNA-in-situ-hybridization 427
RNA polymerase 56, 101, 169, 183
RNase 160, 394, 485
Rous sarcoma virus *see* RSV
RPA 56, 66
RSV (Rous sarcoma virus) 72ff, 73F, 74T, 128
RTK *see* receptor tyrosine kinase
RT-PCR 235f, 431, 438
RUNX 116T, 347
RXR 239

## S

SAβ-GAL 148
SAH (S-adenosylhomocysteine) 174F, 179, 412F, 413
SAM (S-adenosylmethionine) 52, 174F, 179, 353, 412F, 413
SARA 132, 133F
sarcoma 13T, 72, 74T, 80T, 107, 245, 309, 465, 475
satellite 174, 247
scatter factor *see* HGF
scavenging receptor 335, 400
SCC (of the bladder) 291ff, 291H, 305, 343, 347
SCC (of the head and neck) 7, 90B, 132, 261, 297, 336, 424, 481
SCC (of the skin) 58, 90B, 257ff, 258H, 260ff, 297f, 332, 449
SCF (stem cell factor) 189, 222T
schistosoma 6T, 9, 291f, 343
scintigraphy 17
screening 386, 415, 420ff, 450, 463, 471
secondary resistance 324, 457f, 465
second-line therapy 21
γ-secretase 140F, 142
'seed-and-soil' hypothesis 214, 401

selenium 68, 388, 410, 411F
senescence 100T, 102, 106, 119, 125, 130, **145ff**(Ch), 163F, 182, 188, 214, 241, 281, 297, 474
serin/threonine kinase 74T, 86f, 127, 132, 233, 366
SERM 363, 374, 420
serpentine receptor *see* GPCR
serum protein markers 435f
SET domain 181
SFRP 140, 178T, 277F, 278, 281
SH2 domain 85, 120
SHC 85, 86F, 128
SHH 116T, 139F, 141, 189
SHH pathway 109, 116T, 138, **141ff**, 139F, **262ff**, 263F, 280, 298, 306
short patch repair 49T, 52, 53F, 62, 413
SHP 134, 349
side effects 325, 403, 423, 453, 471f *see also* adverse reactions
signal transducer proteins 73
signature (in expression profiling) 206, 380ff
silencer 31
silent mutation 27
simian virus 40 *see* SV40
SIN3A 102
SINE 174
single nucleotide polymorphism *see* SNP
singlet oxygen 6, 9, 10B
siRNA 464, 485
sister chromatid (exchange) 59, 64, 368
skin 189, 239, 257F, 299, 425, 454, 459, 474
  type 266
skin cancer 1, 7, 58, 79T, **255ff**(Ch), 404ff, 475
SKP2 80T, 318
slipping (of DNA polymerase) 31, 48f, 283
SMAC/diablo 151ff, 151F
SMAD 116T, 132, 133F, 274T, 280ff
SMO 79T, 109, 116T, 139F, 141f, 262ff, 263F
SNP 36F, 39, 96
SOCS 116T, 134, 134F, 178T, 332
soft tissue cancers 1, 21, 32, 64, 83 *see also* GIST, sarcoma
somatic hypermutation 229f
somatic mutation 26, 44, 68
Sonic hedgehog *see* SHH
SOS 85f, 86F, 234
SOX 300, 347f
SP1 361
S-phase 19, 178
spleen 248
splicing
  altered 201
  alternative 30, 202, 208, 249f
  mutation 29f, 107, 346

sporadic cancer 43, 93ff, 94F, 250f, 288B, 306B
squamous cell carcinoma *see* SCC
SRC (protein kinase) 74T, 78, 87, 128f, 129F, 132, 200, 349, 350F
SRC1 (co-activator) 361, 363
SRD5A1 see 5α-reductase
SRE (serum response element) 86F
SRF (serum response factor) 86F, 124
stable disease 471
staging (of cancers) 17, 382, 427f, 439ff, 471
STAT factors 116T, 132ff, 134F, 331f, 375
statins 468
stem cell 138ff, 149, 160, 168, **187ff**, 214, 234, 256, 262, 286, 298, 331, 347, 358, 391
  niche 189, 279
stem cell transplantation 4, 227, 233, 238, 417, 429, 449, 465, 475, 478
steroid hormones 359ff, 392
steroid hormone receptor (superfamily) 239, 359ff, 360F, 390, 392
STI-571 *see* imatinib
stomach cancer 1f, 6, 9, 12, 96, 273, 282, **342ff**(Ch), 394, 404, 407, 415
  diffuse type 200, 344ff, 344H
  intestinal type 344ff, 344H
stroma (cells) 131, 187, **195ff**, 210ff, **212ff**, 221, 244, 281, 286, 335, 358f, 373f, 385, 396, 398ff
stromal reaction 195ff, 380
SUFU 116T, 139F, 262
sulfotransferase 293, 294T
sulindac 286
sulphoraphane 412
SUMO 104F, 239
superoxide 10
surgery 3ff, 17, 21ff, 197, 245, 295f, 310, 328, 342, 378ff, 385f, 419f, 421ff, 442ff, 449, 465, 481, 486
surrogate parameter 437
survival growth factors 123
survivin 152f, 158, 322, 456
SV40 7f, 46, 108, 143, 149, 481
sweat glands 137

T
t(2;8) 224T, 227ff
t(2;3) 224T
t(3;8) 312T, 315f
t(3;14) 224T
t(3;21) 236
t(3;22) 224T
t(5;17) 240
t(7;9) 143, 224T
t(8;14) 224T, 227ff, 430F
t(8;16) 224T

# KEYWORD INDEX

t(8;22) 224T, 227ff
t(9;22) 223, 224T
t(11;14) 224T
t(11;15) 237
t(11;17) 224T, 240
t(14;18) 150, 224T
t(15;17) 224T, 237, 432
T-ALL 143
TAM see NOTCH
tamoxifen 363, 363F, 374, 420
tankyrase 160, 160F
T-antigen (large, small, of SV40) 108, 143, 149
TAP 211
target (therapy) 83, 266, 296, 380, 429, 446, 450ff, **459ff**, 479ff
tat (of HIV) 76, 192B
taurine 68, 69T
taxols 450, 452
TBP 101
TCC see bladder cancer
T-cell 8, 143, 150, 156, 168, 210ff, 221ff, 257, 259, 325, 330, 338, 350, 475ff, 478F, 481
T-cell acute leukemia see T-ALL
T-cell receptor (TCR) 62, 168, 224T
TCF 138F, 277f, 484
telangiectasia 65
telomerase 16, 160F, **160ff**, 188ff, 230, 297
telomere 35, 39T, 64, 68, **159ff**, 160F, 165B, 474
telomeric fusions 36, 161
teratocarcinoma 189, 214
terminal differentiation 14, 17, **146ff**, 182, 231, 238, 248, 256, 260, 302, 323, 348, 389, 466
testes 158, 189, 390
testicular cancer 1, 3, 21, 26, 42, 44F, 61, 79T, 108, 150, 173, 449, 451, 456f
testosterone 363F, 390ff, 391F, 423
tetracycline 203
TFIIH 56
TGFα 79T, 83, 83F, **131f**, 320, 322, 334, 375f, 376T, 392, 400
TGFβ 100, 116T, **131f**, 133F, 202, 204, 211, 268, 280ff, 304, 320, 347, 399f, 476
TGFβ receptors (TGFBR) 132f, 133F, 280ff, 284, 306, 322, 347, 400
therapy see cancer therapy
thiopurine methyltransferase see TPMT
thrombospondin see TSP1
thrombopoetin 222
thymi(di)ne 50ff, 51F, 353, 412, 451
thymidine-cytosine photoproducts see photoproducts
thymidine kinase 483, 483F
thymidine-thymidine dimers see photoproducts

thymidylate synthetase (TS) 40T 412F, 413, 443f, 443T, 452F, 458
thyroid cancer 7, 22, 33, 33F, 79T, 97
tight junction 197f, 198F, 290
TIMP 205f, 206F
tissue homeostasis 105, 139ff, 147, 149, 156, **185ff**, 212, 358ff, 410
tissue regeneration 9, 105, 256, 259, 295, 331, 347, 350
TNFα 14, 137, 153ff, 212, 349
T-loop 159, 160F, 205
TNFR 137, 154F, 154ff
TNFRSF 130, 135, **153ff**, 153F, 154F, 157
TNM classification 18
tobacco smoke 2, 5f, 40T, 259, 292f, 352, 355B, 388, 404, 409, 424
tocopherol see vitamin E
toll-like receptors 135, 136F
topoisomerase 378, 459
   inhibitor 62, 443, 451T, 452, 453F
TP53 29f, 38T, 40T, 42, 50, 64ff, 93T, 100T, **101ff**, 102F, 103F, 104F, 112B, 116T, 119, 122, 125, **129ff**, 150ff, 155, 158, 159ff, 187f, 208, 210, 214, 226, 231, 235f, 239, 246, 252, 258ff, 264, 267f, 280ff, 285, 297ff, 303ff, 318, 322, 324, 331, 336ff, 345, 347, 352f, 366, 366T, 393, 404, 429, 439, 455ff, 461f, TP53ff, 480F, 482F
TP73 29
TP73L 29
TPA 127
TPMT 443T, 444
TRADD 153F
TRAIL 154F, 155, 157
transcription 21, 56, 127, 175, 338, 484
transcriptional activator (factor) 73ff, 74T, 80T, 101f, 124, 132ff, 135ff, 169, 180f, 183ff, 222, 225, 240, 249, 339, 359ff, 395f
transcriptional repressor 181, 249
transcription factor cascade 169, 347f
transcription initiation factors (basal transcription factors) 14, 169, 361
transfection 81
transferrin 319
transgenic mice 97, 340B, 473
transient amplification compartment 256
transition (base) 27
translational research 296, 427, 487
translesional repair see bypass repair
translocation see chromosomal translocation
transposition 31
transversion (base) 27
trastuzumab 378f, 382, 441, 461, 471
TRC8 315
βTRCP 277f, 317f, 332
TRF1/2 159f, 160F

# KEYWORD INDEX

trichloro-ethylene 6, 6T, 41
trichothiodistrophy 58
TRK see receptor tyrosine kinase
TRK (oncogene) 128
TS see thymidylate synthetase
TSC1/2 38T, 119ff, 306B, 312, 312T
TSP1 103F, 105, 209
TTD-4 58
Tuberin see TSC1
tuberous sclerosis 38T, 121f, 312, 312T
Turcot syndrome 273ff, 274T
tumor 13T
    benign 13T
    grade 13T
    malignant 13T
    stage 13T
tumor antigen 269, 475ff
tumor necrosis factor α
    see TNFα
tumor suppressor (gene) 37, 42ff, 46, 83, **91ff**(Ch), 119, 121, 141ff, 163, 177, 182, 236, 261, 264, 276, 284, 300ff, 314, 317, 322, 339, 346, 365ff, 394, 417, 464, 479ff, 486
    classification **109ff**
    cloning 288B, 306B
    cooperativity 302
tumor virus 31, 46
two-hit model see Knudson model
tyrosinase 266ff
tyrosine kinase 74T, 77, 79T, **85ff**, 133, 233, 464T, 465, 470ff, 473ff
tyrosine phosphatases 85
tyrphostins 470

## U

ubiquitination 61, 125, 130, 137
ubiquitin ligase 80T, 102, 277f, 317f, 318T, 332, 370
UDPGT 292F, 293, 443f, 443T
UGH see uracil-N-glycohydrolase
UGTA see UDPGT
ultrasound 17, 309
uniparental disomy 250
uPA 202f, 203F, 379, 400f, 441
uPAR 201T, 203, 203F
uranium 7
uracil 51, 51F, 175
uracil-N-glycohydrolase (UGH) 49T, 51, 413
urease 349
uridine 51
urokinase see uPA
uroplakin 290, 298
urothelial cancer 309 see also bladder cancer
urothelium 19F, 148, 290ff, 290H, 298ff, 309, 347, 358

UTR (untranslated region) 27
UV 5ff, 6T, 40T, 57f, 103, 103F, 117, 258ff, 259F, 266f, 269, 334, 404ff, 416
    absorption by DNA 55
    categorization 258

## V

vacA 349ff
vaccination 338, 407f, 415, 464, 477f
vasculogenesis 207, 212
V(D)J joining 168, 229, 231 see also recombinase
VDR 127, 360F, 395T
vegetables 287, 343, 353, 355B, 408ff
VEGF 187, **208ff**, 208T, 209T, 319, 400f
VEGFR 208ff, 209F, 469
v-erbA 75
v-erbB 73f, 77, 77F, 374
v-fms 73
v-fos 73, 80
VHL 38T, 93T, 96, 178T, 261, 312T, **316ff**, 316F, 317F, 318T
vinblastine 452
vincristine 452
viral genome 31
virus 31, 147, 461
    cytolytic 464
    integration 75ff, 112B, 339
    oncolytic 481ff, 486
    receptors 482f
vitamin(s) 388, 425
vitamin A 68f, 239f, 411F
vitamin $B_{12}$ 40T, 179, 412f
vitamin C 68f, 69T, 353, 410, 411F
vitamin D 411F
vitamin E 68f, 69T, 410, 411F
v-jun 73
v-myb 73
v-myc 73ff, 76F, 229
von Hippel-Lindau syndrome 38T, 93T, 187, 210, 312T, 316ff see also VHL
v-raf 73
v-rel 137
v-src 72f, 78, 80, 128
v-sis 74T

## W

WAGR syndrome 246T, 247
'watchful waiting' 421
Werner syndrome 37, 39T, 64f, 67T, 165B see also WRN
WHO 5f
Wilms tumor 26, 42, 171, 182, 214, 243ff(Ch), 245H, 308, 335, 429, 449
'window of opportunity' 248

WNT factors 79T, 84, 116T, 138F, 139ff, 189, 276f, 376
WNT pathway 109, 116T, 122, 130, **138ff**, 138F, 265, **273ff**, 277F, 298, 331ff, 333F, 345f, 355B, 376, 392, 434f, 436F, 484
WNT/Ca2+ pathway 141
WNT/polarity pathway 141
wound repair 158, 185f, 185F, 202, 207, 212, 257
WRN 39T, 49T, 63f, 66, 67T, 162, 165B
WT1 246T, 246ff, 249F

## X

X-chromosome 391
    inactivation 168, 170ff, 175, 180
xenografts 16, 473
Xeroderma pigmentosum 37, 39T, 58ff, 67T, 258, 404, 416
XIAP 152f
X-inactivation center 172
XIST 172
XP (A-G) 49T, 56ff, 67T
X-rays 17
XRCC1 40T, 52
XRCC4 63

## Y

Y-chromosome 172, 311

yes 128

## Z

ZD1839 *see* gefitinib
zinc (ion) 203
zinc finger 249f, 360